ENDOCRINE DISRUPTION MODELING

QSAR in Environmental and Health Sciences

Series Editor

James Devillers
*CTIS-Centre de Traitement de
l'Information Scientifique
Rillieux La Pape, France*

Aims & Scope

The aim of the book series is to publish cutting-edge research and the latest developments in QSAR modeling applied to environmental and health issues. Its aim is also to publish routinely used QSAR methodologies to provide newcomers to the field with a basic grounding in the correct use of these computer tools. The series is of primary interest to those whose research or professional activity is directly concerned with the development and application of SAR and QSAR models in toxicology and ecotoxicology. It is also intended to provide the graduate and postgraduate students with clear and accessible books covering the different aspects of QSARs.

Published Titles

Endocrine Disruption Modeling, *James Devillers,* 2009

QSAR in Environmental and Health Sciences

ENDOCRINE DISRUPTION MODELING

Edited by

James Devillers

CRC Press
Taylor & Francis Group
Boca Raton   London   New York

CRC Press is an imprint of the
Taylor & Francis Group, an **informa** business

CRC Press
Taylor & Francis Group
6000 Broken Sound Parkway NW, Suite 300
Boca Raton, FL 33487-2742

First issued in paperback 2017

ISBN 13: 978-1-138-11191-2 (pbk)
ISBN 13: 978-1-4200-7635-6 (hbk)

Library of Congress Cataloging-in-Publication Data

Endocrine disruption modeling / editor, James Devillers.
     p. ; cm. -- (QSAR in environmental and health sciences)
  Includes bibliographical references and index.
  ISBN 978-1-4200-7635-6 (hardcover : alk. paper)
    1. Endocrine disrupting chemicals--Research--Methodology. 2. QSAR (Biochemistry) I. Devillers, J. (James), 1956- II. Series: QSAR in environmental and health sciences.
    [DNLM: 1. Endocrine Disruptors. 2. Quantitative Structure-Activity Relationship. WK 102 E5477 2009]

RA1224.2.E625 2009
615'.190072--dc22   2009002784

Visit the Taylor & Francis Web site at
http://www.taylorandfrancis.com

and the CRC Press Web site at
http://www.crcpress.com

Contents

Series Introduction

The correlation between the toxicity of molecules and their physicochemical properties can be traced to the 19th century. Indeed, in a French thesis entitled *Action de l'alcool amylique sur l'organisme* (Action of amyl alcohol on the body), which was presented in 1863, by A. Cros before the Faculty of Medicine at the University of Strasbourg, an empirical relationship was made between the toxicity of alcohols and their number of carbon atoms as well as their solubility. In 1875, Dujardin-Beaumetz and Audigé were the first to stress the mathematical character of the relationship between the toxicity of alcohols and their chain length and molecular weight. In 1899, Hans Horst Meyer and Fritz Baum, at the University of Marburg, showed that narcosis or hypnotic activity was in fact linked to the affinity of substances to water and lipid sites within the organism. At the same time at the University of Zurich, Ernest Overton came to the same conclusion, providing the foundation of the lipoid theory of narcosis. The next important step was made in the 1930s, by Lazarev in St. Petersburg who first demonstrated that different physiological and toxicological effects of molecules were correlated with their oil–water partition coefficient through formal mathematical equations in the form: $\log C = a \log P_{oil/water} + b$. Thus, the Quantitative Structure-Activity Relationship (QSAR) discipline was born. Its foundations were definitively fixed in the early 1960s, by the seminal works contributed by C. Hansch and T. Fujita. Since that period, the discipline has gained tremendous interest, and now the QSAR models represent key tools in the development of drugs as well as in the hazard assessment of chemicals. The new REACH (Registration, Evaluation, Authorization, and Restriction of Chemicals) legislation on substances, which recommends the use of QSARs and other alternative approaches instead of laboratory tests on vertebrates, clearly reveals that this discipline is now well established and is an accepted practice in regulatory systems.

In 1993, the journal *SAR and QSAR in Environmental Research* was launched by Gordon and Breach to focus on all of the important works published in the field and to provide an international forum for the rapid publication of SAR (Structure-Activity Relationship) and QSAR models in (eco)toxicology, agrochemistry, and pharmacology. Today, the journal, which is now owned by Taylor & Francis and publishes twice more issues per year, continues to promote research in the QSAR field by favoring the publication of new molecular descriptors, statistical techniques, and original SAR and QSAR models. This field continues to grow rapidly, and many subject areas that require larger development are unsuitable for publication in a journal due to space limitations.

This prompted us to develop a series of books entitled *QSAR in Environmental and Health Sciences* to act in synergy with the journal. I am extremely grateful to Colin Bulpitt and Fiona Macdonald for their enthusiasm and invaluable help in making the project become a reality.

This book is the first of the series. Since it is also the first dedicated to endocrine disruption modeling, it clearly illustrates the aim of the series, which is to bring together the opinions of different experts to discuss important new topics in QSAR.

At the time of going to press, two books are in the pipeline. One deals with three-dimensional (3D) QSAR methods, and the other focuses on the topological description of molecules. I gratefully acknowledge Hilary Rowe for her willingness to assist me in the development of this series.

James Devillers

Acknowledgments

I am extremely grateful to the authors of the chapters for their acceptance to participate in this book and for preparing such valuable contributions. To ensure the scientific quality and clarity of the book, each chapter was sent to two referees for review. I would like to thank all of the referees for their useful comments. Finally, I would like to thank Patricia Roberson and the entire publication team at Taylor & Francis for making the publication of this book possible.

Contributors

Richard D. Beger
Division of Systems Toxicology
National Center for Toxicological
    Research (NCTR)
U.S. Food and Drug Administration
Jefferson, Arkansas

Andrew R. Brown
AstraZeneca U.K. Limited
Brixham Environmental Laboratory
Brixham, Devon, United Kingdom

Dan A. Buzatu
Division of Systems Toxicology
National Center for Toxicological
    Research (NCTR)
U.S. Food and Drug Administration
Jefferson, Arkansas

Daniel M. Consoer
Department of Chemistry and
    Biochemistry
University of Minnesota Duluth
Duluth, Minnesota

Albert R. Cunningham
James Graham Brown Cancer Center
    and Departments of Medicine and
    Pharmacology and Toxicology
University of Louisville
Louisville, Kentucky

Suzanne L. Cunningham
Crestwood, Kentucky

James Devillers
Centre de Traitement de l'Information
    Scientifique (CTIS)
Rillieux La Pape, France

Jean-Christophe Doré
Muséum National d'Histoire Naturelle
Paris, France

Jean-Pierre Doucet
Interfaces, Traitements, Organisation et
    Dynamique des Systèmes (ITODYS)
Université Paris
Paris, France

Hong Fang
Z-Tech Corporation, an ICF
    International Company at National
    Center for Toxicological Research/
    U.S. Food and Drug Administration
    (NCTR/FDA)
Jefferson, Arkansas

Toshio Fujita
Department of Agricultural Chemistry
Kyoto University
Kyoto, Japan

Seena A. Iype
James Graham Brown Cancer Center
    and Departments of Medicine and
    Pharmacology and Toxicology
University of Louisville
Louisville, Kentucky

Nathalie Marchand-Geneste
Université de Bordeaux
Bordeaux, France

Ovanes Mekenyan
Laboratory of Mathematical Chemistry
University "Prof. As. Zlatarov"
Bourgas, Bulgaria

Yoshiaki Nakagawa
Division of Applied Life Sciences
Kyoto University
Kyoto, Japan

Marjana Novič
Laboratory of Chemometrics
National Institute of Chemistry
Ljubljana, Slovenia

Annick Panaye
Interfaces, Traitements, Organisation et
    Dynamique des Systèmes (ITODYS)
Université Paris
Paris, France

Grace H. Panter
AstraZeneca U.K. Limited
Brixham Environmental Laboratory
Brixham, Devon, United Kingdom

Roger Perkins
Z-Tech Corporation, an ICF
    International Company at National
    Center for Toxicological Research/
    U.S. Food and Drug Administration
    (NCTR/FDA)
Jefferson, Arkansas

Jean-Marc Porcher
Institut National de l'Environnement
    Industriel et des Risques (INERIS)
Unité d'évaluation des Risques
    Écotoxicologiques
Verneuil en Halatte, France

Andrew M. Riddle
AstraZeneca U.K. Limited
Brixham Environmental Laboratory
Brixham, Devon, United Kingdom

Rossitsa Serafimova
Laboratory of Mathematical Chemistry
University "Prof. As. Zlatarov"
Bourgas, Bulgaria

Daniel M. Sheehan
Division of Systems Toxicology
National Center for Toxicological
    Research (NCTR)
U.S. Food and Drug Administration
Jefferson, Arkansas

Leming Shi
Division of Systems Toxicology
National Center for Toxicological
    Research (NCTR)
U.S. Food and Drug Administration
Jefferson, Arkansas

Weida Tong
Division of Systems Toxicology
National Center for Toxicological
    Research (NCTR)
U.S. Food and Drug Administration
Jefferson, Arkansas

Marjan Vračko
Laboratory of Chemometrics
National Institute of Chemistry
Ljubljana, Slovenia

Jon G. Wilkes
Division of Systems Toxicology
National Center for Toxicological
    Research (NCTR)
U.S. Food and Drug Administration
Jefferson, Arkansas

Paul F. Robinson
AstraZeneca U.K. Limited
Brixham Environmental Laboratory
Brixham, Devon, United Kingdom

1 In Silico Methods for Modeling Endocrine Disruption

James Devillers

CONTENTS

ABSTRACT

A number of xenobiotics released into the environment have the potential to disturb the normal functioning of the endocrine system. These chemicals termed "endocrine disrupting chemicals" act by mimicking or antagonizing the normal functions of natural hormones and may pose serious threats to the reproductive capability and development of living species. Batteries of laboratory tests exist for detecting these chemicals. However, due to time and cost limitations, they cannot be used for all chemicals that can be found in ecosystems. *In silico* approaches are particularly suited to overcome these problems. The principles and interest of these computational tools are briefly presented.

KEYWORDS

Computational methods
Endocrine disruptors
Environmental contamination
Xenobiotics

1.1 ENDOCRINE FUNCTION DISORDERS IN WILDLIFE AND HUMANS

In a huge number of species, including human, endocrine signaling is involved in pivotal physiological functions such as reproduction, embryo development, and growth [1–3]. Hormones trigger such complex functions by interacting with their

1

receptors, which are present at a nuclear as well as cellular level, in various organs and tissues as part of a complex feedback system [4]. Any disruption of this balance can yield injury in the physiological status of the whole organism, especially during the early stages of development. If the regulatory role of the endocrine system is disrupted, adverse effects on the functioning of the reproductive, nervous, and immune systems may occur [4].

Endocrine disruption became an issue when associations were made between chemical contaminations of the environment and the onset of diseases, reproductive failure, and death of wildlife species [5]. Thus, for example, until the late 1950s, sparrowhawks (*Accipiter nisus* L) were common and widespread in Britain but they suddenly showed a marked decline in numbers, almost disappearing from some districts. Their population crash followed the widespread introduction of organochlorine pesticides, namely, aldrin, dieldrin, and dichlorodiphenyltrichloroethane (DDT) [6]. Such declines in bird populations occurred not only in Britain, but extended over much of Europe, North America, and Australia wherever highly hydrophobic organochlorine pesticides were commonly used [6–8].

Another well-known example of endocrine disruption in wildlife, also discovered accidentally, deals with the effect of tributyltin (TBT) on the mollusks. Although neogastropod mollusks are gonochoristic (that is, sexes are separate), in 1970, Blaber [9] observed that many female dogwhelk (*Nucella lapillus*) had a penis-like structure behind the right tentacle. Shortly thereafter, Smith [10] reported similar reproductive abnormalities in the American mud snail (*Ilyanassa obsoleta*; formerly *Nassarius obsoletus*) on the Connecticut coast and coined the term "imposex" to describe this superimposition of male characters onto females [11]. At this stage, no link had been made with pollution, but 10 years later, Smith [12,13] showed that levels of imposex in *I. obsoleta* were elevated close to marinas where the mollusks were exposed to antifouling paints including TBT.

The detection of reproductive abnormalities and failures in American alligators (*Alligator mississippiensis*) from contaminated lakes in Florida at the end of the 1980s was also unexpected, the populations in Florida being protected and controlled since these reptiles were listed as endangered species in the early 1970s [14]. Whereas many alligator populations were healthy, in one location, Lake Apopka, a dramatic decline was observed. Lake Apopka, one of Florida's largest lakes, became polluted with dicofol and DDT metabolites from several sources. A combination of agricultural runoff due to historically intensive activity around the lake and a 1980 pesticide spill from the former Tower Chemical Company resulted in several adverse health outcomes for the lake's wildlife, mainly in the population of *A. mississippiensis*. Thus, for example, the female alligators exhibited abnormal ovarian morphology, and the males presented poorly organized testes and abnormally small phalli. About 80% to 95% of the alligator eggs at Lake Apopka failed to hatch. These adverse effects and others in relation to the endocrine system were linked to the chemical contaminations [14–17].

The ability of environmental contaminants to influence reproduction and development of invertebrates and vertebrates in the aquatic and terrestrial ecosystems is now well documented (see, for example, [18–22] and references therein). Unfortunately, adverse effects of xenobiotics on the reproductive system of humans have also been clearly demonstrated [4,23–28]. The historical example of diethylstilbestrol (DES) clearly illustrates the situation. DES, a synthetic form of estrogen, was commonly given to pregnant women in the United States, Europe, and Australia to prevent spontaneous abortions from 1948 until 1971, when its use for this purpose was banned [18]. Daughters born to women who took DES in the first months of their pregnancy may have structural abnormalities in their vagina, cervix, or uterus. They may also have difficulties in pregnancy, such as infertility, miscarriage, ectopic pregnancy, and premature births. As young adults, these women also may suffer from vaginal clear-cell adenocarcinomas [29,30]. Maternal ingestion of DES during early pregnancy also may result in an increased incidence of malformations of testes, the development of epididymal cysts, and impaired sperm quality in male offspring [30,31].

A variety of pesticides can lead to significant adverse health effects for both farm workers and consumers. Thus, for example, chlordecone (also termed kepone) is an organochlorine insecticide that was massively used for controlling banana crop pests in the French Caribbean islands. Suspected to be a carcinogen, it was banned in 1993, although there are indications of illegal uses after this date. Consequently, the environment is highly contaminated by this persistent pollutant known to have impact on human health [32]. Indeed, neurological illness involving tremor and nervousness but also abnormalities of the livers and testes were first described in the 1970s, after an excessive and uncontrolled occupational exposure to chlordecone in a small chemical plant in Virginia as well as the exposure of the population in the vicinity of the plant [33]. Chlordecone, which is also known to interfere with reproduction and to be fetotoxic to experimental animals [34,35], is suspected to be at the origin of the increased rates of prostate cancer, congenital malformations, and infertility in the French Caribbean islands of Martinique and Guadeloupe [35,36]. In the same way, it has been shown that simultaneous maternal consumption of soy, including the phytoestrogen genistein, and the fungicide vinclozolin, such as can occur in a nonorganic vegetarian diet, might result in an increase in hypospadias frequency [37], and precocious puberty as well as altered breast development were noted in young girls living in agricultural environments highly contaminated by pesticides [38]. All these xenobiotics, which directly or indirectly interact on the functioning of the endocrine system of the invertebrates and vertebrates, are termed "endocrine disrupting chemicals" (EDCs).

1.2 WHAT IS AN ENDOCRINE DISRUPTING CHEMICAL (EDC)?

An EDC is an exogenous substance or mixture that alters the functions of the endocrine system and consequently causes adverse health effects in an intact organism, its progeny, or its (sub)population [4]. EDCs are believed to exert their effects by (1) mimicking normal hormones such as estrogens and androgens, (2) antagonizing hormones; (3) altering the pattern of synthesis and metabolism of hormones; (4) modifying hormone receptor levels; or (5) interfering in other signaling systems, which are indirectly in relation with the endocrine system, such as the immune and nervous systems [8,39].

Although the EDCs are generally less potent than the hormones, such as estradiol, it is now admitted that they additively act with them [40], even if deviations to

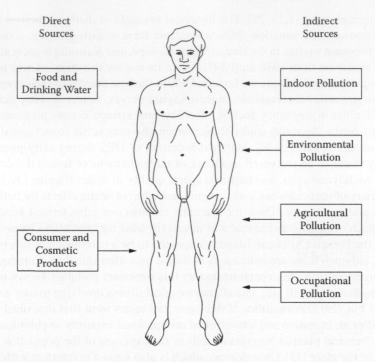

Direct
Sources

Indirect
Sources

Food and
Drinking Water

Indoor Pollution

Environmental
Pollution

Agricultural
Pollution

Consumer and
Cosmetic
Products

Occupational
Pollution

FIGURE 1.1 Sources of human exposure to EDCs.

this rule have been found [41]. In the same way, some EDCs can induce an inverted U-shaped dose-response relationship, resulting from low-dose response stimulation [42,43]. In other words, lower doses induce more profound effects than higher doses. This type of behavior challenges the elaboration of methods for the hazard and risk assessment of EDCs.

Numerous xenobiotics, provided from direct and indirect sources of contaminations (Figure 1.1) have the ability to interfere with the endocrine system. These EDCs present a high variety of structures and physicochemical properties. They include persistent organic pollutants (POPs) such as the p,p'-DDT and its metabolites, the dioxins, and the polychlorinated biphenyls (PCBs), agrochemicals, biocides, and industrial organic chemicals such as bisphenol A and several phthalates. Unfortunately, chemicals included in cosmetics, toiletries, and foods can also be EDCs. Thus, for example, parabens (p-hydroxybenzoic acid esters), which are the most commonly used preservatives in cosmetics, toiletries, pharmaceuticals, and foods due to their rather low human toxicity and their effective antimicrobial activity, show an endocrine disrupting activity [44]. Some ultraviolet (UV)-absorbing chemicals that are widely employed in sunscreens and various cosmetic products for protection against UV radiation present an estrogenic activity [45]. Last, if a good dietary intake of phytoestrogens might be protective against some cancers, in high doses, especially in vegetarian diets, the adverse effects may outweigh their beneficial ones [37,46,47].

During the last three decades, substantial efforts have been made to develop various *in vivo* and *in vitro* tests allowing the detection of the potential endocrine effects of chemicals [48–54]. The challenge with the EDCs is the need to evaluate thousands of chemical structures for their potential impact on the different endocrine targets. Unfortunately, due to time and cost limitations, the task is unrealistic. Computational (eco)toxicology in its various forms offers a practical approach to meet this challenge but can also allow us to better understand the mechanisms of action of the EDCs as well as their environmental fate.

1.3   MODELING EDCs

A mathematical model is a simplified representation of a system. Generally, a model has an information input, an information processor, and an output of expected result(s). Its construction requires making some assumptions about the essential structure of the studied system and about the relationships existing between its constitutive elements. This means that to make a model, certain compromises have to be made with reality. In other words, a model will never be able to fully reproduce a real-world system. There exist different categories of models that are classified according to the nature and/or functions of their output, the nature of their constitutive equations, and so on [55,56]. Regarding the EDCs, it is obvious that the most interesting models are those able to predict the endocrine activity of chemicals in a quantitative or qualitative way and to explain the mechanisms of action of EDCs. The different models presented in this book belong to these categories. They are rooted on various methodologies and deal with different endpoints. Rather than catalog all the existing models on EDCs in this book, the main axes of research in the field are presented. Moreover, even though the first motivation of the book is the modeling of the adverse effects of EDCs on wildlife and humans, it is noteworthy that computational techniques are increasingly used to design new endocrine therapeutic agents such as selective estrogen receptor modulators (SERMs) and selective androgen receptor modulators (SARMs) (for example, [57–61]; see also Chapters 9 and 11). In the same way, modeling approaches are also widely used in the search for chemicals acting on the endocrine system of insects for the development of insecticides disrupting their development and metamorphosis (for example, [62–65]; see also Chapter 13, this volume).

Despite their differences, all these models share common characteristics. First, they are built for a specific endocrine target. This makes sense when the goal of the modeling process is to produce a new therapeutic agent or a new insecticide. Conversely, this is more annoying when the objective of the model is the detection of xenobiotics that could adversely affect the endocrine system because the targets are multiple. It is worthy to note that recently, attempts have been made to predict the endocrine disruption profile of chemicals from a unique model [66]. There is a need for such models or for collections of models allowing for the prediction of the multifaceted aspects of the endocrine disruption phenomenon. All these models present an applicability domain that depends on the diversity and quality of the information used for their design and that has to be respected for securing reliable simulation results. Last, the endocrine disruption models suffer from a lack of activity data for facilitating and securing their validation, and for extending their domain of application. This lack of experimental data represents the main hurdle in the development of models for different endpoints [39] (see Chapter 14, this volume).

Even if the case studies presented in the different chapters clearly show that there exist powerful models allowing for the prediction of the endocrine disruption potential of chemicals and for insights to be provided into some of their mechanisms of action, it is obvious that endocrine disruption modeling is still in its infancy; hence, there is a need to pursue modeling efforts in various directions to be able, in the future, to correctly detect and simulate this complex activity.

REFERENCES

[1] J.C. Wingfield, M.E. Visser, and T.D. Williams, *Introduction. Integration of ecology and endocrinology in avian reproduction: A new synthesis*, Phil. Trans. R. Soc. B 363 (2008), pp. 1581–1588.

[2] J. Lintelmann, A. Katayama, N. Kurihara, L. Shore, and A. Wenzel, *Endocrine disruptors in the environment*, Pure Appl. Chem. 75 (2003), pp. 631–681.

[3] G.A. LeBlanc, *Crustacean endocrine toxicology: A review*, Ecotoxicol. 16 (2007), pp. 61–81.

[4] D. Caserta, L. Maranghi, A. Mantovani, R. Marci, F. Maranghi, and M. Moscarini, *Impact of endocrine disruptor chemicals in gynaecology*, Human Reprod. Update 14 (2008), pp. 59–72.

[5] J.P. Sumpter and A.C. Johnson, *Lessons from endocrine disruption and their application to other issues concerning trace organics in the aquatic environment*, Environ. Sci. Technol. 39 (2005), pp. 4321–4332.

[6] I. Newton and I. Wyllie, *Recovery of a sparrowhawk population in relation to declining pesticide contamination*, J. Appl. Ecol. 29 (1992), pp. 476–484.

[7] P. Olsen, P. Fuller, and T.G. Marples, *Pesticide-related eggshell thinning in Australian raptors*, Emu 93 (1993), pp. 1–11.

[8] C.M. Markey, B.S. Rubin, A.M. Soto, and C. Sonnenschein, *Endocrine disruptors: From Wingspread to environmental developmental biology*, J. Steroid Biochem. Molec. Biol. 83 (2003), pp. 235–244.

[9] S.J.M. Blaber, *The occurrence of a penis-like outgrowth behind the right tentacle in spent females of* Nucella lapillus *(L.)*, Proc. Malacol. Soc. Lond. 39 (1970), pp. 231–233.

[10] B.S. Smith, *Sexuality in the American mud snail*, Nassarius obsoletus *Say*, Proc. Malacol. Soc. Lond. 39 (1971), pp. 377–378.

[11] P. Matthiessen and P.E. Gibbs, *Critical appraisal of the evidence for tributyltin-mediated endocrine disruption in mollusks*, Environ. Toxicol. Chem. 17 (1998), pp. 37–43.

[12] B.S. Smith, *Reproductive anomalies in stenoglossan snails related to pollution from marinas*, J. Appl. Toxicol. 1 (1981), pp. 15–21.

[13] B.S. Smith, *Male characteristics on female mud snails caused by antifouling bottom paints*, J. Appl. Toxicol. 1 (1981), pp. 22–25.

[14] L.J. Guillette, T.S. Gross, G.R. Masson, J.M. Matter, H.F. Percival, and A.R. Woodward, *Developmental abnormalities of the gonad and abnormal sex hormone concentrations in juvenile alligators from contaminated and control lakes in Florida*, Environ. Health Prespect. 102 (1994), pp. 680–688.

[15] J.C. Semenza, P.E. Tolbert, C.H. Rubin, L.J. Guillette, and R.J. Jackson, *Reproductive toxins and alligator abnormalities at Lake Apopka, Florida*, Environ. Health Prespect. 105 (1997), pp. 1030–1032.

[16] M.R. Milnes, D.S. Bermudez, T.A. Bryan, M.P. Gunderson, and L.J. Guillette, *Altered neonatal development and endocrine function in* Alligator mississippiensis *associated with a contaminated environment*, Biol. Reprod. 73 (2005), pp. 1004–1010.

[17] P.M. Lind, M.R. Milnes, R. Lundberg, D. Bermudez, J. Orberg, and L.J. Guillette, *Abormal bone composition in female juvenile American alligators from a pesticide-polluted lake (Lake Apopka, Florida)*, Environ. Health Prespect. 112 (2004), pp. 359–362.

[18] T. Colborn, F.S. vom Saal, and A.M. Soto, *Developmental effects of endocrine-disrupting chemicals in wildlife and humans*, Environ. Health Prespect. 101 (1993), pp. 378–384.

[19] L.G. Parks, C.S. Lambright, E.F. Orlando, L.J. Guillette, G.T. Ankley, and L.E. Gray, *Masculinization of female mosquitofish in Kraft mill effluent-contaminated Fenholloway River water is associated with androgen receptor agonist activity*, Toxicol. Sci. 62 (2001), pp. 257–267.

[20] G. Toft, T.M. Edwards, E. Baatrup, and L.J. Guillette, *Disturbed sexual characteristics in male mosquitofish* (Gambusia holbrooki*) from a lake contaminated with endocrine disruptors*, Environ. Health Prespect. 111 (2003), pp. 695–701.

[21] C. Sonne, P.S. Leifsson, R. Dietz, E.W. Born, R.J. Letcher, L. Hyldstrup, F.F. Riget, M. Kirkegaard, and D.C.G. Muir, *Xenoendocrine pollutants may reduce size of sexual organs in East Greenland polar bears (*Ursus maritimus*)*, Environ. Sci. Technol. 40 (2006), pp. 5668–5674.

[22] T. Soin and G. Smagghe, *Endocrine disruption in aquatic insects: A review*, Ecotoxicology 16 (2007), pp. 83–93.

[23] J. Toppari, J.C. Larsen, P. Christiansen, A. Giwercman, P. Grandjean, L.J. Guillette, B. Jégou, T.K. Jensen, P. Jouannet, N. Keiding, H. Leffers, J.A. McLachlan, O. Meyer, J. Müller, E. Rajpert-De Meyts, T. Scheike, R. Sharpe, J. Sumpter, and N.E. Skakkebaek, *Male reproductive health and environmental xenoestrogens*, Environ. Health Prespect. 104 Sup. 4 (1996), pp. 741–803.

[24] H.R. Pohl, J.G.M. van Engelen, J. Wilson, and A.J.A.M. Sips, *Risk assessment of chemicals and pharmaceuticals in the pediatric population: A workshop report*, Regul. Toxicol. Pharmacol. 42 (2005), pp. 83–95.

[25] P.D. Gluckman and M.A. Hanson, *Evolution, development and timing of puberty*, Trends Endocrinol. Metabol. 17 (2006), pp. 7–12.

[26] J. van Oostdam, S.G. Donaldson, M. Feeley, D. Arnold, P. Ayotte, G. Bondy, L. Chan, E. Dewaily, C.M. Furgal, H. Kuhnlein, E. Loring, G. Muckle, E. Myles, O. Receveur, B. Tracy, U. Gill, and S. Kalhok, *Human health implications of environmental contaminants in Artic Canada: A review*, Sci. Total Environ. 351–352 (2005), pp. 165–246.

[27] G.S. Prins, L. Birch, W.Y. Tang, and S.M. Ho, *Developmental estrogen exposures predispose to prostate carcinogenesis with aging*, Reprod. Toxicol. 23 (2007), pp. 374–382.

[28] R.R. Newbold, E. Padilla-Banks, R.J. Snyder, T.M. Phillips, and W.N. Jefferson, *Developmental exposure to endocrine disruptors and the obesity epidemic*, Reprod. Toxicol. 23 (2007), pp. 290–296.

[29] A.L. Herbst, H. Ulfelder, and D.C. Poskanzer, *Adenocarcinoma of the vagina: Association of maternal stilbestrol therapy with tumor appearance in young women*, N. Engl. J. Med. 284 (1971), pp. 878–881.

[30] M. Bibbo, W.B. Gill, F. Azizi, R. Blough, V.S. Fang, R.L. Rosenfield, G.F. Schumacher, K. Sleeper, M.G. Sonek, and G.L. Wied, *Follow-up study of male and female offspring of DES-exposed mothers*, Obstet. Gynecol. 49 (1977), pp. 1–8.

[31] O.V. Martin, T. Shialis, J.N. Lester, M.D. Scrimshaw, A.R. Boobis, and N. Voulvoulis, *Testicular dysgenesis syndrome and the estrogen hypothesis: A quantitative meta-analysis*, Environ. Health Prespect. 116 (2008), pp. 149–157.

[32] G. Bocquené and A. Franco, *Pesticide contamination of the coastline of Martinique*, Mar. Pollut. Bull. 51 (2005), pp. 612–619.

[33] C.W. Heath, *Environmental pollutants and the epidemiology of cancer*, Environ. Health Prespect. 27 (1978), pp. 7–10.

[34] B. Hammond, B.S. Katzenellenbogen, N. Krauthammer, and J. McConnell, *Estrogenic activity of the insecticide chlordecone (kepone) and interaction with uterine estrogen receptors*, Proc. Natl. Acad. Sci. 76 (1979), pp. 6641–6645.

[35] C. Dubuisson, F. Héraud, J.C. Leblanc, S. Gallotti, C. Flamand, A. Blateau, P. Quenel, and J.L. Volatier, *Impact of subsistence production on the management options to reduce the food exposure of the Martinican population to chlordecone*, Regul. Toxicol. Pharmacol. 49 (2007), pp. 5–16.

[36] D. Belpomme, *Rapport d'Expertise et d'Audit Externe Concernant la Pollution par les Pesticides en Martinique. Conséquences Agrobiologiques, Alimentaires et Sanitaires et Proposition d'un Plan de Sauvegarde en Cinq Points*, ARTAC, Paris, 2007.

[37] M.L.B. Vilela, E. Willingham, J. Buckley, B.C. Liu, K. Agras, Y. Shiroyanagi, and L.S. Baskin, *Endocrine disruptors and hypospadias: Role of genistein and the fungicide vinclozolin*, Urology 70 (2007), pp. 618–621.

[38] E.A. Guillette, C. Conard, F. Lares, M. Guadalupe Aguilar, J. McLachlan, and L.J. Guillette, *Altered breast development in young girls from an agricultural environment*, Environ. Health Prespect. 114 (2006), pp. 471–475.

[39] J. Devillers, N. Marchand-Geneste, A. Carpy, and J.M. Porcher, *SAR and QSAR modeling of endocrine disruptors*, SAR QSAR Environ. Res. 17 (2006), pp. 393–412.

[40] E. Silva, N. Rajapakse, and A. Kortenkamp, *Something from "nothing" — Eight weak estrogenic chemicals combined at concentrations below NOECs produce significant mixture effects*, Environ. Sci. Technol. 36 (2002), pp. 1751–1756.

[41] N. Rajapakse, E. Silva, M. Scholze, and A. Kortenkamp, *Deviation from additivity with estrogenic mixtures containing 4-nonylphenol and 4-tert-octylphenol detected in the E-SCREEN assay*, Environ. Sci. Technol. 38 (2004), pp. 6343–6352.

[42] F.S. vom Saal, B.G. Timms, M.M. Montano, P. Palanza, K.A. Thayer, S.C. Nagel, M.D. Dhar, V.K. Ganjam, S. Parmigiani, and W.V. Welshons, *Prostate enlargement in mice due to fetal exposure to low doses of estradiol or diethylstilbestrol and opposite effects at high doses*. Proc. Natl. Acad. Sci. 94 (1997), pp. 2056–2061.

[43] L. Li, M.E. Andersen, S. Heber, and Q. Zhang, *Non-monotonic dose-response relationship in steroid hormone receptor-mediated gene expression*, J. Molec. Endocrinol. 38 (2007), pp. 569–585.

[44] S. Oishi, *Effects of propyl paraben on the male reproductive system*, Food Chem. Toxicol. 40 (2002), pp. 1807–1813.

[45] P.Y. Kunz and K. Fent, *Estrogenic activity of UV filter mixtures*, Toxicol. Appl. Pharmacol. 217 (2006), pp. 86–99.

[46] C.F. Skibola and M.T. Smith, *Potential health impacts of excessive flavonoid intake*, Free Rad. Biol. Med. 29 (2000), pp. 375–383.

[47] J.P. Antignac, R. Cariou, B. Le Bizec, and F. André, *New data regarding phytoestrogens content in bovine milk*, Food Chem. 87 (2004), pp. 275–281.

[48] T. Zacharewski, In vitro *bioassays for assessing estrogenic substances*, Environ. Sci. Technol. 31 (1997), pp. 613–623.

[49] B.E. Gillesby and T.R. Zacharewski, *Exoestrogens: Mechanisms of action and strategies for identification and assessment*, Environ. Toxicol. Chem. 17 (1998), pp. 3–14.

[50] F. Eertmans, W. Dhooge, S. Stuyvaert, and F. Comhaire, *Endocrine disruptors: Effects on male fertility and screening tools for their assessment*, Toxicol. In Vitro 17 (2003), pp. 515–524.

[51] H.P. Gelbke, M. Kayser, and A. Poole, *OECD test strategies and methods for endocrine disruptors*, Toxicology 205 (2004), pp. 17–25.

[52] R. Lepage and C. Albert, *Fifty years of development in the endocrinology laboratory*, Clin. Biochem. 39 (2006), pp. 542–557.

[53] H.P. Gelbke, A. Hofmann, J.W. Owens, and A. Freyberger, *The enhancement of the subacute repeat dose toxicity test OECD TG 407 for the detection of endocrine active chemicals: Comparison with toxicity tests of longer duration*, Arch. Toxicol. 81 (2007), pp. 227–250.

[54] W. Lilienblum, W. Dekant, H. Foth, T. Gebel, J.G. Hengstler, R. Kahl, P.J. Kramer, H. Schweinfurth, and K.M. Wollin, *Alternative methods to safety studies in experimental animals: Role in the risk assessment of chemicals under the new European chemical legislation (REACH)*, Arch. Toxicol. 82 (2008), pp. 211–236.

[55] A. Ford, *Modeling the Environment: An Introduction to System Dynamics Modeling of Environmental Systems*, Island Press, Washington, 1999.

[56] N. Bouleau, *Philosophies des Mathématiques et de la Modélisation. Du Chercheur à l'Ingénieur*, L'Harmattan, Paris, 1999.

[57] E.F.F. da Cunha, R.C.A. Martins, M.G. Albuquerque, and R.B. de Alencastro, LIV-3D-*QSAR model for estrogen receptor ligands*, J. Mol. Model. 10 (2004), pp. 297–304.

[58] S. Mukherjee, A. Saha, and K. Roy, *QSAR of estrogen receptor modulators: Exploring selectivity requirements for ERα versus ERβ binding of tetrahydroisoquinoline derivatives using E-state and physicochemical parameters*, Bioorg. Med. Chem. Lett. 15 (2005), pp. 957–961.

[59] J. Chen, J. Kim, and J.T. Dalton, *Discovery and therapeutic promise of selective androgen receptor modulators*, Molec. Interv. 5 (2005), pp. 173–188.

[60] M.A. Perera, D. Yin, D. Wu, K.K. Chan, D.D. Miller, and J. Dalton, In vivo *metabolism and final disposition of a novel nonsteroidal androgen in rats and dogs*, Drug Metab. Disp. 34 (2006), pp. 1713–1721.

[61] W. Gao and J.T. Dalton, *Ockham's razor and selective androgen receptor modulators (SARMs): Are we overlooking the role of 5α-reductase?*, Molec. Interv. 7 (2007), pp. 10–13.

[62] X. Qian, *Molecular modeling study on the structure-activity relationship of substituted dibenzoyl-1-tert-butylhydrazines and their structural similarity to 20-hydroxyecdysone*, J. Agric. Food Chem. 44 (1996), pp. 1538–1542.

[63] L. Dinan, R.E. Hormann, and T. Fijimoto, *An extensive ecdysteroid CoMFA*, J. Comput-Aided Molec. Design 13 (1999), pp. 185–207.

[64] L. Dinan, *Ecdysteroid structure-activity relationships*, in *Studies in Natural Products Chemistry*, A.U. Rahman, ed., vol. 29, Elsevier, Amsterdam, 2003, pp. 3–71.

[65] C.E. Wheelock, Y. Nakagawa, T. Harada, N. Oikawa, M. Akamatsu, G. Smagghe, D. Stefanou, K. Iatrou, and L. Swevers, *High-throughput screening of ecdysone agonists using a reporter gene assay followed by 3-D QSAR analysis of the molting hormonal activity*, Bioorg. Med. Chem. 14 (2006), pp. 1143–1159.

[66] J. Devillers, N. Marchand-Geneste, J.C. Doré, J.M. Porcher, and V. Poroikov, *Endocrine disruption profile analysis of 11,416 chemicals from chemometrical tools*, SAR QSAR Environ. Res. 18 (2007), pp. 181–193.

2 Mechanisms of Endocrine Disruptions
A Tentative Overview

Jean-Marc Porcher, James Devillers,
and Nathalie Marchand-Geneste

CONTENTS

ABSTRACT

The term "endocrine disrupting chemicals" (EDCs) is used to define a structurally diverse class of synthetic and natural compounds that possess the ability to alter various components of the endocrine system and potentially induce adverse effects in exposed individuals and populations.

Research on these compounds has revealed that they act through a variety of receptor-mediated and nonreceptor-mediated mechanisms to modulate different components of the endocrine system. Receptor-mediated mechanisms have received the most attention, and EDCs are typically identified as compounds that interact with receptors and thus act as agonists or antagonists of endogenous hormones. Nevertheless, research has clearly shown that EDCs can act at multiple sites via multiple mechanisms of action. For example, growing evidence shows that they may modulate the activity or expression of steroidogenic enzymes or interact with hormone transport and metabolism. EDCs are generally described as substances with (anti)estrogenic or (anti)androgenic effect, but other targets have been evidenced such as the thyroid and immune system. In addition, EDCs alter a wide variety of behaviors, including sexual and other reproductive behaviors, social behaviors, and learning and other cognitive abilities.

Because of cross talk between different components of the endocrine systems, effects may occur unpredictably in endocrine target tissues other than the system predicted to be affected. Moreover, critical periods of reproductive and nervous system development during early life stages are especially sensitive to hormonal disruption. Although there is considerable information on the early molecular events involved in EDC response, there is very little knowledge concerning the relationship between those molecular events and adverse health effects. This knowledge gap is perhaps the most limiting factor in our ability to evaluate exposure–response relationships. For most associations reported between exposure to EDCs and a variety of biological outcomes, the mechanisms of action are poorly understood. This makes it difficult to distinguish between direct and indirect effects and primary versus secondary effects of exposure to EDCs. It also indicates that considerable caution is necessary in extrapolating from *in vitro* data to *in vivo* effects, in predicting effects from limited *in vivo* data, and in extrapolating from experimental data to human or wildlife situations.

It is beyond the scope of this chapter to describe all the possible disruption events of the endocrine systems of vertebrates; instead the focus will be on little known or new modes of action through which EDCs might act.

KEYWORDS

Endocrine disrupting chemicals (EDCs)
Interactions
Mechanisms
Nuclear receptors
Steroidogenesis

2.1 INTRODUCTION

It is now well established that a number of environmental contaminants (from agricultural, industrial, and household origins) are able to disturb the normal physiology and endocrinology of organisms. These substances, termed "endocrine disrupting chemicals" (EDCs), have been defined as "an exogenous substance or mixture that alters function(s) of the endocrine system and consequently causes adverse health effects in an intact organism or its progeny or (sub)population" [1]. Concerns regarding exposure to these EDCs are due primarily to adverse effects observed in wildlife, but the increased incidence of certain endocrine-related human diseases is also generating concern.

Endocrine systems are found in most varieties of animal life and play an essential role in many and varied functions of an organism (for example, growth and development, tissue function, and metabolism). Considerable homology exists in the endocrinology of vertebrates; hence, toxicants that alter endocrine function in one species are likely to produce adverse effects in another. However, there are significant differences between some species in endocrine function that warrant consideration for further interspecies extrapolations. Although the hormones, hormone synthesis, and their receptors are highly conserved, the role of specific hormones in reproductive function and development can vary greatly. Classically, the endocrine system was considered as made up of glands that secrete hormones, and receptors that detect and react to the hormones. Hormones are released by glands and travel throughout the body, acting as chemical messengers. Recently, our concept of "endocrine" has been broadened by the discovery of other chemical regulators, such as chemicals secreted into the blood by neurons, and include growth factors, paracrine and autocrine regulators, second messengers, and transcription factors [1].

Endocrine disruptors are commonly considered to be compounds that mimic or block the transcriptional activation elicited by naturally circulating hormones by binding to steroid hormone receptors.

EDCs encompass a variety of natural (phytoestrogens, toxins, natural hormones released by animals) or man-made substances including some pesticides (including vinclozolin, organochlorine insecticides such as endosulfan, DDT and its derivatives), pharmaceuticals (including ethinylestradiol), plastic additives, and a number of industrial chemicals (including polychlorinated biphenyls, dioxins, and phthalates).

Chemicals can exert their effect through a number of different mechanisms. They may mimic the biological activity of a hormone by binding to a cellular receptor, leading to an unwarranted response by initiating the cell's normal response to the naturally occurring hormone at the wrong time or to an excessive extent (agonist effect). They may bind to the receptor but not activate it, so the presence of the chemical on the receptor will prevent binding of the natural hormone (antagonist effect). They may bind to transport proteins in the blood, thus altering the amounts of natural hormones that are present in the circulation. They may interfere with the metabolic processes in the body, affecting the synthesis or breakdown rates of the natural hormones (Figure 2.1).

It is beyond the scope of this chapter to describe all the possible disruptions of the endocrine systems of vertebrates; instead, the focus will be on little known or new modes of action through which EDCs might act.

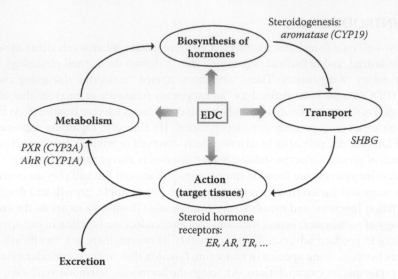

FIGURE 2.1 Hormonal dynamics and endocrine disruption.

Because of cross talk between different components of the endocrine systems, effects may occur unpredictably in endocrine target tissues other than the system predicted to be affected. Moreover, critical periods of reproductive and nervous system development during early life stages are especially sensitive to hormonal disruption. Although there is considerable information on the early molecular events involved in EDC response, there is very little knowledge concerning the relationship between those molecular events and adverse health effects. This knowledge gap is perhaps the most limiting factor in our ability to evaluate exposure–response relationships.

2.2 RECEPTOR-MEDIATED MECHANISMS

The first characterized mechanism of action for endocrine disruptors is to act directly as ligands to steroid hormone nuclear receptors (NRs), in particular, estrogen, androgen, and thyroid hormone nuclear receptors. Nuclear receptors are a class of proteins found within the interior of cells that are responsible for sensing the presence of hormones. In response, these receptors work in concert with other proteins to regulate the expression of specific genes. Nuclear receptors have the ability to directly bind to DNA and regulate the expression of adjacent genes; hence, these receptors are classified as transcription factors [2]. The regulation of gene expression happens when ligand binding to a nuclear receptor results in a conformation change, which in turn activates the receptor, resulting in up-regulation of gene expression. Nuclear receptors may be classified according to either mechanisms [3–5] or homology [6,7].

Schematically, NRs may be classified into four classes according to their dimerization and DNA-binding properties. Class I NRs include the known steroid receptors, which in absence of ligand are located in the cytosol. Hormone binding triggers dissociation of heat shock proteins (HSPs), homodimerization, and translocation to the nucleus where it binds to a specific sequence of DNA half-sites organized as inverted repeats. The nuclear receptor DNA complex in turn recruits other proteins that are responsible for transcription of downstream DNA into mRNA, which is eventually translated into protein, which results in change in cell function. The class II NRs, in contrast, are retained in the nucleus regardless of the binding status. Class II receptors heterodimerize with retinoid X receptor (RXR) and characteristically bind to direct repeat. Exclusive of the steroid hormones, this group includes all other known ligand-dependent receptors. In the absence of ligand, type II NRs are often complexed with corepressor proteins. Ligand binding to the nuclear receptor causes dissociation of corepressor and recruitment of coactivator proteins. Additional proteins including RNA polymerase are then recruited to the NR/DNA complex, which transcribe DNA into mRNA. Class III receptors bind primarily to direct repeats as homodimers. Class IV receptors typically bind to extended core sites as monomers. Most of the orphan receptors fall into class III and IV categories [4].

Sequence alignment and phylogenetic tree construction resulted in a classification of the human NR family into six evolutionary groups of unequal size [3,7]. Group 1 contains thyroid (TRs), retinoic acids (RARs), vitamin D (VDR) and peroxisome proliferator-activated receptors (PPARs), as well as orphan receptors such as RORs, Rev-erbs, CAR, PXR LXRs, and others. Group 2 includes retinoid X (RXRs), COUP-TF, and HNF-4 receptors. Group 3 includes the steroid receptors with estrogen (ERs), glucuronid (GRs), progesterone (PRs), and androgen (ARs) receptors, as well as estrogen-related receptors (ERRs). Group 4 contains the nerve growth factors-induced clone B group of orphan receptors (NGFI-B, NURR1, and NOR1). Group 5 is another small group that includes the steroidogenic factor 1 and the receptors related to the *Drosophila* FTZ-F1. Group 6 contains only the GCNF1 receptor, which does not fit well into any other subfamilies. A correlation exists between the DNA-binding and dimerization abilities of each given NR and its phylogenetic position, which is not the case for ligand-binding activity.

Nuclear receptors form a superfamily of phylogenetically related proteins, with 21 genes in the complete genome of *Drosophila melanogaster,* 48 in humans, 68 in the teleost fish (*Fugu rubripes*), and more than 270 genes in the nematode worm *Caenorhabditis elegans* [8]. Comparison of these gene families has provided and continues to provide valuable information regarding the origin and function of specific NR family members. The *C. elegans* NR set displays a high degree of duplication and divergence relative to both *Drosophila* and human sets. Conversely, all 68 *Fugu* receptors have a clear human homolog and are expressed, indicating that the majority of the additional *Fugu* receptors are likely to be functional [9].

Nuclear receptors share a common structural organization (Figure 2.2) including five domains [8,10]. The N-terminal region (A/B domain) is highly variable and contains at least one constitutionally active transactivation region (AF-1) whose action is independent of the presence of the ligand, and several autonomous transactivation domains. A/B domains are variable in length, from less than 50 to more than 500 amino acids. The most conserved region is DNA binding domain (DBD, C domain), which notably contains a short motif involved in dimerization of nuclear receptor and two zinc-fingers, which bind to specific sequences of DNA called hormone response element (HRE). A less conserved flexible hinge region (D domain) contains

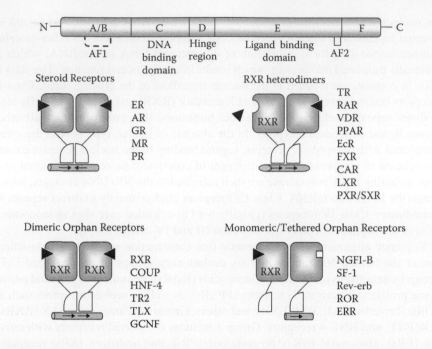

FIGURE 2.2 Domain organization of nuclear receptor superfamily. The six domains (A–F) of nuclear receptors: variable N-terminal region (A/B) contains ampf-I activation functions, DNA-binding domain (region C), variable length hinge region (region D), ligand-binding domain (region E), and variable length C-terminal domain (region F). (Adapted from J.M. Olefsky, *J. Biol. Chem.*, 276, 36863, 2001. With permission.)

the nuclear localization signal (NLS) influencing intracellular trafficking and sub-cellular distribution. Its flexibility can provide DBD rotation along LBD by 180° that is important for the interaction of receptor dimmers with asymmetric HREs. The ligand-binding domain (LBD, E domain) is moderately conserved in sequence and highly conserved in structure between the various nuclear receptors. The E domain is responsible for many functions, mostly ligand induced, notably the AF-2 transactivation function, a strong dimerization interface, another NLS, and often a repression function. Nuclear receptors may or may not contain an extremely variable final C-terminal domain (F domain).

NRs are highly regulated DNA-binding transcription factors that control transcription via several distinct mechanisms, which include both activation and repression activities. After site-specific DNA binding, their final transcription activity depends on the set of associated proteins, the coactivators and corepressors, interacting with them. The functions of these coregulators are varied and include chromatin remodeling (making the target gene either more or less accessible to transcription) or bringing function to stabilize the binding of other regulatory proteins. The binding of agonist ligands to nuclear receptors induces a conformation of the receptor that preferentially binds coactivator proteins. These proteins often have an intrinsic histone acetyltransferase (HAT) activity that weakens the association of histones to

DNA, and therefore promotes gene transcription. The binding of antagonist ligands to nuclear receptors, in contrast, induces a conformation of the receptor that preferentially binds corepressor proteins. These proteins in turn recruit histone deacetylases (HDACs), which strengthen the association of histones to DNA and therefore repress gene transcription. Depending on the receptor involved, the chemical structure of the ligand, and the tissue that is being affected, nuclear receptor ligands may display dramatically diverse effects ranging in a spectrum from agonism to antagonism to inverse agonism. NR signaling is remarkably complex because many receptors respond to cellular signals through ligand-dependent or ligand-independent mechanisms and because many accessory coregulators dictate cell-specific transcriptional responses to a given receptor [3].

The activity of endogenous ligands when bound to their cognate receptors is normally to up-regulate gene expression. This stimulation of gene response by the ligand is referred to as an agonist response. The agonist effect of the endogenous hormones can also be mimicked by xenobiotics. Agonist ligands work by inducing a conformation of the receptor that favors coactivator binding. Other nuclear receptor ligands have no apparent effect on gene transcription in the absence of an endogenous ligand. However, they block the effect of an agonist through competitive binding to the same binding site in the nuclear receptor. These ligands are referred to as antagonists. Antagonist ligands work by inducing a conformation of the receptor that prevents coactivators and promotes corepressor binding. Finally, some nuclear receptors promote a low level of gene transcription in the absence of agonists (also referred to as basal or constitutive activity). Xenobiotics that reduce this basal level of activity in nuclear receptors are known as inverse agonists.

A number of ligands that work through nuclear receptors display an agonist response in some tissues and an antagonistic response in other tissues. Selective receptor modulator (SRM) describes the ability of a ligand to manifest receptor agonist activity in some tissues but blocks receptor activity in other tissues [11,12]. The mechanism of action of SRMs may vary depending on the chemical structure of the ligand and the receptor involved; however, it is thought that many SRMs work by promoting a conformation of the receptor that is closely balanced between agonism and antagonism. In tissues where the concentration of coactivator proteins is higher than corepressors, the equilibrium is shifted in the agonist direction. Conversely, in tissues where corepressors dominate, the ligand behaves as an antagonist [13]. The properties of SRMs are due in part to unique ligand-induced conformational changes in the hormone receptor that affect the subsequent tissue-specific recruitment of other nuclear factors required for ligand-induced gene expression [14]. The antiestrogenic drug tamoxifen is an example of a selective estrogen receptor modulator, but structurally diverse natural and synthetic compounds can have selective receptor modulating activity

The most common mechanism of nuclear receptor action involves direct binding of the nuclear receptor to a DNA hormone response element. This mechanism is referred to as transactivation. However, some nuclear receptors not only have the ability to directly bind to DNA, but also to other transcription factors. This binding often results in deactivation of the second transcription factor in a process known as transrepression [15].

It has been observed that some effects from the application of hormones such as estrogens occur within minutes, which is inconsistent with the classical genomic mechanism of nuclear receptor action. It has been hypothesized that there are variants of nuclear receptors that are membrane associated instead of being localized in the cytosol or nucleus. These nongenomic mechanisms or membrane-initiated signals are activated quickly and are refractory to transcription and translation inhibitors [16–19].

From a toxicological point of view, two superfamilies of nuclear signaling molecules have gained enormous attention in the past years, in particular because our understanding of the molecular basis of their activation and downstream effects has been greatly increasing. These superfamilies include the steroid hormone receptors and PAS receptors. The first family includes (each with different subtypes) the peroxisome proliferator-activated receptor (PPAR), the thyroid hormone receptor (TR), the estrogen receptor (ER), the androgen receptor (AR), and others. The second family includes the aryl hydrocarbon receptor (AhR).

A number of nuclear receptors, referred to as orphan receptors, have no known or at least generally agreed upon endogenous ligands. Some of these receptors, such as FXR, LXR, and PPAR, bind a number of metabolic intermediates such as fatty acids, bile acids, and sterols with relatively low affinity. These receptors hence may function as metabolic sensors. Other nuclear receptors, such as CAR and PXR, appear to function as xenobiotic sensors up-regulating the expression of cytochrome P450 enzymes that metabolize these xenobiotics.

2.2.1 Estrogen Receptors as a Target for Endocrine Disruption

Initial concern regarding EDCs was focused on environmental estrogens. Estrogenic compounds exert pleiotropic effects in wildlife and humans, and endogenous estrogens, like 17β-estradiol, regulate growth and development of their target tissue. A vast range of both natural and man-made steroidal and nonsteroidal compounds has been investigated for estrogen activities. Structural studies of the ligand-binding domains of nuclear hormone receptors have provided a wealth of information on the nature of ligand-binding and its role in receptor activation. Such studies have revealed numerous structural motifs that are able to bind to ER and exert either estrogenic or antiestrogenic activities.

ER ligands, known to produce distinct biological effects, induce distinct conformational changes in the receptors, providing a strong correlation between ER conformation and biological activity [20].

In humans, two estrogen receptors, alpha and beta, mediate the action of endogenous and environmental estrogens, including effects on cell proliferation and cancer induction. Each ER activates ERE containing target genes in the presence of estrogen. In most ERE contexts, ERβ tends to be a weaker activator than ERα, and the weaker activation is dominant in cells with both receptors [21,22]. Moreover, one role of ERβ is to modulate ERα transcriptional activity, and thus the relative expression level of the two isoforms will be a key determinant of cellular responses to agonists and antagonists [14]. All three possible dimers (ERα or ERβ homodimers and ERα/ERβ heterodimers) are able to bind to ERE and, consequently, both homodimers and heterodimers can induce gene expression. ERβ homodimers appear to be less active than ERα/ERβ heterodimers, and evidence exists that ERβ decreases activity of ERα when cotransfected to human cells [14]. The ligand-receptor complex induces differential gene expression depending on the target cell. It has to be able to interact with coactivators and to displace corepressors to modify chromatin structure and induce gene transcription.

The ERs, α and β (and possibly γ), in fish, alone or in combination, may control different subsets of genes. Research indicates that E2 mimics may bind differently to these receptors, acting as agonist in one case and antagonist in another. This complicates the issue of determining the risks of environmental exposure [23].

The receptor has a general tolerance for binding appropriately substituted phenols and other classes of lipophilic compounds. Estradiol, estrone, and estriol are the principal endogenous mammalian estrogens. Because endogenous estrogens are hormones produced naturally and are essential for development, health, and reproduction, and because estrogen pharmaceuticals are administered for their hormonal effect in regulating fertility and for menopausal hormone replacement, these compounds are not typically thought of as potential endocrine disruptors in humans. Nevertheless, there is evidence that some of these compounds can find their way into rivers and streams and can affect aquatic wildlife [24]. Estrogens from natural sources include natural products in food, such as soy isoflavones and whole grain lignans, as well as microbial products and components from wood. Aside from pharmaceutical, man-made estrogen ligands can be found in industrial products, such as alkyl phenols from nonionic detergents, bisphenols from plastics, indicator dye impurities, polymer chemicals, and chlorinated aromatics and pesticides [12].

The interior of the ligand-binding pocket exhibits a considerable degree of plasticity, and reshapes itself around the contours of the ligand. It is not surprising, then, that the nuclear hormone receptors, with their large, flexible, and hydrophobic ligand-binding pockets, are able to interact with ligands that span a wide range of sizes and structures. The affinity with which exogenous ligands bind to these receptors depends, of course, on their structure and their functional nature, and the degree to which they deviate in size, shape, functional nature, and hydrophobilicity from the natural high-affinity ligands. The relationship between ligand structure and binding affinity in nuclear hormone receptors, however, is not always a smooth one where predictions can be made with confidence. In some cases, small changes in structure and stereochemistry can have large effects on ligand binding, yet molecules of different size and shape can have comparable affinities [12].

The functions of estrogens in cellular metabolism and signaling have been increasingly expanded beyond the now classical and conventional functions mediated by nuclear receptors. For example, E2 also modulates gene expression by an indirect mechanism that involves the interaction of ER with other transcription factors which, in turn, bind their cognate DNA element. In this case, ER modulates the activities of transcription factors such as the activator protein (AP-1), nuclear factor-κB (NF-κB), and stimulating protein-1 (SP-1), by stabilizing DNA-protein complexes and/or recruiting coactivators [25]. In addition, E2 binding to ER may exert rapid actions that start with the activation of a variety of signal transduction pathways. The cell context-specific environment (for example, differentiation, ER level, and ER coexpression) has an impact on the integration of rapid signaling by E2 from the membrane and on subsequent nuclear transcription. This leads to different signal cascades, different gene expression in response to the same hormone, and different cell biological outcomes. The debate about the contribution of different ER-mediated signaling pathways to coordinate the expression of specific sets of genes is still open [25]. The physiological effect of ER action induced by a specific ligand is therefore not solely defined by the ligand structure, but rather by the interaction of the receptor-ligand complex with various modulating factors. The interplay between ligand-bound ER and these factors determines the response depending on the target cell. This explains the fact that *in vitro* tests are suitable for the identification of potentially endocrine-active compounds but cannot be used to deduce the risk of endocrine-related adverse effects [26].

Many estrogens have been identified using *in vitro* assays (for example, ER binding, cell proliferation, and transcriptional activation), and several also display estrogenic activity *in vivo*, including methoxychlor, chlordecone, octylphenol, nonylphenol, bisphenol A, genistein, and ethynyl estradiol [1]. Other chemicals that have shown evidence *in vitro* of estrogenic activity have not shown similar evidence in *in vivo* systems, and so caution is warranted in interpreting *in vitro* results without *in vivo* confirmation.

2.2.2 Androgen Receptors as a Target for Endocrine Disruption

The androgen receptor (AR) is a member of NR subfamily 3C and, alone or in combination, plays a role in some of the most fundamental aspects of physiology such as immune function, metabolism, development, and reproduction. Multiple signaling pathways have been established. One main signaling pathway is via direct DNA binding and transcriptional regulation of responsive genes. Another is via protein–protein interactions, mainly with other transcription factors, to regulate gene expression patterns. Both pathways can up-regulate and down-regulate gene expression. Both pathways require ligand activation of the receptor and interplay with multiple protein factors such as chaperone and coregulator proteins [27]. The formation of a productive AR transcriptional complex requires the functional and structural interaction of the AR with its coregulators. In the last decade, a great amount and ever-increasing number of proteins have been proposed to possess AR coactivating or corepressing characteristics. A description of the AR regulation is beyond the scope of this chapter, but an overview of the androgen coregulators can be found in Heemers and Tindall [28]. In addition to the classical AR genomic pathway, rapid nongenomic effects of androgens emerge and are now generally accepted as contributing to the physiological effects of the steroids [17,29].

Although mammals are believed to possess a single AR [30], some piscivorous species have two distinct androgen receptors (ARs) with different characteristics. In the Atlantic croaker, *Micropogonias undulatus*, AR1 was identified in the brain and had high-affinity binding sites for testosterone. AR2 was found in the ovary and had high-affinity binding sites for 5α-dihydrotestosterone (DHT) [31]. It has physiochemical properties similar to those of other vertebrate ARs. AR2 has high-affinity binding for a broad spectrum of natural and synthetic androgens, including 17α-methyl-5α-dihydrotestosterone. The cytosolic AR2 interacts with heat shock proteins in a similar manner to other steroid receptors. In contrast, AR1 is highly specific for only a few androgens and does not interact with heat shock proteins in the usual manner [31].

A number of chemicals used in agriculture have been identified as precursors to environmental antiandrogens. Antiandrogenic chemicals alter sex differentiation by several different mechanisms. The cellular and molecular mechanisms of action of the antiandrogenic fungicide vinclozolin are some of the most thoroughly characterized. The ability of vinclozolin metabolites M1 and M2 to inhibit AR-dependent gene expression has been demonstrated both *in vitro* and *in vivo*. In addition, neither vinclozolin nor its metabolites display affinity for ER, although they do have a weak affinity for progesterone receptor [32,33]. Several other toxic substances have been shown to display AR-antagonist activity, including the DDT metabolite *p,p'*-DDE [34], methoxychlor active metabolite HPTE [35,36], the organophosphate fenitrothion [37], and the dicarboximide fungicide procymidone [38]. Linuron is a urea-based herbicide that displays a weak affinity for the AR, but the reproductive malformations induced in male rat offspring indicate that it may alter mammalian sex differentiation by more than one mechanism of action [39]. Moreover, data indicate that *in utero* exposure to linuron preferentially impairs testosterone-mediated, rather than DHT-mediated, reproductive development [40]. More recently, *in vivo* antiandrogenic activity of both linuron and fenitrothion was evidenced in teleost fish (*Gasterosteus aculeatus*) by using spiggin measurement in the kidney [41]. Moreover, Kang et al. [42] used the rodent Hershberger assay to demonstrate that procymidone may act as a stronger androgen receptor antagonist than vinclozolin, linuron, and *p,p'*-DDE.

In vertebrates such as fish, sexual dimorphism has been demonstrated to be affected by both estrogens and androgens [43]. Reproductive abnormalities, delayed sexual maturation, and modification of secondary sexual characteristics have been reported in fish exposed to pulp and paper mill effluents [44–49]. The androgen precursor androstenedione was identified in the river water and it was postulated that phytosterols from pine tree oil were metabolized to androstene-like compounds by *Mycobacterium* sp. bacteria [50]. Other studies have confirmed the presence of androgen activity in the river downstream of a paper mill, but several compounds might contribute to the effect and were not clearly identified [51]. However, bioactive substances originate from wood and are derived from lignin and terpenoids [45].

Another potential source for environmental androgens is excreted anabolic steroids administered to cattle to improve beef production [52]. The anabolic steroid trenbolone acetate is rapidly converted *in vivo* into 17ß-trenbolone, an active metabolite with androgen activity.

Recently, Chen et al. [53] reported a new category of endocrine disruption. In this report, the data presented on triclocarban (TCC; 3,4,4'-trichlorocarbanilide), a common ingredient in personal-care products, suggest that the bioactivity of endogenous hormones (testosterone) may be amplified by exposure to commercial personal-care products containing sufficient levels of TCC.

2.2.3 THYROID RECEPTORS AS A TARGET FOR ENDOCRINE DISRUPTION

In addition to the well-documented estrogen and androgen pathways, other hormonal systems are susceptible to disruption. Thyroid axis represents one potential target for environmental chemicals, and many different thyroid disruptions are possible.

In contrast to steroid hormone receptors, thyroid hormone receptors bind DNA in the absence of hormone, usually leading to transcriptional repression. Hormone binding is associated with a conformational change in the receptor that causes it to function as a transcriptional activator.

Mammalian thyroid hormone receptors are encoded by two genes, designated alpha and beta. Further, the primary transcript for each gene can be alternatively spliced, generating different alpha and beta receptor isoforms. Currently, four different thyroid hormone receptors are recognized: alpha-1, alpha-2, beta-1, and beta-2.

The different forms of thyroid receptors have patterns of expression that vary by tissue and by developmental stage. For example, almost all tissues express the alpha-1, alpha-2, and beta-1 isoforms, but beta-2 is synthesized almost exclusively in the hypothalamus, the anterior pituitary, and the developing ear. Receptor alpha-1 is the first isoform expressed in the conceptus, and there is a profound increase in expression of beta receptors in the brain shortly after birth. Interestingly, the beta receptor preferentially activates expression from several genes known to be important in brain development (for example, myelin basic protein), and up-regulation of this particular receptor may thus be critical to the well-known effects of thyroid hormones on development of the fetal and neonatal brain.

The presence of multiple forms of the thyroid hormone receptor, with tissue and stage-dependent differences in their expression, suggests an extraordinary level of complexity in the physiologic effects of thyroid hormone.

Thyroid hormone receptors bind to short repeated sequences of DNA called thyroid or T3 response elements (TREs) as monomers, as homodimers, or as heterodimers with the retinoid X receptor (RXR), another member of the nuclear receptor superfamily that binds 9-*cis* retinoic acid. The heterodimer affords the highest affinity binding and is thought to represent the major functional form of the receptor.

Thyroid hormone receptors bind to TRE DNA regardless of whether they are occupied by T3. However, in general, binding of thyroid hormone receptor alone to DNA leads to repression of transcription, whereas binding of the thyroid hormone-receptor complex activates transcription.

In vertebrates, thyroid hormones are essential for postembryonic development, such as establishing the central nervous system in mammals and metamorphosis in amphibians. The regulation of thyroid hormone delivery to tissues and cells during development and in the adult represents a very complex web of feedback systems, providing redundant and compensatory regulatory responses to maintain thyroid hormone signaling in the face of specific deficiencies in circulating levels of thyroid hormones. Environmental factors and xenobiotics can perturb this web at various points of regulation, inducing a variety of responses ([54] for a review). However, more research is needed to truly understand the degree of thyroid disruption induced by a particular toxicant that is necessary to be considered either "compensatory/adaptive" or "adverse" with respect to noncancer endpoints. Changes in thyroid hormone levels due to chemical exposure are generally considered adverse, but the thyroid field is still new, and continued research on new endpoints of thyroid hormone action is needed to link the changes in thyroid hormone levels to doses of chemicals and clear downstream actions. Moreover, toxicants interfering with the hypothalamic–pituitary–thyroid

(HPT) axis by different mechanisms appear to produce different effects on the relationship among the various endpoints within the thyroid endocrine system.

In mammals, thyroid hormones T3 and T4 are secreted from the thyroid gland under the control of thyroid-stimulating hormone (TSH or thyrotropin), which, in turn, is controlled in part by thyrotropin-releasing hormone (TRH) from the hypothalamus. The thyroid hormone triiodothyronine (T3) regulates gene expression by binding to a high-affinity nuclear receptor. Thyroid hormone receptors (TRs) bind DNA as heterodimers with the retinoid X-receptor. In general, T3-stimulated genes are involved in gluconeogenesis, glycogenolysis, lipogenesis, cell proliferation, and apoptosis, and genes repressed by T3 include some involved in insulin signal transduction, cell immunity, extracellular matrix structure, cell architecture, protein glycosylation, and mitochondrial functions [55].

Numerous environmentally relevant chemicals, including polychlorinated biphenyls (PCBs), pesticides, bisphenol A, polybrominated diphenyl ethers (PBDEs), and metals, exert acute or chronic effects on the thyroid cascade in various species, including rodents, fish, and amphibians [56–61]. However, the mechanisms underlying thyroid changes and their physiological consequences are usually poorly understood, because the thyroid cascade may respond indirectly and it has considerable capacity to compensate for abuses that otherwise would disrupt thyroid hormone homeostasis.

2.2.4 Other Nuclear Receptors

Recently, the PPAR/RXR system has also been identified as a target for endocrine disruption through organotin agents used as marine antifouling agents [62,63]. Furthermore, orphan nuclear receptors such as the pregnane X-receptor (PXR) and the constitutive androstane receptor (CAR), which are important regulators in the adaptation of chemical stress, are activated by a variety of ligands [64].

Pregnane X receptor (PXR), peroxisome proliferator activated receptors (PPARs), and the constitutive androstane receptor (CAR) have received particular attention because they bind to drugs and xenobiotics, which in turn activate the expression of genes involved in phase I [65], phase II [66], and phase III [67] of biotransformation pathways.

Moreover, PPARs, PXR, and vitamin D receptors act as sensors for various molecules encountered by the body on a daily basis. The effects of these ligands can be understood by the fact that numerous genes involved in the cellular processes, such as homeostasis, growth, and defense against microbes, are under the control of these NRs.

2.2.4.1 Peroxisome Proliferator-Activated Receptor/ Retinoid X Receptor (PPAR/RXR) System

The sequencing of mammalian genomes indicated that there are only three peroxisome proliferator-activated receptor (PPAR) isotypes [68]. In contrast to steroid hormone receptors, which act as homodimers, transcriptional regulation by PPARs requires heterodimerization with the RXR, which belongs to the same receptor superfamily [69,70]. When activated by a ligand, the dimer modulates transcription via binding to a specific DNA sequence element called a peroxisome proliferator response element (PPRE) in the promoter region of target genes. Transcriptional control by PPAR/RXR heterodimers also requires interaction with coregulator complexes. Thus, selective action of PPARs *in vivo* results from the interplay at a given time between expression levels of each of the three PPAR and RXR isotypes, affinity for a specific promoter PPRE, and ligand and cofactor availabilities.

PPARs exhibit broad, isotype-specific tissue expression patterns, and many cellular and systemic roles have been attributed to these receptors, such as cellular differentiation and development and metabolism of carbohydrates, lipids, and proteins.

PPARα is expressed at high levels in organs that carry out significant catabolism of fatty acids, such as brown adipose tissue, liver, heart, kidney, and intestine [71]. PPARβ/δ has the broadest expression pattern, and the levels of expression in certain tissues depend on the extent of cell proliferation and differentiation. PPARγ is expressed as two isoforms, of which PPARγ2 is found at high levels in the adipose tissues, whereas PPARγ1 has a broader expression pattern. Transcriptional regulation by PPARs requires heterodimerization with the RXR. Consistent with its distribution in tissues with high catabolic rates of fatty acids and high peroxisomal activity, the major role of PPARα is the regulation of energy homeostasis [72].

PPARβ/δ is necessary for placental and gut development and is also involved in the control of energy homeostasis [73–77]. In addition, PPARβ/δ has an important role in the control of cell proliferation, differentiation, and survival and is involved in tissue repair [78–81].

PPARγ plays a major role in adipose tissue differentiation and in maintaining adipocyte-specific functions, such as lipid storage in the white adipose tissue and energy dissipation in the brown adipose tissue [82–86]. Furthermore, it is required for the survival of differentiated adipocytes [87]. In addition, PPARγ is involved in glucose metabolism through an improvement of insulin sensitivity and thus represents a molecular link between lipid and carbohydrate metabolism [88–92].

A wide variety of natural or synthetic compounds were identified as PPAR ligands. Among the synthetic ligands, the lipid-lowering drugs, fibrates, and the insulin sensitizers, thiazolidinediones, are PPARα and PPARγ agonists, respectively, which underscore the important role of PPARs as therapeutic targets [68]. PPARα was discovered as the receptor mediating hepatic peroxisome proliferation and carcinogenesis in rodents in response to a wide range of chemicals including pesticides, industrial solvents, and plasticizers [93,94]. Recently, Feige et al. [95] demonstrated how monoethyl-hexyl-phthalate (MEHP), a metabolite of the most abundantly used phthalate (diethyl-hexyl-phthalate, DEHP), selectively activates different PPARγ target genes and promotes adipogenesis. These studies highlight some key mechanisms of endocrine disruption actions of DEHP through its metabolite MEHP and suggest that the metabolic function of PPARγ can be targeted by a subclass of endocrine disruptors defined as metabolic disruptors.

Another example of the implication of PPARs in endocrine disruption is given by organotin compounds, which have been widely used as agricultural fungicides, rodent repellents, and molluscicides and in antifouling paints for ships and fishing nets. In aquatic invertebrates, particularly marine gastropods, organotin compounds, such as tributyltin (TBT) and triphenyltin (TPT), induce irreversible sexual abnormality in females, which is termed "imposex" at very low concentrations. Imposex has been established as a form of endocrine disruption caused by elevated testosterone levels, leading to masculinization in organotin-exposed females [96–98]. Using *in vitro* systems for human nuclear receptor activation, Kanayama et al. [63] found that TBT and TPT were potential agonists of $RXR\alpha$ and $PPAR\gamma$. In addition, these compounds induced the transactivation function of $RXR\alpha$ and $PPAR\gamma$ in mammalian cell culture, indicating that these organotin compounds function as $RXR\alpha$ and $PPAR\gamma$ agonists in mammalian cells. As the gene expression of human aromatase is regulated by the activation of $PPAR\gamma$ and/or RXR, the aromatase expression regulated by organotin compounds may involve the activation of $PPAR\gamma$ and RXR because the aromatase expression pattern induced in the human placenta and ovary by activation of $PPAR\gamma$ and/or RXR is similar to that induced by organotin compounds.

2.2.4.2 Pregnane X Receptor (PXR)

Two related nuclear receptors, the pregnane X receptor (PXR) and the constitutive androstane receptor (CAR), act as xenobiotic sensors that protect the body from a multitude of xenobiotics and play a central role in the metabolism and clearance of steroids and toxic endogenous lipids [99].

Like other type II nuclear receptors, when activated, it forms a heterodimer with the retinoid X receptor and binds to hormone response elements on DNA, which elicit the expression of gene products. The primary function of PXR is to sense the presence of xenobiotics and in response to up-regulate the expression of proteins involved in the detoxification and clearance of these substances from the body. CAR nuclear hormone receptors act in concert with PXRs to detoxify xenobiotics.

One of the primary targets of PXR activation is the induction of CYP3A4, an important phase I oxidative enzyme that is responsible for the metabolism of many drugs. In addition, PXR up-regulates the expression of phase II conjugating enzymes such as glutathione-S-transferase and phase III transport uptake.

The detoxification system consists of microsomal cytochrome P450 enzymes (CYPs) and other oxidizing and hydroxylating enzymes (phase I response), conjugation enzymes such as glucuronosyl- and sulfotransferases (phase II response), and membrane-bound drug pumps such as MDR1 (phase III) that function in a concerted fashion to inactivate and clear chemical compounds (reviewed in [100]). The same system is also utilized to metabolize endogenous compounds such as steroids, bile acids, thyroid hormones, retinoids, cytokines, and fatty acids. One characteristic of xenobiotic metabolizing enzymes and transporters is their inducibility by their substrates. This allows enhanced production of these proteins only as needed. These compounds also induce a variety of other metabolic enzymes and transporters.

From the available studies, it is likely that many industrial and natural endocrine-active substances bind and activate PXR, CAR, or both. PXR and CAR most likely play a protective role against most of these substances by promoting their detoxification and clearance. However, in rare but relevant cases, PXR and CAR may inadvertently promote their deleterious effects. First, detoxification reactions, such as CYP3A induction, are also known to activate certain substrates to carcinogenic and cytotoxic products. For example, the first step in aflatoxin metabolism creates a reactive adduct for DNA, enabling its activity as a hepato-carcinogen. Thus, activation of PXR or CAR may enhance the toxicity of some environmental chemicals. Second, because steroids and thyroid hormones are metabolized and typically inactivated by enzymes such as CYPs, UGTs, and SULTs, constitutive activation of PXR or CAR by environmental chemicals could alter endocrine systems. In fact, the chronic activation of PXR in transgenic mouse livers results in increased corticosterone in serum and urine and, by extension, most likely stimulates the production of gonadal steroids [101]. Furthermore, prototypical chemical activators for PXR or CAR induce increased metabolism and decreased level of thyroid hormones, consequential increase in thyroid-stimulating hormone, and thyroid hypertrophy in rats [102]. Further studies are required to clarify the effect of chronic activation or inactivation of the xenobiotic receptors on global endocrine physiology and disease progression [99].

2.2.5 ARYL HYDROCARBON RECEPTOR (AHR)

The aryl hydrocarbon receptor (AhR) is a member of the basic helix loop-helix (bHLH) PER-ARNT-SIM (PAS) family of nuclear transcription factors that heterodimerize with another PAS family member, the AhR nuclear translocator (ARNT), to mediate gene expression. The mechanism of AhR-mediated gene transactivation is similar to that of DNA binding proteins that belong to the steroid receptor family. AhR mediates the biological effects of polycyclic aromatic hydrocarbons (PAHs), polychlorinated dibenzodioxins (PCDDs), and polychlorinated dibenzofurans (PCDFs), by-products from incomplete combustion of fossil fuels, wood, and other organic substances [103].

AhR was originally identified and characterized as a result of its central role in the vertebrate response to many planar aromatic hydrocarbons. In this capacity, AhR binds exogenous ligand and transcriptionally activates a battery of enzymes that promote metabolic transformation and excretion of the xenobiotic from the organism. This AhR-mediated pathway is commonly viewed as an adaptive response toward these xenobiotics. The activated AhR/ARNT heterodimer complex binds to its cognate DNA sequences, termed "xenobiotic response elements" (XREs), and activates the expression of AhR target genes, such as cytochrome P4501A1 (CYP1A1), and CYP1B1 [104]. This toxicity mediated by AhR receptor remains the most clearly understood aspect of AhR biology (see [105] for review). More recently, efforts have been made to understand the physiology of the AhR under normal cellular conditions (in the absence of xenobiotics) on cell proliferation and differentiation, endogenous mechanism of activation, gene regulation, tumor development, cell motility and migration, and so on [106]. Interestingly, significant differences in the metabolism of EDCs can result in market species differences in response to these chemicals.

Modulation of estrogen receptor signaling by association with activated AhR has been demonstrated [107]. Agonist-activated AhR/ARNT heterodimer directly associates with estrogen receptors ERα and ERβ. This association results in the recruitment of unliganded ER and coactivator p300 to estrogen-responsive gene promoters, leading to activation of transcription and attenuated estrogenic effects. Several studies have reported that activated AhR inhibits the expression of E2-induced genes

[108]. The precise molecular mechanisms for this cross talk are unclear and may be a combination of several different mechanisms: direct inhibition by the activated AhR/ARNT heterodimer through binding to inhibitory XRE present in ER target genes; squelching of shared coactivators, including ARNT; synthesis of an unknown inhibitory protein; increased proteosomal degradation of ER; and altered estrogen synthesis/metabolism through increase in aromatase and cytochrome P450 1A1 and 1B1 expression [109]. Active AhR can also redirect ER from ER target genes to AhR target genes, suggesting that AhR can regulate ERα protein levels and, consequently, estrogenic responses [110].

2.3 ENDOCRINE DISRUPTION BY MODULATING STEROID HORMONE METABOLISM

2.3.1 STEROIDOGENIC ENZYMES

The ability of xenobiotics to disrupt steroidogenesis and the mechanisms by which these compounds interfere with the function of steroidogenic enzymes is a relatively unexplored area of endocrine toxicology. Nevertheless, key enzymes involved in steroid hormone synthesis and metabolism are being considered as important targets for EDCs. The cytochrome P450 enzymes responsible for the highly specific reactions in the steroid biosynthesis pathway are particularly gaining interest as molecular targets, given their key role in the formation of various highly potent endogenous steroid hormones. It is possible for certain chemicals to cause or contribute to hormonal disruption by interfering with the function of key enzymes involved in steroid synthesis and breakdown [111]. Steroidogenic enzymes are responsible for the biosynthesis from cholesterol of various steroid hormones, including glucocorticoids, mineralocorticoids, progestins, androgens, and estrogens. They consist of several specific cytochrome P450 enzymes (CYPs), hydroxysteroid deshydrogenases (HSDs), and steroid reductases [112]. The enzyme that has received the most attention with regard to enzyme modulation and endocrine disruptors has been aromatase (CYP19) that converts androgens to estrogens (Figure 2.3) [113]. The interference of EDCs with aromatase CYP19 expression or activity and the consequences for reproduction of teleost fish have been recently reviewed [114], and it was shown that expression and/or activity of aromatase CYP19 genes may be affected through a variety of mechanisms. Nevertheless, all the steroidogenic enzymes are potential targets for disruption. Table 2.1 [115–138] synthesizes the main examples of steroidogenic enzyme disruptions in humans and fish.

2.3.2 METABOLISM OF STEROID HORMONES

Increasing or decreasing of steroid metabolism contributes to the detrimental effect of EDCs. In humans and rodents, 50% or more of the drugs and pesticides currently used induce the expression of the hepatic enzyme cytochrome P450 (CYP)3A [139]. Numerous steroids, including testosterone, 17β-estradiol, progesterone, and androstenedione, are metabolized by CYP3A [139]. It is therefore conceivable that wildlife or

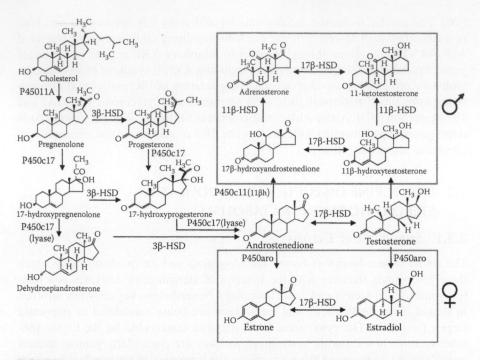

FIGURE 2.3 Steroid synthesis of the sex hormones and the steroidogenic enzymes in fish gonads (*Onchorhynchus mykiss* and *Tilapia sparrmanii*). Note that biosynthesis pathways for pregnenolone and androstenedione are the same in humans. 3β-HSD: 3β-hydroxysteroid dehydrogenase (Δ4/5-isomerase); P450c17: cytochrome P450/C$_{17-20}$lyase; 17β-HSD: 17β-hydroxysteroid dehydrogenases; 11-βHSD: 11β-hydroxysteroid dehydrogenases; P450aro: cytochrome P450 19; P450c11 (11 β-H): 11 βhydroxylase. (Adapted from J. F. Baroiller, Y. Guigen, and A. Fostier, *Cell. Mol. Life Sci.*, 55, 910, 1999. With permission.)

humans exposed to contaminated environments would exhibit elevated CYP3A activity that leads to increased clearance of such steroids as testosterone and 17β-estradiol from the plasma, as reported in chickens [140,141] and humans [142]. Transcription of CYP3A, as well as of other hepatic enzymes involved in the biotransformation of testosterone, pharmaceutical agents, and xenobiotics, appears to be regulated, at least in part, by SXR and related nuclear transcription factors [143,144]. Many vertebrates exhibit species-specific, sexually dimorphic patterns of hepatic enzyme activity that appear to be regulated by sex steroids and/or growth hormone [145]. For example, differences in the activity of the hepatic enzymes responsible for the biotransformation of sex steroids exist among alligators collected from contaminated and relatively uncontaminated sites in Florida [146,147]. The observed differences in plasma sex steroid concentrations could be caused, in part, by differences in hepatic clearance of these sex steroids [148] as demonstrated in other species [149,150].

Interestingly, several members of the nuclear receptor superfamily are known to be degraded through the ubiquitin-proteasome pathway in a ligand-dependent pathway [151]. Inhibition of the ubiquitin-proteasome pathway down-regulates the transcriptional activity of nuclear-steroid receptors such as progesterone receptor [152] and

TABLE 2.1
Examples of Environmental Chemicals Interacting with Steroidogenic Enzymes

Enzyme	Substance	Type of Effect	Reference
AMPc	Bisphenol A/octylphenols	Inhibitor	[127]
	Lindane	Inhibitor	[130]
	Atrazine	Inductor	[132]
StAR Protein	Barbiturates	Inhibitor	[120]
	Lindane	Inhibitor	[137]
	"Roundup"	Inhibitor	[136]
	AhR Agonists (TCDD)	Inhibitor	[115]
P450scc	Cadmium chloride	Inhibitor/inductor	[118,123,207]
	Lead	Inhibitor	[121]
	Ketoconazole	Inhibitor	[122]
	AhR agonists	Inhibitor	[115]
3β-HSD	Cadmium chloride	Inhibitor	[123]
	Lithium chloride	Inhibitor	[119]
	Monobutyltin	Inhibitor	[125]
	Tributyltin	Inhibitor	[125]
	Isoflavonoids	Inhibitor	[128]
P450c17	Isoflavonoids	Inhibitor	[128]
	Aroclor 1260 (Askarel®)	Inhibitor	[117]
17β-HSD / 17KSR	Lithium chloride	Inhibitor	[119]
	Dicofol	Inhibitor	[134]
	Dibutyltin	Inhibitor	[134]
5α-reductase	Dicofol	Inhibitor	[134]
	Atrazine	Inhibitor	[134]
	Organotins	Inhibitor	[134]
	Phathalates	Inhibitor	[134]
	TCDD	Inhibitor	[129]
	Vinclozolin*	Inhibitor	[124]
Aromatase	Fenarimol	Inhibitor	[135]
	Prochloraz	Inhibitor	[116]
	Propiconazole	Inhibitor	[132]
	Endosulfan	Inhibitor	[138]
	Atrazine	Inductor	[133]
	Tributyltin	Inhibitor/inductor	[126,131]

* 2-[[(3,5-dichlorophenyl)-carbamoyl]-oxy]-2-methyl-3-butenoic acid and 3',5'-dichloro-2-hydroxy-2-methylbut-3-enanilide are the main la 5α-reductase inhibitor degradation products of vinclozolin.

AR [153,154]. ERα undergoes different rates of proteasome-mediated degradation in the presence of ER agonists, antagonists, and selective ER modulators, demonstrating that transcriptional activity can be affected by modulating receptor stability [155]. This leads to the hypothesis that EDCs could act on proteasome-mediated degradation of nuclear receptors or coregulatory proteins to directly affect the magnitude and duration of normal hormone responses, thereby causing endocrine disrupting effects [151].

2.4 HORMONE TRANSPORT

In most vertebrate species, sex steroid hormones circulate in the plasma predominantly bound to a specific high-affinity sex hormone–binding globulin (SHBG) and low-affinity proteins such as corticosteroid-binding proteins and albumin (see [156] for review). In the blood, typically 97% to 99% of total estrogens and androgens are carried bound to these proteins. Although the functions of the SHBG are not fully understood, it is believed to be involved in regulating circulating endogenous sex steroids as well as cellular signal transduction to nuclear steroid receptors in sex steroid–sensitive tissues (see [157] for review). Interestingly, binding proteins have been reported to bind several endocrine disruptors [158–162] and to alter the biological activity of natural and synthetic estrogens [163–165]. The consequence is that modulation of SHBG constitutes a new indirect pathway for regulating sex steroid action. In practice, the main regulators for SHBG production are the sex steroids as well as other regulators of SHBG production.

2.5 TRANSGENERATIONAL CONSEQUENCES OF EDC EXPOSURE

Recent studies have demonstrated the ability of EDCs to have epigenetic effects (DNA methylation) [166–170]. When these changes occur during certain stages of development, they are permanent and can be inherited by offspring. The term "epigenetic" means outside conventional genetic [171] and was coined by the developmental biologist Conrad H. Waddington [172]. Epigenetics is the process by which the genotype of an organism interacts with the environment to produce its phenotype. It provides a framework to explain the source of variations in individual organisms [173] and also explains what makes cells, tissues, and organs different despite the identical nature of the genetic information in every cell in the body. Currently, epigenetics is defined as the molecular phenomena that regulate gene expression without alterations to DNA sequence [174]. The most studied epigenetic modification is DNA methylation of CpG nucleotides that are essential for mammalian development [175,176]. In most studies, increased DNA methylation is associated with gene silencing, and decreased methylation is associated with gene activation.

These epigenetic mechanisms help explain the transgenerational effects of some hormonally active chemicals. Diethylstilbestrol (DES) during pregnancy results in vaginal adenocarcinoma in female offspring in humans [177] and mice [178]. Female offspring of mice exposed to DES themselves express this same rare genital tract cancer [179]. Newbold and colleagues showed that specific rare testes cancers are also expressed and therefore transmitted to the male offspring of females treated *in utero* with DES [169,180]. Rats treated with the estrogenic pesticide methoxychlor or the antiandrogenic fungicide vinclozolin during pregnancy produce male offspring that have decreased sperm capacity and fertility [166]. Remarkably, the compromised fertility is passed through the adult male germ line for four generations. The authors demonstrated altered patterns of DNA methylation in germ cells of generations two and three. Interestingly, individuals initially are fertile, but with age, fertility is reduced. This study is an elegant demonstration of epigenetic alteration in genes that is apparently important to reproductive function, by two kinds of EDCs in age-dependent manner. In addition to research linking EDCs and epigenetic and reproductive diseases, a growing body of information suggests that epigenetic effects might extend to gender differences in brain and behavior with potential evolutionary significance [181].

Other environmental contaminants such as 2,3,7,8-tetrachlorodibenzo-*p*-dioxin (TCDD) [182], polychlorinated biphenyls (PCBs) [183,184], phthalate esters [167], and chlorine disinfection by-products [185] also affect the reproductive system or induce tumor development by altering DNA methylation and steroid hormone metabolism and signaling.

2.6 CROSS TALK BETWEEN COMPONENTS OF ENDOCRINE SYSTEMS

Cross talk between nuclear receptor–mediated and other signal-transduction pathways is an important aspect of nuclear receptor–based regulation. Phosphorylation, for example, has been extensively studied, but other types of modifications such as ubiquitylation and acethylation have been demonstrated [186].

Another type of nuclear receptor cross talk that has only recently been recognized is the "nongenomic" action of several nuclear receptors. It is now clear that some nongenomic actions of nuclear receptor ligands are apparently mediated through membrane receptors that are not part of the nuclear receptor superfamily.

2.7 MULTIPLE AND CRITICAL PERIODS OF EXPOSURES

2.7.1 Exposure to Multiple Chemicals

Most research resources are devoted to single chemical studies with an almost complete neglect of mixture studies. A contributing factor to this imbalance is no doubt the inaccessibility of theoretical concepts in mixture toxicology and the resulting uncertainty as to how to proceed experimentally [187]. Mixture studies are perceived to be challenging, both conceptually and experimentally, but despite these difficulties, many articles on combination effects of EDCs have been published (reviewed in [187,188]). As estrogenic chemicals have been the focus of most of the work on EDCs, it is not surprising that this group of substances has been the topic of the majority of EDC mixture studies. Nevertheless, some studies have been performed on AR antagonists and to a lesser extent on mixtures of thyroid-disrupting chemicals. Globally, combined effects of EDCs belonging to the same category (for example, estrogenic, antiandrogenic or thyroid-disrupting agents) can be predicted by using dose addition. This applies to a wide range of endpoints reflecting various levels of hormone actions in a variety of organisms [189–194]. Combinations of EDCs are able to produce significant effects, even when each chemical is present in low doses that individually do not induce observable effects. However, comparatively little work has been performed with mixtures of different classes of endocrine disruptors, such as estrogenic agent combined with antiestrogenic chemicals, or EDCs combined with other toxicants. Perhaps the best-known example of effect modulation is the inhibitory effect of AhR agonists on the action of estrogenic chemicals.

2.7.2 DOSE-RESPONSE RELATIONSHIPS

A number of specific factors influence the dose of an EDC that reaches the target cells to produce a response. These factors include route of administration, distribution, metabolism, rate of clearance, plasma transport, cell uptake, affinity for the receptor subtype in the cell, and interaction of the ligand receptor complex with tissue-specific factors comprising the transcriptional apparatus.

The physiological and the environmentally relevant dose ranges typically fall well below the toxicological dose range based on using established protocols for examining acute toxic effects of chemicals. Furthermore, toxicology assumes that it is valid to extrapolate linearly from high doses over a very wide dose range to predict responses at doses within the physiological range of receptor occupancy for an EDC; however, because of receptor-mediated response saturate, this assumption is incorrect. Furthermore, receptor-mediated responses can first increase and then decrease as dose increases, contradicting the assumption that dose-response relationships are monotonic. Exogenous hormones modulate a system that is physiologically active and thus is already above threshold, contradicting the traditional toxicological assumption of thresholds for endocrine responses to EDCs [195,196].

2.7.3 FETAL AND DEVELOPMENTAL BASIS OF ADULT DISEASE

The timing of exposure to exogenous hormonally active substances is critical to the outcome of that exposure, with early lifetime exposures particularly detrimental because they produce permanent effects [197]. It is generally accepted that EDCs have the greatest impact when exposure occurs during development [198,199]. During fetal life, endogenous hormones regulate the differentiation and growth of cells, and developmental processes appear to have evolved to be exquisitely sensitive to changes in hormone concentrations. Those life stages most vulnerable to ED are the prenatal and early postnatal periods, because these are times when organ and neural systems are changing most rapidly. Pubertal and perimenopausal periods may also be sensitive windows of exposure because of the changing hormonal effects during these periods [200]. Numerous strong examples of the fetal/developmental hypothesis can be found in the literature for a variety of systems and species: male infertility in rats [201], metabolism and obesity [202,203], mouse genital tracts [178], mammary glands [204], and so on.

2.7.4 MULTIPLE EFFECTS INDUCED BY A SINGLE COMPOUND

It is actually assumed that a single chemical may disrupt the endocrine system via different mechanisms of action implying different targets. A good example is constituted by an *in vitro* study on 200 pesticides for estrogen and androgen receptor activities [205]. In this study, 29 pesticides had both hERα and hERβ agonistic activities, and 34 pesticides possessed both estrogenic and antiandrogenic activities indicating pleiotropic effects on hER and hAR. A recent review, summarizing the endocrine properties of 127 pesticides, confirmed the complexity of the responses induced even with a single compound [206].

2.8 CONCLUSIONS

The potential mechanisms underlying the effects of EDCs are incredibly diverse, making studies on their biological effects daunting. Endogenous hormones act through several mechanisms. The classical mechanism of action for hormones such as estrogens, androgens, thyroid, and progesterone involves the binding of the hormone to the receptor, the interaction of this hormone-receptor complex with other cofactors in a cell, and the activation or inactivation of transcription of a target gene. More recently, membrane steroid hormone receptors using different intracellular signaling pathways have been identified. An important consideration is that a unique ligand activates a diversity of target receptors and signaling mechanisms and may interact with completely different complements of cofactors depending upon the phenotype of the target cell.

Hormone signaling also involves the synthesis, degradation, or inactivation of the hormones by specific enzymes, any or all of which may be targeted by EDCs. Another consideration is that the endogenous hormones, particularly estrogens, androgens, and thyroid, bind to proteins in blood that reduce their bioavailability. EDCs may not bind to the same binding proteins, thereby increasing their bioavailability relative to endogenous hormones. Furthermore, if EDCs are not as rapidly metabolized as endogenous steroid hormones, they may remain bioavailable far longer and get incorporated to body burden, generally the fat tissues, as most EDCs are lipophilic.

Another challenge for endocrine disruption is to understand how extremely low doses of EDCs can exert potent effects on endocrine and homeostatic systems and why EDCs exert nonlinear dose-response curves, often U or inverted U in shape [195,197]. The overall shape of the dose-response curve thus reflects the cumulative action of EDCs upon a range of targets. It is now clear that low-dose EDC exposure, particularly at vulnerable development windows, can have long-term consequences on later health.

Related to the low-dose phenomenon is the question of whether thresholds for different EDCs actually exist. Environmental toxicological protocols continue to utilize a single dose of a single chemical at different concentrations, seeking the lowest dose at which no adverse effects are observed in the animal subject. This no observed adverse effect level (NOAEL) for the chemical in question is then used as a threshold dose for risk assessment. However, a power-analysis study revealed that thresholds may not exist for estrogenic EDCs, as any amount of the exogenous steroidal agent automatically exceeds the organism's threshold [196].

Various chemicals will have a combination of effects on the endocrine system, as they may act as steroid receptor (ant)agonists, steroidogenic enzyme inducers/inhibitors, and, via other less well-understood mechanisms, cause net effects on the endocrine system that will be highly concentration and endpoint dependent. Thus, it is to be expected that for many EDCs more than one mechanism will play a role, inevitably resulting in complex dose-response relationships for many different endocrine parameters.

Furthermore, EDCs rarely occur for a single substance, and for the most part, the environment is contaminated by complex mixtures.

Nevertheless, research has clearly shown that EDCs can act at multiple sites via multiple mechanisms of action. Because of cross talk between different components of the endocrine systems, effects may occur unpredictably in endocrine target tissues other than the system predicted to be affected. All these factors confirm that to make the prediction from *in vitro* and *in vivo* experiments to risk analysis for human and wildlife is extremely complex and indeed hazardous.

ACKNOWLEDGMENT

This study was granted by the French Ministry of Ecology and Sustainable Development (PNETOX-N°24-B/2004-N°CV 04000175).

REFERENCES

[1] IPCS, *Global assessment of the state-of-the-science of endocrine disruptors*, in *International Program on Chemical Safety*, IPCS, 2002, pp. 180.

[2] J.M. Olefsky, *Nuclear receptor minireview series*, J. Biol. Chem. 276 (2001), pp. 36863–36864.

[3] P. Germain, B. Staels, C. Dacquet, M. Spedding, and V. Laudet, *Overview of nomenclature of nuclear receptors*, Pharmacol. Rev. 58 (2006), pp. 685–704.

[4] D.J. Mangelsdorf, C. Thummel, M. Beato, P. Herrlich, G. Schutz, K. Umesono, B. Blumberg, M. Kastner, M. Mark, P. Chambon, and R.M. Evans, *The nuclear receptor superfamily: The second decade*, Cell 83 (1995), pp. 835–839.

[5] N. Novac and T. Heinzel, *Nuclear receptors: Overview and classification*, Curr. Drug Targets Inflamm. Allergy 3 (2004), pp. 335–346.

[6] V. Laudet, *Evolution of the nuclear receptor superfamily: Early diversification from an ancestral orphan receptor*, J. Mol. Endoc. 19 (1997), pp. 207–226.

[7] Nuclear Receptor Nomenclature Committee, *A unified nomenclature system for the nuclear receptor superfamily*, Cell 97 (1999), pp. 161–163.

[8] M. Robinson-Rechavi, H. Escriva Garcia, and V. Laudet, *The nuclear receptor superfamily*, J. Cell Sci. 116 (2003), pp. 585–586.

[9] J.M. Maglich, J.A. Caravella, M.H. Lambert, T.M. Willson, J.T. Moore, and L. Ramamurthy, *The first completed genome sequence from a teleost fish* (Fugu rubipes) *adds significant diversity to the nuclear receptor superfamily*, Nucleic Acids Res. 31 (2003), pp. 4051–4058.

[10] A.N. Smirnov, *Nuclear receptors: Nomenclature, ligands, mechanisms of their effects on gene expression*, Biochem. (Moscow) 67 (2002), pp. 957–977.

[11] B.S. Katzenellenbogen and J.A. Katzenellenbogen, *Defining the "S" in SERMs*, Science 295 (2002), pp. 2380–2381.

[12] J.A. Katzenellenbogen and R. Muthyala, *Interactions of exogenous endocrine active substances with nuclear receptors*, Pure Appl. Chem. 75 (2003), pp. 1797–1817.

[13] C.A. Smith and B.W. O'Malley, *Coregulator function: A key to understanding tissue specificity of selective receptor modulators*, Endocrinol. Rev. 25 (2004), pp. 45–71.

[14] J.M. Hall and D.P. McDonnell, *The estrogen receptor β-isoform (ERβ) of the human estrogen receptor modulates ERα transcriptional activity and is a key regulator of the cellular response to estrogens and antiestrogens*, Endocrinology 140 (1999), pp. 5566–5578.

[15] G. Pascual and C.K. Glass, *Nuclear receptors versus inflammation: Mechanisms of transrepression*, Trends Endocrinol. Metab. 17 (2006), pp. 321–327.

[16] L. Björnström and M. Sjöberg, *Estrogen receptor-dependent activation of AP-1 via non-genomic signalling*, Nucl. Recept. 2 (2004), pp. 3.

[17] R.M. Lösel, E. Falkenstein, M. Feuring, A. Schultz, H.C. Tillmann, K. Rossol-Haseroth, and M. Wehling, *Nongenomic steroid action: Controversies, questions, and answers*, Physiol. Rev. 83 (2003), pp. 965–1016.

[18] D. Zivadinovic, B. Gametchu, and C.S. Watson, *Membrane estrogen receptor-α levels in MCF-7 breast cancer cells predict cAMP and proliferation responses*, Breast Cancer Res. 7 (2005), pp. 101–112.

[19] F. Acconcia and R. Kumar, *Signaling regulation of genomic and nongenomic functions of estrogen receptors*, Cancer Lett. 238 (2006), pp. 1–14.

[20] L.A. Paige, D.J. Christensen, H. Grøn, J.D. Norris, E.B. Gottlin, K.M. Padilla, C.-Y. Chang, L.M. Ballas, P.T. Hamilton, D.P. McDonnell, and D.M. Fowlkes, *Estrogen receptor (ER) modulators each induce distinct conformational changes in ERα and ERβ*, Proc. Natl. Acad. Sci. USA 96 (1999), pp. 3999–4004.

[21] P.J. Kushner, P. Webb, R.M. Uht, M.M. Liu, and R.H. Price, *Estrogen receptor action through target genes with classical and alternative response elements*, Pure Appl. Chem. 75 (2003), pp. 1757–1769.

[22] K. Pettersson, F. Delaunay, and J.A. Gustafsson, *Estrogen receptor β acts as a dominant regulator of estrogen signaling*, Oncogene 19 (2000), pp. 4970–4978.

[23] M.B. Hawkins, J.W. Thornton, D. Crews, J.K. Skipper, A. Dotte, and P. Thomas, *Identification of a third distinct estrogen receptor and reclassification of estrogen receptors in teleosts*, Proc. Natl. Acad. Sci. USA 97 (2000), pp. 10751–10756.

[24] J.P. Sumpter, *Xenoendocrine disrupters — Environmental impacts*, Toxicol. Lett. 102–103 (1998), pp. 337–342.

[25] M. Marino, P. Galluzzo, and P. Ascenzi, *Estrogen signaling multiple pathways to impact gene transcription*, Curr. Genomics 7 (2006), pp. 497–508.

[26] S.O. Mueller, *Xenoestrogens: Mechanisms of action and detection methods*, Anal. Bioanal. Chem. 378 (2004), pp. 582–587.

[27] D.M. Lonard and B.W. O'Malley, *Expanding functional diversity of the coactivators*, Trends Biochem. Sci. 30 (2005), pp. 126–132.

[28] H.V. Heemers and D.J. Tindall, *Androgen receptor (AR) coregulators: A diversity of functions converging on and regulating the AR transcriptional complex*, Endocrinol. Rev. 28 (2007), pp. 778–808.

[29] F. Rahman and H.C. Christian, *Non-classical actions of testosterone: An update*, Trends Endocrinol. Metab. 18 (2007), pp. 371–378.

[30] C.A. Quigley, A. De Bellis, K.B. Marschke, M.K. El-Awady, E.M. Wilson, and F.S. French, *Androgen receptor defects: Historical, clinical, and molecular perspectives*, Endocrinol. Rev. 16 (1995), pp. 271–321.

[31] T.S. Sperry and P. Thomas, *Characterization of two nuclear androgen receptors in Atlantic croaker: Comparison of their biochemical properties and binding specificities*, Endocrinology 140 (1999), pp. 1602–1611.

[32] W.R. Kelce, E. Monosson, M.P. Gamcsik, S.C. Laws, and L.E. Gray, *Environmental hormone disruptors: Evidence that vinclozoline developmental toxicity is mediated by antiandrogenic metabolites*, Toxicol. Appl. Pharmacol. 126 (1994), pp. 275–285.

[33] W.R. Kelce and E.M. Wilson, *Environmental antiandrogens: Developmental effects, molecular mechanisms, and clinical implications*, J. Mol. Med. 75 (1997), pp. 198–207.

[34] W.R. Kelce, C.R. Stone, S.C. Laws, L.E. Gray, J.A. Kemppainen, and E.M. Wilson, *Persistent DDT metabolite p,p'-DDE is a potent androgen receptor antagonist*, Nature 375 (1995), pp. 581–585.

[35] K.W. Gaido, L.S. Leonard, S.C. Maness, J.M. Hall, D.P. McDonnell, B. Saville, and S. Safe, *Differential interaction of the methoxychlor metabolite 2,2-bis-(p-hydroxyphenyl)-1,1,1trichlorethane with estrogen receptors α and β*, Endocrinology 140 (1999), pp. 5746–5753.

[36] S.C. Maness, D.P. McDonnell, and K.W. Gaido, *Inhibition of androgen receptor-dependent transcriptional activity by DDT isomers and methoxychlor in HepG2 human hepatoma cells*, Toxicol. Appl. Pharmacol. 151 (1998), pp. 135–142.

[37] H. Tamura, S.C. Maness, K. Reischmann, D.C. Dorman, L.E. Gray, and K.W. Gaido, *Androgen receptor antagonism by the organophosphate insecticide fenitrothion*, Toxicol. Sci. 60 (2001), pp. 56–62.

[38] J. Ostby, W.R. Kelce, C. Lambright, C.J. Wolf, P. Mann, and L.E. Gray, *The fungicide procymidone alters sexual differentiation in the male rat by acting as an androgen-receptor antagonist* in vivo *and* in vitro, Toxicol. Ind. Health 15 (1999), pp. 80–93.

[39] C. Lambright, J. Ostby, K. Bobseine, V. Wilson, A.K. Hotchkiss, P.C. Mann, and L.E. Gray, *Cellular and molecular mechanisms of action of linuron: An antiandrogenic herbicide that produces reproductive malformations in male rats*, Toxicol. Sci. 56 (2000), pp. 389–399.

[40] B.S. McIntyre, N.J. Barlow, D.G. Wallace, S.C. Maness, K.W. Gaido, and P.M.D. Foster, *Effect of in utero exposure of linuron on androgen-dependent reproductive development in male Crl:CD(SD)BR rat*, Toxicol. Appl. Pharmacol. 167 (2000), pp. 87–99.

[41] I. Katsiadaki, S. Morris, C. Squires, M.R. Hurst, J.D. James, and A.P. Scott, *Use of the three-spined stickleback (Gasterosteus aculeatus) as a sensitive* in vivo *test for detection of environmental antiandrogens*, Environ. Health Perspect. 114 (2006), pp. 115–121.

[42] I.H. Kang, H.S. Kim, J.H. Shin, T.S. Kim, H.J. Moon, I.Y. Kim, K.S. Choi, K.S. Kil, Y.I. Park, M.S. Dong, and S.Y. Han, *Comparison of anti-androgenic activity of flutamide, vinclozolin, procymidone, linuron, and p,p'-DDE in rodent 10-day Hershberger assay*, Toxicology 199 (2004), pp. 145–159.

[43] G.T. Ankley and J.P. Giesy, *Endocrine disruptors in wildlife: A weight of evidence perspective*, in *Principles and Processes for Assessing Endocrine Disruption in Wildlife*, R. Kendall, R. Dickerson, W. Suk, and J.P. Giesy, eds., SETAC Press, Pensacola, Florida, 1998, pp. 349–368.

[44] M.M. Gagnon, D. Bussieres, J.J. Dodson, and P.V. Hodson, *White sucker (Catostomus commersoni) growth and sexual maturation in pulp mill contaminated and reference rivers*, Environ. Toxicol. Chem. 14 (1995), pp. 317–327.

[45] L.M. Hewitt, J.L. Parrott, and M.E. McMaster, *A decade of research on the environmental impacts of pulp and paper mill effluents in Canada: Sources and characteristics of bioactive substances*, J. Toxicol. Environ. Health Part B: Crit. Rev. 9 (2006), pp. 341–356.

[46] K.R. Munkittrick, M.E. McMaster, L.H. McCarthy, M.R. Servos, and G.J. Van der Kraak, *An overview of recent studies on the potential of pulp-mill effluents to alter reproductive parameters in fish*, Toxicol. Environ. Health Part B Crit. Rev. 1 (1998), pp. 347–371.

[47] K.R. Munkittrick, C.B. Portt, G.J. Van der Kraak, I.R. Smith, and D.A. Rokosh, *Impact of bleached kraft mill effluent on population characteristics, liver MFO activity, and serum steroid levels of a Lake Superior white sucker* (Catastomus commersoni) *population*, Can. J. Fish. Aquat. Sci. 48 (1991), pp. 1371–1380.

[48] E. Neuman and P. Karas, *Effects of pulp mill effluent on a Baltic coastal fish community*, Water Sci. Technol. 20 (1988), pp. 95–106.

[49] G. Van der Kraak, T. Zacharewski, D.M. Janz, M.B. Sanders, and J.W. Gooch, *Comparative endocrinology and mechanisms of endocrine modulation in fish and wildlife*, in *Principles and Processes for Evaluating Endocrine Disruption in Wildlife*, R. Kendall, R. Dickerson, J.P. Giesy, and W.P. Suk, eds., SETAC Press, Pensacola, 1998, pp. 97–119.

[50] R. Jenkins, R.A. Angus, H. McNatt, W.M. Howell, J.A. Kemppainen, M. Kirk, and E.M. Wilson, *Identification of androstenedione in a river containing paper mill effluent*, Environ. Toxicol. Chem. 20 (2001), pp. 1325–1331.

[51] L. Parks, C. Lambright, E. Orlando, L. Guillette, G.T. Ankley, and L.E. Gray, *Masculinization of female mosquitofish in kraft mill effluent-contaminated Fenholloway river water is associated with androgen receptor agonist activity*, Toxicol. Sci. 62 (2001), pp. 257–267.

[52] V.S. Wilson, C. Lambright, J. Ostby, and L.E. Gray, *In vitro and in vivo effects of 17 beta trenbolone: A feedlot effluent contaminant*, Toxicol. Sci. 70 (2002), pp. 202–211.

[53] J.G. Chen, K.C. Ahn, N.A. Gee, M.I. Ahmed, A.J. Duleba, L. Zhao, S.J. Gee, B.D. Hammock, and B.L. Lasley, *Triclocarban enhances testosterone action: A new type of endocrine disruptor?* Endocrinology 149 (2008), pp. 1173–1179.

[54] OECD, *Detailed revue paper on thyroid hormone disruption assays*, in *OECD Series on Testing and Assessment*, 2006, pp. 1–434.

[55] C.B. Harvey and G.R. Williams, *Mechanism of thyroid hormone action*, Thyroid 12 (2002), pp. 441–446.

[56] S.B. Brown, R.E. Evans, L. Vandenbyllardt, K.W. Finnson, V.P. Palace, A.S. Kane, A.Y. Yarechewski, and D.C.G. Muir, *Altered thyroid status in lake trout* (Salvelinus namaycush) *exposed to co-planar 3,3',4,4',5-pentachlorobiphenyl*, Aquat. Toxicol. 67 (2004), pp. 75–85.

[57] M.S. Christian and N.A. Trenton, *Evaluation of thyroid function in neonatal and adult rats: The neglected endocrine mode of action*, Pure Appl. Chem. 75 (2003), pp. 2055–2068.

[58] M. Kaneko, R. Okada, K. Yamamoto, M. Nakamura, G. Mosconi, A.M. Polzonetti-Magni, and S. Kikuyama, *Bisphenol A acts differently from and independently of thyroid hormone in suppressing thyrotropin release from the bullfrog pituitary*, Gen. Comp. Endocrinol. 155 (2008), pp. 574–580.

[59] Y. Kudo, K. Yamauchi, H. Fukazawa, and Y. Terao, *In vitro and* in vivo *analysis of the thyroid system-disrupting activities of brominated phenolic and phenol compounds in Xenopus laevis*, Toxicol. Sci. 92 (2006), pp. 87–95.

[60] V.M. Richardson, D.F. Staskal, D.G. Ross, J.J. Diliberto, M.J. DeVito, and L.S. Birnbaum, *Possible mechanisms of thyroid hormone disruption in mice by BDE 47, a major polybrominated diphenyl ether congener*, Toxicol. Appl. Pharmacol. 226 (2008), pp. 244–250.

[61] L.H. Tseng, M.H. Li, S.S. Tsai, C.W. Lee, M.H. Pan, W.J. Yao, and P.C. Hsu, *Developmental exposure to decabromodiphenyl ether (PBDE 209): Effects on thyroid hormone and hepatic enzyme activity in male mouse offspring*, Chemosphere 70 (2008), pp. 640–647.

[62] F. Grün, H. Watanabe, Z. Zamanian, L. Maeda, K. Arima, R. Cubacha, D.M. Gardiner, J. Kanno, T. Iguchi, and B. Blumberg, *Endocrine-disrupting organotin compounds are potent inducers of adipogenesis in vertebrates*, Mol. Endocrinol. 20 (2006), pp. 2141–2155.

[63] T. Kanayama, N. Kobayashi, S. Mamiya, T. Nakanishi, and J. Nishikawa, *Organotin compounds promote adipocyte differentiation as agonists of the peroxisome proliferator-activated receptor γ/retinoid X receptor pathway*, Mol. Pharmacol. 67 (2005), pp. 766–774.

[64] X.C. Kretschmer and W.S. Baldwin, *CAR and PXR: Xenosensors of endocrine disrupters?* Chem. Biol. Interact. 155 (2005), pp. 111–128.

[65] B. Goodwin, M.R. Redinbo, and S.A. Kliewer, *Regulation of CYP3a gene transcription by the pregnane X receptor*, Annu. Rev. Pharmacol. Toxicol. 42 (2002), pp. 1–23.

[66] J. Zhou, J. Zhang, and W. Xie, *Xenobiotic nuclear receptor-mediated regulation of UDP-glucuronosyl-transferases*, Curr. Drug Metab. 6 (2005), pp. 289–298.

[67] C. Xu, C.Y. Li, and A.N. Kong, *Induction of phase I, II and III drug metabolism/transport by xenobiotics*, Arch. Pharm. Res. 28 (2005), pp. 249–268.

[68] L. Michalik, J. Auwerx, J.P. Berger, V. Krishna Chatterjee, C.K. Glass, F.J. Gonzales, P.A. Grimaldi, T. Kadowaki, M.A. Lazar, S. O'Rahilly, C.N.A. Palmer, J. Plutzky, J.K. Reddy, B.M. Spiegelman, B. Staels, and W. Wahli, *International union of pharmacology. LXI. Peroxisome proliferator-activated receptors*, Pharmacol. Rev. 58 (2006), pp. 726–741.

[69] H. Keller, C. Dreyer, J. Medin, A. Mahfoudi, K. Ozato, and W. Wahli, *Fatty acids and retinoids control lipid metabolism through activation of peroxisome proliferator-activated receptor-retinoid X receptor heterodimers*, Proc. Natl. Acad. Sci. USA 90 (1993), pp. 2160–2164.

[70] S.A. Kliewer, K. Umesono, D.J. Noonan, R.A. Heyman, and R.M. Evans, *Convergence of 9-cis retinoic acid and peroxisome proliferator signalling pathways through heterodimer formation of their receptors*, Nature 358 (1992), pp. 771–774.

[71] S. Mandard, M. Muller, and S. Kersten, *Peroxisome proliferator-activated receptor α target genes*, Cell. Mol. Life Sci. 61 (2004), pp. 393–416.

[72] P. Lefebvre, G. Chinetti, J.C. Fruchart, and B. Staels, *Sorting out the roles of PPAR α in energy metabolism and vascular homeostasis*, J. Clin. Invest. 116 (2006), pp. 571–580.

[73] Y. Barak, D. Liao, W. He, E.S. Ong, M.C. Nelson, J.M. Olefsky, R. Boland, and R.M. Evans, *Effects of peroxisome proliferator-activated receptor delta on placentation, adiposis, and colorectal cancer*, Proc. Natl. Acad. Sci. USA 99 (2002), pp. 303–309.

[74] K. Nadra, S.I. Anghel, E. Joye, N.S. Tan, S. Basu-Modak, D. Trono, W. Wahli, and B. Desvergne, *Differentiation of trophoblast giant cells and their metabolic functions are dependent on peroxisome proliferator-activated receptor β/δ*, Mol. Cell. Biol. 26 (2006), pp. 3266–3281.

[75] J.M. Peters, S.S. Lee, W. Li, J.M. Ward, O. Gavrilova, C. Everett, M.L. Reitman, L.D. Hudson, and F.J. Gonzales, *Growth, adipose, brain, and skin alterations resulting from targeted disruption of the mouse peroxisome proliferator-activated receptor beta (delta)*, Mol. Cell. Biol. 20 (2000), pp. 5119–5128.

[76] F. Varnat, B. Bordier-ten Heggeler, P. Grisel, N. Bouquard, I. Corthésy-Theulaz, W. Wahli, and B. Desvergne, *PPAR β/δ regulates Paneth cells differentiation via controling the Hedgedhog signaling pathway*, Gastroenterology 131 (2006), pp. 538–553.

[77] Y.X. Wang, C.H. Lee, S. Tiep, R.T. Yu, J. Ham, H. Kang, and R.M. Evans, *Peroxisome-proliferator-activated receptor δ activates fat metabolism to prevent obesity*, Cell 113 (2003), pp. 159–170.

[78] N. Di-Poi, N.S. Tan, L. Michalik, W. Wahli, and B. Desvergne, *Antiapoptotic role of PPARβ in keratinocytes via transcriptional control of the Akt1 signaling pathway*, Mol. Cell 10 (2002), pp. 721–733.

[79] E. Letavernier, J. Perez, E. Joye, A. Bellocq, B. Fouqueray, J.P. Haymann, D. Heudes, W. Wahli, B. Desvergne, and L. Baud, *Peroxisome proliferator-activated receptor beta/delta exerts a strong protection from ischemic acute renal failure*, J. Am. Soc. Nephrol. 16 (2005), pp. 2395–2402.

[80] L. Michalik and W. Wahli, *Involvment of PPAR nuclear receptors in tissue injury and wound repair*, J. Clin. Invest. 116 (2006), pp. 598–606.

[81] N.S. Tan, L. Michalik, N. Noy, R. Yasmin, C. Pacot, M. Heim, B. Flühmann, B. Desvergne, and W. Wahli, *Critical roles of PPAR beta/delta in keratinocyte response to inflammation*, Genes Dev. 15 (2001), pp. 3263–3277.

[82] W.M. He, Y. Barak, A. Hevener, P. Olson, D. Liao, J. Le, M. Nelson, E. Ong, J.M. Olefsky, and R.M. Evans, *Adipose-specific peroxisome proliferator-activated receptor gamma knockout causes insulin resistance in fat and liver but not in muscle*, Proc. Natl. Acad. Sci. USA 100 (2003), pp. 15712–15717.

[83] H. Koutnikova, T.A. Cock, M. Watanabe, S.M. Houten, M.F. Champy, A. Dierich, and J. Auwerx, *Compensation by the muscle limits the metabolic consequences of lipodystrophy in PPAR γ hypomorphic mice*, Proc. Natl. Acad. Sci. USA 100 (2003), pp. 14457–14462.

[84] E.D. Rosen, C.J. Walkey, P. Puigserver, and B.M. Spiegelman, *Transcriptional regulation of adipogenesis*, Genes Dev. 14 (2000), pp. 1293–1307.

[85] P. Tontonoz, E. Hu, R.A. Graves, A.I. Budavari, and B.M. Spiegelman, *mPPAR γ 2: Tissue-specific regulator of an adipocyte enhancer*, Genes Dev. 8 (1994), pp. 1224–1234.

[86] P. Tontonoz, J.B. Kim, R.A. Graves, and B.M. Spiegelman, *ADD1: A novel helix-loop-helix transcription factor associated with adipocyte determination and differentiation*, Mol. Cell. Biol. 13 (1993), pp. 4753–4759.

[87] T. Imai, R. Takakuwa, S. Marchand, E. Dentz, J.M. Bornert, N. Messaddeq, O. Wendling, M. Mark, B. Desvergne, W. Wahli, P. Chambon, and D. Metzger, *Peroxisome proliferator-activated receptor γ is required in mature white and brown adipocytes for their survival in the mouse*, Proc. Natl. Acad. Sci. USA 101 (2004), pp. 4543–4547.

[88] N. Kubota, Y. Terauchi, H. Miki, H. Tamemoto, T. Yamauchi, K. Komeda, S. Satoh, R. Nakano, C. Ishii, T. Sugiyama, K. Eto, Y. Tsubamoto, A. Okuno, K. Murakami, H. Sekihara, G. Hasegawa, M. Naito, Y. Toyoshima, S. Tanaka, K. Shiota, T. Kitamura, T. Fujita, O. Ezaki, S. Aizana, R. Nagai, K. Tobe, S. Kimura, and T. Kadowaki, *PPAR γ mediates high-fat diet-induced adipocyte hypertrophy and insulin resistance*, Mol. Cell 4 (1999), pp. 597–609.

[89] J. Rieusset, F. Touri, L. Michalik, P. Escher, B. Desvergne, E. Niesor, and W. Wahli, *A new selective peroxisome proliferator-activated receptor γ antagonist with antiobesity and antidiabetic activity*, Mol. Endocrinol. 16 (2002), pp. 2628–2644.

[90] E.D. Rosen, P. Sarraf, A.E. Troy, G. Bradwin, K. Moore, D.S. Milstone, B.M. Spiegelman, and R.M. Mortensen, *PPAR γ is required for the differentiation of adipose tissue in vivo and in vitro*, Mol. Cell 4 (1999), pp. 611–617.

[91] B.D. Savage, G.D. Tan, C.L. Acerini, S.A. Jebb, M. Agostini, M. Gurnell, R.L. Williams, A.M. Umpleby, E.L. Thomas, J.D. Bell, A.K. Dixon, F. Dunne, R. Boiani, S. Cinti, A. Vidal-Ruig, F. Karpe, V.K.K. Chatterjee, and S. O'Rahilly, *Human metabolic syndrome resulting from dominant-negative mutations in the nuclear receptor peroxisome proliferator-activated receptor γ*, Diabetes 52 (2003), pp. 910–917.

[92] Z. Wu, E.D. Rosen, R. Brun, S. Hauser, G. Adelmant, A.E. Troy, C. McKeon, G.J. Darlington, and B.M. Spiegelman, *Cross-regulation of C/EBP α and PPAR γ controls the transcriptional pathway of adipogenesis and insulin sensitivity*, Mol. Cell 3 (1999), pp. 151–158.

[93] I. Issemann and S. Green, *Activation of a member of the steroid-hormone receptor superfamily by peroxisome proliferators*, Nature 347 (1990), pp. 645–650.

[94] D.Y. Lai, *Rodent carcinogenicity of peroxisome proliferators and issues on human relevance*, J. Environ. Sci. Health C Environ. Carcinog. Ecotoxicol. Rev. 22 (2004), pp. 37–55.

[95] J.N. Feige, L. Gelman, D. Rossi, V. Zoete, R. Métivier, C. Tudor, S.I. Anghel, A. Grosdidier, C. Lathion, Y. Engelborghs, O. Michielin, W. Wahli, and B. Desvergne, *The endocrine disruptor monoethyl-hexyl-phthalate is a selective peroxisome proliferator-activated receptor γ modulator that promotes adipogenosis*, J. Biol. Chem. 282 (2007), pp. 19152–19166.

[96] C. Bettin, J. Oehlmann, and E. Stroben, *TBT-induced imposex in marine neogastero-pods in mediated by an increasing androgen level*, Helgolander Meeresuntersuchungen 50 (1996), pp. 299–317.

[97] T. Nakanishi, *Potential toxicity of organotin compounds via nuclear receptor signal-ing in mammals*, J. Health Sci. 53 (2007), pp. 1–9.

[98] N. Spooner, P.E. Gibbs, G.W. Bryan, and L.J. Goad, *The effects of tributyltin upon ste-roid titers in the female dogwhelk, Nucella lapillus, and the development of imposex*, Mar. Environ. Res. 32 (1991), pp. 37–49.

[99] J. Sonoda and R.M. Evans, *Biological function and mode of action of nuclear xenobi-otic receptors*, Pure Appl. Chem. 75 (2003), pp. 1733–1742.

[100] F. Oesch and M. Arand, *Xenobiotic metabolism*, in *Toxicology*, H. Marquardt, S.G. Schafer, R. McClellan, and F. Welsch, eds., Academic Press, New York, 1999, pp. 83–109.

[101] W. Xie, M.F. Yeuh, A. Radominska-Pandya, S.P. Saini, Y. Negishi, B.S. Bottroff, G.Y. Cabrera, R.H. Tukey, and R.M. Evans, *Control of steroid, heme, and carcinogen metabolism by nuclear pregnane X receptor and constitutive androstane receptor*, Proc. Natl. Acad. Sci. USA 100 (2003), pp. 4150–4155.

[102] C.D. Klaassen and A.M. Hood, *Effects of microsomal enzyme inducers on thyroid fol-licular cell proliferation and thyroid hormone metabolism*, Toxicol. Pathol. 29 (2001), pp. 34–40.

[103] S. Safe, *Modulation of gene expression and endocrine response pathways by 2,3,7,8-tetrachlorodibenzo-p-dioxin and related compounds*, Pharmacol. Ther. 67 (1995), pp. 247–281.

[104] O. Hankinson, *The aryl hydrocarbon receptor complex*, Annu. Rev. Pharmacol. Toxicol. 35 (1995), pp. 307–340.

[105] A.B. Okey, *An aryl hydrocarbon receptor odyssey to the shores of toxicology: The Deichmann lecture, international congress of toxicology-XI*, Toxicol. Sci. 98 (2007), pp. 5–38.

[106] R. Barouki, X. Coumoul, and P. M. Fernandez-Salguero, *The aryl hydrocarbon receptor, more than a xenobiotic-interacting protein*, FEBS Lett. 581 (2007), pp. 3608–3615.

[107] F. Ohtake, K.I. Takeyama, T. Matsumoto, H. Kitagawa, Y. Yamamoto, K. Nohara, C. Tohyama, A. Krust, J. Mimura, P. Chambon, J. Yanagisawa, Y. Fujii-Kuriyama, and S. Kato, *Modulation of oestrogen receptor signaling by association with activated dioxin receptor*, Nature 423 (2003), pp. 545–550.

[108] S. Safe and M. Wormke, *Inhibitory aryl hydrocarbon receptor-estrogen recep-tor alpha cross-talk and mechanisms of action*, Chem. Res. Toxicol. 16 (2003), pp. 807–826.

[109] J. Matthews and J.A. Gustafsson, *Estrogen receptor and aryl hydrocarbon receptor signaling pathways*, Nucl. Recept. Signal. 4 (2006), p. e016.

[110] J. Matthews, B. Wihlen, J. Thomsem, and J.A. Gustafsson, *Aryl hydrocarbon recep-tor-mediated transcription: Ligand-dependant recruitment of estrogen receptor α to 2,3,7,8-tetrachlorodibenzo-p-dioxin-responsive promoters*, Mol. Cell. Biol. 25 (2005), pp. 5517–5328.

[111] J.T. Sanderson, *The steroid hormone biosynthesis pathway as a target for endocrine-disrupting chemicals*, Toxicol. Sci. 94 (2006), pp. 3–21.

[112] W.L. Miller, *Molecular biology of steroid-hormone synthesis*, Endocrinol. Rev. 9 (1988), pp. 295–318.

[113] J.F. Baroiller, Y. Guigen, and A. Fostier, *Endocrine and environmental aspects of sex differentiation in fish*, Cell. Mol. Life Sci. 55 (1999), pp. 910–931.

[114] K. Cheshenko, F. Pakdel, H. Segner, O. Kah, and R.I.L. Eggen, *Interference of endocrine disrupting chemicals with aromatase CYP19 expression or activity, and consequences for reproduction of teleost fish*, Gen. Comp. Endocrinol. 155 (2008), pp. 31–62.

[115] N. Aluru, R. Renaud, J.F. Leatherland, and M.M. Vijayan, *Ah receptor-mediated impairment of interrenal steroidogenesis involves StAR protein and P450scc gene attenuation in rainbow trout*, Toxicol. Sci. 84 (2005), pp. 260–269.

[116] H.R. Andersen, A.M. Vinggaard, T.H. Rasmussen, I.M. Gjermandsen, and E.C. Bonefeld-Jorgensen, *Effects of currently used pesticides in assays for estrogenicity, androgenicity, and aromatase activity* in vitro, Toxicol. Appl. Pharmacol. 179 (2002), pp. 1–12.

[117] S.A. Andric, T.S. Kostic, S.M. Dragisic, N.L. Andric, S.S. Stojilkovic, and R.Z. Kovacevic, *Acute effects of polychlorinated biphenyl-containing and -free transformer fluids on rat testicular steroidogenesis*, Environ. Health Perspect. 108 (2000), pp. 955–959.

[118] P.J. Chedrese, M.R. Rodway, C.L. Swan, and C. Gillio-Meina, *Establishment of a stable steroidogenic porcine granulosa cell line*, J. Mol. Endoc. 20 (1998), pp. 287–292.

[119] P.K. Ghosh, N.M. Biswas, and D. Ghosh, *Effect of lithium-chloride on spermatogenesis and testicular steroidogenesis in mature albino-rats — Duration dependent response*, Life Sci. 48 (1991), pp. 649–657.

[120] P.M. Gocze, I. Szabo, Z. Porpaczy, and D.A. Freeman, *Barbiturates inhibit progesterone synthesis in cultured Leydig tumor cells and human granulosa cells*, Gynecol. End. 13 (1999), pp. 305–310.

[121] B.M. Huang, H.Y. Lai, and M.Y. Liu, *Concentration dependency in lead-inhibited steroidogenesis in MA-10 mouse Leydig tumor cells*, J. Toxicol. Environ. Health Part A 65 (2002), pp. 557–567.

[122] P.B. Kan, M.A. Hirst, and D. Feldman, *Inhibition of steroidogenic cytochrome-P-450 enzymes in rat testis by ketoconazole and related imidazole antifungal drugs*, J. Steroid Biochem. Mol. Biol. 23 (1985), pp. 1023–1029.

[123] M. Kawai, K.F. Swan, A.E. Green, D.E. Edwards, M.B. Anderson, and M.C. Henson, *Placental endocrine disruption induced by cadmium: Effects on P450 cholesterol side-chain cleavage and 3 β-hydroxysteroid dehydrogenase enzymes in cultured human trophoblasts*, Biol. Reprod. 67 (2002), pp. 178–183.

[124] W.R. Kelce, E. Monosson, M.P. Gamcsik, S.C. Laws, and L.E. Gray, *Environmental hormone disruptors — Evidence that vinclozolin developmental toxicity is mediated by antiandrogenic metabolites*, Toxicol. Appl. Pharmacol. 126 (1994), pp. 276–285.

[125] M.J. McVey and G.M. Cooke, *Inhibition of rat testis microsomal 3 β-hydroxysteroid dehydrogenase activity by tributyltin*, J. Steroid Biochem. Mol. Biol. 86 (2003), pp. 99–105.

[126] T. Nakanishi, J. Kohroki, S. Suzuki, J. Ishizaki, Y. Hiromori, S. Takasuga, N. Itoh, Y. Watanabe, N. Utoguchi, and K. Tanaka, *Trialkyltin compounds enhance human CG secretion and aromatase activity in human placental choriocarcinoma cells*, J. Clin. Endocrinol. Metab. 87 (2002), pp. 2830–2837.

[127] H. Nikula, T. Talonpoika, M. Kaleva, and J. Toppari, *Inhibition of hCG-stimulated steroidogenesis in cultured mouse Leydig tumor cells by bisphenol A and octylphenols*, Toxicol. Appl. Pharmacol. 157 (1999), pp. 166–173.

[128] S. Ohno, S. Shinoda, S. Toyoshima, H. Nakazawa, T. Makino, and S. Nakajin, *Effects of flavonoid phytochemicals on cortisol production and on activities of steroidogenic enzymes in human adrenocortical H295R cells*, J. Steroid Biochem. Mol. Biol. 80 (2002), pp. 355–363.

[129] B.L. Roman, R.J. Sommer, K. Shinomiya, and R.E. Peterson, *In utero and lactational exposure of the male-rat to 2,3,7,8-Tetrachlorodibenzo-p-dioxin — Impaired prostate growth and development without inhibited androgen production*, Toxicol. Appl. Pharmacol. 134 (1995), pp. 241–250.

[130] A.M. Ronco, K. Valdes, D. Marcus, and M. Llanos, *The mechanism for lindane-induced inhibition of steroidogenesis in cultured rat Leydig cells*, Toxicology 159 (2001), pp. 99–106.

[131] M. Saitoh, T. Yanase, H. Morinaga, M. Tanabe, Y.M. Mu, Y. Nishi, M. Nomura, T. Okabe, K. Goto, R. Takayanagi, and H. Nawata, *Tributyltin or triphenyltin inhibits aromatase activity in the human granulosa-like tumor cell line KGN*, Biochem. Biophys. Res. Commun. 289 (2001), pp. 198–204.

[132] J.T. Sanderson, J. Boerma, G.W.A. Lansbergen, and M. van den Berg, *Induction and inhibition of aromatase (CYP19) activity by various classes of pesticides in H295R human adrenocortical carcinoma cells*, Toxicol. Appl. Pharmacol. 182 (2002), pp. 44–54.

[133] J.T. Sanderson, W. Seinen, J.P. Giesy, and M. van den Berg, *2-chloro-s-triazine herbicides induce aromatase (CYP19) activity in H295R human adrenocortical carcinoma cells: A novel mechanism for estrogenicity?* Toxicol. Sci. 54 (2000), pp. 121–127.

[134] R. Thibaut and C. Porte, *Effects of endocrine disrupters on sex steroid synthesis and metabolism pathways in fish*, J. Steroid Biochem. Mol. Biol. 92 (2004), pp. 485–494.

[135] A.M. Vinggaard, C. Hnida, V. Breinholt, and J.C. Larsen, *Screening of selected pesticides for inhibition of CYP19 aromatase activity in vitro*, Toxicol. In Vitro 14 (2000), pp. 227–234.

[136] L.P. Walsh, C. McCormick, C. Martin, and D.M. Stocco, *Roundup inhibits steroidogenesis by disrupting steroidogenic acute regulatory (StAR) protein expression*, Environ. Health Perspect. 108 (2000), pp. 769–776.

[137] L.P. Walsh and D.M. Stocco, *Effects of lindane on steroidogenesis and steroidogenic acute regulatory protein expression*, Biol. Reprod. 63 (2000), pp. 1024–1033.

[138] H. Morinaga, T. Yanase, M. Nomura, T. Okabe, K. Goto, N. Harada, and H. Nawata, *A benzimidazole fungicide, benomyl, and its metabolite, carbendazim, induce aromatase activity in a human ovarian granulose-like tumor cell line (KGN)*, Endocrinology 145 (2004), pp. 1860–1869.

[139] G.G. Gibson, N.J. Plant, K.E. Swales, A. Ayrton, and W. El-Sankary, *Receptor-dependent transcriptional activation of cytochrome P450A genes: Induction mechanisms, species differences, and interindividual variation in man*, Xenobiotica 32 (2002), pp. 165–206.

[140] S.W. Chen, P.J. Dziuk, and B.M. Francis, *Effect of four environmental toxicants on plasma Ca and estradiol 17β and hepatic P450 in laying hens*, Environ. Toxicol. Chem. 13 (1994), pp. 789–796.

[141] S.W. Chen, B.M. Francis, and P.J. Dziuk, *Effect of concentration of mixed-function oxidase on concentration of estrogen, rate of egg lay, eggshell thickness, and plasma calcium in laying hens*, J. Anim. Sci. 71 (1993), pp. 2700–2707.

[142] A. Bammel, K. Vandermee, E.E. Ohnhaus and W. Kirch, *Divergent effects of different enzyme-inducing agents on endogenous and exogenous testosterone*, Eur. J. Clin. Pharmacol. 42 (1992), pp. 641–644.

[143] B. Blumberg, W. Sabbagh, H. Juguilon, J. Bolado, C.M. Van Meter, E.S. Ong, and R.M. Evans, *SXR, a novel steroid and xenobiotic-sensing nuclear receptor*, Genes Dev. 12 (1998), pp. 3195–3205.

[144] W. Xie, J.L. Barwick, M. Downes, B. Blumberg, C.M. Simon, M.C. Nelson, B.A. Neuschwander-Tetri, E.M. Bruntk, P.S. Guzelian, and R.M. Evans, *Humanized xenobiotic response in mice expressing nuclear receptor SXR*, Nature 206 (2000), pp. 435–439.

[145] J.A. Gustafsson, *Regulation of sexual dimorphism in rat liver*, in *The Differences between Sexes*, R.V. Short and E. Balaban, eds., Cambridge University Press, Cambridge, 1994, pp. 231–241.

[146] M.P. Gunderson, G.A. LeBlanc, and L.J. Guillette, *Alterations in sexually dimorphic biotransformation of testosterone in juvenile American alligators* (Alligator mississippiensis) *from contaminated lakes*, Environ. Health Perspect. 109 (2001), pp. 1257–1264.

[147] M.P. Gunderson, E. Oberdorster, and L.J. Guillette, *Phase I and II enzyme activities in juvenile alligators* (Alligator mississipiensis) *collected from three sites in the Kissimmee-Everglades drainage, Florida (USA)*, Comp. Biochem. Physiol. C Toxicol. Pharmacol. 139 (2004), pp. 39–46.

[148] L.J. Guillette, *Endocrine disrupting contaminants — Beyond the dogma*, Environ. Health Perspect. 114 (2006), pp. 9–12.

[149] V.S. Wilson and G.A. LeBlanc, *Endosulfan elevates testosterone biotransformation and clearence in CD-1 mice*, Toxicol. Appl. Pharmacol. 148 (1998), pp. 158–168.

[150] V.S. Wilson and G.A. Leblanc, *The contribution of hepatic inactivation of testosterone to the lowering of serum testosterone levels by ketoconazole*, Toxicol. Sci. 54 (2000), pp. 128–137.

[151] M.M. Tabb and B. Blumberg, *New modes of action for endocrine-disrupting chemicals*, Mol. Endocrinol. 20 (2006), pp. 475–482.

[152] H. Syvala, A. Vienonen, Y.H. Zhuang, M. Kivineva, T. Ylikomi, and P. Tuohimaa, *Evidence for enhanced ubiquitin-mediated proteolysis of the chicken progesterone receptor by progesterone*, Life Sci. 63 (1998), pp. 1505–1512.

[153] H.K. Lin, S. Altuwaijri, W.J. Lin, P.Y. Kan, L.L. Collins, and C. Chang, *Proteasome activity is required for androgen receptor transcriptional activity via regulation of androgen receptor nuclear translocation and interaction with coregulators in prostate cancer cells*, J. Biol. Chem. 277 (2002), pp. 36570–36576.

[154] L. Sheflin, B. Keegan, W. Zhang, and S.W. Spaulding, *Inhibiting proteasomes in human HepG2 and LNCaP cells increases endogenous endocrine receptor levels*, Biochem. Biophys. Res. Commun. 276 (2000), pp. 144–150.

[155] A.L. Wijayaratne and D.P. McDonnell, *The human estrogen receptor-α is an ubiquinated protein whose stability is affected differentially by agonists, antagonists and selective estrogen receptor modulator*, J. Biol. Chem. 276 (2001), pp. 35684–35692.

[156] U. Westphal, *Sex steroid-binding protein (SBP)*, in *Steroid-Protein Interactions II. Monographs on Endocrinology*, Springer-Verlag, New York, 1986, pp. 198–264.

[157] N. Fortunati, *Sex hormone-binding globulin: Not only a transport protein. What news is around the corner?* J. Endoc. Inv. 22 (1999), pp. 223–234.

[158] H. Déchaud, C. Ravard, F. Claustrat, A. Brac de la Perrière, and M. Pugeat, *Xenoestrogen interaction with human sex hormone-binding globulin (hSHBG)*, Steroids 64 (1999), pp. 328–334.

[159] M.E. Martin, M. Haourigui, C. Pelissero, C. Benassayag, and E.A. Nunez, *Interactions between phytoestrogens and human sex steroid binding protein*, Life Sci. 58 (1996), pp. 429–436.

[160] S.R. Milligan, O. Khan, and M. Nash, *Competitive binding of xenobiotic oestrogens to rat alpha-fetoprotein and to sex steroid-binding proteins in human and rainbow trout* (Oncorhyncus mykiss) *plasma*, Gen. Comp. Endocrinol. 112 (1998), pp. 89–95.

[161] K.E. Tollefsen, *Interaction of estrogen mimics, singly and in combination, with plasma sex steroid-binding proteins in rainbow trout* (Oncorhynchus mykiss), Aquat. Toxicol. 56 (2002), pp. 215–225.

[162] K.E. Tollefsen, J. Ovrevik, and J. Stenersen, *Binding of xenoestrogens to the sex steroid-binding protein in plasma from Arctic chaff* (Salvelinus alpinus L.), Comp. Biochem. Physiol. C Toxicol. Pharmacol. 139 (2004), pp. 127–133.

[163] S.F. Arnold, B.M. Collins, M.K. Robinson, L.J. Guillette, and J.A. McLachlan, *Differential interaction of natural and synthetic estrogens with extracellular binding proteins in yeast estrogen screen*, Steroids 61 (1996), pp. 642–646.

[164] D.A. Crain, N. Noriega, P.M. Vonier, S.F. Arnold, J.A. McLachlan, and L.J. Guillette, *Cellular bioavailability of natural hormones and environmental contaminants as a function of serum and cytosolic binding factors*, Toxicol. Ind. Health 14 (1998), pp. 261–273.

[165] S.C. Nagel, F.S. vom Saal, and W.V. Welshons, *The effective free fraction of estradiol and xenoestrogens in human serum measured by whole cell uptake assays: Physiology of delivery modifies estrogenic activity*, Proc. Soc. Exp. Biol. Med. 217 (1998), pp. 300–309.

[166] M.D. Anway, A.S. Cupp, M. Uzumcu, and M.K. Skinner, *Epigenetic transgenerational actions of endocrine disruptors and male fertility*, Science 308 (2005), pp. 1466–1469.

[167] P.M.D. Foster, *Disruption of reproductive development in male rat offspring following* in utero *exposure to phthalate esters*, Int. J. Androl. 29 (2006), pp. 140–147.

[168] S.M. Ho, W.Y. Tang, J. Belmonte de Frausto, and G.S. Prins, *Developmental exposure to estradiol and bisphenol A increases susceptibility to prostate carcinogenesis and epigenetically regulates phosphodiesterase type 4 variant 4*, Cancer Res. 66 (2006), pp. 5624–5632.

[169] R.R. Newbold, E. Padilla-Banks, and W.N. Jefferson, *Adverse effects of the model environmental estrogen diethylstilbestrol are transmitted to subsequent generations*, Endocrinology 147 (2006), pp. S11–S17.

[170] W.Y. Tang and S.M. Ho, *Epigenetic reprogramming and imprinting in origins of disease*, Rev. Endocr. Metab. Disord. 8 (2007), pp. 173–182.

[171] R. Jaenisch and A. Bird, *Epigenetic regulation of gene expression: How the genome integrates intrinsic and environmental signals*, Nat. Genet. 33 (2003), pp. 245–254.

[172] L. Van Speybroeck, *From epigenesis to epigenetics: The case of W. H. Waddington*, Ann. N.Y. Acad. Sci. 981 (2002), pp. 61–81.

[173] A. Akhtar and G. Cavalli, *The epigenome network of excellence*, PLoS Biol. 3 (2005), pp. e177.

[174] P.A. Jones and D. Takai, *The role of DNA methylation in mammalian epigenetics*, Science 293 (2001), pp. 1068–1070.

[175] R. Holliday, *DNA methylation and epigenetic mechanisms*, Cell Biophys. 15 (1989), pp. 15–20.

[176] E. Li, T.H. Bestor, and R. Jaenisch, *Targeted mutation of the DNA methyl-transferase gene results in embryonic lethality*, Cell 69 (1992), pp. 915–926.

[177] A.L. Herbst, H. Ulfelder, and D.C. Poskanzer, *Adenocarcinoma of the vagina association of the maternal stilbestrol therapy, with tumor appearance in young women*, New Engl. J. Med. 284 (1971), pp. 878–881.

[178] J. McLachlan, R. Newbold, and B. Bullock, *Long-term effects on the female mouse genital tract associated with prenatal exposure to diethylstilbestrol*, Cancer Res. 40 (1980), pp. 3988–3999.

[179] R.R. Newbold, R.B. Hanson, W.N. Jefferson, B.C. Bullock, J. Haseman, and J.A. McLachlan, *Increased tumors but uncompromised fertility in the female descendants mice exposed developmentally to diethylstilbestrol*, Carcinogenesis 19 (1998), pp. 1655–1663.

[180] R.R. Newbold, R.B. Hanson, W.N. Jefferson, B.C. Bullock, J. Haseman, and J.A. McLachlan, *Proliferative lesions and reproductive tract tumors in male descendants of mice exposed developmentally to diethylstilbestrol*, Carcinogenesis 21 (2000), pp. 1355–1363.

[181] D. Crews and J.A. McLachlan, *Epigenetics, evolution, endocrine disruption, health, and disease*, Endocrinology 147 (2006), pp. S4–S10.

[182] Q. Wu, S. Ohsako, R. Ishimura, J.S. Suzuki, and C. Tohyama, *Exposure of mouse preimplantation embryos to 2,3,7,8-tetrachlorodibenzo-p-dioxin (TCDD) alters the methylation status of imprinted genes H19 and Igf2*, Biol. Reprod. 70 (2004), pp. 1790–1797.

[183] D. Desaulniers, G.H. Xiao, K. Leingartner, I. Chu, B. Musicki, and B.K. Tsang, *Comparisons of brain, uterus, and liver mRNA expression for cytochrome P450s, DNA methyltransferase-1, and cathecol-o-methyltransferase in prepurbertal female Sprague-Dawley rats exposed to a mixture of aryl hydrocarbon receptor agonists*, Toxicol. Sci. 86 (2005), pp. 175–184.

[184] J.A. McLachlan, E. Simpson, and M. Martin, *Endocrine disrupters and female reproductive health*, Best Pract. Res. Clin. Endocrinol. Metab. 20 (2006), pp. 63–75.

[185] L. Tao, W. Wang, L. Li, P.K. Kramer, and M.A. Pereira, *DNA hypomethylation induced by drinking water disinfection by-products in mouse and rat kidney*, Toxicol. Sci. 87 (2005), pp. 344–352.

[186] M. Fu, C. Wang, J. Wang, X. Zhang, T. Sakamaki, Y.G. Yeung, C. Chang, T. Hopp, S.A.W. Fuqua, E. Jaffray, R.T. Hay, J.J. Palvimo, O.A. Jänne, and R.G. Pestell, *Androgen receptor acetylation governs trans activation and MEKK1-induced apoptosis without affecting* in vitro *sumoylation and trans-repression function*, Mol. Cell. Biol. 22 (2002), pp. 3373–3388.

[187] A. Kortenkamp, *Ten years of mixing cocktails: A review of combination effects of endocrine-disrupting chemicals*, Environ. Health Perspect. 115 Suppl 1 (2007), pp. 98–105.

[188] A. Kortenkamp and R. Altenburger, *Synergisms with mixtures of xenoestrogens: A reevaluation using the method of isoboles*, Sci. Total Environ. 221 (1998), pp. 59–73.

[189] J.V. Brian, C.A. Harris, M. Scholze, T. Backhaus, P. Booy, M. Lamoree, G. Pojana, N. Jonkers, T. Runnalls, A. Bonfa, A. Marcomini, and J.P. Sumpter, *Accurate prediction of the response of freshwater fish to a mixture of estrogenic chemicals*, Environ. Health Perspect. 113 (2005), pp. 721–728.

[190] G.D. Charles, C. Genning, T. Zacharewski, B.B. Gollapudi, and E.W. Carney, *Assessment of interactions of diverse ternary mixtures in an estrogen receptor-α reporter assay*, Toxicol. Appl. Pharmacol. 180 (2002), pp. 11–21.

[191] K.M. Crofton, E.S. Craft, J.M. Hedge, C. Gennings, J.E. Simmons, R.A. Carchman, W.H. Carter, and M.J. DeVito, *Thyroid-hormone-disrupting chemicals: Evidence for dose-dependent additivity or synergism*, Environ. Health Perspect. 113 (2005), pp. 1549–1554.

[192] Y. Le Page, M. Scholze, O. Kah, and F. Pakdell, *Assessment of xenoestrogens using three distinct estrogen receptors and the zebrafish brain aromatase gene in a highly responsive glial cell system*, Environ. Health Perspect. 114 (2006), pp. 752–758.

[193] K.L. Thorpe, R.I. Cummings, T.H. Hutchinson, M. Scholze, G. Brighty, J.P. Sumpter, and C.R. Tyler, *Relative potencies and combination effects of steroidal estrogens in fish*, Environ. Sci. Technol. 37 (2003), pp. 1142–1149.

[194] H. Tinwell and J. Ashby, *Sensitivity of the immature rat uterotrophic assay to mixtures of estrogens*, Environ. Health Perspect. 112 (2004), pp. 575–582.

[195] W.V. Welshons, K.A. Thayer, B.M. Judy, J.A. Taylor, E.M. Curran, and F.S. vom Saal, *Large effects from small exposures. I. Mechanisms for endocrine-disrupting chemicals with estrogenic activity*, Environ. Health Perspect. 111 (2003), pp. 994–1006.

[196] D.M. Sheehan, *No-threshold dose-response curves for nongenotoxic chemicals: Findings and applications for risk assessment*, Environ. Res. 100 (2006), pp. 93–99.

[197] A.C. Gore, J.J. Heindel, and R.T. Zoeller, *Endocrine disruption for endocrinologists (and others)*, Endocrinology 147 (2006), pp. S1–S3.

[198] H.A. Bern, K.T. Mills, D. Hatch, P.L. Ostrander, and T. Iguchi, *Altered mammary responsiveness to estradiol and progesterone in mice exposed neonatally to diethylstilbestrol*, Cancer Lett. 63 (1992), pp. 117–124.

[199] T. Colborn and C. Clement, *Chemically induced alterations in sexual and functional development: The wildlife/human connection*, Princeton Scientific, Princeton, NJ, 1992.

[200] M.S. Golub, C.E. Hogrefe, S.L. Germann, and C.P. Jerome, *Endocrine disruption in adolescence: Immunologic, hematologic, and bone effects in monkeys*, Toxicol. Sci. 82 (2004), pp. 598–607.

[201] M.D. Anway and M.K. Skinner, *Epigenetic transgenerational actions of endocrine disruptors*, Endocrinology 147 (2006), pp. S43–S49.

[202] J.J. Heindel, *Endocrine disruptors and the obesity epidemic*, Toxicol. Sci. 76 (2003), pp. 247–249.

[203] R.R. Newbold, E. Padilla-Banks, R.J. Snyder, T.M. Phillips, and W.N. Jefferson, *Developmental exposure to endocrine disruptors and the obesity epidemic*, Reprod. Toxicol. 23 (2007), pp. 290–296.

[204] S.E. Fenton, *Endocrine-disrupting compounds and mammary gland development: Early exposure and later life consequences*, Endocrinology 147 (2006), pp. S18–S24.

[205] H. Kojima, E. Katsura, S. Takeuchi, K. Niiyama, and F. Kobayashi, *Screening for estrogen and androgen receptor activities in 200 pesticides by in vitro reporter gene assays using chinese hamster ovary cells*, Environ. Health Perspect. 112 (2004), pp. 524–531.

[206] R. McKinlay, J.A. Plant, J.N.B. Bell, and N. Voulvoulis, *Endocrine disrupting pesticides: Implications for risk assessment*, Environ. Int. 34 (2008), pp. 168–183.

[207] P.J. Chedrese, M.R. Rodway, C.L. Swan, and C. Gillio-Meina, *Establishment of a stable steroidogenic porcine granulosa cell line*, J. Mol. Endocrinol. 20 (1998), pp. 287–292.

3 Population Dynamics Modeling

A Tool for Environmental Risk Assessment of Endocrine Disrupting Chemicals

Andrew R. Brown, Paul F. Robinson, Andrew M. Riddle, and Grace H. Panter

CONTENTS

ABSTRACT

Environmental risk assessment (in Europe) and equivalent processes including ecological risk assessment (United States) and ecosystem risk assessment (Japan) are collectively referred to in this chapter as ERA. They each involve the scientific analysis and characterization of environmental hazards (frequently chemicals) based on the predicted likelihood and level of exposure, versus the predicted severity of any consequent adverse effects. The discovery of "endocrine disruption" in wildlife due to natural hormones and synthetic mimics (endocrine disrupting chemicals, EDCs) in the last two decades has challenged the established ERA process, because very low exposure levels may illicit adverse effects including nonmonotonic responses in exposed organisms (Section 3.1.2). Accepting that it is impossible to legislate to protect every organism, especially when some natural losses are inevitable (for example, due to climatic stress or predation), the aims of ERA are to ensure the sustainability of populations to protect species and maintain the structure and functioning of ecosystems in which they live.

Ultimately, ERA requires clear understanding and communication of risks and objective decision making on the part of risk managers. This is based on an assessment of the most relevant data or facts (assessment endpoints) concerning the predicted behavior, fate, and effects of chemicals in the environment. A number of mathematical models are available to risk assessors, which have enormous potential in helping to relate chemical properties to potential behavior and to understand empirical data concerning exposure, uptake, and effects experienced by individual organisms (see Chapters 6, 7, 11, 13, this volume). Population modeling can offer additional help in linking individual exposure to population effect, extrapolating from laboratory to field, but more importantly by providing a means of communicating risk and uncertainty to risk managers in a form that they can clearly understand and apply. By focusing on ecologically relevant effects and scales, population-level ERA can aid regulatory decision making and help focus finite resources on managing ecologically significant risks (Section 3.1.1). This chapter illustrates how population dynamics modeling can support the established ERA process. Case studies are presented concerning endocrine disruption in fish (Section 3.3).

KEYWORDS

Deterministic model
ERA
Fish
Population dynamics model
Stochastic model

3.1 INTRODUCTION TO POPULATION DYNAMICS MODELING

Fish population dynamics models reflecting the life history of fish in the wild have been shown to be highly applicable in ecological risk assessments [1–3]. Many applications of models have been developed since the 1950s for the assessment of fish populations, to assess stocks, potential yield, sustainability, and predation [4–6]. Also, effects such as habitat fragmentation, nutrient enrichment, and physical hazards such as power plant intakes have been studied using population models [7,8]. The application of models to assess the potential adverse effects of chemicals on fish populations is more recent. Barnthouse et al. [9] used a logistic model to investigate the effects of toxic contaminants to fish populations, and Schaaf et al. [10] and Barnthouse [1] applied Leslie matrix models to simulate the effects of pollution on fish populations with different life-history strategies. Delay differential equation (DDE) models have also been used to assess the effects of chemical stressors on fish populations [3,11] but have more generally been applied in agriculture and forestry [12,13], as well as in the conservation of terrestrial fauna [14]. Individual-based (IB) models are now generally widely used in many branches of species and population modeling and have been applied to assess the effects of poor environmental quality on fish populations [15–18].

3.1.1 THE APPLICATION OF POPULATION DYNAMICS MODELING IN ENVIRONMENTAL RISK ASSESSMENT

The ultimate aim of environmental risk assessment [19] and, similarly, ecological risk assessment [20] and ecosystem risk assessment [21], is to determine, with a reasonable degree of certainty, whether or not an environmental hazard (for example, chemical "x" at X concentration) is environmentally safe. Traditionally this involves relating the chemical's predicted environmental concentration (PEC) with its predicted effects threshold, as determined by laboratory-based ecotoxicology studies. Risk is highlighted if toxicity is observed at exposure concentrations approaching or below the PEC. In the former case, when the risk is not so clear-cut, experimental uncertainties surrounding the environmental fate and behavior of chemicals can be addressed relatively easily by measuring their environmental concentrations. However, field measurement of the environmental effects of chemical "x" is not so straightforward because there may be many other stressors or factors that can either compound or ameliorate the effects of the chemical being assessed. Hence, it is very difficult to link cause and effect in the field. The detection of specific biomarkers, such as the expression of a gene, or the induction of an enzyme or other immune response, may help in this regard. However, gathering a significant body of evidence confirming adverse effects (versus a "normal" baseline) in potentially vulnerable species in the wild may not be practical or financially viable. One possible compromise is to simulate environmental effects with greater realism than in the laboratory using mesocosms [22,23], although some concerns have been raised regarding exaggerated fish population growth in miniature experimental ecosystems [24]. Failing that, computational population modeling can be used to make the link between laboratory data and potential field effects.

> Population dynamics modelling is the simulation of changes in population numbers or biomass, reflecting the distribution of individuals of different ages or developmental stages within defined populations, and biological and environmental processes influencing those changes.

As well as adding value in the interpretation of effects data in higher-tier risk assessment, population modeling has a potentially important role in helping to streamline the ERA process. This stems from the need to gather scientifically robust data on exposure and effects, while also optimizing the use of finite experimental and economic resources and, from an ethical standpoint, minimizing the use of vertebrates [25]. Once set up and validated, population models are relatively quick and cheap to run, allowing a wide range of exposure and effects scenarios to be simulated. This is one obvious reason why predictive modeling (for example, population dynamics) is considered useful in ERA [26,27].

However, the main synergies between ERA and population modeling may be derived from the improved use of environmental test data — that is, the inclusion of multiple effects endpoints in risk predictions made using population models, as opposed to the traditional emphasis on the most sensitive endpoint or lowest observed effect concentration (LOEC). Stochastic models, which account for the variability between individuals in a population, can be used to identify those effects endpoints that are most influential on the trajectory (success or failure) of that population over successive generations. This can be achieved by sensitivity analysis. Consequently, judgments regarding the importance of effects endpoints showing statistical significance versus those considered to be ecologically relevant can be avoided for the most part. This is an issue that has challenged traditional ERA for decades.

Risk managers essentially need the following from an ERA:

- To determine the likelihood and severity of any adverse environmental effect resulting from the hazard (chemical "x")
- To be able to demonstrate the validity of data and predictions, highlighting any assumptions or limitations including any uncertainties in a clear and transparent way

Population modeling can also provide them with the following:

- The interpretation/extrapolation of laboratory (ecotoxicological) effects data via the provision of more meaningful, environmentally relevant predictions of the sustainability of wild populations.

- A better understanding of the overall mechanism of toxicity of a chemical, helping to make the link between cause and effect via the identification of critical parameters affecting population sustainability.
- The integration of the effects of the chemical (or chemicals) with other background stressors or environmental variables, predicting the overall net effect on a population.

3.1.2 THE APPLICATION OF POPULATION DYNAMICS MODELING IN THE ASSESSMENT OF ENDOCRINE DISRUPTING CHEMICALS

Endocrine disruption is a phenomenon that was indentified in the field, for example, in marine invertebrates such as the dog whelk (*Nucella lapillus*) [28]. Their exposure to trace, sublethal concentrations of the antifouling ingredient tri-butyl tin was observed to induce a condition of imposex (the growth of male genitalia in females) that affected their reproductive ability and resulted in localized impacts on populations inhabiting harbor areas [29]. Similar impacts were observed in other species' populations in response to their exposure to other endocrine disrupting chemicals (EDCs) (Western gull [30], Pacific oyster [31], common seal [32], alligator [33], Florida panther [34], mosquito fish [35], roach [36]) and brought to the attention of the public in Theo Colborn's book *Our Stolen Future* [37]. In each case, adverse ecological effects were first observed in the field, rather than in the laboratory (that is, all chemicals in question evaded traditional ERA).

> The basic definition of an EDC is an exogenous substance that causes adverse health effects in an intact organism, or its progeny, consequent to changes in endocrine function [43].

Predicting the sublethal impacts of EDCs on fish populations has become a possibility, because in the past 10 years there have been major advances in knowledge regarding reproductive ecotoxicology and endocrine disruption [38–40]. Experimental data are becoming more readily available following the inclusion of abbreviated fish reproduction studies in the regulatory testing of chemicals including the "base set" for pharmaceuticals [41]. This will be further enhanced by the introduction of a 21-day fish reproduction test for EDC screening, which is currently under development on behalf of the Organization for Economic Cooperation and Development (OECD) [42]. Such screens will complement recent testing guidelines concerning the ERA of chemicals, for example, pharmaceuticals in Europe, which recognize the peculiarities of EDCs and require assessment of their potential fate and effects (including fish embryo-larval developmental effects), irrespective of the quantities released into the environment [41].

Endocrine disruption is an effect rather than a mode of action. EDCs may interfere with the synthesis, secretion, transport, binding, action, or elimination of natural hormones in the body that are responsible for the maintenance of normal cell metabolism. In order to model the effects of EDCs on populations, it is necessary to appreciate how these physiological effects are manifested in an individual's growth and development, reproductive capacity, and behavior.

The following are challenges EDCs present to environmental risk assessment, with particular reference to fish:

- EDCs are highly potent and can elicit adverse developmental, reproductive, or behavioral effects at very low environmental concentrations [38].
- Effects may follow a nonmonotonic dose response, making them difficult to detect in typical laboratory range-finding studies — that is, low exposure concentrations may elicit an entirely different response to that observed at high concentrations [44,45].
- The consequences of endocrine disruption may be multifaceted (for example, as well as affecting sexual differentiation in fish, including sex reversal, intersex may be combined with alterations in sexual maturation, sex ratio, male and female fertility, behavior, and so forth) [46–49].
- The degree of intersex tends to increase with the age of an individual, which may be related to susceptibility as well as exposure [50].
- Some fish are hermaphrodites and undergo sex change naturally in response to changes in temperature or food availability [51].
- Preexposure of embryos may infer greater resistance or greater sensitivity to continued exposure of juvenile fish [52,53].
- Endocrine effects may be passed on to future generations via the accumulation of some EDCs in the egg yolk [54] or via epigenetic changes such as methylations [55].

3.1.3 INCREASING REGULATORY FOCUS ON POPULATION-LEVEL ASSESSMENT

Environmental regulations ultimately aim to protect ecosystem structure and function, but this tends to be an aspirational or implicit goal rather than an explicit goal [56,57]. Current guidelines stipulate the measurement of ecotoxicity endpoints such as individual survival, growth, and fecundity, which are intended to reflect population-level impact within a particular species. However, it is rarely stated how this level of assessment should be achieved or what level of impact is acceptable. In a recent review of chemical regulations in North America, only one set of state regulations (for Oregon) concerning chemical risk assessment was found to specifically require population-level ERA. This set a tolerance limit for adverse effects on <20% of wildlife populations [56].

Recent advances in population-level risk assessment have been made by government, industry, and academia. For example, over the past two or three decades, the environmental regulation of pesticides (for example, via North America's Federal Insecticide, Fungicide, and Rodenticide Act [FIFRA] and Europe's Pesticides Directive 91/414/EEC) has required a tiered approach to product testing, in some cases culminating in the assessment of population- and community-level effects in field studies [5,58] or artificial mesocosms [24,59,60]. In turn, population modeling has been used to broaden some of these risk assessments by extrapolating the results to simulate wider agricultural field situations [61,62].

Initiatives concerned with the advancement and acceptance of population-level ERA within the existing regulatory framework for chemicals in general are currently being funded via the European Chemistry Council's Long-Range Research Initiative [63] and other industry consortia [64]. As part of this work, a critical review of computational models used in a population-level ERA context, published between 2000 and 2007, identified 41 different models and several inconsistencies in their description and application. Model validation was another important issue that was raised [65] (see Section 3.2.3).

3.1.4 INTEGRATING POPULATION DYNAMICS MODELING INTO ERA

A reasonable first step toward integrating ERA and population modeling is to compare basic ERA needs (inputs) with potential model outputs. Conducting this simple exercise immediately illustrates many complementary areas (Table 3.1).

TABLE 3.1
Potential Overlaps between ERA (Effects Assessment) and Population Dynamics Modeling

ERA Effects Assessment Needs (Inputs)	Population Modeling Aims (Outputs)
1. Identify relevant test species and effects assessment endpoints along with their normal baseline/ reference values.	1a. Define the range of possible trajectories of potentially affected "assessment" or "meta-populations" (for example, a population or subpopulation within a defined river catchment or reach, under normal, reference conditions).
	1b. Define acceptable, minimum viable population size [66] as a limit for specifying an unacceptable level of population effect.
2. Assess departure from normal "control" conditions due to simulated chemical exposure — identify effects threshold.	2. Assess the likelihood of an adverse population effect — that is, expressed as a range of interval decline probabilities from 0 to 1 that the population will decline by a critical amount (for example, 25%, 50%, or 75%) following chemical exposure at an expected concentration and duration (interval).
3. Evaluate the range of available effects data (endpoints). Interpretation relies on expert judgment to evaluate statistical significance of effects versus environmental relevance.	3. Identify the most influential input (effect) parameter(s) affecting the trajectory of the population (by sensitivity or elasticity analysis).
4. Place the observed effects data in the context of other environmental data — comparative risk assessment.	4. Assessment of the importance of other relevant environmental variables — abiotic (for example, temperature, resource availability, other chemicals) and biotic (for example, competition and predation).
5. Determine the likelihood and severity of an adverse environmental effect, presenting results clearly and transparently (for example, dose-response relationships to facilitate communication of risk to risk managers).	5. Graphical or tabular presentation of results including interval decline probabilities (and population recovery rates for time-limited exposure scenarios).

3.2 POPULATION DYNAMICS MODELS

3.2.1 SELECTING THE APPROPRIATE MODEL

It is important to appreciate the capabilities of different types of population dynamics models and hence their suitability for particular risk assessments. Deterministic age- or stage-structured models can be used to simulate incremental changes in mean survivorship and fecundity of individual females from birth to death and ultimately calculate the intrinsic rate of population increase. They are relatively easy to parameterize and use, satisfying many practical considerations concerning data availability and utility [67–70].

At the other end of the spectrum, individual-based (IB) models (also known as agent-based models) can account for the continuous development and interaction of individuals throughout their lifetime and may range considerably in terms of their complexity. However, added complexity provides increased scope for including additional assessment endpoints such as individual behavior [71], as well as the effects of environmental factors (for example, temperature [72], pesticide exposure [16,62], and intersex effects due to steroidal estrogens [73,74]). The predictive capabilities are summarized for these various types of models in Table 3.2.

TABLE 3.2
Capabilities of Different Population Dynamics Models

		Predictive Capabilities	
Model Type	Minimum Unit	Deterministic	Stochastic
1. Matrix	Group of individuals ascribed to a discrete age class or stage, females only	Population decline, population recovery, intrinsic rate of population increase (r) predicted in discrete time steps (usually annual)	Relative importance of vital rates
2. DDE	Group of individuals ascribed to a discrete age class or stage, females only	Population decline, population recovery, intrinsic rate of population increase (r) predicted in continuous time steps (daily)	Interval decline risk, most influential effect parameter
3. IB	Individuals, including males and females	Population decline, population recovery, intrinsic rate of population increase (r) predicted in continuous time	Most influential effect parameter

Notes: Matrix: Leslie Matrix models (see P.H. Leslie, *Biometrika*, 35, 213, 1945 [75]); DDE: Delay Differential Equation models (see R.M. Nisbet, in *Structured Population Models in Marine Terrestrial and Freshwater Systems,* Chapman & Hall, New York, 1997 [76]); IB: individual-based models (see W. van Winkle, H.I. Jager, S.F. Railsback, B.D. Holcomb, T.K. Studley, and J.E. Baldrige, *Ecol. Model.*, 110, 175, 1998 [77]).

The suitability of different population dynamics models in ERA will depend on the questions being posed. These define the necessary level of complexity of the model, which is also limited by computer processing power, the availability of input data (control and effects), and the required level of user training [56]. In order that model outputs meet or exceed specific ERA requirements, models should be properly calibrated and validated (Section 3.2.3). This ensures that model outputs accurately quantify population viability in terms of rate of decline and recovery, interval decline probability, extinction risk, and so forth. However, very few published population models have been validated due to the lack of suitable field effects data [61,62]. The specificity of different chemical effects and the costs associated with validation are major contributory factors. The transparency of models is also of key importance in determining their usefulness in ERA. Modeling assumptions need to be explicitly stated and the implications understood by the risk assessor. The adoption of guidelines describing population models clearly and consistently (for example, overview, design concept, details [78]) should increase the availability and reliability of models and facilitate their uptake and use in ERA.

3.2.1.1 Deterministic or Stochastic Simulations?

Matrix or DDE models simulate population numbers as a whole and are basically deterministic — that is, a model run gives a unique answer for a given set of life-history and exposure effects parameters. Figure 3.1 shows the structure of a deterministic run output for one year, showing the different processes taking place and the influence on population numbers.

However, there is uncertainty in the normal vital rates (of survival and fecundity) and effects responses (for example, for perch in Windermere fecundity of year 2+ to 6+ fish varies with a mean value of 25,905 eggs and a standard deviation [sd] of 3,776 eggs [67]). To allow for this, the model is often set in a Monte Carlo (MC) framework to generate a stochastic output by computing many different runs, where each of the parameters for a run is chosen at random from the statistical distribution of the parameter, for example,

$$\text{fecundity} = \text{mean} + \alpha.\text{sd}$$

where α is a random number from a standard normal distribution (or other statistical distribution). The results for each run fall between the maximum and minimum curves shown in Figure 3.2. The deterministic curve on this plot is obtained using the mean values of all the parameters.

Often the MC is run 1,000 times and a statistical analysis of the results is carried out to give information such as the probability of a 50% population decline due to exposure. Also, if the exposure is stopped or reduced, the model will then predict the time for recovery to say 95% of the original population numbers [67].

An IB model simulates each fish individually and is inherently stochastic in that the variability of life-history parameters and effects parameters is incorporated into one model run. The output from the model looks like the deterministic prediction from the matrix or DDE model, but it is produced allowing for all the variability

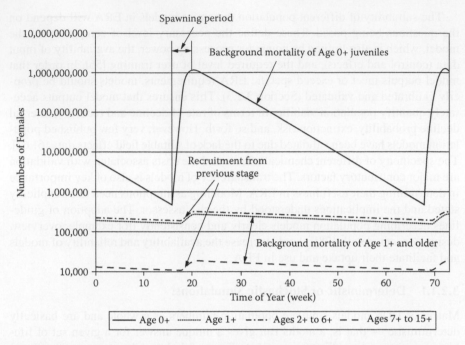

FIGURE 3.1 The variation throughout the annual spawning cycle in mean perch population numbers in the North Basin of Windermere. (From A.R. Brown et al., *Ecol. Model.*, 189, 387, 2005. With permission.)

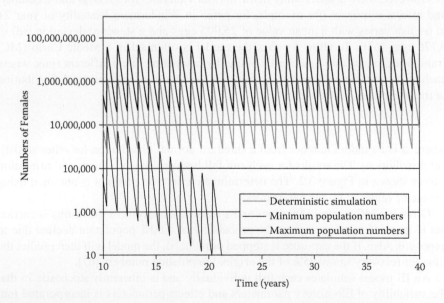

FIGURE 3.2 Maximum, deterministic, and minimum predicted population numbers for perch exposed to steroidal estrogens in Windermere. (Adapted from A.R. Brown et al., *Ecol. Model.*, 189, 387, 2005. With permission.)

between individuals. At a deeper level, the responses of the "weaker" and "stronger" fish can be assessed and the statistical distribution of the population at different times (for example, before exposure and at different times during exposure) can be output. The IB model framework can incorporate other types of variability that would not be considered in matrix or DDE models, such as different behaviors induced by exposure (for example, males behaving more aggressively during the spawning season) [48].

A single deterministic model prediction is generally not that useful on its own. These simple runs could be used to compare and rank the potential magnitude of effects on the population of different exposures to give an idea of which is the worst case. A stochastic prediction gives more robust information, because it allows for variability in the population and gives statistical output that can be used to assess the probability of risk. Further discussions on deterministic versus stochastic models can be found elsewhere [79–81].

3.2.2 MODEL DESIGN AND PARAMETERIZATION

3.2.2.1 Design

Model design should incorporate those parameters (assessment endpoints) that are pertinent to a particular risk assessment and accurately reflect their interrelationship. These parameters include vital, intrinsic (life-history) traits such as survival, growth, and reproduction, as well as extrinsic (environmental) variables such as chemical exposure and temperature (Figure 3.3).

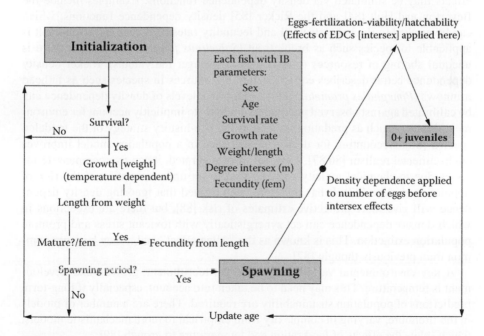

FIGURE 3.3 The iteration of an individual-based model "ROACH" (described in more detail in Section 3.3.3).

3.2.2.1.1 Control Simulation

A life stage of a group of fish or an individual has a range of attribute information associated with it, including sex, age, maturity, survival, growth rate, weight and length, fecundity (females), male fertility, and spawning behavior. The essential parameters or "vital rates" are survival and fecundity.

The life cycle of the modeled species is divided into discrete phases, which for instance could be based on changes in mortality or growth rate. On moving into a new life stage, the vital rates are updated with new statistical distributions representing the natural variability of this life stage. Other vital rates (e.g., fecundity) can be derived from primary attributes or parameters (e.g., length).

For early life stages (for example, age 0+), it may be necessary to estimate parameters such as survival based on stock recruitment curves or back-calculation methods, which assume stable recruitment [82]. Another parameter that is usually difficult to quantify is the immigration/emigration rate. This is invariably accounted for in estimates of annual stock-recruitment, which also includes age 0+ survival. However, modern molecular and landscape population genetics techniques provide a potential means of estimating immigration/emigration rates (or average dispersal distance) based on assessments of genetic distance and gene flow between neighboring populations [83].

Most populations are limited in number by factors such as availability of food and space. Population growth is limited by reductions in the production of offspring, rate of offspring survival, rate of adult survival, and rate of maturation. These effects may be simulated via density dependence functions. Examples include the Beverton and Holt [84] and the Ricker [85] density dependence functions, which can be applied to the basic survival and fecundity rates. Whereas Beverton–Holt is applicable to species such as brook trout (*Salvelinus fontinalis*), in which there is unequal sharing of resources (biased toward stronger individuals), Ricker density dependence better describes equal sharing of resources in species such as fathead minnows (*Pimephales promelas*) [69]. Appropriate levels of density dependence can be calibrated against observed population numbers to implicitly account for environmental factors such as predation, as well as the life-history strategy of the modeled species [86]. Accounting for density dependence in a population model improves environmental realism [86,87]. However, if exaggerated, it can overcompensate for losses due to chemical toxicity and can lead to an underestimation of the risk of population decline [87]. It has therefore been argued that ignoring density dependence will give more protective estimates of risk [88], but there are exceptions in which density dependence can act synergistically with toxicant stress and promote population extinction. This is known as the "Allee effect" and it may be more common than previously thought [87].

A key environmental variable affecting fish recruitment, growth, and development is temperature. This may need to be taken into account, especially if long-term predictions of population sustainability are required. There are a number of models in the literature, varying in complexity from empirical to more mechanistic models, which relate the effects of food ration and temperature to growth [89].

Control (and effects) simulation results may be expressed as total population number, numbers of individuals in specific age/stage classes, or in terms of total or age-specific biomass per unit area. Biomass is a particularly useful measure of population size, because it can be related directly to environmental carrying capacity (that is, maximum supportable fish stock) [85]. Numerical abundance may be misleading in this regard, due to the temporary, unsustainable explosion in numbers of fish fry immediately after spawning. Therefore, if results are expressed in terms of abundance, a reliable, consistent approach is to focus on prespawning population numbers [3].

3.2.2.1.2 Effects Simulation

At the most basic level, the effects of individual EDCs can be simulated by altering the vital rates (survival and fecundity) in the control model based on observed laboratory effects data. These "life-table response experiments" (LTREs) involve predicting net replacement rate (R_o), population growth rate (λ) or the intrinsic rate of population increase ($r = \ln \lambda$) and determining the exposure concentration at which there is zero increase ($r = 0$, $\lambda = 1$) [91]. These predictions (made using matrix models) may be further enhanced by directly relating the model response (measured in eigenvalues) to the perturbation (chemical exposure) via a transfer function [92].

Although predictions of ecotoxicity based on laboratory studies are enhanced via traditional LTREs [86,91,93], they ignore key ecological processes, such as density dependence, and lack environmental realism. On the other hand, by default, density-independent models have less parameter uncertainty associated with them [94]. There is no escaping the uncertainty introduced from "reading-across" effects from surrogate laboratory species to identical or other species in the wild. Having said that, extrapolating chronic data from one related species to another (however distant) is better than employing acute data for the same species or family in population modeling [95].

For EDCs, it may be necessary to account for the mode of action of the chemical and how this manifests itself in terms of individual development, sexual differentiation, and behavior. The emergence of "intelligent testing," incorporating the measurement of these and other more specific biomarker endpoints, favors the development of more mechanistic IB models. These may include toxicokinetic/dynamic submodels accounting for chemical uptake, receptor binding, and elimination [96,97]. However, it should be appreciated that describing these complex effect mechanisms in mathematical terms is a major challenge. Complex models can be perceived as less certain or reliable. For this reason, it is best to avoid overcomplexity and concentrate on simulating the principal toxic effect mechanisms. A universal rule is to treat the results from all population dynamics models as a guide for decision making and not to accept them as the final answer [98].

3.2.2.2 Parameterization

Having selected an appropriate type of population model (for example, matrix, DDE, or IB), it is necessary to parameterize it with control data for the species in question.

TABLE 3.3
Parameterization of Population Dynamics Models

| | Input Control Data | | Input Effects |
Model Type	Intrinsic (Life History)	Extrinsic (Environmental)	Data
1. Matrix	Mean (sd) annual survival, fecundity	Ignored [91] or implicit in life-history data [3]	Mean (sd) survival, fecundity
2. DDE	Mean (sd) daily survival, fecundity	Implicit in life-history data [67]	Mean (sd) survival, fecundity, growth (via submodel)
3. IB	Daily survival, growth, development, fecundity, male fertility, behavior	Environmental variables include temperature, habitat, and food availability, within and between species interaction [75]	Survival, growth, development, fecundity, male fertility, behavior, physiological effects

The most basic data are laboratory control data [91,93], whereas the most realistic are life-history data reflecting baseline (control) conditions in the wild [67]. Generic life-history data for a wide range of fish species can be sourced from databases such as www.fishbase.org/search.php and www.freshwaterlife.org/. For case-specific risk assessments, life-history data may be acquired from wider peer-reviewed literature or, if necessary, derived from local fisheries data [67].

Details concerning the design and parameterization of model types 1 to 3 (Table 3.3) were previously described [3,72] and are summarized in the case studies presented in Section 3.3.

Suitable effect data may be obtained relatively easily from laboratory or field-based studies (see Table 3.5). The development of the Organization for Economic Cooperation and Development (OECD) 21-day fish reproduction test for EDC screening [42] and other intelligent testing protocols incorporating informative biomarker endpoints will benefit the risk assessment of chemicals with specific modes of action (Type IV compounds) as defined by Verhaar et al. [99]. It is likely that the uptake of population-level ERA will help propagate these changes in the design of higher-tier ecotoxicological studies. This will be justified if the accuracy and ecological relevance of risk predictions are improved, providing an alternative approach to the use of overly conservative safety factors in traditional organism-level ERA.

3.2.3 MODEL CALIBRATION AND VALIDATION

3.2.3.1 Calibration

Calibration is the process that ensures that the model predictions of the control population reflect actual field data (or surrogate mesocosm or laboratory data). This is required because there are three major sources of uncertainty associated with population dynamics models: natural variability in the environment (temporal and spatial heterogeneity and genetic variability within a population), potential errors in the estimation of model input parameters [94], and potential errors in the model structure [100].

Natural variability within a population is normally accounted for in the life-history data used to parameterize the control model (Section 3.2.2).

Errors in model structure result from oversimplification of ecological processes and the limitations of mathematics in describing these processes [98]. Therefore, the design and calibration of population dynamics models should be based on a clear conceptual model that identifies the relevant processes and suitable data (assessment endpoints) linking exposure and effect [101].

Errors in the estimation of model input parameters mainly concern stock recruitment and the contribution of under-yearling survival versus immigration/emigration; density dependence [102]; and spawning behavior [47]; including the proportion of individuals contributing to spawning [103].

One or several of these parameters may require adjustment as part of the calibration process.

3.2.3.2 Validation

Validation concerns the checking of effects predictions versus actual effects, ideally observed in the field. Simple models representing constant exposure (both spatially and temporally) to a single chemical are obviously easiest to validate, and it is feasible that this may be done in the laboratory. However, such models lack environmental realism. In reality, chemical exposure and the recipient organisms or populations are in a constant state of flux. As mentioned previously, biological variability is typically included in population dynamics models within the "control" life-history input data. Variable chemical exposure can also be incorporated in temporally or spatially explicit models [61]. However, their inclusion presents a greater challenge in terms of validation. In a recent critical review of 41 population dynamics models used in ERA, 30 models made no reference to validation [64]. This may be because suitable field effects data are lacking for many compounds. Having said that, there are numerous examples of multigeneration laboratory studies [104], mesocosm studies [59,60], and field studies [49,58] in the literature, which could be used to validate existing population dynamics models.

The modeling of open systems such as rivers presents a greater challenge in terms of validation than does the modeling of enclosed systems such as lakes, in which contaminants and populations are confined. The work reported by Palace et al. [47] and Blanchfield et al. [48], looking at the effects on fathead minnows in lake 260 from the Canadian Experimental Lakes Area during an experiment where EE2 was dosed into the lake, is a prime example in which detailed population effects measurements were made. However, in many cases, full validation is not possible, especially for models used to predict long-term population extinction probability [104].

TABLE 3.4

Overview of Four Models with Potential Applications in Risk Assessment

Model Type	Model Name	Purpose (Application)	References
1. Matrix	BROOK (Brook trout)	Assess long-term population impact of and recovery from continuous EDC exposure	[3]
2. DDE	MINNOW (Fathead minnow)	Assess long-term and short-term population impact of and recovery from continuous and periodic (for example, seasonal pesticide applications) EDC exposure	[3]
3. DDE	PERCH (Eurasian perch)	Assess the susceptibility of perch populations in Windermere pre- and postdisease outbreak to EDC exposure	[67]
4. IB	ROACH (Roach)	Assess long-term population impact related to intersex in males from continuous steroid estrogen exposure (as part of a program to devise Environmental Quality Standards in U.K. rivers)	[73]

3.2.4 EXAMPLES OF MODEL APPLICATIONS

Despite the lack of validated population dynamics models and the fact that they are "nothing more than abstractions of reality," they "simplify systems so that defensible decisions can be made" [105]. There are many potential applications for population dynamics models in an ERA context. These were reviewed elsewhere [65,102].

Selecting the most appropriate population dynamics model depends on the type of input data available and the regulatory goal or "question being asked." The following examples, which feature in a series of case studies in Section 3.3, illustrate the potential applications for different types of models (Table 3.4).

3.3 CASE STUDIES

Laboratory and field effects data for putative EDCs including methoxychlor (MXC), nonylphenol (NP), estrone (E1) 17-β estradiol (E2) and 17-α ethinylestradiol (EE2) concerning individual fish survival, growth, and reproduction (Table 3.5), have been translated into predictions of population impact on the brook trout (*Salvelinus fontinalis*), fathead minnow (*Pimephales promelas*), perch (*Perca fluviatilis*), and roach (*Rutilus rutilus*). The selected case studies represent a range of chemical exposure scenarios, from seasonal, short-term, to continuous, long-term exposures, whereas the selected

TABLE 3.5
Examples of Laboratory and Field Effects Data Used in Case Studies

Compound	Exposure	Effects (% of Control)	Fish Species
Nonylphenol (NP) results from the degradation of detergents in sewage treatment. Ambient river concentrations range from <2 to 300 mg l⁻¹ [108,118].	Low (1 μg l⁻¹)	Age 0+ survival = 73%	*Salmo trutta* — brown trout [106] (reduction in survival related to growth[a])
		Age >0+ survival = 100%	*Oncorhynchus mykiss* — rainbow trout [107]
			Lepomis macrochirus — bluegill sunfish [108]
			Pimephales promelas — fathead minnow [109]
		New adult fecundity = 105%	[107]
		Older adult fecundity = 87%	[106]
		Growth[a] = 95%	[107]
			Rutilus rutilus — roach [110]
	High (50 μg l⁻¹)	Age 0+ survival = 73%	Reduction in survival related to growth[a] [106]
		Age >0+ survival = 95%	[107,108]
		Fecundity = 125%	[107]
		Growth[a] = 80%	[107,110]
Methoxychlor (MXC) is used in the control of black fly larvae in rivers each spring, summer, and autumn. Pulse concentrations range from 2 to 200 μg l⁻¹ [112].	Low (500 μg l⁻¹)	Age 0+ survival = 73%	*Salmo trutta* — brown trout [106] (reduction in survival related to growth[a])
		Age >0+ survival = 95%	*Oncorhynchus mykiss* — rainbow trout [111,112]
		New adult fecundity = 95%	*Catostomus commersoni* — white sucker [52]
			Jordinella floridae — flagfish [113]
		Older adult fecundity = 87%	[111]
		Growth[a] = 95%	[52,112]
	High (10,000 μg l⁻¹)	Survival (all ages) = 60%	*Oncorhynchus mykiss* — rainbow trout [111,114]
		New adult fecundity = 75%	[52,113]
		Older adult fecundity = 55%	[106] (reduction in fecundity related to growth[a])
		Growth[a] = 80%	[111,114]

(continued)

TABLE 3.5 (CONTINUED)
Examples of Laboratory and Field Effects Data Used in Case Studies

Compound	Exposure	Effects (% of Control)	Fish Species
Ethinylestradiol (EE2), a female contraceptive, is >10 times as potent as the natural hormone 17-β estradiol. Ambient river concentrations range from <0.02 to 15 ngl^{-1} [119].	Low (0.001 µg l^{-1})	Survival (all ages) = 100%	*Pimephales promelas* — fathead minnow [115–117]
		New adult fecundity = 53%	[115]
		Older adult fecundity = 32%	[106] (reduction in fecundity related to growth[a])
		Growth[a] = 80%	[115]
	High (0.01 µg l^{-1})	Survival (all ages) = 100%	[115–117]
		Adult fecundity (all ages) = 0%	[115]
		Growth[a] = 80% to 90%	[115–117]

Note: Exposure ranges (low to high) relate to environmental exposure concentrations.

[a] Growth effects translated into effects on survival and fecundity.

species populations exhibit contrasting ecological life histories (strategies) and varying demographic structures that influence their susceptibilities to these exposures.

3.3.1 COMPARING THE EFFECTS OF TWO EDCs ON TWO FISH SPECIES WITH DIFFERENT LIFE-HISTORY STRATEGIES

3.3.1.1 Overview and Model Setup

The first case study, adapted from Brown et al. [3], illustrates how the ecological impact of different EDCs, methoxychlor (MXC) and nonylphenol (NP), on fish may vary from one species to another — the fathead minnow (*Pimephales promelas*) and the brook trout (*Salvelinus fontinalis*). The former species is representative of the cyprinidae, the world's largest family of fish [51], and is commonly used in regulatory studies to provide effects data for ERA. Another widely used test species is the rainbow trout (*Oncorhynchus mykiss*). This is representative of another major family of fish, the salmonidae, to which the brook trout also belongs.

Published life-history data were obtained for natural populations of fathead minnow in Horseshoe Lake, Minnesota [120], and brook trout in Hunt Creek, Michigan [121], to parameterize basic control models for these species. These control models were run until they achieved steady state. Then we superimposed survival, growth, and reproductive effects data from laboratory-based studies involving these or other related fish species (Table 3.5) to assess the impact of differential exposure of NP and MXC on population sustainability and recovery. Both the MINNOW model and the BROOK model simulate the population dynamics of female fish, based on the assumption that they limit fish reproduction rather than males [51,122].

The two modeled species have very different life-history strategies. The majority of female brook trout mature at age 1+. They are k-strategists, exhibiting low egg production (38 to 303 eggs per female), high hatchling survival, and subsequent survival rates (0.42 survival probability from age 0+ to age 1+), leading to inherently stable recruitment. This, coupled with a relatively long life span of up to 5 years and a population growth rate (net replacement rate) of $R_0 = 1.13$ conferred stable population numbers in Hunt Creek [121], whose carrying capacity was limited according to the Beverton–Holt density dependence function [104]. On the other hand, female fathead minnows are r-strategists, maturing rapidly within 10 to 14 weeks (under optimal environmental conditions), exhibiting higher egg production (255 to 2,400 eggs per female), lower hatching success, and lower survival rates (0.15 survival probability from age 0+ to age 1+). Their overall population growth rate in Horseshoe Lake marginally exceeded $R_0 = 1$ [120], with a limited carrying capacity according to the Ricker density dependence function [85]. These density dependence functions were used as calibration parameters to limit fecundity in the BROOK and MINNOW models in order to simulate total estimated population numbers (preceding spawning) and population growth rates in Hunt Creek [121] and Horseshoe Lake [120], respectively.

3.3.1.2 Results and Discussion

The BROOK model predicted that continuous NP exposure would not pose a risk to brook trout; in fact, high exposure concentrations of 50 µg l^{-1} could even enhance population numbers by up to 20%. On the other hand, the MINNOW model predicted that the same concentration of NP would cause a population decline of 15% [3]. Considering the stochasticity in both models, there is still a possibility that NP could cause a decline in both populations. The greater sensitivity of the MINNOW model to NP — that is, the higher probability of a 25% reduction in the population — stems from the relatively low lifetime reproductive capacity of individuals. The majority of fathead minnows spawn only once, whereas brook trout may have up to three spawning seasons, over which time fecundity increases almost exponentially. Generally speaking, in brook trout, increasing fecundity appears to more than compensate for the observed reduction in survival rates (73% to 95%) of laboratory controls (Table 3.5). This is not the case in fathead minnows, which supports the suggestion that life-history characteristics such as life span are important determinants of a species' risk from exposure to estrogens and estrogen mimics [49].

With a time step of 1 day, the MINNOW model is capable of resolving the effects of chemicals with discrete seasonal applications such as MXC, whereas the BROOK model, with an annual time step, is not appropriate. Our predictions indicated that MXC applications may potentially cause a 10% reduction in fathead minnow population size, but only if concentrations greatly exceed maximal predicted environmental exposure concentrations. Therefore, MXC presents a lower risk to the sustainability of fathead minnow populations than NP, mainly because NP exposure can be assumed to occur throughout the year. However, to put these predictions into context, the exposure simulations for neither MXC nor NP caused the fathead minnow population to decline by more than 50%. This level of effect is less than the natural variations observed in Horseshoe Lake [101] but greater than the 20% limit stated in one set of U.S. state regulations, albeit for Oregon rather than Minnesota [56].

3.3.2 ACCOUNTING FOR THE EFFECTS OF OTHER ENVIRONMENTAL FACTORS IN CONJUNCTION WITH CHEMICAL EXPOSURE

3.3.2.1 Overview and Model Setup

The second case study, adapted from Brown et al. [67], illustrates how the ecological impact of two EDCs, nonylphenol (NP) and ethinylestradiol (EE2), can vary within the same species and also within the same population, following significant changes in life-history strategy and demographic structure with time. Such changes may be linked to fishing pressure or environmental factors, such as temperature and food availability, affecting annual recruitment success, or major selection pressures (for example, those associated with disease outbreaks). The subject of this case study, the population of perch (*Perca fluviatilis*) in Windermere, United Kingdom, has been affected by each of these "pressures" in the past 50 years. This highlights the importance of defining the "assessment" or "meta-population" not only spatially (for example, within a lake, river reach, or catchment), but also temporally. This is especially important given predictions of global climate change.

Although perch is not an accepted test species nor a close relation to those species commonly used in regulatory studies, there are several reasons to commend it as an environmental monitor or sentinel species. It has a wide geographical distribution extending from Great Britain through Northern Europe, including Scandinavia, to Northeast Asia. Despite this, individuals have a narrow home range [123], which, along with their position as predators in the aquatic food chain, means that perch could be particularly susceptible to point-source chemical discharges [67].

Extensive life-history data for perch inhabiting the North Basin of Windermere [124–128] were used to parameterize the PERCH control model. These data represented two distinct phases of stable recruitment and contrasting demographic structures preceding and following the outbreak of "perch disease" (ferunculosis). Between 1966 and 1976, environmental conditions favored high recruitment success and growth rates, and fish had life spans of up to 15 years. Following the disease outbreak in 1976, recruitment was temporarily impaired and, subsequently, the number of age classes was reduced to five or less.

Using the Ricker density dependence function [85] to limit individual fecundity, the pre- and postdisease control models were calibrated against estimates of total population size based on mark-and-recapture data over these periods. Laboratory effects data concerning survival, growth, and reproduction in other fish species (Table 3.5) were superimposed on both control models to assess the impact of differential exposure of NP and EE2 on population sustainability and recovery. Like the MINNOW and BROOK models, the PERCH model simulated the population dynamics of female fish.

3.3.2.2 Results and Discussion

Sustained high-level exposure of EE2 (0.01 µg l^{-1}) had a high probability of causing the extinction of the confined fish population in Windermere due to reproductive failure in the majority of individuals [67]. However, such an exposure scenario is unlikely. Far greater uncertainty surrounds the prediction of effects due to low-level exposure of fish populations. In the case of EE2, statistically significant effects observed in laboratory studies (that is, fecundity lowered by 30%) did not translate into significant population effects according to the PERCH model predictions. This is not surprising, as the control life-history data showed high natural variability in terms of individual vital rates, even during the selected periods of stable recruitment. Our work suggests that for a decline in the perch population numbers to be significant in Windermere, it would have to be substantially more than 50%. Our model predictions indicated that the postdisease population was more vulnerable than the predisease population and had a significantly greater probability of declining by 50% following single chemical exposure. However, both populations were shown to be equally likely to decline following combined chemical exposure to NP and EE2, assuming additivity of effects [129]. This supports the "risk cup" concept, taking into account the effects of mixtures as well as other cumulative background stressors in the environment [2].

The PERCH model showed that the rate of recovery in population numbers was a more sensitive measure in terms of differentiating the effects of low and high chemical exposure, as well as the vulnerability of populations with contrasting demographic structures, histories, or levels of background stress. However, properties such as genetic diversity within a population may take considerably longer to recover than its demographics (especially if it is confined), and reduction in genetic diversity (as opposed to reduction in fecundity) could have a far more pronounced effect, especially if the fitness of remaining individuals is eroded, eventually giving way to a sudden population crash [67].

3.3.3 USING FIELD BIOMARKER RESPONSES IN MALE AND FEMALE FISH TO PREDICT POPULATION LEVEL EFFECTS

3.3.3.1 Overview

The third case study concerns the assessment of the impact of steroidal estrogens estrone (E1), 17-β estradiol (E2) and the synthetic mimic 17-α ethinylestradiol (EE2) on roach populations in U.K. rivers to help enable the derivation of statutory

Environmental Quality Standards for these substances under the European Water Framework Directive. Roach (*Rutilus rutilus*) is an abundant freshwater fish species in many European rivers and lakes [18,130,131], particularly around treated sewage discharges, which are the main sources of steroidal estrogens in surface waters [46].

The evidence of endocrine disruption in wild fish populations (particularly roach) is substantial in terms of the occurrence of intersex males (possessing varying degrees of feminine characteristics) in rivers with elevated concentrations of steroidal estrogens [36,38,132]. However, the consequences of endocrine disruption, including intersex, on the sustainability of the affected populations have been heavily debated for some time [133,134]. Therefore, the assessment of endocrine disruption in roach on a catchment scale (EDCAT 7) is being coordinated by the U.K. Environment Agency (EA) and the Department for the Environment Fisheries and Rural Affairs (DEFRA). This assessment focuses on a series of spawning studies involving wild fish, in which the contribution of intersex fish (as percent parentage) is assessed. Data from these studies are being used to validate a roach population dynamics model (ROACH), which will then be used to extrapolate fisheries data combined with biomarker assessments of intersex incidence and severity for roach in U.K. rivers, to predict population impact.

> Intersex, in gonochoristic (discrete sex) fish, refers to the alteration in normal sexual differentiation leading to the co-occurrence of male and female characteristics in the gonad and/or reproductive tract. It may be accompanied by: alteration of sex steroid hormone levels, delay in maturation rate; change in phenotypic sex ratio; reduction in male fertility; reduction in female fecundity; alteration in spawning behaviour. (Adapted from Jobling et al., p. 516 [46])

The ROACH model is an IB model, which incorporates the variability of ED effects in wild roach populations, including intersex and the consequences on male fertility and maturation rate, as well as female fecundity. This modeling framework also provides scope for the inclusion of behavioral effects reflecting the social dominance of individuals, particularly in the act of spawning.

3.3.3.2 Control Model Setup

It is assumed that individuals progress through age classes distinguished, among other things, by differences in annual survival rates. These were estimated from the U.K. Environment Agency's fisheries data, as well as wider published data for the ROACH model. It is accepted that the survival rate after maturity is fairly stable [135] and that females generally grow faster, mature later, and live longer than males [136].

A submodel for a dependency on external water temperature has also been included in the ROACH model; a base set of quintile temperatures for a U.K. river was used in the control model. The ROACH growth model was calibrated using the von Bertalanffy equation [137] (Figure 3.4). A repeating cyclic state was achieved in the control model (with the application of a single set of annual quintile temperatures), which predicts rapid growth in the spring and summer, with minimal growth during the autumn and winter months, reflecting the observed seasonal pattern. It should be noted that these

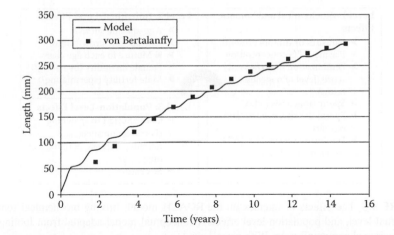

FIGURE 3.4 Comparison of ROACH model growth with the von Bertalanffy curve. For mature roaches, absolute fecundity (F) was related to fork length (L) in centimeters by the relationship log (F) = a.log (L)-b, where "a" and "b" are constants. (See R.H.K. Mann, *J. Fish Biol.*, 5, 707, 1973).

types of growth models do not account for the availability of food and necessary ration for optimal growth; therefore, we assumed that ration is not limiting.

Density dependence in the ROACH model was expressed as a multiplier that reduces the absolute fecundity according to the Beverton–Holt function, whose parameters were derived from EA fisheries stock recruitment data [138]. This density dependence model [84] was adopted because roach exhibit phenotypic plasticity in growth and maturation, leading to asymmetries in cohort size spectra — that is, some fish acquire more resources than others and this asymmetry increases as they grow larger [139,140].

3.3.3.3 Effects Parameterization

The ROACH model considers each of the individuals within a population inhabiting a specific or U.K. average lowland river and can allow for individual susceptibilities conferred by chance or inherent variation within a population.

Although there is a range of qualitative information on the histological condition of male roach associated with intersex, there is little in the literature quantifying the effects on male fertility. We concentrated on the results from the work by Jobling et al. [46], because quantitative effects on male fertility (and female fecundity) are reported for a range of intersex severities found in UK rivers. These effects are characterized in terms of the percent reduction in the number of males spermiating, reduction in percent survival (taken as percent hatch in the model), and percent reduction in fecundity of females (Figure 3.5). During spawning it is assumed that sperm is in excess [51,122], and we assumed that each sperm has an equal chance of fertilizing an egg. Thus, the effect of intersex on male sperm production in terms of the contribution of individual males to spawning can be assumed to be proportionate to the fraction

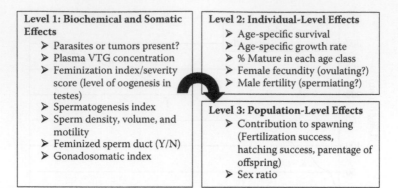

FIGURE 3.5 The effects of intersex on the ROACH model: linking biochemical/somatic, individual-level, and population-level effects. (Conceptual model adapted from Jobling and Burn, personal communication. With permission.)

of roach in each intersex category (based on feminization index). Then, using the percent survival factor (with uncertainty), the proportion of eggs that actually hatch was calculated. It is important to note that sperm competition was not quantified in the laboratory data used to estimate the fertility of intersex males [46] and therefore the above assumption may not be valid. Quantifying differences in the motility and viability of sperm from different classes of intersex fish in the wild may be equally if not more important than understanding the effects of EDCs on social dominance and estimating the contribution of individual males to spawning. It is intended that these data will be provided by the EDCAT 7 program in due course.

3.3.3.4 Results and Discussion

To date, our effects predictions concentrate on population effects associated with intersex in roach, mediated by changes in male fertility and female fecundity. Effects relating, for example, to reduced survival or delayed maturation in intersex males have so far been ignored. Conservative risk assessment dictates that these should be taken into account. Nevertheless, current model predictions serve to illustrate that quantitative biomarkers of intersex in roach may be related to detectable population effects for this species.

So far, we simulated the effects on male fertility and female fecundity assuming a range of intersex severities are associated with adult males such that 25% are allocated randomly to each of the four classes or levels of intersex (Table 3.6). This is a crude way of simulating differential exposure and age-related intersex occurrence (that is, intersex increasing with age [50]), and our preliminary results indicate a 20% reduction in postspawning population numbers (Figure 3.6).

There are potentially a number of other factors relating directly to roach and the ED effects of steroidal estrogens that are not accounted for in the above simulation, and this could positively or negatively affect population sustainability (Table 3.7).

Future experimental and field data (from the EDCAT 7 programme) quantifying the relative contribution of intersex roach to spawning will be used to further develop

TABLE 3.6
Individual Parameters Associated with Each Intersex Class

Level of Intersex in Males (Testes)	Percent Fecundity for Corresponding Females	Percent Males Spermiating	Percent Hatching and Fry Surviving
All male features	100	97.6	45 (st. dev. = 1.7)
Ovarian cavity	100	97.0	45 (st. dev. = 1.7)
Some oocytes	100	82.6	45 (st. dev. = 1.7)
Severely feminized	80 (st. dev. = 3)	66.7	15 (st. dev. = 1.7)

Source: Adapted from S. Jobling et al., *Biol. Reprod.*, 67, 519–521, 2002. (With permission.)

and validate the ROACH model. Additional data (Table 3.5) will also be incorporated. As mentioned previously, the purpose of the model is to relate endocrine disruption, including intersex incidence and severity, to population-level impact in river catchments in order to determine the adequacy of Environmental Quality Standards for steroidal estrogens (E1, E2, and EE2). However, caution should be used when attempting to extrapolate between biomarker effects such as intersex and population-level effects, and that there is no substitute for traditional effects endpoints including survival, growth, development, and reproduction in population-level ecological risk assessment [40,67].

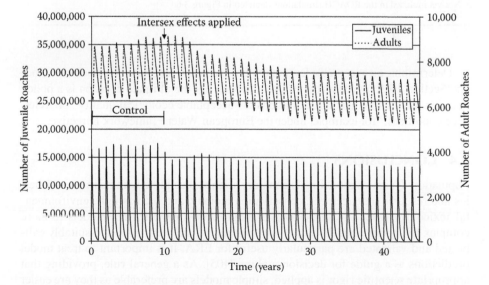

FIGURE 3.6 Roach population trajectory in an arbitrary river with varying intersex severity within the population (25% in each intersex class and 25% are unaffected).

TABLE 3.7
Factors That Could Positively or Negatively Affect Population Sustainability

Positive Factors	Negative Factors
[a]Increasing annual mean temperatures (assuming they are tolerable) could have positive effects on fish growth and relative fecundity	Reduction in male fertility and female fecundity [46]
	[a]Reduced survival in intersex males
	[a]Delayed maturation in intersex males [141]
	[a]Complete blockage of the sperm duct (reported in some severe intersex cases) would prevent sperm release [36]
	[a]Spawning behavior upset by increased male and female aggression [48,142]; altered courtship behavior, mate competition [143]
	[a]Reduced sperm competition [46]

Positive and Negative

• General behavior including avoidance of noxious contaminants versus tendency to feed around organically enriched STP effluent discharges [46]

• Immigration and emigration may reinvigorate or deplete populations by adding or removing numbers of individuals and or genetic diversity [144]

Note: The model domain (equating to an initial adult population of 5,000 fish, as illustrated in Figure 3.6) is assumed to be large enough to minimize the influence of immigration and emigration, but if a significant number of more or less impacted fish exchange with fish from the model domain, this could significantly affect predictions.

[a] Not yet included in the ROACH simulations depicted in Figure 3.6.

Determining an acceptable level of impact and risk is facilitated by stochastic models (Section 3.2.1.1) such as the ROACH model, but the ultimate decision is a policy decision, taking into account definitions of "acceptable ecological quality" or "reference conditions," for example, under the European Water Framework Directive.

3.4 CONCLUSIONS

Population dynamics modeling is a mature science discipline with its roots in fisheries and resource management, conservation biology, and increasingly, environmental toxicology [102]. Available models range from simple matrix models to more complex IB models, but those that are transparent, simple to use, and suitably calibrated and validated are particularly useful for ERA. It is important to treat model predictions as a guide for decision making [105]. As a general rule, providing that appropriate scientific rigor is applied, simple models are preferable as they are easier to parameterize and are less prone to compounding errors. Ultimately, model selection depends on the question being asked. For example, simple, stage-based models may be adequate for determining the inherent sensitivities of different species to chemicals, including EDCs, as illustrated in case study 1. This information may then be used to target appropriate laboratory testing or field monitoring effort as part of a refined risk assessment.

The utility of population dynamics models in regulatory decision making stems from their ability to quantify risk (for example, quasi-extinction risk), integrate multiple effects endpoints (rather than focus on the most sensitive endpoint or lowest observed effect concentration [LOEC]), extrapolate from laboratory measurements to predicted field effects, and optimize the use of finite experimental and economic resources [25–27,56]. Some models are specifically aimed at elucidating the subtle reproductive and transgenerational effects of EDCs on populations and their impact on population sustainability [54], whereas others may be used to assess the population-level impact of particular EDC effects such as intersex [73] (case study 3).

In addition, population dynamics modeling enables the combined risk assessment of multiple stressors (e.g., EDC mixtures) and can take into account other existing or historical selective pressures (e.g., disease) acting on a population, as illustrated in case study 2. Stochastic models, which incorporate the inherent variability among individuals in a population, are able to highlight critical effects parameters (via sensitivity analysis) and determine the probability that the population will become extinct or fall below a predefined level. The acceptable level of risk (for example, probability of population extinction) is down to a relatively "straightforward" policy decision and may range from zero for critically endangered species to perhaps 0.2 to 0.5 (20% to 50%) for more ubiquitous species. The crux is defining "What is the minimum level below which a population is no longer sustainable?" This question is addressed using population viability analysis [66], and the answer depends on the species or population to which it pertains. In the final analysis, the results from population dynamics models should be interpreted in the context of the range of normal reference conditions for the population (or metapopulation) in question. However, natural variations in population numbers frequently involve population numbers rising and falling by as much as 50% in successive generations, as highlighted in case studies 1 and 2. This degree of tolerance is not typically included within traditional ERA, potentially leading to overconservatism.

Validation of models for predicting the effects of chemicals on fish populations is a key issue. Simple models representing constant exposure, both spatially and temporally, to a single chemical are obviously easiest to validate (for example, in the laboratory), but they lack environmental realism. In reality, chemical exposure and the recipient organisms or populations are in a constant state of flux. Biological variability is typically included within the "control" model via the input of life-history data. Variable chemical exposure can also be incorporated in temporally or spatially explicit models [61]. The modeling of open systems such as rivers presents a greater challenge in terms of validation than enclosed systems such as lakes, in which contaminants and populations are confined. However, full validation is invariably not possible, especially for models used to predict long-term population extinction probability [104].

The population dynamics models presented as case studies in this chapter are autecology models and as such do not specifically consider species interactions, although competition and predation are included implicitly in life-history data used to parameterize the control models in each case. Alternative ecological modeling approaches such as food chain or ecosystem models may be more applicable than those presented here, especially if a chemical has the potential to bioaccumulate and biomagnify up the food chain. In cases such as these, it may also be advisable to consider physiological responses, including the metabolism of the compound via toxicokinetic- and toxicodynamic-based modeling approaches as described elsewhere in this book (see Chapter 4).

An interesting point highlighted by case study 2 is the fact that properties such as genetic diversity within a population may take considerably longer to recover than its demographics (especially if it is confined), and reduction in genetic diversity (as opposed to reduction in fecundity) could have a far more pronounced effect [67,144]. This kind of effects assessment is currently beyond the scope of population dynamics modeling.

Despite its considerable benefits, the main drawback concerning population-level ERA is that it is more complex than existing organism-level ERA. Therefore, its future use will depend on the availability of suitable modeling tools, input data, and guidance and training for risk assessors and risk managers [56]. The emergence of versatile model-making software and their intercalibration with a wide range of model scenarios and templates [146] will facilitate this greatly by providing a common platform for population dynamics modeling, which is readily available to risk assessors.

REFERENCES

[1] L.W. Barnthouse, *Population level effects*, in Ecological Risk Assessment, Suter, ed., Lewis, Chelsea, MI, 1993, pp. 247–274.

[2] W.E. Schaaf, D.S. Peters, L. Coston-Clements, D.S. Vaughan, and C.W. Krouse, *A simulation model of how life history strategies mediate pollution effects on fish populations*, Estuaries 16 (1993), pp. 697–702.

[3] A.R. Brown, A.M. Riddle, N.L. Cunningham, T.J. Kedwards, N. Shillabeer, and T.H. Hutchinson, *Predicting the effects of endocrine disruptors on fish populations*, Human Ecol. Risk Assess. 9 (2003), pp. 761–788.

[4] J.A. Wilson, J. French, P. Kleban, S.R. McKay, and R. Townsend, *Chaotic dynamics in a multiple species fishery: A model of community predation*. Ecol. Model. 58 (1991), pp. 303–322.

[5] L.W. Barnthouse, *Modelling ecological risks of pesticide application: A review of available approaches*, Oak Ridge National Laboratory, for the Office of Pesticide Programs, U.S. Environmental Protection Agency, contract DE-AC05-84OR21400, 1996.

[6] A. Aubone, *Threshold for sustainable exploitation of an age-structured fishery stock*, Ecol. Model. 173 (2004), 95–107.

[7] J.K. Summers, *Simulating the indirect effects of power plant entrainment losses on an estuarine ecosystem*, Ecol. Model. 49 (1989), pp. 31–47.

[8] K. Morita and A. Yokota, *Population viability of stream-resident salmonids after habitat fragmentation: A case study with white-spotted charr* (Salvelinus leucomaenis) *by an individual-based model*, Ecol. Model. 155 (2002), pp. 85–94.

[9] L.W. Barnthouse, G.W. Suter, A.E. Rosen, and J.J. Beauchamp, *Estimating responses of fish populations to toxic contaminants*, Environ. Toxicol. Chem. 6 (1987), pp. 811–824.

[10] W.E. Schaaf, D.S. Peters, D.S. Vaughan, L. Coston-Clements, and C.W. Krouse, *Fish population responses to chronic and acute pollution: The influence of life history strategies.* Estuaries 10 (1987), pp. 267–275.

[11] S.N. Wood, *Semi-parametric population models*, Aspects Appl. Biol. 53 (1999), pp. 41–49.

[12] M.F. Acevedo, D.L. Urban, and H.H. Shugart, *Models of forest dynamics based on roles of tree species*, Ecol. Model. 87 (1996), pp. 267–284.

[13] K. Louie, G.C. Wake, G. Lambert, A. MacKay, and D. Barker, *A delay model for the growth of ryegrass-clover mixtures: Formulation and preliminary simulations*. Ecol. Model. 155 (2002), pp. 31–42.

[14] D. Schley and M.A. Bees, *Delay dynamics of the slug* Deroceras reticulatum, *an agricultural pest*, Ecol. Model. 162, (2003), pp. 177–198.

[15] K.A. Rose, E.S. Rutherford, D.S. McDermot, J.L. Forney, and E.L. Mills, *Individual-based model of yellow perch and walleye populations on Oneida Lake*, Ecol. Monogr. 62 (1999), pp. 127–154.

[16] W. van Winkle, K.A. Rose, and R.C. Chambers, *Individual-based approach to fish population dynamics: An overview*, Trans. Am. Fish Soc. 122 (1993), pp. 397–403.

[17] J.S. Jaworska, K.A. Rose, and A.L. Brenkert, *Individual-based modelling of PCB effects on young-of-the-year largemouth bass in southeastern U.S. reservoirs*. Ecol. Model. 99 (1997), pp. 113–135.

[18] F. Holker and B. Breckling, *A spatio-temporal individual-based fish model to investigate emergent properties at the organismal and the population level*, Ecol. Model. 186 (2005), pp. 406–426.

[19] European Chemicals Bureau, *Technical Guidance Document in support of Commission Directive 93/67/EEC on risk assessment for new notified substances and Commission Regulation (EC) No. 1488/94 on risk assessment for existing substances* Part 2, 2006.

[20] U.S. EPA, *Guidelines for ecological risk assessment*, U.S. EPA, Washington, DC, EPA/630/R-95/002Fa, 1998a.

[21] METI, *Risk Assessment of Endocrine Disrupters*, February 2002 (revised May 2003), Ministry of Economy, Trade and Industry, Japan, 2003.

[22] W.H. Clements and P.M. Kiffney, *Assessing contaminant effects at higher levels of biological organisation.* Environ. Toxicol. Chem. 13 (1994), pp. 357–359.

[23] A. Joern and K.D. Hoagland, *In defense of whole-community bioassays for risk assessment*, Environ. Toxicol. Chem. 15 (1996), pp. 407–409.

[24] P.J. Campbell, D.J.S Arnold, T.C.M. Brock, N.J. Grandy, W. Heger, F. Heimbach, S.J. Maund, and M. Streloke, *Guidance document on higher-tier aquatic risk assessment for pesticides (HARAP)*, SETAC-Europe, Brussels, 1999.

[25] S. Bradbury, T.C.J. Feijtel, and C.J. van Leeuwen, *Meeting the scientific needs of ecological risk assessment in a regulatory context*, Environ. Sci. Technol. 38 (2004), pp. 463–470.

[26] P. Calow, P. Armitage, P. Boon, P. Chave, E. Cox, A. Hildrew, M. Learner, L. Maltby, G. Morris, J. Seager, and B. Whitton, *Ecological issues no. 1: River water quality*. British Ecological Society, Field Studies Council, 1990.

[27] S.J. Maund, T.N. Sherratt, T. Stickland, J. Biggs, P. Williams, N. Shillabeer, and P.C. Jepson, *Ecological considerations in pesticide risk assessment for aquatic ecosystems*, Pestic. Sci. 49 (1997), pp. 185–190.

[28] P.E. Gibbs, P.L. Pascoe, and G.R. Burt, *Sex change in the female dog whelk* Nucella lapillus, *induced by tributyltin from anti-fouling paints*, J. Mar. Biol. Ass. U.K. 68 (1988), pp. 715–731.

[29] G.W. Bryan, P.E. Gibbs, and G.R. Burt, *A comparison of the effectiveness of tri-n-butyltin chloride and five other organotin compounds in promoting the development of imposex in the dog whelk,* Nucella lapillus, J. Mar. Biol. Ass. U.K. 68 (1988), pp. 733–744.

[30] D.M. Fry and C.K. Toone, *DDT-induced feminisation of gull embryos*, Science 213 (1981), pp. 922–924.

[31] C.L. Alzieu, J. Sanjauna, J.P. Deltreil, and M. Borel, *Tin contamination in Arcachon bay: Effects on oyster shell anomalies*, Mar. Pollut. Bull. 17 (1986), pp. 494–498.

[32] P.J.H. Reijnders, *Reproductive failure in common seals feeding on fish from polluted coastal waters*, Nature 342 (1986), pp. 456–457.

[33] L.J. Guillette Jr., T.S. Gross, G.R. Masson, J.M. Matter, H.F. Percival, and A.R. Woodward, *Developmental abnormalities of the gonad and abnormal sex hormone concentrations in juvenile alligators from contaminated and control lakes in Florida*, Environ. Health Perspect. 102 (1994), pp. 680–688.

[34] J. Raloff, *The gender benders*, Sci. News 145 (1994), pp. 24–27.

[35] S.A. Bortone and W.P. Davis, *Fish intersexuality as an indicator of environmental stress,* Bioscience 44 (1994), pp. 165–172.

[36] S. Jobling, M. Nolan, C.R. Tyler, G. Brighty, and J.P. Sumpter, *Widespread sexual disruption in wild fish*, Environ. Sci. Technol. 32 (1998), pp. 2498–2506.

[37] T. Colborn, D. Dumanoski, and J. Peterson Myers, *Our Stolen Future: Are We Threatening Our Fertility, Intelligence, and Survival? A Scientific Detective Story*, Dutton, New York, 1996.

[38] J.P. Sumpter and A.C. Johnson, *Lessons from endocrine disruption and their application to other issues concerning trace organics in the aquatic environment*, Environ. Sci. Technol. 39 (2005), pp. 4321–4332.

[39] D.E. Hinton, S.W. Kullman, R.C. Hardman, D.C. Voz, P.J. Chen, M. Carney, and D.C. Bencic, *Resolving mechanisms of toxicity while pursuing ecotoxicological relevance?* Mar. Pollut. Bull. 51 (2005), pp. 635–648.

[40] T.H. Hutchinson, G.T. Ankley, H. Segner, and C.R. Tyler, *Screening and testing for endocrine disruption in fish — biomarkers as "signposts not traffic lights" in risk assessment*, Environ. Health Perspect. 114 (Suppl 1) (2006), pp. 106–114.

[41] European Medicines Agency, *Pre-authorisation evaluation of medicines for human use. Guideline on the environmental risk assessment of medicinal products for human use*, EMEA/CHMP/SWP/4447/00, 2006.

[42] J.W. Owens, *Report of eight 21-day fish endocrine screening assays with additional test substances for phase-3 of the OECD validation program: Studies with octylphenol in the fathead minnow* (Pimephales promelas) *and zebrafish* (Danio rerio) *and with sodium pentachlorophenol and androstenedione in the fathead minnow* (Pimephales promelas), OECD, 2007. Available at www.oecd.org/dataoecd/62/36/38784404.pdf.

[43] *European Workshop on the impacts of endocrine disruptors on human health and wildlife*, Report of the proceedings, Weybridge, United Kingdom, 1996.

[44] C.W. Schmidt, *The lowdown on low-dose endocrine disruptors*, Environ. Health Perspect. 109 (2001), pp. 420–421.

[45] L. Lin, M.E. Andersen, S. Heber, and O. Zhang, *Non-monotonic dose–response relationship in steroid hormone receptor-mediated gene expression*, J. Molec. Endocrinol. 38 (2007), pp. 569–585.

[46] S. Jobling, S. Coeya, J.G. Whitmore, D.E. Kime, K.J.W. van Look, B.G. McAllister, N. Beresford, A.C. Henshaw, G. Brighty, C.R. Tyler, and J.P. Sumpter, *Wild intersex roach* (Rutilus rutilus) *have reduced fertility*, Biol. Reprod. 67 (2002), pp. 515–524.

[47] V. Palace, R.E. Evans, K. Wautier, L. Vandenbyllardt, W. Vandersteen, and K. Kidd, *Induction of vitellogenin and histological effects in wild fathead minnows from a lake experimentally treated with the synthetic estrogen, ethynylestradiol*, Can. J. Water Qual. 37 (2002), pp. 637–650.

[48] P. Blanchfield, A. Majewski, V. Palace, K. Kidd, and K. Mills, *Reproductive and population-level impacts of a synthetic estrogen on the fathead minnow*, poster from the SETAC conference, Austin, TX, 2003.

[49] K.A. Kidd, P.J. Blanchfield, K.H. Mills, V.P. Palace, R.E. Evans, J.M. Lazorchak, and R.W. Flick, *Collapse of a fish population after exposure to a synthetic estrogen*, Proc. Natl. Acad. Sci. USA 104 (2007), pp. 8897–8901.

[50] K.E. Liney, S. Jobling, J.A. Shears, P. Simpson, and C.R. Tyler, *Assessing the sensitivity of different life stages for sexual disruption in Roach* (Rutilus rutilus) *exposed to effluents from wastewater treatment works*, Environ. Health Perspect. 113 (2005), pp. 1299–1307.

[51] G.S. Helfman, B.B. Collette, and D.E. Facey, *The diversity of fish*. Blackwell Science, London, 1997.

[52] D.A. Holdway, E.J. Kemp, and D.G. Dixon, *Acute toxicity of methoxychlor to larva; white sucker* (Catostomus commersoni) *as modified by age, food availability and egg pre-exposure*, Can. J. Fish. Aquat. Sci. 44 (1987), pp. 227–232.

[53] D.A. Holdway and D.G. Dixon, *Effects of methoxychlor exposure of flagfish egg* (Jordanella floridae) *on hatchability, juvenile methoxychlor tolerance and whole-body levels of tryptophan, serotonin and 5-hydroxyindoleacetic acid*, Water Res. 20 (1986), pp. 893–897.

[54] K. McTavish, H. Stech, and F. Stay, *A modelling framework for exploring the population-level effects of endocrine disruptors*, Environ. Toxicol. Chem. 17 (1998), pp. 58–67.

[55] M.D. Anway, A.S. Cupp, M. Uzumcu, and M.K. Skinner, *Epigenetic transgenerational actions of endocrine disruptors and male fertility*, Science 308 (2005), pp. 1466–1469.

[56] G.R. Biddinger, P. Calow, P. Delorme, G. Harris, B. Hope, B.-L. Lin, M.T. Sorensen, and P. van den Brink, *Managing risk to ecological populations*, in *Population-Level Ecological Risk Assessment*, L.W. Barnthouse, W.R. Munns Jr., and M.T. Sorensen, eds., Taylor and Francis, New York, 2008, pp. 7–40.

[57] F. De Laender, K.A.C. De Schamphelaere, P.A. Vanrolleghem, and C.R. Janssen, *Comparison of different toxic effect sub-models in ecosystem modelling for ecological effects assessments and water standard setting*, Ecotoxicol. Environ. Safety 69 (2008), pp. 13–23.

[58] A.F.W. Schroer, J.D.M. Belgers, C.M. Brock, A.M. Matser, S.J. Maund, and P.J. van den Brink, *Comparison of laboratory single species and field population-level effects of the pyrethroid insecticide cyhalothrin on freshwater invertebrates*, Arch. Environ. Contam. Toxicol. 46 (2004), pp. 324–335.

[59] K. Knauer, S. Maise, G. Thoma, U. Hommen, and J. Gozalez-Valero, *Long-term variability of zooplankton populations in aquatic mesocosms*, Environ. Toxicol. Chem. 24 (2005), pp. 1182–1189.

[60] G.H.P. Arts, L.L. Buijse-Bogdan, J.D.M. Belgers, C.H. va Rhene-Kersten, R. Winngaarden, I. Roessink, S.J. Maund, P.J. van den Brink, and T.C.M. Brock, *Ecological impact in ditch mesocosms of simulated spray drift from a crop protection program for potatoes*, Int. Environ. Assess. Manage. 2 (2006), pp. 105–125.

[61] C.J. Topping, T.S. Hansen, T.S. Jensen, J.U. Jepsen, F. Nikolajsen, and P. Odderskær, *ALMaSS, an agent-based model for animals in temperate European landscapes*, Ecol. Model. 167 (2003), pp. 65–82.

[62] P. van den Brink, H. Baveco, J. Verboom, and F. Heimbach, *An individual-based approach to model spatial population dynamics of invertebrates in aquatic ecosystems after pesticide contamination*, Environ. Toxicol. Chem. 26 (2007), pp. 2226–2236.

[63] P. van den Brink, *Population modelling: A tool for Environmental Risk Assessment*, CEFIC Long-range Research Initiative, ECO10 (in progress).

[64] V. Grimm, P. Chapman, P. Thorbek, F. Heimbach, J. Wogram, P. van den Brink, and V. Forbes, *Ecological models in support of regulatory risk assessments of pesticides: Developing a strategy for the future*, LEMTOX Workshop (SETAC), 2007.

[65] V. Grimm, P. Thorbeck, A. Schmolke, and P. Chapman, *State-of-the-art of ecological modelling for pesticide risk assessment: a critical review,* in *Population-level ecological risk assessment,* in *Ecological models in support of regulatory risk assessments of pesticides: developing a strategy for the future: Proceedings of the LEMTOX workshop Leipzig 2007,* V.E. Forbes, U. Hommen, P. Thorbek, F. Heimbach, P.J. Van den Brink, J. Wogram, H-H. Thulke, and V. Grimm, eds., SETAC series, CRC/Taylor& Francis, New York, (in press).

[66] M.L. Shaffer, Minimum population sizes for species conservation, Bioscience 31 (1981), pp. 131–134.

[67] A.R. Brown, A.M. Riddle, I.J. Winfield, J.M. Fletcher, and J.B. James, *Predicting the effects of endocrine disrupting chemicals on healthy and disease impacted populations of perch* (Perca fluviatilis), Ecol. Model. 189 (2005), pp. 377–395.

[68] V. Demyanov, S.N. Wood, and T.J. Kedwards, *Improving ecological impact assessment by statistical data synthesis using process-based models,* Appl. Stats. 55 (2006), pp. 41–62.

[69] M. Spencer and S. Ferson, *RAMAS- Ecotoxicology version 1.0a.* Applied Biomathematics, Setauket, NY, 1997.

[70] S. Tuljapurkar and H. Caswell, *Structured population models in marine terrestrial and freshwater systems,* Chapman & Hall, New York, 1997.

[71] D.L. De Angelis, D.K. Cox, and C.C. Coutant, *Cannabilism and size dispersal in young-of-the-year large mouth bass: experiment and model.* Ecol. Model. 8 (1980), pp. 133–148.

[72] P.F. Robinson, A.R. Brown, A.M. Riddle, and I.J. Winfield, *Modelling the combined effects of EDCs and temperature on perch* (Perca fluviatilis) *in Windermere,* SETAC, Lille, France, 2005.

[73] P.F. Robinson, A.M. Riddle, A.R. Brown, and M. Gross-Sorokin, *A model to predict the effects of intersex on roach* (Rutilus rutilus) *populations in English rivers,* SETAC, Porto, Portugal, 2007.

[74] J.A. Tyler and K.A. Rose, *Individual variability and spatial heterogeneity in fish population models,* Rev. Fish Biol. Fisheries 4 (1994), pp. 91–123.

[75] P.H. Leslie, *On the use of matrices in certain population mathematics,* Biometrika 35 (1945), pp. 213–245.

[76] R.M. Nisbet, *Delay-differential equations for structured populations,* in *Structured Population Models in Marine Terrestrial and Freshwater Systems,* S. Tuljapurkar and H. Caswell, eds., Chapman & Hall, New York, 1997.

[77] W. van Winkle, H.I. Jager, S.F. Railsback, B.D. Holcomb, T.K. Studley, and J.E. Baldrige, *Individual-based model of sympatric populations of brown and rainbow trout for instream flow assessment: Model description and calibration,* Ecol. Model. 110 (1998), pp. 175–207.

[78] V. Grimm, U. Berger, F. Bastiansena, S. Eliassen, V. Ginot, J. Giske, J. Goss-Custard, T. Grand, S.K. Heinz, G. Huse, A. Huth, J.U. Jepsen, C. Jørgensenc, W.M. Mooij, B. Muller, G. Pe'er, C. Piou, S.F. Railsback, A.M. Robbins, M.M. Robbins, E. Rossmanith, N. Ruger, E. Strand, S. Souissi, R.A. Stillman, R. Vabø, U. Visser, and D.L. DeAngelis, *A standard protocol for describing individual-based and agent-based models,* Ecol. Model. 198 (2006), pp. 115–126.

[79] L. Maltby, T.J. Kedwards, V.E. Forbes, K. Grasman, J.E. Kammenga, W.R. Munns Jr., A.H. Ringwood, J.S. Weis, and S.N. Wood, *Linking individual-level responses and population-level consequences,* in *Ecological Variability: Separating Natural From Anthropogenic Causes of Ecosystem Impairment,* D.J. Baird and G.A. Burton, eds., Penascola, FL, SETAC, 2001, pp. 27–82.

[80] R.A. Pastorok, S.M. Bartell, S. Ferson, and L.R. Ginzburg, *Ecological modelling in risk assessment: Chemical effects on populations, ecosystems and landscapes,* CRC Press, Boca Raton, FL, 2002.

[81] W.R. Munns Jr., J. Gervais, A.A Hoffman, U. Hommen, D. Nacci, M. Nakamaru, R. Sibly, and C.J. Topping, *Modelling approaches to population-level ecological risk assessment*, in *Population-Level Ecological Risk Assessment*, L.W. Barnthouse, W.R. Munns Jr., and M.T. Sorensen, eds., Taylor and Francis, New York, 2008, pp. 179–210.

[82] D.S. Vaughan and S.B. Saila, *A method for determining mortality rates using the Leslie matrix*, Trans. Amer. Fish. Soc. 105 (1976), pp. 380–383.

[83] S. Manel, M.K. Scwartz, G. Luikart, and P. Taberlet, *Landscape genetics: Combining landscape ecology and population genetics*, Trends Ecol. Evolut. 18 (2003), pp. 189–197.

[84] R.J.H. Beverton and S.J. Holt, *On the dynamics of exploited fish populations*, (Great Britain) Ministry of Agriculture, Fisheries and Food. Fish. Invest. (series 2) 19 (1957), pp. 500–533.

[85] W.E. Ricker, *Computation and interpretation of biological statistics of fish populations*, Bulletin 191 of the Fish Research Board of Canada, Ottawa, Ontario, 1975.

[86] A. Grant, *Population consequences of chronic toxicity: Incorporating density dependence into the analysis of life table response experiments*, Ecol. Model. 105 (1998), pp. 325–335.

[87] S.J. Moe, *Density dependence in ecological risk assessment*, in *Population-Level Ecological Risk Assessment*, L.W. Barnthouse, W.R. Munns Jr., and M.T. Sorensen, eds., Taylor and Francis, New York, 2008, pp. 69–92.

[88] V.E. Forbes, P. Calow, and R.M. Sibly, *Toxicant impacts on density-limited populations: A critical review of theory, practice and results*, Ecol. Appl. 11 (2001), pp. 1249–1257.

[89] T.J. Pitcher and P.D.M. MacDonald, *Two models of seasonal growth*. J. Appl. Ecol. 10 (1973), pp. 599–606.

[90] S.P. Cox, S.J.D. Martell, C.J. Walters, T.E. Essington, J.F. Kitchell, C. Boggs, and I. Kaplan, *Reconstructing ecosystem dynamics in the central Pacific Ocean, 1952–1998: Estimating population biomass and recruitment of tunas and billfishes*, Can. J. Fish. Aquat. Sci. 59 (2002), pp. 1724–1735.

[91] E.P. Grist, N.C. Wells, P. Whitehouse, G. Brighty, and M. Crane, *Estimating the effects of 17α-ethinylestradiol on populations of the fathead minnow* Pimephales promelas*: Are conventional toxicological endpoints adequate?* Environ. Sci. Technol. 37 (2003), pp. 1609–1616.

[92] D.J. Hodgson and S. Townley, *Linking management changes to population dynamic responses: The transfer function of a projection matrix perturbation*, J. Appl. Ecol. 41 (2004), pp. 1155–1161.

[93] D.H. Miller and G.T. Ankley, *Modelling impacts on populations: Fathead minnow* (Pimephales promelas) *exposure to the endocrine disruptor 17β-trenbolone as a case study*, Ecotoxicol. Environ. Safety 59 (2004), pp. 1–9.

[94] R.V. O'Neill, R.H. Gardener, S.W. Christensen, W. van Winkle, J.H. Carney, and J.B. Mankin, *Some effects of parameter uncertainty in density-independent and density-dependent models for fish populations*, Can. J. Fish. Aquat. Sci. 38 (1980), pp. 91–100.

[95] L.W. Barnthouse, G.W. Suter, and A.E. Rosen, *Risks of toxic contaminants to exploited fish populations: Influence of life history data uncertainty and exploitation intensity*, Environ. Toxicol. Chem. 9 (1990), pp. 297–311.

[96] T.P. Traas, K.A. Stab, P.R.G. Kramer, W.P. Cofino, and T. Aldenberg, *Modelling and risk assessment of tributyltin accumulation in the food web of a shallow freshwater lake*, Environ. Sci. Technol. 30 (1996), pp. 1227–1237.

[97] T.P. Traas, A.P. van Wezel, J.L.M. Hermens, M. Zorn, A.G.M. van Hattum, and C.J. van Leeuwen, *Environmental quality criteria for organic chemicals predicted from internal effect concentrations and a food web model*, Environ. Toxicol. Chem. 23 (2004), pp. 2518–2527.

[98] J.A. Gervais and H.M. Regan, *What conservation biology and natural resource management can offer population-level ecological risk assessment*, in *Population-Level Ecological Risk Assessment*, L.W. Barnthouse, W.R. Munns Jr., and M.T. Sorensen, eds., Taylor and Francis, New York, 2008, pp. 129–150.

[99] H.J.M. Verhaar, C.J. van Leeuwen, and J.L.M. Hermens, *Classifying environmental pollutants 1: Structure-activity relationships for prediction of aquatic toxicity*, Chemosphere 25 (1992), pp. 471–491.

[100] D.S. Vaughan, *An age structured model of yellow perch in western Lake Erie,* in *Quantitative Population Dynamics*, International Cooperative, Fairland, MD, 1981, pp. 189–219.

[101] U.S. EPA, *Endocrine disrupter screening programme: Proposed statement of policy.* Federal Register 63 (1998b), pp. 71542–71568.

[102] L.W. Barnthouse, W.R. Munns Jr., and M.T. Sorensen,. *Population-level ecological risk assessment*. SETAC series, CRC/Taylor & Francis, New York, 2008.

[103] I.R. Franklin and R. Frankham, *How large must populations be to retain evolutionary potential?* Anim. Conserv. 1 (1998), pp. 69–73.

[104] B.W. Brook, J.J. O'Grady, A.P. Chapman, M.A. Burgman, H.R. Ackayaka, and R. Frankham,. *Predictive accuracy of population viability analysis in conservation biology*, Nature 6776 (2000), pp. 385–387.

[105] F.L. Bunnell, *Alchemy and uncertainty: What are good models?* USDA Forest Service Pacific NW Research Station, Portland, OR, General Technical Report PNW-GTR-232, 1989.

[106] J.M. Elliott, *Numerical changes and population regulation in young migratory trout* Salmo trutta *in a lake district stream, 1966–1983*, J. Animal Ecol. 53 (1984), pp. 327–350.

[107] L.A. Ashfield, T.G. Pottinger, and J.P. Sumpter, *Exposure of female juvenile rainbow trout to alkyl-phenolic compounds results in modification to growth and ovosomatic index*, Environ. Toxicol. Chem. 17 (1998), pp. 679–686.

[108] K. Lieber, J.A. Gangl, T.D. Corry, L.J. Heinis, and F.S. Stay, *Lethality and bio-accumulation of 4-nonylphenol in bluegill sunfish in littoral enclosures*, Environ. Toxicol. Chem. 18 (1999), pp. 394–400.

[109] S.R. Miles-Richardson, S.L. Pieren, K.M. Schols, V.J. Kramerc, E.M. Snyderd, S.A. Snyderd, J.A. Renderb, S.D. Fitzgerald, and J.P. Giesy, *Effects of waterborne exposure to 4-nonylphenol and nonylphenol ethoxylate on secondary sex characteristics and gonads of fathead minnows* (Pimephales promelas), Env. Res. A. 80 (1999), pp. 122–137.

[110] C.R. Tyler and E.J. Routledge, *Oestrogenic effects in fish in English rivers with evidence of their causation*, Pure Appl. Chem. 70 (1998), pp. 1795–1804.

[111] M. Krisfalusi, V.P. Eroschenko, and J.G. Cloud, *Methoxychlor and estradiol-17β affect alevin rainbow trout* (Onchorhynchus mykiss) *mortality, growth and pigmentation*, Bull. Environ. Contam. Toxicol. 61 (1998), pp. 519–526.

[112] T.A. Heming, E.J McGuinness, L.M. George, and K.A. Blumhagen, *Effects of pulsed and spiked exposure to methoxychlor on early life-stages of rainbow trout*, Bull. Environ. Contam. Toxicol. 40 (1988), pp. 764–770.

[113] D.A. Holdway and D.G. Dixon, *Acute toxicity of pulse-dosed methoxychlor on juvenile american flagfish* (Jordinella floridae) *as modified by age, food availability*, Aquatic Toxicol. 6 (1985), pp. 243–250.

[114] M. Krisfalusi, V.P. Eroschenko, and J.G. Cloud, *Exposure of juvenile rainbow trout* (Onchorhynchus mykiss) *to methoxychlor results in a dose-dependent decrease in growth and survival but does not alter male sexual differentiation*, Bull. Environ. Contam. Toxicol. 60 (1998), pp. 659–666.

[115] R. Länge, T.H. Hutchinson, C.P. Croudace, F. Siegmund, H. Schweinfurth, P. Hampe, G.H. Panter, and J.P. Sumpter, *Effects of the synthetic estrogen 17α-ethinylestradiol on the life-cycle of the fathead minnow* (Pimephales promelas), Environ. Toxicol. Chem. 20 (2001), 1216–1227.

[116] S. Jobling, D. Casey, T. Rodgers-Gray, J. Oehlmann, U. Schulte-Oehlmann, S. Pawlowski, T. Baunbeck, A.P. Turner, and C.R. Tyler, *Comparative responses of molluscs and fish to environmental estrogens and an estrogenic effluent*, Aquat. Toxicol. 65 (2003), pp. 205–220.

[117] G.H. Panter, *The oestrogenicity of steroids and steroid conjugates to fish*, Ph.D. diss., Brunel University, U.K., 1998.

[118] P. Whitehouse, M. Wilkinson, J.K. Fawell, and A. Sutton, *Proposed environmental quality standards for nonylphenol: EQS aquatic-toxicity freshwater seawater in water*, U.K. EA RandD Technical Report P42 by WRC, 1998.

[119] G.W. Ahern and R. Briggs, *The relevance of the presence of certain synthetic steroids in the aquatic environment*, J. Pharm. Pharmacol. 41 (1989), pp. 735–736.

[120] D. Isaak, *The ecological life history of the fathead minnow*, Ph.D. diss., Minnesota University, 1961.

[121] J.T. McFadden, G.R. Alexander, and D.S. Shetter, *Numerical changes and population regulation in brook trout* (Savelinus fontinalis). J. Fish. Res. Board Can. 24 (1967), pp. 1425–1459.

[122] R.R. Warner, *Sperm allocation in coral reef fish, strategies for coping with demands on sperm production*, Bioscience 47 (1997), pp. 561–564.

[123] J.F. Craig, *The biology of perch and related fish*, Croom Helm, London and Sydney, 1987.

[124] J.F. Craig, *Growth and production of the 1955 to 1972 cohorts of perch,* Perca fluviatilis *L., in Windermere*, J. Animal Ecol. 49 (1980), pp. 291–315.

[125] E.D. Le Cren, *Perch* (Perca fluviatilis) *and pike* (Esox lucius) *in Lake Windermere from 1940 to 1985; studies in population dynamics*, Can. J. Fish. Aquat. Sci. 44 (1987), pp. 216–228.

[126] C.A. Mills, M.A. Hurley, *Long-term studies on the Windermere populations of perch* (Perca fluviatilis), *pike* (Esox lucius) *and arctic charr* (Salvelinus alpinus), Freshwater Biol. 23 (1990), pp. 119–136.

[127] I.J. Winfield, D.G. George, J.M. Fletcher, and D.P. Hewitt, *Environmental factors influencing the recruitment and growth of underyearling perch* (Perca fluviatilis) *in Windermere North Basin, U.K., from 1966 to 1990*, in *Proceedings of NATO Advanced Research Workshop on Management of Lakes and Reservoirs during Global Change*, D.G. George, J.G. Jones, P. Puncochar, C.S. Reynolds, and D.W. Sutcliffe, eds., Kluwer Academic, Dordrecht, 1998, pp. 245–261.

[128] I.J. Winfield, J.M. Fletcher, D.P. Hewitt, and J.B. James, *Long-term trends in the timing of the spawning season of Eurasian perch* (Perca fluviatilis) *in the north basin of Windermere, U.K.*, in *Proceedings of Percis III: The Third International Percid Fish Symposium*, T.P. Barry and J.A. Malison, eds., University of Wisconsin Sea Grant Institute, Madison, WI, 2004, pp. 95–96.

[129] S.A.L.M. Kooijman, *Toxic effects as process perturbations*, in *The Analysis of Aquatic Toxicity Data*, S.A.L.M. Kooijman and J.J.M. Bedaux, eds., VU University Press, Amsterdam, 1996, pp. 25–28.

[130] P.S. Maitland and R.N. Campbell, *Freshwater Fishes*, Harper Collins, U.K., 1992.

[131] C.E. Davies, J. Shelley, P.T. Harding, I.F.G. McLean, R. Gardiner, and G. Pierson, *Freshwater fishes in Britain the species and their distribution*, Harley Books, England, 2004.

[132] C.R. Tyler, S. Jobling, and J.P. Sumpter, *Endocrine disruption in wildlife: A critical review of the evidence*, Crit. Rev. Toxicol. 28 (1998), pp. 319–361.

[133] L.D. Arcand-Hoy and W.H. Benson, *Fish reproduction: An ecologically relevant indicator of endocrine disruption.* Environ. Toxicol. Chem. 17 (1998), pp. 49–57.

[134] P.M. Campbell and T.H. Hutchinson, *Wildlife and endocrine disrupters: Requirements for hazard identification*, Environ. Toxicol. Chem. 17 (1998), pp. 127–135.

[135] R.H.K. Mann, *Observations on the age, growth, reproduction and food of the roach* Rutilus rutilus *in two rivers in southern England*, J. Fish Biol. 5 (1973), pp. 707–736.

[136] L.A. Vøllestad and J.H. L'Abée-Lund, *Reproductive biology of stream-spawning roach,* Rutilus rutilus, Env. Biol. Fishes 18 (1987), pp. 219–227.

[137] L. von Bertalanffy, *A quantitative theory of organic growth*, Human Biol. 10 (1938), pp. 181–213.

[138] F. Eley and R. Sedgwick, *Modelling the potential impacts of endocrine disruption on fish populations: Statistical model*, U.K. Environment Agency Science Report, SC060002/TR1, 2008.

[139] B. Breckling, U. Middelhoff, and H. Reuter, *Individual-based models as tools for ecological theory and application: Understanding the emergence of organisational properties in ecosystems*, Ecol. Model. 194 (2005), pp. 102–113.

[140] J. Ylikarjula, M. Heino, U. Dieckmann, and K. Veijo, *Does density-dependent individual growth simplify dynamics in age-structure populations? A general model applied to Perch* (Perca fluviatilis), Interim Report IR-01-025, International Institute for Applied Systems Analysis, Austria, 2001.

[141] S. Jobling, D. Sheahan, J.A. Osborne, P. Matthiessen, and J.P. Sumpter, *Inhibition of testicular growth in rainbow trout* (Oncorhynchus mykiss) *exposed to estrogenic alkylphenolic chemicals*, Environ. Toxicol. Chem. 15 (1996), pp. 194–202.

[142] T. Coe, P.B. Hamilton, D. Hodgson, G.C. Paull, J.R. Stevens, K. Sumner, and C.R. Tyler, An environmental estrogen alters reproductive hierarchies, disrupting sexual selection in group-spawning fish, Environ. Sci. Technol. 42 (2008), pp. 5020–5025.

[143] M.A. Gray, K.L. Teather, and C.D. Metcalfe, *Reproductive success and behavior of Japanese medaka* (Oryzias latipes) *exposed to 4-tert-octylphenol*, Environ. Toxicol. Chem. 18 (1999), pp. 2587–2594.

[144] D. Nacci and A.A. Hoffmann, *Genetic variation in population-level ecological risk assessment*, in *Population-Level Ecological Risk Assessment*, L.W. Barnthouse, W.R. Munns Jr., and M.T. Sorensen, eds., Taylor and Francis, New York, 2008, pp. 93–112.

[145] S.F. Railsback, S.L. Lytinen, and S.K. Jackson, *Agent-based simulation platforms: Review and development recommendations*, Simulation 82 (2006), pp. 609–623.

4 Application of Pharmacokinetic Modeling to Understand the Mechanism of Action of Endocrine Disrupting Chemicals

James Devillers

CONTENTS

ABSTRACT

Physiologically based pharmacokinetic (PBPK) modeling is now well recognized for mechanistically simulating and predicting the absorption, distribution, metabolism, and excretion of chemicals in an organism. In this review, the basic concepts of this modeling technique are highlighted. Then, different examples of applications focusing on chemicals affecting the functioning of the endocrine system are presented. Last, some advantages and limitations of PBPK models are briefly discussed.

KEYWORDS

Absorption, Distribution, Metabolism, and Excretion (ADME)
Bisphenol A
PBPK model
Perchlorate
Pharmacokinetics

4.1 INTRODUCTION

Much of the research in toxicology and ecotoxicology is focused on elucidating the mechanisms through which the chemicals exert their adverse effects [1]. Basically, the toxicological effect of a compound to living species depends on its physico-chemical properties as well as its concentration and the time of exposure. To fully understand and predict the toxicity of a chemical, it is necessary to consider both its pharmacokinetics (the detailed mechanisms by which chemicals are distributed from the external environment to target tissues) and pharmacodynamics (the detailed mechanisms by which target tissue doses are transformed into adverse biological responses) [2]. This can be done by means of modeling approaches, the combination of pharmacokinetic and pharmacodynamic models allowing us to predict the dose response of an organism to a chemical, based on the exposure concentrations [3].

In this chapter, an attempt is made to review the main physiologically based pharmacokinetic (PBPK) models [4] that have been designed to study endocrine processes with a special emphasis on those dealing with endocrine disruption mechanisms, because they are the most widely used for modeling these adverse effects. After introducing the concepts fundamental to this approach, different examples of applications are presented focusing on chemicals that adversely affect the endocrine system of living species. Last, the advantages and limitations of this modeling technique are briefly discussed.

4.2 PHYSIOLOGICALLY BASED PHARMACOKINETIC (PBPK) MODELING

The idea behind PBPK modeling is to describe via mathematical equations the physicochemical, biochemical, and physiological processes that determine the pharmacokinetics of a substance within an organism — namely, its absorption, distribution, metabolism, and excretion (ADME) [5]. The first step in the modeling process consists in dividing the studied organism into compartments, each representing a tissue or an organ. The physiological characteristics of the compartments and the behavior of the studied chemical between and within them are described by means of differential and algebraic equations that can be solved numerically to obtain the time-course concentrations of the chemical and its metabolites in the compartments [4,6–8].

Such a PBPK model contains a collection of parameters that have to be determined for specific simulations. These parameters can be broadly classified into two classes:

those in relation to the studied organism as well as the modeled endpoint, and those directly linked to the chemical under study. The first class includes basic physiological parameters such as organ volumes, blood flow rates, ventilation rates, and surface areas for the permeation processes as well as specific parameters such as pH values at different locations. The second class contains chemical-dependent parameters such as organ/blood partition coefficients, permeability parameters for different cross-membrane transports, and kinetic constants for the active processes [4,9].

As previously indicated, PBPK models allow the conversion of an exposure concentration to tissue dose, which can be used as input in a pharmacodynamic (PD) model to predict a biological response such as binding activity.

4.3 MODELING ENDOCRINE DYSFUNCTIONS

The endocrine system operates through a complex series of events triggered by glands and their messengers, called hormones. These substances are conveyed by the bloodstream and control many fundamental functions in the organisms such as homeostasis, reproduction, and development. From a modeling point of view, the endocrine system can be viewed as a complex compartmental system; hence, PBPK modeling appears particularly suited to model endocrine functions as well as endocrine dysfunctions. Rather than cataloging all the models in the field, some relevant case studies are summarized in the following paragraphs.

4.3.1 PBPK MODELING OF BISPHENOL A

Bisphenol A (4,4'-isopropylidene-2-diphenol; BPA) is a monomer constituting the starting material for the manufacture of polycarbonate plastics and epoxy resins used in many consumer products, including food-ware, baby formula bottles, water carboys, food can linings, and dental composite fillings and sealants [10]. BPA has been shown to leach from these materials due to incomplete polymerization and to degradation of the polymers by exposure to high temperatures, occurring under normal conditions of use [11]. Obviously, significant amounts of BPA are also released into the atmosphere during production. As a result, exposure to BPA in the general population is widespread. Unfortunately, there has been concern about the estrogenic potential of BPA [12] and several organs have been found to be affected by prenatal exposure to BPA [10,13,14].

Shin et al. [15] developed a PBPK model to simulate the tissue distribution and blood pharmacokinetics of BPA in nonpregnant rats and humans. The model, which included a lot of compartments, was basically constructed to predict the steady-state levels of BPA in blood and various tissues observed in rats after multiple intravenous injections. The PBPK model was further applied to predict blood and various BPA tissue levels in a 70-kg human after a single intravenous injection (5-mg dose) and multiple oral administrations to steady state (100-mg doses every 24 hours). This model was highly criticized by Teeguarden and coworkers [16] who proposed a more realistic PBPK model for BPA and BPA glucuronide (BPAG) to simulate blood and uterine concentrations after exposure to BPA by oral and intravenous routes. The model was initially developed for adult male and female rats and was later extended to humans from data obtained from literature and experiments in male and female volunteers. A nested model structure was used, with submodels for BPA and BPAG comprising the overall model. The BPA model consisted of the following compartments: blood, uterus, liver, the lumen of the gastrointestinal (GI) tract, and a body compartment representing the remaining perfused tissues. In this model, BPA distributed to and from a nonmetabolizing body compartment. The other tissues were formulated the same as the body compartment, but they were described with additional terms representing processes affecting tissue concentrations of BPA (metabolism, protein or ER binding, uptake) [16]. Intravenous- and oral-route blood kinetics of BPA in rats and oral-route plasma and urinary elimination kinetics in humans were well described by the model. Simulations of rat oral-route BPAG pharmacokinetics were of lower quality. In the absence of BPA binding to plasma proteins, simulations showed high ER occupancy at doses without uterine effects. In contrast, the correlation between receptor occupancy and uterine response was stronger when the free available concentration of BPA was restricted by plasma protein binding. These results highlighted the importance of including plasma binding in BPA PBPK models [16].

In 2007, Kawamoto et al. [17] proposed a PBPK model for BPA in pregnant mice using oral administration. The model consisted of two analogous kinetic models for BPA and its metabolites connected to each other at the liver compartment in the metabolism phase. Because the brain is one of the targets of BPA, it was represented as a separate compartment in the model. For the sake of simplicity, the yolk sac and chorioallantoic placentas were modeled as a single placenta. Two gastrointestinal compartments, small and large intestines, and the gallbladder compartment were considered to model the enterohepatic circulation observed in a previous experimental study conducted by these authors [18]. Interestingly, the model was constructed from pharmacokinetic data obtained in experiments performed by the authors on pregnant mice following a single oral administration of BPA. The schematic diagram of the PBPK model for BPA on pregnant mice is shown in Figure 4.1, and the main physiological and physicochemical parameter values used in the model are listed in Table 4.1. Model simulations were compared with the experimental pharmacokinetic data on pregnant mice after a single oral administration (10 mg/kg), resulting in good consistency over the whole period after the administration regarding the concentrations in all tissues. A satisfying consistency was observed for BPA and its metabolite concentrations, as well as total concentrations in maternal blood and the liver. The model also described the rapid transfer of BPA through the placenta to the fetus and the slow disappearance from fetuses. The simulated time course after three-time repeated oral administrations of BPA by the model fitted the experimental data. Good consistency with experimental data was also obtained with a ten-time lower dose of BPA (1 mg/kg) [17].

4.3.2 PBPK Modeling of Perchlorate

Ammonium perchlorate is an oxidizer used in the propulsion systems of solid fuel rockets and missiles. It is also employed in the manufacture of flares, fireworks,

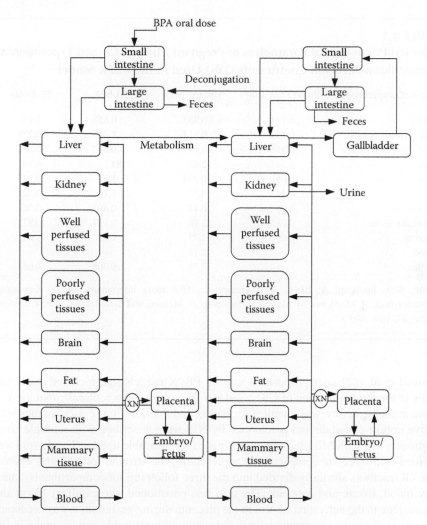

FIGURE 4.1 The physiologically based pharmacokinetic (PBPK) model for bisphenol A (BPA) in pregnant mice. (Adapted from Y. Kawamoto, W. Matsuyama, M. Wada, J. Hishikawa, M. Pui Ling Chan, A. Nakayama, and S. Morisawa, *Toxicol. Appl. Pharmacol.*, 224, 182, 2007.)

and automobile airbags. Past use and disposal practices have resulted in important environmental contaminations yielding the production of the soluble anion (ClO_4^-) that is toxic to aquatic species [19] as well as to humans when found in drinking water. Human health concerns arise from the fact that ClO_4^-, being similar in shape to iodide (I^-), is able to bind to the sodium-iodide symporter (NIS), thus reducing the amount of iodide taken up into the thyroid [20].

A rather simple PBPK model was first derived by Fisher et al. [21] for describing the disposition of perchlorate in the adult male rat from limited experimental data.

TABLE 4.1
**Selected Physiological Parameters of Pregnant Mice (d 15) and Experimental
Tissue/Blood Partition Coefficients (TBs) Used in the PBPK Model**

Organ (Compartment)	Blood Flow (l/h)	Volume (ml)	TB-BPA	TB-BPAG
Kidney	0.154	0.733	0.858	3.18
Well perfused tissues	0.120	0.518	1.43	0.271
Poorly perfused tissues	0.948	33.1	0.682	0.387
Liver	0.344	2.92	384	6.76
Brain	0.056	0.733	1.34	0.125
Fat	0.147	3.77	1.16	0.22
Uterus	0.017	0.44	0.693	0.581
Mammary tissue	0.085	2.18	0.957	0.27
Placenta	0.097	1.73	0.88	0.68
Blood	—	2.11	—	—
Fetuses	—	7.74	0.308	0.058

Notes: BPA, bisphenol A; BPAG, BPA glucuronide. (For more information, see Y. Kawamoto,
W. Matsuyama, M. Morikawa, M. Morita, M. Sugimoto, N. Manabe, and S. Morisawa, *Toxicol. Environ.
Chem.*, 87, 199, 2005.)

Clewell et al. [22] later proposed a suite of PBPK models that included important
steps in the mode of action of ClO_4^- and key life stages for perchlorate toxicity. The
basic model structure was identical for the rat and human (Figure 4.2). Tissues with
active uptake of iodide and perchlorate by NIS were described with multiple compartments and with Michaelis-Menten kinetics for saturable uptake. The thyroid was
divided into three subcompartments representing the stroma, follicle, and colloid.
The GI tract was similarly divided into the three following subcompartments: capillary blood, tissue, and contents. The skin was partitioned into capillary blood and
tissue. Due to the activation of NIS in the placenta during gestation, it was necessary
to include two subcompartments in the placenta with saturable active uptake in the
gestation model. In the same way, the activation of anion channels in the milk and
NIS in the mammary gland required three compartments to be considered for the
mammary gland in the lactation model — namely, capillary blood, mammary tissue,
and milk, with active uptake into both the mammary gland and milk [22]. Active
uptake was also included in the mammary gland of the pregnant rat. It is noteworthy
that flow-limited compartments were also included for the liver, kidney, and fat. The
I^- and ClO_4^- models were broadly the same assuming that similar size and charge
of the anions would yield similar kinetics. Both models were based on the premise
that only free anions would be available for uptake via diffusion or symporter. The
models differed only in that ClO_4^- was described with two compartments in the
serum to account for the binding of the anion to serum proteins, such as albumin
[22]. The kinetic parameters used in this suite of PBPK models are listed in Table 4.2
and Table 4.3.

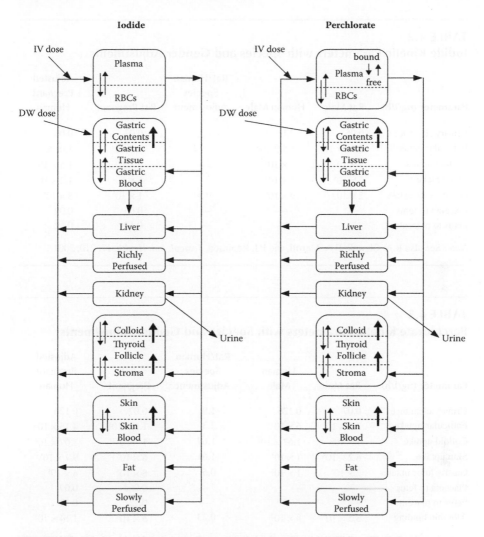

FIGURE 4.2 Basic physiologically based pharmacokinetic (PBPK) model structure for iodide and perchlorate kinetics in male rats and humans. (Adapted from R.A. Clewell, E.A. Merrill, and P.J. Robinson, *Toxicol. Ind. Health*, 17, 210, 2001.)

The gestation and lactation rat models were able to describe perchlorate distribution in serum and thyroid across drinking water doses ranging from 0.01 to 10 mg/kg/day. The models were also able to reproduce I⁻ concentrations in the adult, fetal, and neonatal rats as well as to describe the transfer kinetics during pregnancy and lactation. Preliminary PBPK models for human gestation and lactation were designed from the rat models. They were able to predict serum, thyroid, fetal, milk, and neonatal time-course data from several literature studies at different time points throughout gestation and lactation [22]. The suite of PBPK models was then used to derive more accurate models [23–26].

TABLE 4.2
Iodide Kinetic Parameters with Species and Gender Adjustments

Parameter (ng/l/h)	Rat Male	Human Male	Rat/Human Species Adjustment	Rat Pregnant	Adjusted Pregnant Human
Urinary clearance	0.05	0.1	2	0.03	0.06
Follicular uptake	5.94×10^4	1.5×10^5	2.52	4×10^4	1.01×10^5
Colloid uptake	4×10^7	1×10^8	2.5	6×10^7	1.5×10^8
Skin uptake	5×10^5	1×10^6	2	6×10^4	1.2×10^5
Gastric juice uptake	2×10^6	9×10^5	0.5	1×10^6	5×10^5
Placenta to fetus	—	—	—	0.06	0.09
Fetus to placenta	—	—	—	0.12	0.12

Note: See also R.A. Clewell, E.A. Merrill, and P.J. Robinson, *Toxicol. Ind. Health*, 17, 210, 2001.

TABLE 4.3
Perchlorate Kinetic Parameters with Species and Gender Adjustments

Parameter (ng/l/h)	Rat Male	Human Male	Rat/Human Species Adjustment	Rat Pregnant	Adjusted Pregnant Human
Urinary clearance	0.07	0.126	1.8	0.07	0.126
Follicular uptake	2.6×10^3	6×10^3	2.3	1.8×10^3	4.15×10^3
Colloid uptake	1×10^4	1.67×10^4	1.67	1×10^4	1.67×10^4
Skin uptake	6.2×10^5	1×10^6	1.61	6×10^5	9.7×10^5
Gastric juice uptake	2×10^5	1×10^5	0.5	8×10^5	4×10^5
Placenta to fetus	—	—	—	0.065	0.09
Fetus to placenta	—	—	—	0.12	0.12
Albumin binding	3.5×10^3	8×10^2	0.23	5×10^3	1.14×10^3

Note: See also R.A. Clewell, E.A. Merrill, and P.J. Robinson, *Toxicol. Ind. Health*, 17, 210, 2001.

4.4 ADVANTAGES AND LIMITATIONS OF PBPK MODELS FOR ENDOCRINE-ACTIVE CHEMICALS

The case studies presented in the previous paragraphs clearly highlight the advantages of the PBPK models. Basically, they allow the study of the absorption, distribution, metabolism, and excretion (ADME) of chemicals, which are pivotal activities to determine for drugs and xenobiotics. More generally, the PBPK models appear particularly suited to summarize and rationalize information, to hypothesize and validate mechanisms of action, to easily integrate time and dynamics of organism development in the modeling process, and to extrapolate to lower doses, across routes, and species [27–31]. These models basically consist of different interrelated compartments, which represent

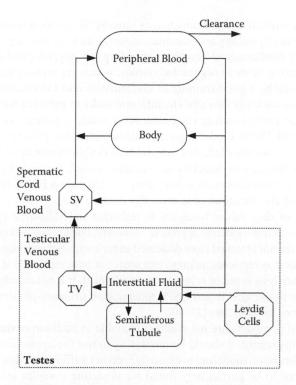

FIGURE 4.3 Physiological compartments included in the rat testicular–hypothalamic–pituitary axis model. (Adapted from H.A. Barton and M.E. Andersen, *Toxicol. Sci.*, 45, 174, 1998.)

the different organs of the studied organism. The level of description of the organs and physiological processes directly depends on the studied phenomenon. Thus, for example, a pharmacokinetic model was developed by Barton and Andersen [32] to describe the dose-response behaviors of chemicals that interact with the testicular–hypothalamic–pituitary axis of male rats and affect its reproductive functions. Consequently, the model focused only on this organ. Indeed, it consisted of compartments for the testes, peripheral venous and arterial blood, and the remainder of the body (Figure 4.3). The testes were divided into compartments including seminiferous tubules, interstitial tissue made up of interstitial fluid, and Leydig cells. Briefly, this model was able to describe behaviors of the central axis regulation of testosterone synthesis in response to perturbations such as castration, testosterone replacement, and exposure to antiandrogen [32].

Some modeling approaches, such as QSARs (Quantitative Structure-Activity Relationships) [33], are based on an accurate topological and physicochemical description of the chemicals, but this is not the case of the PBPK models for which the biological and physiological parameters are the key input parameters. Consequently, PBPK models can be successfully used to study the ADME of simple as well as complex chemical mixtures [34–36].

However, PBPK modeling shows some limitations. The design of a PBPK model needs first to identify the key organs, tissues, and mechanisms involved in the

ADME of the chemical and phenomenon of interest. To do so, it is required to have a good training in physiology and pharmacokinetics. In a second step, it is necessary to create all the mathematical equations encoding the physiological and biological processes occurring in these organs and tissues, which are represented by compartments. Consequently, a good training in mathematics and informatics is also compulsory to algorithmically describe the different tasks to perform within the model and compute the corresponding equations. As a result, it appears obvious that the design of a sound PBPK model requires intimate interdisciplinary collaborations. If this condition is not satisfied, this can yield the design of wrong or faulty models, which are often difficult to detect by individuals. Indeed, it seems logical that if an interdisciplinary communication is necessary to elaborate a PBPK model, this collaboration would also be required to critically analyze it.

The number of data values necessary to elaborate and validate a PBPK model is another hurdle with this modeling approach. Moreover, due to time and cost constraints, very often they are not obtained from dedicated experiments but are collected from various sources. This may represent an important source of bias, especially if an uncertainty or a sensitivity analysis is not or is incorrectly performed. It is noteworthy that attempts have been made to use QSAR and QSPR (quantitative structure-property relationship) models to generate missing data [37–40].

Last, the PBPK models are not readily amenable to addition or deletion of compartments. In this regard, it should be interesting to test the agent-based models [41] in endocrine disruption modeling because they do not suffer from this limitation and they have proved to be particularly suited for modeling complex systems (see, for example, [42–44]; see also Chapter 3, this volume).

4.5 CONCLUSIONS

Physiologically based pharmacokinetic (PBPK) modeling is now commonly used for mechanistically simulating and predicting the ADME of a substance at the tissue/organ/organism level. PBPK models allow individuals to be distinguished within a group because they are able to account for differences related to age, gender, race, and so on. This is particularly interesting for the simulation of a biological process within a specific population such as children or pregnant women. This is the reason why PBPK models have found successful applications for modeling endocrine activity. The combination of PBPK models with pharmacodynamic (PD) models allows for the transformation of target tissue doses into adverse biological responses. Even if the PBPK models show some limitations, they represent very promising tools for elucidating the mechanism of action of endocrine disrupting chemicals.

REFERENCES

[1] G.C. Becking, *Use of mechanistic information in risk assessment for toxic chemicals*, Toxicol. Lett. 77 (1995), pp. 15–24.

[2] M.E. Andersen, *Development of physiologically based pharmacokinetic and physiologically based pharmacodynamic models for application in toxicology and risk assessment*, Toxicol. Lett. 79 (1995), pp. 35–44.

[3] B. Reisfeld, A.N. Mayeno, M.A. Lyons, and R.S.H. Yang, *Physiologically based pharmacokinetic and pharmacodynamic modeling*, in *Computational Toxicology: Risk Assessment for Pharmaceutical and Environmental Chemicals*, S. Ekins, ed., Wiley, Hoboken, NJ, 2007, pp. 33–69.

[4] M.B. Reddy, R.S.H. Yang, H.J. Clewell, and M.E. Andersen, *Physiologically Based Pharmacokinetics Modeling: Science and Applications*, Wiley, Hoboken, NJ, 2005.

[5] M. Dickins and H. van de Waterbeemd, *Simulation models for drug disposition and drug interactions*, Drug Discov. Today BioSilico 2 (2004), pp. 38–45.

[6] H.W. Leung and D.J. Paustenbach, *Physiologically based pharmacokinetic and pharmacodynamic modeling in health risk assessment and characterization of hazardous substances*, Toxicol. Lett. 79 (1995), pp. 55–65.

[7] K.C.T. Hoang, *Physiologically based pharmacokinetic models: Mathematical fundamentals and simulation implementations*, Toxicol. Lett. 79 (1995), pp. 99–106.

[8] M.E. Andersen, *Toxicokinetic modeling and its applications in chemical risk assessment*, Toxicol. Lett. 138 (2003), pp. 9–27.

[9] W. Schmitt and S. Willman, *Physiology-based pharmacokinetic modeling: Ready to be used*, Drug Discov. Today Technol. 2 (2005), pp. 125–132.

[10] M.V. Maffini, B.S. Rubin, C. Sonnenschein, and A.M. Soto, *Endocrine disruptors and reproductive health: The case of bisphenol-A*, Molec. Cell. Endocrinol. 254–255 (2006), pp. 179–186.

[11] J.E. Biles, T.P. McNeal, T.H. Begley, and H.C. Hollifield, *Determination of bisphenol-A in reusable polycarbonate food contact plastics and migration to food simulating liquid*, J. Agric. Food Chem. 45 (1997), pp. 3541–3544.

[12] G.G. Kuiper, J.G. Lemmen, B. Carlsson, J.C. Corton, S.H. Safe, P.T. van der Saag, B. van der Burg, and J.A. Gustafsson, *Interaction of estrogenic chemicals and phytoestrogens with estrogen receptor beta*, Endocrinol. 139 (1998), pp. 4252–4263.

[13] H. Kabuto, M. Amakawa, and T. Shishibori, *Exposure to bisphenol A during embryonic/fetal life and infancy increases oxidative injury and causes underdevelopment of the brain and testis in mice*, Life Sci. 74 (2004), pp. 2931–2940.

[14] C.M. Markey, E.H. Luque, M.M. Munoz de Toro, C. Sonnenschein, and A.M. Soto, *In utero exposure to bisphenol A alters the development and tissue organization of the mouse mammary gland*, Biol. Reprod. 65 (2001), pp. 1215–1223.

[15] B.S. Shin, C.H. Kim, Y.S. Jun, D.H. Kim, B.M. Lee, C.H. Yoon, E.H. Park, K.C. Lee, S.Y. Han, K.L. Park, H.S. Kim, and S.D. Yoo, *Physiologically based pharmacokinetics of bisphenol A*, J. Toxicol. Environ. Health A 67 (2004), pp. 1971–1985.

[16] J.G. Teeguarden, J.M. Waechter, H.J. Clewell, T.R. Covington, and H.A. Barton, *Evaluation of oral and intravenous route pharmacokinetics, plasma protein binding, and uterine tissue dose metrics of bisphenol A: A physiologically based pharmacokinetic approach*, Toxicol. Sci. 85 (2005), pp. 823–838.

[17] Y. Kawamoto, W. Matsuyama, M. Wada, J. Hishikawa, M. Pui Ling Chan, A. Nakayama, and S. Morisawa, *Development of a physiologically based pharmacokinetic model for bisphenol A in pregnant mice*, Toxicol. Appl. Pharmacol. 224 (2007), pp. 182–191.

[18] Y. Kawamoto, W. Matsuyama, M. Morikawa, M. Morita, M. Sugimoto, N. Manabe, and S. Morisawa, *Disposition of bisphenol A in pregnant mice and fetuses after a single and repeated oral administration*, Toxicol. Environ. Chem. 87 (2005), pp. 199–213.

[19] J.W. Park, J. Rinchard, F. Liu, T.A. Anderson, R.J. Kendall, and C.W. Theodorakis, *The thyroid endocrine disruptor perchlorate affects reproduction, growth, and survival of mosquitofish*, Ecotoxicol. Environ. Safety 63 (2006), pp. 343–352.

[20] J. Wolff, *Perchlorate and the thyroid gland*, Pharmacol. Rev. 50 (1998), pp. 89–105.

[21] J. Fisher, P. Todd, D. Mattie, D. Godfrey, L. Narayanan, and K. Yu, *Preliminary development of a physiological model for perchlorate in the adult male rat: A framework for further studies*, Drug Chem. Toxicol. 23 (2000), pp. 243–258.

[22] R.A. Clewell, E.A. Merrill, and P.J. Robinson, *The use of physiologically based models to integrate diverse data sets and reduce uncertainty in the prediction of perchlorate and iodide kinetics across life stages and species*, Toxicol. Ind. Health 17 (2001), pp. 210–222.

[23] R.A. Clewell, E.A. Merrill, K.O. Yu, D.A. Deirdre, A. Mahle, T.R. Sterner, D.R. Mattie, P.J. Robinson, J.W. Fisher, and J.M. Gearhart, *Predicting fetal perchlorate dose and inhibition of iodide kinetics during gestation: A physiologically-based pharmacokinetic analysis of perchlorate and iodide kinetics in rat*, Toxicol. Sci. 73 (2003), pp. 235–255.

[24] E.A. Merrill, R.A. Clewell, J.M. Gearhart, P.J. Robinson, T.R. Sterner, K.O. Yu, D.R. Mattie, and J.W. Fisher, *PBPK predictions of perchlorate distribution and its effect on thyroid uptake of radioiodide in the male rat*, Toxicol. Sci. 73 (2003), pp. 256–269.

[25] R.A. Clewell, E.A. Merrill, K.O. Yu, D.A. Mahle, T.R. Sterner, J.W. Fisher, and J.M. Geaehart, *Predicting neonatal perchlorate dose and inhibition of iodide uptake in the rat during lactation using physiologically-based pharmacokinetic modeling*, Toxicol. Sci. 74 (2003), pp. 416–436.

[26] E.A. Merrill, R.A. Clewell, P.J. Robinson, A.M. Jarabek, J.M. Gearhart, T.R. Sterner, and J.W. Fisher, *PBPK model for radioactive iodide and perchlorate kinetics and perchlorate-induced inhibition of iodide uptake in humans*, Toxicol. Sci. 83 (2005), pp. 25–43.

[27] M.A. Medinsky, *The application of physiologically based pharmacokinetic/pharmacodynamic (PBPK/PD) modeling to understanding the mechanism of action of hazardous substances*, Toxicol. Lett. 79 (1995), pp. 185–191.

[28] F. Welsch, G.M. Blumenthal, and R.B. Conolly, *Physiologically based pharmacokinetic models applicable to organogenesis: Extrapolation between species and potential use in prenatal toxicity risk assessments*, Toxicol. Lett. 82/83 (1995), pp. 539–547.

[29] J.F. Young, *Physiologically-based pharmacokinetic model for pregnancy as a tool for investigation of developmental mechanisms*, Comput. Biol. Med. 28 (1998), pp. 359–364.

[30] M.E. Andersen, R.B. Conolly, E.M. Faustman, R.J. Kavlock, C.J. Portier, D.M. Sheehan, P.J. Wier, and L. Ziese, *Quantitative mechanistically based dose-response modeling with endocrine-active compounds*, Environ. Health Prespect. 107 (1999), pp. 631–638.

[31] G. Ginsberg, D. Hattis, and B. Sonawane, *Incorporating pharmacokinetic differences between children and adults in assessing children's risks to environmental toxicants*, Toxicol. Appl. Pharmacol. 198 (2004), pp. 164–183.

[32] H.A. Barton and M.E. Andersen, *A model for pharmacokinetics and physiological feedback among hormones of the testicular-pituitary axis in adult male rats: A framework for evaluating effects of endocrine active compounds*, Toxicol. Sci. 45 (1998), pp. 174–187.

[33] W. Karcher and J. Devillers, *Practical Applications of Quantitative Structure-Activity Relationships (QSAR) in Environmental Chemistry and Toxicology*, Kluwer Academic, Dordrecht, The Netherlands, 1990.

[34] S. Haddad, M. Béliveau, R. Tardif, and K. Krishnan, *A BPPK modeling-based approach to account for interactions in the health risk assessment of chemical mixtures*, Toxicol. Sci. 63 (2001), pp. 125–131.

[35] J.E. Dennison, M.E. Andersen, I.D. Dobrev, M.M. Mumtaz, and R.S.H. Yang, *PBPK modeling of complex hydrocarbon mixtures: Gasoline*, Environ. Toxicol. Pharmacol. 16 (2004), pp. 107–119.

[36] R.S.H. Yang and J.E. Dennison, *Initial analyses of the relationship between "thresholds" of toxicity for individual chemicals and "interaction thresholds" for chemical mixtures*, Toxicol. Appl. Pharmacol. 223 (2007), pp. 133–138.

[37] F.M. Parham, M.C. Kohn, H.B. Matthews, C. DeRosa, and C.J. Portier, *Using structural information to create physiologically based pharmacokinetic models for all polychlorinated biphenyls. I. Tissue:blood partition coefficients*, Toxicol. Appl. Pharmacol. 144 (1997), pp. 340–347.

[38] F.M. Parham and C.J. Portier, *Using structural information to create physiologically based pharmacokinetic models for all polychlorinated biphenyls*, Toxicol. Appl. Pharmacol. 151 (1998), pp. 110–116.

[39] M. Béliveau, R. Tardif, and K. Krishnan, *Quantitative structure-property relationships for physiologically based pharmacokinetic modeling of volatile organic chemicals in rats*, Toxicol. Appl. Pharmacol. 189 (2003), pp. 221–232.

[40] M. Béliveau, J. Lipscomb, R. Tardif, and K. Krishnan, *Quantitative structure-property relationships for interspecies extrapolation of the inhalation pharmacokinetics of organic chemicals*, Chem. Res. Toxicol. 18 (2005), pp. 475–485.

[41] E. Bonabeau, *Agent-based modeling: Methods and techniques for simulating human systems*, Proc. Nat. Acad. Sci. USA 99 (2002), pp. 7280–7287.

[42] D. Walker, S. Wood, J. Southgate, M. Holcombe, and R. Smallwood, *An integrated agent-mathematical model of the effect of intercellular signalling via the epidermal growth factor receptor on cell proliferation*, J. Theor. Biol. 242 (2006), pp. 774–789.

[43] K. Bentley, H. Gerhardt, and P.A. Bates, *Agent-based simulation of notch-mediated tip cell selection in angiogenic sprout initialisation*, J. Theor. Biol. 250 (2008), pp. 25–36.

[44] H. Devillers, J.R. Lobry, and F. Menu, *An agent-based model for predicting the prevalence of* Trypanosoma cruzi *I and II in their host and vector populations*, J. Theor. Biol. 255 (2008), pp. 307–315.

5 Comparative Modeling Review of the Nuclear Hormone Receptor Superfamily

Nathalie Marchand-Geneste and James Devillers

CONTENTS

ABSTRACT

Endocrine disrupting chemicals (EDCs) are either natural or synthetic chemicals that directly or indirectly enhance (agonist effect) or inhibit (antagonist effect) the action of hormones that bind to specific nuclear receptors (NRs). NRs are important transcriptional regulators involved in diverse physiological functions such as control of homeostasis, cell differentiation, and embryonic development. In the absence of NR X-ray crystal structures, computer-aided molecular modeling becomes an alternative tool to provide *in silico* predictions of three-dimensional (3D) models of such receptors. Moreover, advances in comparative (homology) modeling algorithms, in addition to the increase in experimental protein structure information, allow the generation of homology models for a significant portion of known genomic protein sequences. The aim of this chapter is to provide an overview of the 3D homology models of the NR superfamily. Such models are of considerable value, allowing for the understanding of EDC–receptor interactions and the generation of Structure-Activity Relationship (SAR) or 3D Quantitative Structure-Activity Relationship (QSAR) models. The strengths and weaknesses of the homology models applied to the EDCs are also analyzed.

KEYWORDS

Comparative modeling
Homology model
Nuclear receptor
Protein Data Bank (PDB)
Sequence identity
Three-Dimensional Quantitative Structure-Activity Relationship (3D QSAR)
X-ray

5.1   INTRODUCTION

Three-dimensional protein structures are key to a detailed understanding of the molecular basis of protein functions. Techniques to experimentally determine protein structure (for example, nuclear magnetic resonance [NMR] spectroscopy, X-ray crystallography, and electron microscopy) have made great progress in recent years, and more than 47,000 experimental protein structures have been deposited in the Protein Data Bank (PDB) (www.rcsb.org/pdb) [1,2]. The knowledge of such three-dimensional (3D) structural information can give insight into the key molecular interactions between the EDCs and the nuclear receptor active site and enables us to determine how modifications to the ligand or the protein structure (by site-directed mutagenesis of several amino acids) may potentially affect the binding. Moreover, the experimental binding affinity data of several EDCs can be correlated to calculated binding energy values by using current molecular modeling methods (such as docking simulation).

However, the number of structurally characterized proteins is smaller than the number of known protein sequences in the Swiss-Prot and TrEMBL databases [3], which contain about five million sequence entries. For example, membrane-bound proteins are complex to crystallize in the aqueous environment required for X-ray diffraction experiments; other proteins are insufficiently soluble or too large for NMR studies or are not robust enough to obtain an electron micrograph. Thus, for the majority of protein sequences, there is no experimental structural information available. Therefore, to address this deficit, alternative methods have been developed to give insight into the functional role of a receptor in terms of its 3D structure. Among all the currently available computational approaches, homology (or comparative) modeling is the method that can reliably generate a 3D model for a protein [4]. Other techniques include Quantitative Structure-Activity Relationships (QSARs) and the use of pharmacophores.

This chapter focuses on the 3D homology models of nuclear receptors that are reported in the literature. The advantages and limitations of these different models are discussed.

5.2 NUCLEAR RECEPTOR SUPERFAMILY

5.2.1 NUCLEAR RECEPTOR (NR) FUNCTIONAL DOMAINS

Nuclear receptors (NRs), which belong to different classes [5–8] (see also Chapter 2, this volume), are involved in many important roles in eukaryotic development, differentiation, reproduction, and metabolic homeostasis [9,10]. NRs exhibit a modular structure composed of five domains designated A through F [11] (Figure 5.1). The N-terminal region (A/B domain) varies considerably between the different proteins, both in sequence and size, ranging among classical NRs from 23 amino acids in the vitamin D receptor (VDR) to 602 amino acids in mineralocorticoid receptor (MR). The DNA-binding domain (DBD) including around 70 amino acids (C domain) and the

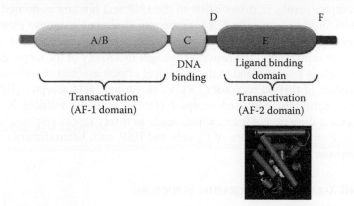

FIGURE 5.1 The structural and functional organization of nuclear receptors. **(See color insert following page 244.)**

ligand-binding domain (LBD) with 250 amino acids (E domain) are the best conserved regions throughout the superfamily [10,12,13]. A linker region known as D domain is located between the DBD and the LBD. In addition to the domains described above, we found the carboxyl-terminal region (F domain), which is not always present. Its function is poorly understood.

The C domain contains a P-box, which is a short motif responsible for DNA-binding specificity and which is involved in the dimerization of nuclear receptors including the formation of both heterodimers and homodimers. The D region serves as a flexible hinge between the C and E domains. It contains the nuclear localization signal. Phosphorylation of the hinge region is coupled with increased transcriptional activation. The LBD or E domain is responsible for the binding of cognate ligand or hormone. This domain also contains a ligand-dependent transactivation function (AF-2) necessary for recruiting transcriptional coactivators interacting with the general transcriptional activation machinery. Most NRs contain a transcriptional activation function (AF-1) in the variable A/B domain. Contrary to the moderately conserved AF-2 domain, the AF-1 is weakly conserved across the NR superfamily and may mediate differential promoter regulation *in vivo*.

5.2.2 NR Mechanism of Signaling

Present in vertebrates, arthropods, and nematodes, NRs control numerous processes involved in development, growth, reproduction, cell differentiation, proliferation, apoptosis, and homeostasis [11]. All NRs modulate gene transcription but in a different way depending on the class to which they belong [9,10,14]. In most cases, they activate transcription in response to the binding of a cognate ligand, generally a small lipophilic molecule. The mechanism of action of the steroid receptors has been characterized in considerable detail [15,16]. These receptors are transformed into active transcriptional factors by the binding of the specific ligand, the appropriate steroid hormone. The proteins are found within the target cells as soluble complexes with heat shock protein (HSP) [17]. Binding of the steroid to its cognate receptor results in dissociation of the HSP and binding as homodimers to DNA. Interaction with specific DNA sequences (hormone response elements) in the proximity of genes results in changes in the rate of transcriptional initiation (stimulation or repression). The resultant changes in activity of the target genes and the corresponding proteins lead to the biological effects associated with hormones. The association of thyroid receptor (TR), VDR, retinoic acid receptor (RAR), and peroxisome proliferator-activated receptor (PPAR) with the retinoid X receptor (RXR) forms a dimeric complex, which leads to direct repeat response elements [9,10,14,18]. Even in the absence of ligands and HSP, such heterodimers are bound to their response element.

5.2.3 NR X-ray Crystallographic Structures

X-ray crystallographic structures of studied NRs (Table 5.1) that were resolved and deposited in the Brookhaven PDB were collected with their PDB accession codes and the chemical name of the ligand cocrystallized (when existing).

TABLE 5.1

Studied Nuclear Receptors and Their Natural Ligands

Receptor Name	Receptor Subtypes	Natural Ligand
Estrogen receptor (ER)	α, β, γ	Estradiol
Androgen receptor (AR)		Testosterone
Progesterone receptor (PR)		Progesterone
Glucocorticoid receptor (GR)		Cortisol
Mineralocorticoid receptor (MR)		Aldosterone
Thyroid receptor (TR)	α, β	Thyroid hormone
Vitamin D receptor (VDR)		Vitamin D
Retinoic acid receptor (RAR)	α, β, γ	Retinoic acid
Peroxisome proliferator activated receptor (PPAR)	α, δ, γ	Fatty acids
Retinoic X receptor (RXR)	α, β, γ	*trans*-Retinoic acid
Estrogen related receptor (ERR)	α, β, γ	—
Constitutive androstane receptor (CAR)		—
Pregnane X receptor (PXR)		—
RAR related orphan receptor (ROR)	α, β, γ	

This information will help the modeler to have a global view of the existing NR crystallographic coordinates in order to select the appropriate template structure for a future comparative modeling of NRs.

5.3 PROTEIN MODELING

The ultimate goal of protein modeling is to predict a structure from its primary sequence with a comparable accuracy to the best results achieved experimentally. This would allow users to safely use generated *in silico* receptor models in all the research areas where only experimental structures provide a solid basis such as in the structure-based drug design and in the analysis of protein function or interactions. Moreover, protein modeling is the best approach to obtain structural information when experimental techniques fail. Homology or comparative modeling uses experimentally determined protein structures to predict the conformation of another protein that has a similar amino acid sequence. The method relies on the observation that in nature the structural conformation of a protein is more highly conserved than its primary sequence and that changes in amino acid sequence typically result in only small changes in the 3D structure [19] In homology modeling, a known sequence with an unknown structure is mapped against a known structure of one or several similar (homologous) proteins. It is claimed that two proteins of similar origin and function present reasonable structural similarity. Although the number of PDB entries is growing rapidly (~13 new entries daily), the 3D structure of only 1% to 2% of all known proteins has been experimentally characterized. The quality of homology models depends on the sequence identity between the protein of known structure (template) and the protein to be modeled (target). To achieve the comparative modeling process, a large number of applications and services are freely available on the Internet [20].

In practice, homology modeling process can be summarized in five steps: (1) fold assignment and template selection, (2) sequence alignment, (3) model generation, (4) loop modeling, and (5) model optimization and validation [21]. These different steps are discussed in the following paragraphs.

5.3.1 FOLD ASSIGNMENT AND TEMPLATE SELECTION

In the fold assignment step, a set of template proteins of known 3D structure that are related to the target sequence of unknown structure is identified. The template 3D structure must be determined by reliable experimental methods (for example, X-ray crystallography and NMR) and published as an atomic coordinate "PDB" file available from the PDB Web site [1,2]. Several computerized search methods are available to assist in identifying homologs [20]. Next, a sequence database of proteins with known structures is searched with the target sequence. Template identification is performed using sequence identity search algorithms or threading techniques [22], which scan sequence and structure databases, such as PDB, structural classification of proteins (SCOP) [23], distance-matrix alignment [24], and Class, Architecture, Topology, and Homology (CATH) [25,26].

5.3.2 SEQUENCE ALIGNMENT

After identifying the template 3D structure(s), the next key step is the accurate alignment of the target sequence to the template sequence(s). An erroneous alignment will lead to the design of an incorrect model. The alignment between two sequences is typically calculated by optimizing an alignment scoring function [27]. Many programs are available to align a number of related sequences, such as CLUSTALW, and the resulting alignment contains useful additional information [28].

5.3.3 MODEL GENERATION

The next step in the homology modeling process is model generation. The target protein structure is built from the substitution, insertion and/or deletion of amino acids in the template protein 3D structure according to the sequence alignment. It is noteworthy that experimentally determined protein structures (templates) are not perfect. Most of the errors found range from poor electron density in the X-ray diffraction map to human errors when generating the atomic coordinate PDB file. To build a good model, the selection of the template with the fewest errors is required. To help in such a task, the PDBREPORT database [28] freely available at www.cmbi. nl/gv/pdbreport can be very useful.

5.3.4 LOOP MODELING

Generally, the alignment between the model and the template sequence contains gaps. There are two main approaches to loop modeling: the knowledge-based method and the energy-based method. The former technique is a database search approach that scans a protein 3D structure database to find sequence segments fitting the anchor core regions. All major molecular modeling programs and servers support this approach (for example, 3D-JIGSAW [30], INSIGHT [31], MODELLER [32], SWISS-MODEL [33], or WHAT IF [34]). In the energy-based method, as in true *ab initio* fold prediction, an energy function is used to evaluate the quality of a loop. Then, this function is minimized, using Monte Carlo [35] or molecular dynamics simulation techniques [36] to reach the best loop conformation.

5.3.5 MODEL OPTIMIZATION AND VALIDATION

Once the comparative model is constructed, an energy minimization process using force field approaches is needed to correct all conformational troubles. A molecular dynamics simulation of the model can be used to follow the motions of the protein on a femtosecond (10^{-15} s) timescale, which mimics the true folding process.

There are many model evaluation programs and servers available on the Internet [20,37]. Two types of checks are generally used: (1) stereochemical quality where the model is more likely to be correct if its structure is the less strained and (2) side-chain environment where the model is more likely to be correct if its atoms are in the most thermodynamically favorable configuration. Moreover, to estimate the conformational space sampled by the homology model, it is useful to calculate the root mean square deviation (RMSD) between the main chain atoms of the model and the most homologous template X-ray structure. The main programs for verifying stereochemistry are PROCHECK [38,39], PROCHECK-NMR [40], SQUID [41], and WHATCHECK [29]. VERIFY 3D [42] can be used to assess the amino acid environment. The model features checked by such programs include bond lengths and angles, peptide bond and side-chain ring planarities, backbone and side-chain torsion angles, chirality, and clashes between nonbonded pairs of atoms. It is important to note that results obtained from all these validation procedures have to be contextualized by comparison with the quality of the template structure. Obviously, it is impossible to expect that a model outperforms its template.

Finally, experimental findings can be valuable solutions to validate the homology model, such as site-directed mutagenesis data, which can give insight into potential structural and functional roles of the respective amino acids.

5.3.6 ERRORS IN HOMOLOGY MODELING

It is crucial for modelers to assess the accuracy of the methods they used. An attempt to address this problem was made by the Critical Assessment of Techniques for Protein Structure Prediction (CASP) and the Critical Assessment of Fully Automated Structure Prediction (CAFASP) experiments [43,44]. EVA is a Web server that automatically assesses secondary structure prediction, residue clashes, fold assignment, and comparative modeling [45].

It is worth mentioning that the homology modeling process is based on template structures that are experimentally determined. However, every experiment, no matter how carefully carried out, may have errors. It is important to identify these errors before the structures are used for research. The X-ray diffraction technique contains several difficult steps that could lead to errors such as the crystallization of the protein. If the crystallographic structure resolution is low, the number of errors is expected to increase and, therefore, these experimental errors will be transferred to the predicted models. Thus, the quality of the template structure selected for modeling is essential. With low-quality templates, homology models are likely to be incorrect.

Furthermore, the correctness of the homology models is dependent on the level of sequence identity between the protein of known structure and the protein to be modeled [46,47]. The accuracy of a model can be estimated by calculating the RMSD with the used template structure. The model can be compared to X-ray determined structures if the percentage sequence identity is greater than 90%. From 50% to 90% identity, the deviation in the modeled atomic coordinates can be equal to 1.5 Å. If the sequence identity drops to 20%, the alignment leads to large errors. If the sequence identity decreases below 15%, structure modeling becomes speculative.

Errors can be due to incorrect alignments between the template and target structure. In some cases, gaps need to be inserted in the target sequence to reach a better alignment with the template. However, if a gap is inserted at a wrong position in the alignment, the quality of the final model will be affected.

5.4 NUCLEAR RECEPTOR MODELS

5.4.1 Estrogen Receptor (ER)

Table 5.2 displays the numerous X-ray crystallographic structures of ERα and β subtypes that are available in the Brookhaven PDB Web site. Diverse ligands are cocrystallized with the receptor leading to agonist and antagonist conformation of the protein. It is important to note that one theoretical model of hERα was deposited in the PDB and is accessible with the 1AKF PDB code [48].

The first 3D homology model of human ER (hER) reported in the literature was built by Lewis et al. [49] based on alignment with the α-antitrypsin sequence of known crystal structure [49,50]. Although the identity and similarity percentages between the template and ER sequences were quite low (11% and 25%, respectively), they performed the alignment and assigned the coordinates of α-antitrypsin amino acids to the target amino acids. The model was minimized, but the stereochemistry of the model was not verified with any of the existing programs available on the Internet (for example, PROCHECK [38,39]).

On the basis of the determined crystal structures of the ligand-binding domain (LBD) of the human retinoic acid receptor-γ (hRARγ) [51], Wurtz et al. [52] developed a 3D molecular model of human estrogen receptor-α (hERα). The multiple sequence alignment between ER LBD of various organisms (human, rat, mouse, chicken, boar, and finch), hRXRα, and hRARγ of known crystal structures revealed 21% identity between hRARγ and hERα. Large errors in the model are expected for 21% of sequence identity. However, the authors were aware of this low identity percentage and took special care with the evaluation of the hERα model. They carried out a literature search to retrieve mutants affecting the binding properties of the human or mouse ER, which exhibits 96% sequence identity for the LBDs. After the energy minimization of the hERα model, a Cα-trace superposition with the hRAR

TABLE 5.2
X-ray Crystallographic Estrogen Receptor (ER) Structures Available in the Protein Data Bank

Receptor	Ligand Cocrystallized	PDB Code, Year, Reference
hERα	17 β-Estradiol	1ERE (1997, [56]);1A52 (1998, [57]); 1AKF (1998, [48]); 1G50 (2001, [66]); 1QKT, 1QKU (2001, (63)]; 1GWR (2002, [67]); 1PCG (2003, [68])
	Raloxifene	1ERR (1997, [56]); 1GWQ (2002, [67])
	Diethylstilbestrol	3ERD (1998, [58])
	4-Hydroxytamoxifen	3ERT (1998, [58]); 1PCG (2005, [69])
	Genistein	1X7R (2004, [70])
	(R,R)-5,11-cis-Diethyl-5,6,11,12-tetrahydrochrysene-2,8-diol	1L2J (2002, [65])
	2-Phenyl-1-[4-(2-Piperidin-1-yl-ethoxy)-phenyl]-1,2,3,4-tetrahydro-isoquinolin-6-ol	1UOM (2003, [71])
	(2S,3R)-2-(4-(2-(Piperidin-1-yl)ethoxy)phenyl)-2,3-dihydro-3-(4-hydroxyphenyl)benzo[b][1,4]oxathiin-6-ol	1SJ0 (2004, [72])
	[5-hydroxy-2-(4-Hydroxyphenyl)-1-benzofuran-7-yl] acetonitrile	1X7E (2004, [73])
	(2E)-3-{4-[(1E)-1,2-Diphenylbut-1-enyl]phenyl} acrylic acid	1R5K (2005, [74])
	(2S,3R)-2-(4-{2-[(3R,4R)-3,4-Dimethylpyrrolidin-1-yl] ethoxy}phenyl)-3-(4-hydroxyphenyl)-2,3-dihydro-1,4-benzoxathiin-6-ol	1XP1(2005, [75])
	(2S,3R)-2-(4-{2-[(3S,4S)-3,4-Dimethylpyrrolidin-1-yl]ethoxy}phenyl)-3-(4-hydroxyphenyl)-2,3-dihydro-1,4-benzoxathiin-6-ol	1XP6 (2005, [75])
	(2S,3R)-3-(4-Hydroxyphenyl)-2-(4-{[(2S)-2-pyrrolidin-1-ylpropyl]oxy}phenyl)-2,3-dihydro-1,4-benzoxathiin-6-ol	1XP9 (2005, [75])
	(2S,3R)-3-(4-Hydroxyphenyl)-2-(4-{[(2R)-2-pyrrolidin-1-ylpropyl]oxy}phenyl)-2,3-dihydro-1,4-benzoxathiin-6-ol	1XPC (2005, [75])
	(1S)-1-{4-[(9aR)-Octahydro-2H-pyrido[1,2-a]pyrazin-2-yl]phenyl}-2-phenyl-1,2,3,4-tetrahydroisoquinolin-6-ol	1XQC (2005, [76])
	(2R,3R,4S)-3-(4-Hydroxyphenyl)-4-methyl-2-[4-(2-pyrrolidin-1-ylethoxy)phenyl]chroman-6-ol	1YIM (2005, [77])
	(2R,3R,4S)-5-Fluoro-3-(4-hydroxyphenyl)-4-methyl-2-[4-(2-piperidin-1-ylethoxy)phenyl]chroman-6-ol	1YIN (2005, [77])

(*continued*)

TABLE 5.2 (CONTINUED)
X-ray Crystallographic Estrogen Receptor (ER) Structures Available in the Protein Data Bank

Receptor	Ligand Cocrystallized	PDB Code, Year, Reference
	6-(4-Methylsulfonyl-phenyl)-5-[4-(2-piperidin-1-ylethoxy)phenoxy]naphthalen-2-ol	2AYR (2005, [78])
	4-[(1S,2S,5S)-5-(Hydroxymethyl)-6,8,9-trimethyl-3-oxabicyclo[3.3.1]non-7-en-2-yl]phenol	1ZKY (2006, [79])
	4-[(1S,2S,5S)-5-(Hydroxymethyl)-8-methyl-3-oxabicyclo[3.3.1]non-7-en-2-yl]phenol	2B1V (2006, [79])
	4-[(1S,2S,5S,9R)-5-(Hydroxymethyl)-8,9-dimethyl-3-oxabicyclo[3.3.1]non-7-en-2-yl]phenol	2FAI (2006, [79])
	4-[(1S,2R,5S)-4,4,8-Trimethyl-3-oxabicyclo[3.3.1]non-7-en-2-yl]phenol	2G44 (2006, [80])
	(3aS,4R,9bR)-4-(4-Hydroxyphenyl)-1,2,3,3a,4,9b-hexahydrocyclopenta[c]chromen-8-ol	2I0J (2006, [81])
	(5R,6S)-6-Phenyl-5-[4-(2-pyrrolidin-1-ylethoxy)phenyl]-5,6,7,8-tetrahydronaphthalen-2-ol	2OUZ (2007, [82])
	(17β)-17-{(E)-2-[2-(Trifluoromethyl)phenyl]vinyl}estra-1(10),2,4-triene-3,17-diol	2P15 (2007, [83])
	4-[(Dimesitylboryl)(2,2,2-trifluoroethyl)amino]phenol	2Q6J (2007, [84])
	(3aS,4R,9bR)-2,2-Difluoro-4-(4-hydroxyphenyl)-1,2,3,3a,4,9b-hexahydrocyclopenta[c]chromen-8-ol	2Q70 (2007, [85])
	(3aS,4R,9bR)-4-(4-Hydroxyphenyl)-6-(methoxymethyl)-1,2,3,3a,4,9b-hexahydrocyclopenta[c]chromen-8-ol	2QE4 (2007, [86])
	N-[(1R)-3-(4-Hydroxyphenyl)-1-methylpropyl]-2-[2-phenyl-6-(2-piperidin-1-ylethoxy)-1H-indol-3-yl]acetamide	2IOG (2007, [87])
	N-[(1R)-3-(4-Hydroxyphenyl)-1-methylpropyl]-2-(2-phenyl-1H-indol-3-yl)acetamide	2IOK (2007, [87])
hERβ	Genistein	1QKM (1998, [60]); 1X7J (2004, [70])
	(R,R)-5,11-cis-Diethyl-5,6,11,12-tetrahydrochrysene-2,8-diol	1L2I (2002, [65])
	5-Hydroxy-2-(4-hydroxyphenyl)-1-benzofuran-7-carbonitrile	1X76 (2004, [73])
	[5-Hydroxy-2-(4-hydroxyphenyl)-1-benzofuran-7-yl]acetonitrile	1X78 (2004, [73])
	2-(3-Fluoro-4-hydroxyphenyl)-7-vinyl-1,3-benzoxazol-5-ol	1X7B (2004, [73])
	2-(4-Hydroxy-phenyl)benzofuran-5-ol	1U9E (2004, [73])

TABLE 5.2 (CONTINUED)
X-ray Crystallographic Estrogen Receptor (ER) Structures Available in the Protein Data Bank

Receptor	Ligand Cocrystallized	PDB Code, Year, Reference
	4-(2-{[4-{[3-(4-Chlorophenyl)propyl]sulfanyl}-6-(1-piperazinyl)-1,3,5-triazin-2-yl]amino}ethyl)phenol	1NDE (2002, [88])
	4-(6-Hydroxy-benzo[d]isoxazol-3-yl)benzene-1,3-diol	1U3Q (2004, [89])
	2-(5-Hydroxy-naphthalen-1-yl)-1,3-benzooxazol-6-ol	1U3R (2004, [89])
	3-(6-Hydroxy-naphthalen-2-yl)-benzo[d]isooxazol-6-ol	1U3S (2004, [89])
	1-Chloro-6-(4-hydroxyphenyl)-2-naphthol	1YY4 (2005, [89])
	3-(3-Fluoro-4-hydroxyphenyl)-7-hydroxy-1-naphthonitrile	1YYE (2005, [90])
	3-Bromo-6-hydroxy-2-(4-hydroxyphenyl)-1H-inden-1-one	1ZAF (2005, [91])
	4-Hydroxytamoxifen	2FSZ (2006, [92])
	(9aS)-4-Bromo-9a-butyl-7-hydroxy-1,2,9,9a-tetrahydro-3H-fluoren-3-one	2GIU (2006, [93])
	(3aS,4R,9bR)-4-(4-Hydroxyphenyl)-1,2,3,3a,4,9b-hexahydrocyclopenta[c]chromen-8-ol	2IOG (2006, [81])
	(3aS,4R,9bR)-2,2-Difluoro-4-(4-hydroxyphenyl)-1,2,3,3a,4,9b-hexahydrocyclopenta[c]chromen-8-ol	2Z4B (2007, [85])
	(3aS,4R,9bR)-4-(4-Hydroxyphenyl)-6-(methoxymethyl)-1,2,3,3a,4,9b-hexahydrocyclopenta[c]chromen-8-ol	2JJ3 (2007, [86])
	4-(4-Hydroxyphenyl)-1-naphthaldehyde oxime	2NV7 (2007, [94])
	(3aS,4R,9bR)-4-(4-Hydroxyphenyl)-1,2,3,3a,4,9b-hexahydrocyclopenta[c]chromen-9-ol	2POG (2007, [95])
	(3aS,4R,9bR)-2,2-Difluoro-4-(4-hydroxyphenyl)-6-(methoxymethyl)-1,2,3,3a,4,9b-hexahydrocyclopenta[c]chromen-8-ol	2QTU (2007, [96])
Rat ERβ	Raloxifene	1QKN (1999, [60])
	N-Butyl-11-[(7R,8R,9S,13S,14S,17S)-3,17-dihydroxy-13-methyl-7,8,9,11,12,13,14,15,16,17-decahydro-6H-cyclopenta[a]phenanthren-7-yl]-N-methylundecanamid	1HJ1 (2001, [97])

LBDs revealed a RMSD of 0.7 Å. Finally, a PROCHECK analysis confirmed that the geometry and the dihedral angles of the model were in a range expected for a good protein structure (for example, more than 98% of the residues are in the most favorable or allowed regions of the Ramachandran plot).

Maalouf and coworkers [48] modeled the LBD of hERα protein by homology to the known crystal structure of the α isoform of human LBD hRXR. The secondary structure of the hERα model was deduced from the sequence alignment with members of the NR superfamily — that is, hRXR, human retinoic acid receptor-γ, and rat thyroid receptor [51,53,54]. The 3D model was refined using molecular dynamics simulation and energy minimization. Then, the model obtained was inspected with PROCHECK. The authors deposited their model in the Brookhaven Protein Data Bank with 1AKF accession code. However, it would have been useful to report identity percentage between hER sequence and the NR sequences used in the alignment.

Mori and coworkers [55] reported the homology model of rainbow trout ER LBD built from the X-ray structure of 17β-estradiol complexed to ERα (1ERE PDB accession code) [56]. The pairwise sequence alignment of the LBDs of 1ERE and rtER was given as well as the homology percentage. But, it was not specified if it was the identity or similarity percentage. It would have been interesting and helpful to perform a multiple alignment with other X-ray crystal structures of ER in complex with 17β-estradiol or diethylstilbestrol (1A52 and 3ERD PDB accession codes, respectively) that were experimentally resolved before this article [57,58]. The rainbow trout estrogen receptor subtype (α, β, or γ) was not specified in the article. The authors did not provide computational details of the energy minimization procedure of the theoretical model. The RMSD value between the target and template atoms LBDs was 0.87 Å. Although this value is good, the major drawback of this modeling approach is that the quality of the homology model was not estimated by any of the existing programs (for example, PROCHECK or WHATCHECK).

A homology model of human estrogen receptor subtype β (hERβ) was built using crystallographic structures of ERα in complex with estradiol (two structures), diethylstilbestrol, tamoxifen, and raloxifene (1ERE, 1ERR, 1A52, 3ERD, and 3ERT PDB accession codes, respectively) [56–59]. It is noteworthy that the authors specified the primary sequence of hERβ used to build the theoretical model (GenBank accession number BAA24953). This information is crucial for anyone wanting to build such a model if the coordinates are not available. Even though the ERα and ERβ LBD sequences hold more than 60% identity, the authors warned of the gaps in the ERα amino acid sequences due to either poor resolution or lack of order within particular domains [56–58]. All the computational details of the refinement procedure are given in the article. Moreover, DeLisle and coworkers [59] analyzed all the possible sources of errors in their comparative model, scanned it for possible clashes between amino acids of the side chains, and assessed it for appropriate rotamers, particularly at the points where random loops were inserted. Finally, to assess the structural accuracy of the ERβ homology model, the crystal structure (resolved at the time of their work by Pike et al. [60]) and model were superimposed by structural alignment and led to an RMSD value less than 1.6 Å, which is correct. Although the model was conscientiously generated, the authors should have performed a validation of their 3D ERβ model with PROCHECK or WHATCHECK programs.

Jacobs et al. [61] generated the 3D structure of human estrogen receptor β (hERβ) ligand-binding domain by homology modeling to the human ERα crystallographic coordinates [56,61]. The sequence alignment was determined based on the article of Brzozowski and coworkers [56] but without any sequence identity and/or similarity information. Moreover, the hERβ crystallographic structure was resolved by Pike et al. [60] in 1999, and was available in the PDB. It is surprising that such a template structure was not used in the sequence alignment. The major drawback relies on the lack of the stereochemical quality and the side-chain environment assessment.

A 3D model of the rainbow trout estrogen receptor-α (rtERα) was reported by Marchand-Geneste et al. [62] based on the human ERα X-ray crystallographic structure (1QKU PDB accession code) [62,63]. This X-ray structure was selected among the other ER crystallographic structures because it was the most complete (no missing residues). All the sequence FASTA codes are given in the publication, and the pairwise sequence alignment revealed 64% identity between rtERα and hERα. Manual and automatic homology modeling procedures were proposed and were compared leading to two analog 3D structures of rtERα. Both models were energy minimized. The superimposition between Cα atoms of both models and the crystal structure yielded acceptable RMSD values, although the automatic procedure generated a model with larger RMSD value than the manual procedure.

Homology models for the LBD of zebrafish estrogen receptors (zfERs) α, β1, and β2 were constructed using X-ray crystal structures of human homologs (1ERE, 1ERR, 3ERD, 3ERT, and 1L2I PDB accession codes for hERα structures and 1L2J and 1QKM PDB files for hERβ structures) [56,58,60,64,65]. Human structures of ER agonist complexes (1ERE, 3ERD, 1L2I) yielded agonist zfERα structures, while antagonist template structures (1ERR and 1ERT) were used to model antagonist zfERα. All zebrafish sequence accession numbers used are given in the article as well as the template accession numbers. The zebrafish homology models were minimized and superimposed to their corresponding human X-ray crystal homologs yielding RMSD values of 0.92 Å for α subtypes and 1.37 Å for β subtypes. Finally, the quality assessment of the models was validated with a PROSA analysis to create energy graphs that provided diagnostic indicators of poorly calculated/folded regions of tertiary structure [47]. A PROCHECK analysis provided scores of overall quality of the models, and Ramachandran plots indicated the most favored regions of amino acids [38,39]. The homology modeling procedure performed by Costache et al. [64] leads to reliable zfER α, β1, and β2 3D models.

5.4.2 Androgen Receptor (AR)

Table 5.3 presents the existing X-ray crystallographic structures of AR for rat, chimpanzee, and human with their corresponding bound ligand that could be used to generate further homology models.

The homology models of human AR DNA-binding domain (DBD) and LBD were determined by McDonald et al. [98] based on rat glucocorticoid receptor (ratGR, 1GLU PDB accession code) and human progesterone receptor (hPR, 1A28) known crystal structures, respectively [98–100]. Pairwise sequence alignments yielded 52% identity between hPR template structure and hAR LBD and 82% identity between

TABLE 5.3
X-ray Crystallographic Androgen Receptor (AR) Structures Available in the Protein Data Bank

Receptor	Ligand Cocrystallized	PDB Code, Year, Reference
hAR	Metribolone	1E3G (2000, [103]); 1XOW, 1XQ3, 2AO6 (2004, [104])
	9α-Fluorocortisol	1GS4 (2002, [105])
	Dihydrotestosterone	1T5Z, 1T63, 1T65, 1XJ7 (2005, [106])
	(R)-Bicalutamide	1Z95 (2005, [107])
	Hydroxyflutamide	2AX6 (2005, [108])
	(S)-3-(4-Fluorophenoxy)-2-hydroxy-2-methyl-N-[4-nitro-3-(trifluoromethyl)phenyl]propanamide	2AX7, 2AX8, 2AXA (2005, [108])
	(R)-3-Bromo-2-hydroxy-2-methyl-N-[4-nitro-3-(trifluoromethyl)phenyl]propanamide	2AX9 (2005, [108])
	Testosterone	2AM9 (2006, [109]); 2Q7I, 2Q7J, 2Q7K, 2Q7L (2007, [110])
	Dihydrotestosterone	2AMA (2006, [109])
	Tetrahydrogestrinone	2AMB (2006, [109])
	6-[bis(2,2,2-Trifluoroethyl)amino]-4-(trifluoromethyl)quinolin-2(1H)-one	2HVC (2006, [111])
	Cyproterone acetate	2OZ7 (2007, [112])
	(5S,8R,9S,10S,13R,14S,17S)-13-[2-[(3,5-Difluorophenyl)methoxy]ethyl]-17-hydroxy-10-methyl-1,2,4,5,6,7,8,9,11,12,14,15,16,17-tetradecahydrocyclopenta[a]phenanthren-3-one	2PNU (2007, [113])
Rat AR	Dihydrotestosterone	1I37, 1I38 (2001, [114])
	(3aR,4S,7R,7aS)-3a,4,7,7a-Tetrahydro-2-(4-nitro-1-naphthalenyl)-4,7-ethano-1H-isoindole-1,3(2H)-dione	1XNN (2005, [115])
	4-[(7R,7aS)-7-Hydroxy-1,3-dioxotetrahydro-1H-pyrrolo[1,2-c]imidazol-2(3H)-yl]-1-naphthonitrile	2IHQ (2006, [116])
	2-Chloro-4-[(7R,7aS)-7-hydroxy-1,3-dioxotetrahydro-1H-pyrrolo[1,2-c]imidazol-2(3H)-yl]-3-methylbenzonitrile	2NW4 (2007, [117])
Chimpanzee AR	Dihydrotestosterone	1T73, 1T74, 1T76, 1T79, 1T7F, 1T7M, 1T7R, 1T7T (2004, [118])

ratGR template structure and hAR DBD. The sequences were aligned directly without the introduction of gaps. Each human AR domain model was energy minimized with a different protocol. In fact, to avoid conformational collapse of the hAR active site model during the energy minimization procedure, testosterone was manually inserted into it. Close contacts and buried hydrophilic side chains were adjusted manually. Finally, the geometry and stereochemistry of final models were checked with PROCHECK, leading to 99% (LBD) and 91% (DBD) of residues in the allowed regions of the Ramachandran plot, and stereochemical parameters that fell within allowable limits for both models.

Models of the human AR LBD were constructed by Poujol and coworkers [101] using the RARγ, hERα, and hPR crystal structures [51,56,100,101]. The authors only described the model based on the hPR crystal structures. The alignment between hPR and hAR revealed 56% identity, suggesting that a reliable model could be built from this template structure. The 3D hAR model generated was energy minimized. Although the statistical details were not reported in the article, the quality of the model was checked by PROCHECK.

Marhefka and coworkers [102] proposed a 3D model for the human AR LBD bound to testosterone using the hPR known crystal structure (1A28) [100,102]. The sequence alignment revealed 56% identity and 88% similarity between the target and the template sequences. The absence of insertions or deletions within these sequences led to the direct mutation of the hPR residues to the corresponding residues of the hAR. Model refinement was performed by multiple molecular dynamics simulations. The major drawback of this model is the lack of quality assessment.

5.4.3 PROGESTERONE RECEPTOR (PR)

The hPR structures resolved by X-ray and deposited in the PDB are gathered in Table 5.4. Among these hPR crystal structures, that reported by Williams and Sigler [100] is the most used to generate theoretical models of NRs.

TABLE 5.4
X-ray Crystallographic Progesterone Receptor (PR) Structures Available in the Protein Data Bank

Receptor	Ligand Cocrystallized	PDB Code, Year, Reference
hPR	Progesterone	1A28 (1998, [100])
	Estradiol	1A52 (1998, [57])
	Metribolone	1E3K (2000, [103])
	Norethindrone	1SQN (2004, [120])
	Mometasone furoate	1SR7 (2004, [120])
	5-(4,4-Dimethyl-2-thioxo-1,4-dihydro-2H-3, 1-benzoxazin-6-yl)-1-methyl-1H-pyrrole-2-carbonitrile	1ZUC (2005, [121])
	4-[(11β,17β)-17-Methoxy-17-(methoxymethyl)-3-oxoestra-4,9-dien-11-yl]benzaldehyde oxime	2OVH, 2OVM (2007, [122])

A progesterone receptor (PR) antagonist homology model was modeled by Jones et al. [119] from the crystal structure of PR complexed to the endogenous agonist progesterone (1A28 PDB accession code) and ERα complexed to the antagonist tamoxifen (3ERT) [58,100,119]. It is surprising that more recent hPR crystallographic structures (Table 5.4) were not selected, although these structures are bound with agonist ligands (except for 2OVH and 2OVM PDB files). Very few computational details are given in this article. No energy minimization and validation of the model built were reported.

5.4.4 GLUCOCORTICOID RECEPTOR (GR)

As few X-ray crystallographic structures of GR (uncomplexed to DNA or other NRs) are available in the PDB (Table 5.5), numerous comparative models were generated. They are briefly presented below.

Lind et al. [123] developed and optimized a homology model of the GR LBD based on the ERα crystal structure [56]. However, no sequence alignment was reported in the article and no identity or similarity percentages were given. Consequently, it is difficult to appreciate the validity of such a model. Moreover, the quality assessment of the constructed model was not performed.

Dey et al. [124] proposed a comparative model of mouse GR (mGR) LBD designed from the X-ray crystal structure of progesterone complexed to its receptor (1A28 PDB accession code) [100]. The authors reported the PDB accession codes of all the X-ray template structures used to align the mouse GR sequence that were extracted from the SWISSPROT sequence data bank. The multiple sequence alignment among mGR, hPR, hAR and hER LDB is given and discussed. The identity and similarity percentages are presented between the target and each template X-ray crystal structure. The refinement protocol of the homology model is well detailed, and the RMSD value between the mGR LBD backbone and hPR is 1.3 Å. The reliability of the theoretical model of mGR LBD was checked with PROCHECK. In conclusion, this mGR homology model is reliable because each step of the homology process has been respected.

Lewis et al. [125] reported the results of homology modeling of the human GR LBD (hGR) based on the LBD of human ERα (1ERE PDB code) [56]. Although the authors proposed a multiple sequence alignment, they did not report sequence identity and similarity information. Nevertheless, they argued that the hERα template structure was chosen for modeling due to the relatively high homology with the hGR

TABLE 5.5
X-ray Crystallographic Glucocorticoid Receptor (GR)
Structures Available in the Protein Data Bank

Receptor	Ligand Cocrystallized	PDB Code, Year, Reference
hGR	Dexamethasone	1M2Z (2002, [127]); 1P93 (2003, [130])
	Mifepristone	1NHZ (2003, [130])

---

sequence. The model obtained was energy minimized using molecular mechanics to obtain a low-energy geometry and stable protein conformation by relaxing the molecule and relieving any unfavorable steric contacts. However, the weakness of their model comes from the absence of any quality assessment (for example, stereochemical quality, side-chain environment, conformational space sampled by the homology model).

Honer et al. [126] created a homology model for the LBD of human GR based on the X-ray crystal structure of the progesterone bound to PR LBD (1A28) [99]. This template was chosen because the pairwise sequence alignment revealed 53% identity and 74% homology between hGR and hPR. The refinement and the quality assessment procedure of the hGR homology model built were not carried out. Moreover, it is surprising that the X-ray crystal structure of hGR mentioned in the article was not used as a template for their alignment and was not compared to their 3D model [127].

In order to understand the potential structural differences between hGRα and hGRβ, Yudt et al. [128] modeled the hGRα and hGRβ LBDs using the coordinates from the progesterone receptor (hPR) (1A28 PDB accession code) [100]. Multiple sequence alignment of hGRα, hGRβ, and several other related receptors (hPR, hAR, hERα, and RXRγ) was carried out. However, no sequence identity between the target sequences and NR sequences was given. Energy minimization and dynamics of the 3D homology models were performed as part of the homology modeling process. The models developed were checked using the PROCHECK and WHAT IF programs to insure that psi and phi angles occurred in the expected regions of the Ramachandran map and that there were no steric bumps or overlaps in the structure. As previously indicated, the X-ray crystal structure of hGR was resolved before the submission of this article [127]. Thus, it would have been interesting to add this 3D structure to the sequence alignment or to superimpose the model built to this X-ray structure.

A homology model of the human GR LBD based on the crystal structure of the human progesterone receptor (1A28 PDB accession code) was reported by von Langen et al. [100,129]. Table 5.5 displays the different crystal structures resolved before the work of von Langen and coworkers [129]. In 2002, Bledsoe and coworkers [127] resolved the crystal structure of hGR LBD bound to dexamethasone and a coactivator motif. Moreover, in 2003, the 3D crystal structures of hGR LDB in complex with the antagonist RU-486 at 2.3 Å resolution and with the agonist dexamethasone ligand together with a coactivator peptide at 2.8 Å was published by Kauppi et al. [130]. The drawback of this study comes from the template selected to build the hGR model. In fact, the hPR template structure chosen is surprising because both crystal structures of hGR were published prior to their article. However, the stereochemical correctness of their hGR model was checked with the PROCHECK program and the Ramachandran plot reported that 97% of the residues were in the most favored or allowed region. Although the authors did not select any of the hGR crystal structures available in the PDB, they performed the superimposition of the hGR homology model built with the X-ray crystal structure of hGR in complex with dexamethasone and found an RMSD value for the backbone atoms of 2.1 Å. The authors reported that such a high value was due to structural differences mainly at the loop regions in the outer parts and to helix 12. However, it is well known that the

TABLE 5.6
X-ray Crystallographic Mineralocorticoid Receptor (MR) Structures
Available in the Protein Data Bank

Receptor	Ligand Cocrystallized	PDB Code, Year, Reference
hMR	Desoxycorticosterone	1Y9R (2005, [135])
	Progesterone	1YA3 (2005, [135]); 2AA5, 2AA6 (2005, [136])
	Corticosterone	2A3I (2005, [137])
	Aldosterone	2AA2 (2005, [136])
	Desoxycorticosterone	2AA7 (2005, [136])
	Cortisone	2AAX (2005, [136])
	Spironolactone	2AB2 (2005, [136]); 2OAX (2007, [138])

helix 12 is involved in the conformational change observed between an agonist and antagonist receptor structure [56].

5.4.5 MINERALOCORTICOID RECEPTOR (MR)

Table 5.6 collects the X-ray crystallographic structures of hMR available from the PDB Web site.

Fagart et al. [131] proposed a homology model of the human MR LBD based on the crystal structure of the human retinoic acid receptor-γ ligand-binding domain (hRARγ, 2LBD PDB accession code) [51]. The major drawback of this model is the low sequence identity shared with the hRARγ sequence (<20%). This template was preferred by the authors because all 11 helices observed in the hRARγ X-ray structure are well matched and represented the anchoring points for the alignment process. Analysis of the homology procedure reveals that the refinement step of the model building was omitted as well as the superimposition of the 3D homology model of hMR with the hRARγ crystal structure, which allows us to assess the quality of the model. Nevertheless, the reported PROCHECK statistics show that more than 97% of the residues in the Ramachandran plot (not displayed in the article) are in the most favored or allowed regions and that side-chain stereoparameters are inside the range. PROSAII analysis gave a Z-score value close to the range observed for RARγ and RXRα structures [132]. These results suggest that the 3D homology model of hMR is of good quality despite the low sequence identity reported. It should have been useful to specify the primary sequence used for the hMR.

Another hMR LDB homology model was generated based on a sequence identity of 56% with hPR [100,133]. The model building was achieved according to the method described by Fagart et al. [131]. All the critical points stressed in the article by Fagart and coworkers are also applicable in this article.

A 3D X-ray structure of progesterone complexed to its receptor was used by Dey and Roychowdhury [134] to develop a homology model of human MR LBD in a similar procedure as mouse GR LBD discussed previously [100]. All the homology modeling steps were followed, yielding a reliable 3D hMR model.

TABLE 5.7
X-ray Crystallographic Thyroid Receptor (TR) Structures Available in the Protein Data Bank

Receptor	Ligand Cocrystallized	PDB Code, Year, Reference
hTRα	[4-(4-Hydroxy-3-isopropylphenoxy)-3,5-dimethylphenyl]acetic acid	1NAV (2003, [140])
	3,5,3′-Triiodothyronine	2H77, 2H79 (2006, [141])
hTRβ	[4-(4-Hydroxy-3-isopropyl-phenoxy)-3,5-dimethylphenyl]-6-azauracil	1N46 (2003, [142])
	[4-(4-Hydroxy-3-isopropylphenoxy)-3,5-dimethylphenyl]acetic acid	1NAX (2003, [140])
	[4-(4-Hydroxy-3-iodo-phenoxy)-3,5-diiodophenyl]acetic acid	1NQ0, 1NQ1 (2003, [143]); 1NQ2 (2003, [144]); 1NUO (2003, [144])
	[4-(3-Benzyl-4-hydroxybenzyl)-3,5-dimethylphenoxy]acetic acid	1Q4X (2003, [145])
	2-[3,5-Dibromo-4-(4-hydroxy-3-{hydroxy[(2-phenylethyl)amino]methyl}phenoxy)phenyl]ethane-1,1-diol	1R6G (2004, [146])
	3,5,3″-Triiodothyronine	1BSX (1998, [54]); 1XZX (2004, [147]); 2H6W (2006, [141])
	3,5,3′,5′-Tetraiodo-l-thyronine	1Y0X (2004, [147])
	3,5-Dibromo-4-(3-isopropyl-phenoxy) benzoic acid	2J4A (2006, [148])

5.4.6 THYROID HORMONE RECEPTOR (TR)

Several X-ray crystallographic structures of TR have been resolved since 1998. Their PDB accession codes are gathered in Table 5.7. Concerning the theoretical models of TR, Schapira et al. [139] used the crystal structure of raloxifene bound to hERα (1ERR PDB accession code) as the template to derive, from the crystallographic structure of agonist bound TRβ ligand-binding domain (1BSX), a predicted model of the thyroid receptor bound to an antagonist [54,56,139]. However, sequence alignment and identity percentage were not reported in this study. The model was refined by a Monte Carlo simulation but neither the quality of the model nor the RMS deviation with the template was checked. Thus, the consistency of this model is doubtful.

5.4.7 VITAMIN D RECEPTOR (VDR)

The 3D VDR crystallographic coordinates for human, rat, and zebrafish, which are available in the PDB, are presented in Table 5.8. Very few models are reported in the literature. The first human vitamin D receptor (hVDR) LBD modeled was based on the crystal structure of the human retinoic acid receptor-γ (hRARγ, 2LBD

TABLE 5.8
X-ray Crystallographic Vitamin D Receptor (VDR) Structures Available in the Protein Data Bank

Receptor	Ligand Cocrystallized	PDB Code, Year, Reference
hVDR	5-{2-[1-(5-Hydroxy-1,5-dimethylhexyl)-7a-methyl-octahydro-inden-4-ylidene]-ethylidene}-4-methylene-cyclohexane-1,3-diol	1DB1 (2000, [151]); 1IE9 (2001, [152])
	5-(2-{1-[1-(4-Ethyl-4-hydroxyhexyloxy)-ethyl]-7a-methyl-octahydro-inden-4-ylidene}-ethylidene)-4-methylene-cyclohexane-1,3-diol	1IE8 (2001, [152])
	Seocalcitol	1S0Z (2004, [153])
	Calcipotriol	1S19 (2004, [153])
	$(1R,3R)$-5-((Z)-2-(($(1R,7aS)$-Hexahydro-1-(((S)-6-hydroxy-6-methylhept-4-yn-2-yl)-7a-methyl-1 H-inden-4(7aH)-ylidene)ethylidene)cyclohexane-1,3-diol	1TXI (2005, [154])
	2α-Propyl-1α,25-dihydroxyvitamin D3	2HAM (2006, [155])
	2α-(3-Hydroxypropoxy)-1α,25-dihydroxyvitamin D3	2HAR (2006, [155])
	2α-Propoxy-1α,25-dihydroxyvitamin D3	2HAS (2006, [155])
	2α-(3-Hydroxypropyl)-1α,25-dihydroxyvitamin D3	2HB7 (2006, [155])
	2α-Methyl-1α,25-dihydroxy-vitamin D3	2HB8 (2006, [155])
Rat VDR	5-{2-[1-(5-Hydroxy-1,5-dimethylhexyl)-7a-methyl-octahydro-inden-4-ylidene]-ethylidene}-2-methylene-cyclohexane-1,3-diol	1RJK (2004, [156])
	5-{2-[1-(5-Hydroxy-1,5-dimethylhexyl)-7a-methyl-octahydro-inden-4-ylidene]-ethylidene}-4-methylene-cyclohexane-1,3-diol	1RK3 (2004, [156])
	5-{2-[1-(1-Methyl-propyl)-7a-methyl-octahydro-inden-4-ylidene]-ethylidene}-2-methylene-cyclohexane-1,3-diol	1RKG (2004, [156])
	5-{2-[1-(5-Hydroxy-1,5-dimethylhexyl)-7a-methyl-octahydro-inden-4-ylidene]-ethylidene}-2-methyl-cyclohexane-1,3-diol	1RKH (2004, [156])
	$(1R,3R,7E,17Z)$-17-(5-Hydroxy-1,5-dimethylhexylidene)-2-methylene-9,10-secoestra-5,7-diene-1,3-diol	2O4J (2007, [157])
Rat VDR	$(1R,3R,7E,17E)$-17-(5-Hydroxy-1,5-dimethylhexylidene)-2-methylene-9,10-secoestra-5,7-diene-1,3-diol	2O4R (2007, [157])
Zebrafish VDR	5-{2-[1-(5-Hydroxy-1,5-dimethyl-hexyl)-7a-methyl-octahydro-inden-4-ylidene]-ethylidene}-4-methylene-cyclohexane-1,3-diol	2HC4 (2007, [158])

TABLE 5.8 (CONTINUED)
X-ray Crystallographic Vitamin D Receptor (VDR) Structures Available in the Protein Data Bank

Receptor	Ligand Cocrystallized	PDB Code, Year, Reference
Zebrafish VDR	(1*R*,3*S*,5*Z*)-5-[(2*E*)-2-[(1*R*,3a*S*,7a*R*)-1-(2,10-Dihydroxy-2, 10-dimethyl-undecan-6-yl)-7a-methyl-2,3,3a,5,6,7-hexah ydro-1*H*-inden-4-ylidene]ethylidene]-4-methylidene-cycl ohexane-1,3-diol	2HCD (2007, [158])
Zebrafish VDR	(1*R*,3*S*,5*Z*)-5-[(2*E*)-2-[(1*S*,3a*S*,7a*S*)-1-(5-Hydroxy-5-methyl-hexa-1,3-diynyl)-7a-methyl-2,3,3a,5,6,7-hexahydro-1*H*-inden-4-ylidene] ethylidene]-4-methylidene-cyclohexane-1,3-diol	2HBH (2007, [159])

PDB accession code) [51,149]. The authors used this template for the following reasons: RAR is a member of the same NR subfamily with VDR and from a sequence alignment with six NRs (hPPARγ, TRα, hRARγ, hRXRα, hERα, and hPR). It has the highest homology percentage with VDR. However, from the multiple sequence alignment proposed, Yamamoto et al. [149] did not report any sequence identity or similarity percentage, which could give an idea about the correctness of the homology model built that is strongly dependent on the level of sequence identity. The 3D homology model of hVDR obtained was energy minimized without and with solvent. Finally, the authors evaluated their model with PROCHECK, and Ramachandran plot analysis reported that 99% of the residues were in the most favored or allowed regions.

The second study concerning the comparative modeling of the rat vitamin D receptor (rat VDR) 3D structure was reported and evaluated by Rotkiewicz et al. [150]. The VDR sequence was aligned with all sequences of the NCBI nonredundant sequence database until convergence of multiple sequence alignment was achieved by using a PSI-BLAST procedure. The use of the whole sequence database improved the sensitivity of the search and the quality of the alignments. Sequence identity between the five NRs and rat VDR was given by the authors in addition to PSI-BLAST scores. Five NRs of known crystallographic structures were found — namely, hPR (1A28), hTR (1BSX), hER (1ERE), hRAR (2LBD), and hPPARγ (4PRG) — and were used as template structures in the modeling of rat VDR. A molecular dynamics minimization procedure of the obtained model was performed. The deviation of the model structure from the templates was on the same level as the deviation between various template structures. This suggests that this 3D rat VDR comparative model is of relatively high accuracy, probably close to the accuracy of crystallographic structures. However, it would have been interesting to check the model quality with PROCHECK to complete the study.

5.4.8 RETINOIC ACID RECEPTOR (RAR)

Although the X-ray structure of RARγ bound to the agonist all *trans*-retinoic acid was available in the PDB with the 2LBD code (Table 5.9), the conformation of the receptor bound to an antagonist was not known [51]. Thus, a model of the antagonist bound structure of human retinoic acid receptor-α (hRARα) was elaborated by Schapira et al. [160] using information derived from tamoxifen bound estrogen receptor-α (3ERT PDB accession code) [58]. The major drawbacks of this model rely on the lack of sequence alignment and sequence identity percentage allowing the modeler to judge the accuracy of the model, as well as the absence of any quality assessment. Later, Schapira and coworkers [161] built a model of the RAR-α agonist-binding pocket from the crystal structure of the RAR-γ ligand-binding domain (2LBD). However, the critical analysis of this model leads to the same drawbacks as previously noted.

A homology model of RARβ ligand-binding domain was built by Lund et al. [162] using known crystal structures of RARγ-LBD as templates (Table 5.9) [51,162–165]. Sequence identity between the β and γ isoforms of RAR was as high as 87%, which made modeling a straightforward procedure. Homology models built were subject to coarse molecular mechanics refinement with AMBER94 force field including the GB/SA solvation term. Models derived from various templates were found to be

---

TABLE 5.9

X-ray Crystallographic Retinoic Acid Receptor (RAR) Structures Available in the Protein Data Bank

Receptor	Ligand Cocrystallized	PDB Code, Year, Reference
hRARβ	4-[(1E)-2-(5,5,8,8-Tetramethyl-5,6,7, 8-tetrahydronaphthalen-2-yl)prop-1-enyl]benzoic acid	1XAP (2004, [166])
hRARγ	Retinoic acid	2LBD (1995, [51]); 3LBD (1998, [163])
	3-Fluoro-4-[2-hydroxy-2-(5,5,8,8-tetramethyl-5,6,7, 8-tetrahydro-naphtalen-2-yl)-acetylamino]-benzoic acid	4LBD (1998, [163])
	6-[Hydroxy-(5,5,8,8-tetramethyl-5,6,7,8-tetrahydro-naphtalen-2-yl)-methyl]-naphtalene-2-carboxylic acid	1FCX (2000, [164])
	6-(5,5,8,8-Tetramethyl-5,6,7,8-tetrahydro-naphtalene-2-carbonyl)-naphtalene-2-carboxylic acid	1FCY (2000, [164])
	4-[3-Oxo-3-(5,5,8,8-tetramethyl-5,6,7,8-tetrahydro-naphthalen-2-yl)-propenyl]-benzoic acid	1FCZ (2000, [164])
	6-[Hydroxyimino-(5,5,8,8-tetramethyl-5,6,7,8-tetrahydro-naphtalen-2-yl)-methyl]-naphtalene-2-carboxylic acid	1FD0 (2000, [164])
	(R)-3-Fluoro-4-[2-hydroxy-2-(5,5,8,8-tetramethyl-5,6,7,8-tetrahydro-naphtalen-2-yl)-acetylamino]-benzoic acid	1EXA (2000, [165])
	(S)-3-Fluoro-4-[2-hydroxy-2-(5,5,8,8-tetramethyl-5,6,7,8-tetrahydro-naphtalen-2-yl)-acetylamino]-benzoic acid	1EXX (2000, [165])

highly consistent. The average RMSD of the positions of α-carbons was around 0.26 Å. However, a quality assessment would have been useful to complete the validation of the models.

5.4.9 Peroxisome Proliferator Activated Receptor (PPAR)

Numerous hPPAR X-ray crystallographic structures of α, γ, and δ subunits have been experimentally reported since 1998. Lewis and Lake [167] built a homology model of the rat PPARα LBD based on the crystal structure of the human retinoic acid X receptor-α (hRXRα) [53] (Table 5.10). The authors reported a multiple sequence alignment between rat, mouse, and human PPARs and hRXR, which exhibited a homology around 50%. However, they did not report the identity percentage between the protein of known structure (hRXR) and the target protein (PPAR), which is the most important percentage with which to evaluate the reliability of the model to build. Although the energy optimization of PPAR models was performed, the stereochemistry quality, the side-chain environment, and the RMSD with the template structure were not checked. These PPAR homology models for rat, mouse, and human were rebuilt in 2002 by Lewis et al. [168] using the hPPARγ crystal structure instead of the hRXRα X-ray structure [169]. As displayed in Table 5.11, note that five X-ray crystal structures were deposited in 1999 in the PDB. It is surprising that these crystallographic structures were not included in the alignment procedure to improve the correctness of the model. Although the multiple sequence alignment between the template sequence of known structure and the target sequence was given, no identity or similarity percentage was reported. The models were refined and a close comparison between the PPARα LBDs and that of the crystal structure template (hPPARγ) was carried out. It is rather surprising that without any RMS deviation evaluation or quality assessment with PROCHECK or WHAT IF programs, the authors concluded that their homology models were likely to bear a high degree of validity.

Blaney [170] used stereodiagrams from the RARγ and RXRα publications to generate the starting structure of the RARγ model as the coordinates of both NRs were not available [51,53]. Using the published sequences of the secondary structural elements, the author built an idealized 3D structure that was sequentially optimized by constrained molecular dynamics. Then, an accurate alignment of three

---

TABLE 5.10

X-ray Crystallographic Retinoic Acid X Receptor (RXR) Structures Available in the Protein Data Bank

Receptor	Ligand Cocrystallized	PDB Code, Year, Reference
hRXRα	—	1LBD (1995, [53]); 1G1U (2000, [171])
	Retinoic acid	1FBY (2000 [172]); 1G5Y (2000, [171])
	Docosa-4,7,10,13,16,19-hexaenoic acid	1MV9 (2002, [173])
	4-[2-(5,5,8,8-Tetramethyl-5,6,7,8-tetrahydro-naphthalen-2-yl)-[1,3]dioxolan-2-yl]-benzoic acid	1MVC, 1MZN (2002, [173])

TABLE 5.11
X-ray Crystallographic Peroxisome Proliferator Activated Receptor (PPAR) Structures Available in the Protein Data Bank

Receptor	Ligand Cocrystallized	PDB Code, Year, Reference
hPPARα	(2S)-2-Ethoxy-3-[4-(2-{4-[(methylsulfonyl)oxy]phenyl}ethoxy)phenyl]propanoic acid	1I7I, 1I7G (2001, [174])
	2-(1-Methyl-3-oxo-3-phenyl-propylamino)-3-{4-[2-(5-methyl-2-phenyl-oxazol-4-yl)-ethoxy]-phenyl}-propionic acid	1K7L (2001, [175])
	N-((2S)-2-({(1Z)-1-Methyl-3-oxo-3-[4-(trifluoromethyl)phenyl]prop-1-enyl}amino)-3-{4-[2-(5-methyl-2-phenyl-1,3-oxazol-4-yl)ethoxy]phenyl}propyl)propanamide	1KKQ (2002, [176])
	(2R,3E)-2-{4-[(5-Methyl-2-phenyl-1,3-oxazol-4-yl)methoxy]benzyl}-3-(propoxyimino)butanoic acid	2NPA (2007, [177])
	2-Methyl-2-(4-{[({4-methyl-2-[4-(trifluoromethyl)phenyl]-1,3-thiazol-5-yl}carbonyl)amino]methyl}phenoxy)propanoic acid	2P54 (2007, [178])
hPPARδ	—	2GWX (1999, [179])
	2-(4-{3-[1-[2-(2-Chloro-6-fluoro-phenyl)-ethyl]-3-(2,3-dichloro-phenyl)-ureido]-propyl}-phenoxy)-2-methyl-propionic acid	1GWX (1999, [179])
	5,8,11,14,17-Eicosapentaenoic acid	3GWX (1999, [179])
	(2S)-2-(4-[2-(3-[2,4-difluorophenyl]-1-heptylureido)ethyl]phenoxy)-2-methylbutyric acid	1Y0S (2000, [180])
	Vaccenic acid	2AWH, 2B50 (2006, [181])
	Heptyl-β-d-glucopyranoside	2BAW (2006, [182])
	(3-{4-[2-(2,4-Dichloro-phenoxy)-ethylcarbamoyl]-5-phenyl-isoxazol-3-yl}-phenyl)-acetic acid	2J14 (2006, [183])
hPPARγ	—	1PRG (1998, [169])
	—	3PRG (1998, [184])
	(5S)-5-[[4-[2-(Methyl-pyridin-2-yl-amino)ethoxy]phenyl]methyl]-1,3-thiazolidine-2,4-dione	2PRG (1998, [169]); 1ZGY (2005, [185])
	(2S,5S)-3-(4-(4-Carboxyphenyl)butyl)-2-heptyl-4-oxo-5-thiazolidine	4PRG (1999, [186])
	(S)-3-(4-(2-Carbazol-9-yl-ethoxy)-phenyl)-2-ethoxy-propionic acid	1KNU (2002, [187])
	(2S)-2-Ethoxy-3-{4-[2-(10H-phenoxazin-10-yl)ethoxy]phenyl}propanoic acid	1NYX (2003, [188])
	2-[(2,4-Dichlorobenzoyl)amino]-5-(pyrimidin-2-yloxy)benzoic acid	1WM0 (2004, [189]); 2Q6S (2007, [190])
	(2S)-(4-Isopropylphenyl)[(2-methyl-3-oxo-5,7-dipropyl-2,3-dihydro-1,2-benzisoxazol-6-yl)oxy]acetate	1ZEO (2005, [191])

TABLE 5.11 (CONTINUED)
X-ray Crystallographic Peroxisome Proliferator Activated Receptor (PPAR) Structures Available in the Protein Data Bank

Receptor	Ligand Cocrystallized	PDB Code, Year, Reference
	2-{5-[3-(7-Propyl-3-trifluoromethylbenzo[d] isoxazol-6-yloxy) propoxy]indol-1-yl}ethanoic acid	2ATH (2005, [192])
	(5-{3-[(6-Benzoyl-1-propyl-2-naphthyl)oxy] propoxy}-1H-indol-1-yl)acetic acid	2F4B (2006, [193])
	1-(3,4-Dimethoxybenzyl)-6,7-dimethoxy-4-{[4-(2-methoxyphenyl)piperidin-1-yl]methyl} isoquinoline	2FVJ (2006, [194])
	3-Fluoro-N-[1-(4-fluorophenyl)-3-(2-thienyl)-1 H-pyrazol-5-yl]benzenesulfonamide	2G0G (2006, [195])
	N-[1-(4-Fluorophenyl)-3-(2-thienyl)-1H-pyrazol-5-yl]-3,5-bis(trifluoromethyl)benzenesulfonamide	2G0H (2006, [195])
	(2S)-3-(1-{[2-(2-Chlorophenyl)-5-methyl-1, 3-oxazol-4-yl]methyl}-1H-indol-5-yl)-2-ethoxypropanoic acid	2GTK (2006, [196])
	3-(4-Methoxyphenyl)-N-(phenylsulfonyl)-1-[3-(trifluoromethyl)benzyl]-1H-indole-2-carboxamide	2HFP (2006, [197])
	[(1-{3-[(6-Benzoyl-1-propyl-2-naphthyl)oxy] propyl}-1H-indol-5-yl)oxy]acetic acid	2HWQ (2006, [198])
	2-[(1-{3-[(6-Benzoyl-1-propyl-2-naphthyl)oxy] propyl}-1H-indol-4-yl)oxy]-2-methylpropanoic acid	2HWR (2006, [198])
	(2R)-2-(4-{2-[1,3-Benzoxazol-2-yl(heptyl)amino] ethyl}phenoxy)-2-methylbutanoic acid	2I4J (2007, [199])
	(2S)-2-(4-{2-[1,3-Benzoxazol-2-yl(heptyl)amino] ethyl}phenoxy)-2-methylbutanoic acid	2I4P, 2I4Z (2007, [199])
	(6aR,10aR)-3-(1,1-Dimethylheptyl)-1-hydroxy-6, 6-dimethyl-6a,7,10,10a-tetrahydro-6H-benzo[c] chromene-9-carboxylic acid	2OM9 (2007, [200])
	(2S)-2-(2-{[1-(4-Methoxybenzoyl)-2-methyl-5-(trifluoromethoxy)-1H-indol-3-yl]methyl} phenoxy)propanoic acid	2Q59 (2007, [190])
	(2S)-2-(3-{[1-(4-Methoxybenzoyl)-2-methyl-5-(trifluoromethoxy)-1H-indol-3-yl]methyl} phenoxy)propanoic acid	2Q5P (2007, [190])
	5-Chloro-1-(4-chlorobenzyl)-3-(phenylthio)-1 H-indole-2-carboxylic acid	2Q5S (2007, [190])
	1-Benzyl-5-chloro-3-(phenylthio)-1H-indole-2-carboxylic acid	2Q61 (2007, [190])
	5-Chloro-1-(3-methoxybenzyl)-3-(phenylthio)-1H-indole-2-carboxylic acid	2Q6R (2007, [190])

human PPAR subtype sequences with that of RARγ and RXRα was performed. A model of hPPARγ was obtained (few computational details are given) and refined. The quality assessment of the models was not carried out. The major weakness of this homology modeling procedure is the absence of any 3D crystallographic coordinates. Moreover, the lack of sequence identity percentage makes the reliability of the model unlikely.

Jacobs et al. [61] built a model of human PPARα by homology modeling using the human PPARγ ligand-binding domain crystallographic coordinates [169]. However, this model has several weaknesses, such as the absence of identity percentage from the sequence alignment using the hPPARγ crystal structure, the RMS deviation between the model and the template backbone atoms which was not given, and the quality of the hPPARα model built which was not evaluated. Moreover, as shown in Table 5.11, four crystallographic structures of hPPARα were already resolved and deposited in the PDB before the article was published.

No homology models of RXR were reported in the literature, and few crystal structures were deposited into the PDB data bank (Table 5.10).

5.4.10 ESTROGEN RELATED RECEPTOR (ERR)

The agonist bound form of estrogen related receptor-α (ERRα) was modeled by Chen et al. [201] using the DES bound structure of the estrogen receptor as a template (3ERD PDB accession code) [58]. The antagonist form was built using the tamoxifen bound form (PDB accession code: 3ERT) [58]. However, only few details are given by the authors concerning the comparative modeling procedure (no sequence alignment, no sequence identity, no minimization process, and no validation).

In the same way, two homology models of estrogen related receptor-γ (ERRγ) were built by Nam et al. [202] based on X-ray crystal structures of ERα in an agonist conformation (3ERD) and in an antagonist conformation (3ERT) [58]. As previously mentioned, few computational details of the homology modeling and minimization procedure are given in the article. The major drawback of this publication is the absence of any stereochemistry verification in order to validate both models.

Suetsugi et al. [203] generated a homology model of hERRα based on the ERRγ free protein X-ray structure (PDB accession code: 1KV6) reported by Greschik et al. [204] as ERRγ shared 57% amino acid sequence identity with the hERRα ligand-binding domain (LBD). However, they found that the comparative model built possessed a very small ligand-binding pocket. More specifically, they reported that the side chain of Phe399 amino acid on helix 11 partially filled the cavity and interfered with the formation of hydrogen bonds between His398 and the ligand. The position of Phe399 led to a tightly packed pocket that is only about half of the size of ERα, as indicated by Greschik et al. [204]. Consequently, they decided not to use this model for the screening of agonists of ERRα. In order to generate a more accurate, diversified, and unbiased ligand-binding pocket of ERRα, three hERα crystal structures with different agonist ligands (DES, (R,R)-5,11-cis-diethyl-5,6,11,12-tetrahydrochrysene-2,8-diol, and 17β-estradiol) were used as templates (PDB codes 3ERD, 1L2I, and 1GWR, respectively) [58,65,67]. The homology model of a mutant ERRα was built from the same procedure. Both structures were energy minimized and their quality was evaluated by checking the stereochemistry, local geometry, solvent accessible surface areas, and side-chain conformational probabilities. They also performed a Ramachandran plot analysis showing that 97.5% of the residues in both models were in the most favored or allowed regions. The authors' critical analysis toward the first model built strengthened the accuracy of their second model.

5.4.11 Constitutive Androstane Receptor (CAR)

Xiao et al. [205] derived a homology model of hCAR using hPXR (1ILH PDB accession code) and hVDR (1DB1 PDB accession code) X-ray structures [151,206]. The sequences of the LBD of hCAR homologs (hPXR and hVDR) were extracted from their PDB files, and multiple sequence alignment yielded, respectively, 43% and 34% of sequence identity with hCAR. Nevertheless, it would have been interesting to include the hPXR X-ray crystal structure unbound (1ILG PDB accession code) into the multiple alignment step [206]. The model was optimized, and several residue side chains around the binding cavity were manually adjusted to obtain side-chain torsion angles similar to those of the templates. The validation of this model yielded 90% of residues in the most favorable region of Ramachandran plot. Finally, the superimposition of the hCAR model with hPXR and hVDR $C\alpha$ backbone was achieved and resulted in RMSD values of 0.25 Å and 0.65 Å, respectively, which suggest that the 3D homology model of hCAR is reliable and of good quality.

Another 3D comparative model of hCAR was proposed by Dussault and coworkers [207]. A multiple sequence alignment with different nuclear receptor sequences (1LBD, 3ERD, 2PRG, 2LBD, 1A28, 1DB1, 1BSX, and 1ILH PDB accession codes) revealed that hPXR and hVDR crystal structures were the most appropriate template for the modeling with 35.1% and 30.8% of sequence identity with hCAR. Although the sequence identity percentage is similar between hPXR and hVDR, the latter was not selected for the following two reasons: (1) the authors noted that steric clashes and exposed hydrophobic residues modified the structure of the two loops connecting H1 and H3, leading to incompatible region conservation with CAR sequence; and (2) the active site was too small to accommodate a potent agonist 1,4-bis[2-(3,5-dichloropyridyloxy)]benzene (TCPOBOP) ligand involved in the study. The refinement procedure was completed but not detailed. The α carbons of hPXR and CAR structural model were superimposed and yielded a RMSD of 3.15 Å. This value suggests that several differences are apparent. In fact, the authors reported that some loops of the template became shorter in the target structure and that three hydrophobic amino acids became smaller in size in the CAR model, which modified the binding ability of the CAR structure. The quality of the model was tested with PROCHECK, which revealed that 82.6% of residues fell within the most favored regions of the Ramachandran plot.

Jacobs et al. [61] reported a 3D homology model of CAR based on the hERα template structure [56]. The model was refined, but the major drawbacks of this study are that the identity or similarity percentages were not reported, the RMS deviation was not specified, and the quality of the 3D model generated was not checked with any of the free available programs such as PROCHECK or WHAT IF.

5.4.12 PREGNANE X RECEPTOR (PXR)

Jacobs et al. [61] built the 3D homology model of human PXR from the hERα crystallographic coordinates [56]. However, the authors ignored both X-ray crystallographic structures of hPXR (1ILG and 1ILH accession codes, Table 5.12) resolved in 2001 by Watkins et al. [206]. These two template structures should have been added into their alignment. Neither identity nor similarity percentage were given from the alignment with the hERα template structure; hence, the validity of the template structure selected is not proven. Finally, the quality of their 3D hPXR model was not checked.

Numerous X-ray crystal structures of PXR are available in the PDB (1ILG, 1ILH, 1M13, 1NRL, 1SKX, and 1O9I PDB accession codes, Table 5.12), but none of them contain an intact LBD. Especially, a flexible loop of 15 amino acids is unresolved in these crystal structures. In the absence of ligands, PXR can repress gene expression by interacting with transcriptional corepressors, such as the silencing mediator for retinoid and thyroid hormone receptor (SMRT). From this observation, Wang et al. [208] proposed a homology model of PXR in complex with SMRT containing two interacting domains (ID1 and ID2). The apo-form of PXR (1ILG PDB code) was chosen as the template structure to build the model of PXR and the X-ray crystal structure of SMRT-ID2 bound to PPARα (1KKQ PDB accession code) to generate the final model of PXR/SMRT-ID2 complex [176]. Then, the ID2 sequence was mutated to the corresponding ID1 sequence while preserving the remaining structural components of PXR/SMRT-ID2 structure. Both models were refined with energy minimization and molecular dynamics simulations. Finally, the two refined structures of PXR/ SMRT complex were analyzed with PROCHECK, which revealed acceptable quality. The major drawback of this study is the lack of identity sequence percentage.

5.4.13 RAR RELATED ORPHAN RECEPTOR (ROR)

Few crystal structures of ROR have been deposited into the PDB database (Table 5.12). Moreover, only two homology models were found in the literature. Harris et al. [226] constructed a homology model of the RORα using TRβ (1BSX PDB code) as template structure [54]. This structure was preferred to RARγ because the critical region for coactivator interaction showed much higher conservation with TRβ than RARγ. The alignment revealed that the AF2 region was almost identical between RORα and TRβ. However, the authors reported numerous missing side chains and an incomplete Cα backbone between helix 2 and 3 in the template structure. To circumvent such problems, the RORα homology model built was subjected to loop modeling through loop databank. Finally, the model was refined by energy minimization and validated with a Ramachandran, leading to 3% of residues in disallowed regions of the Ramachandran plot. Furthermore, a WHAT IF quality analysis was performed and yielded a robust and reliable RORα homology model.

Kurebayashi et al. [227] built the homology model of the LBD RORγ from the LBD crystal structure of RORβ (1K4W PDB code) [224]. The modeling process was not detailed (no alignment and refinement) and the model quality was not reported.

TABLE 5.12
X-ray Crystallographic Estrogen Related Receptor (ERR), Constitutive Androstane Receptor (CAR), Pregnane X Receptor (PXR), and RAR (Retinoic Acid Receptor) Related Orphan Receptor (ROR) Structures Available in the Protein Data Bank

Receptor	Ligand Cocrystallized	PDB Code, Year, Reference
hERRα	—	1XB7 (2004, [209])
	1-Cyclohexyl-*N*-{[1-(4-methylphenyl)-1 *H*-indol-3-yl]methyl}methanamine	2PJL (2007, [210])
hERRγ	—	1KV6 (2002, [204])
	—	1TFC (2004, [211])
	Diethylstilbestrol	1S9P (2004, [211])
	4-Hydroxytamoxifen	1S9Q (2004, [211]); 2GPV (2006, [212])
	(Z)-4-(1-{4-[2-(Dimethylamino)ethoxy] phenyl}-5-hydroxy-2-phenylpent-1-enyl) phenol	2EWP (2006, [213])
	Bisphenol A	2E2R (2007, [214])
Mouse ERRγ	4-Hydroxytamoxifen	1VJB (2004, [211])
hCAR	(5β)-Pregnane-3,20-dione	1XV9 (2004, [215])
	6-(4-Chlorophenyl)imidazo[2,1-*b*][1,3] thiazole-5-carbaldehyde *O*-(3,4-dichlorobenzyl)oxime	1XVP (2004, [215])
Mouse CAR	3,5-Dichloro-2-{4-[(3,5-dichloropyridin-2-yl) oxy]phenoxy}pyridine	1XLS (2004, [216])
	16,17-Androstene 3 ol	1XNX (2004, [217])
hPXR	—	1ILG (2001, [206])
	[2-(3,5-*di-tert*-Butyl-4-hydroxy-phenyl)-1-(diethoxy-phosphoryl)-vinyl]-phosphonic acid diethlyl ester	1ILH (2001, [206])
	4-Hydroxy-5-isobutyryl-6-methyl-1,3,7-tris-(3-methyl-but-2-enyl)-6-(4-methyl-pent-3-enyl)-bicyclo[3.3.1]non-3-ene-2,9-dione	1M13 (2003, [218])
	[2-(3,5-*di-tert*-Butyl-4-hydroxy-phenyl)-1-(diethoxy-phosphoryl)-vinyl]-phosphonic acid diethlyl ester	1NRL (2003, [219])
	Rifampicin	1SKX (2005, [220])
	N-(2,2,2-Trifluoroethyl)-*N*-{4-[2,2,2-trifluoro-1-hydroxy-1-(trifluoromethyl)ethyl]phenyl} benzenesulfonamide	1O9I (2007, [221])
hRORα	Cholesterol	1N83 (2002, [222])
	Cholest-5-en-3-yl hydrogen sulfate	1S0X (2004, [223])

(*continued*)

TABLE 5.12 (CONTINUED)
X-ray Crystallographic Estrogen Related Receptor (ERR), Constitutive Androstane Receptor (CAR), Pregnane X Receptor (PXR), and RAR (Retinoic Acid Receptor) Related Orphan Receptor (ROR) Structures Available in the Protein Data Bank

Receptor	Ligand Cocrystallized	PDB Code, Year, Reference
Rat RORβ	Stearic acid	1K4W (2001, [224])
	Retinoic acid	1N4H (2003, [225])
	7-(3,5-*di-tert*-Butylphenyl)-3-methylocta-2,4,6-trienoic acid	1NQ7 (2003, [225])

5.5 CONCLUSIONS

In this chapter, an attempt was made to review the homology models dealing with the nuclear hormone receptor superfamily. These models are well suited to better understand the mechanisms of action of the endocrine disruptor chemicals in relation with their 3D structure. This is particularly important because numerous studies have shown that the design of powerful SAR and QSAR models allowing us to estimate the endocrine disruption potential of chemicals needed the use of 3D molecular descriptors.

Despite an impressive bibliographical investigation, the number of models found in the literature is limited, even if it is interesting to note that most of the types of receptors have been subject to modeling investigations. Conversely, our review clearly reveals that a lot of them are of limited value, often suffering from a lack of validation and critical analysis. It seems surprising that existing resources, easily available via the Internet, are not considered, but they should be helpful in the design and refinement of the models.

The potentialities of homology modeling applied to the understanding of endocrine disruption mechanisms are still not fully exploited. Undoubtedly, a lot of work remains to be done in this field.

REFERENCES

[1] J. Westbrook, Z. Feng, S. Jain, T.N. Bhat, N. Thanki, V. Ravichandran, G.L. Gilliland, W. Bluhm, H. Weissig, D.S. Greer, P.E. Bourne, and H.M. Berman, *The Protein Data Bank: Unifying the archive*, Nucleic Acids Res. 30 (2002), pp. 245–248.

[2] J. Westbrook, Z. Feng, L. Chen, H. Yang, and H.M. Berman, *The Protein Data Bank and structural genomics*, Nucleic Acids Res. 31 (2003), pp. 489–491.

[3] B. Boeckmann, A. Bairoch, R. Apweiler, M.C. Blatter, A. Estreicher, E. Gasteiger, M.J. Martin, K. Michoud, C. O'Donovan, I. Phan, S. Pilbout, and M. Schneider, *The SWISS-PROT protein knowledgebase and its supplement TrEMBL in 2003*, Nucleic Acids Res. 31 (2003), pp. 365–370.

[4] A. Tramontano, R. Leplae, and V. Morea, *Analysis and assessment of comparative modelling predictions in CASP4*, Proteins 45 (2001), pp. 22–38.

[5] V. Giguère, N. Yang, P. Segui, and R.M. Evans, *Identification of a new class of steroid hormone receptors*, Nature 331 (1988), pp. 91–94.

[6] F. Chen, Q. Zhang, T. McDonald, M.J. Davidoff, W. Bailey, C. Bai, Q. Liu, and C.T. Caskey, *Identification of two hERR2-related novel nuclear receptors utilizing bioinformatics and inverse PCR*, Gene 228 (1999), pp. 101–109.

[7] H Hong, L. Yang, and M.R. Stallcup, *Hormone-independent transcriptional activation and coactivator binding by novel orphan nuclear receptor ERR3*, J. Biol. Chem. 274 (1999), pp. 22618–22626.

[8] D.J. Heard, P.L. Norby, J. Holloway, and H. Vissing, *Human ERRγ, a third member of the estrogen receptor-related receptor (ERR) subfamily of orphan nuclear receptors: Tissue-specific isoforms are expressed during development and in the adult*, Mol. Endocrinol. 14 (2000), pp. 382–392.

[9] R.C.J. Ribeiro, P.J. Kushner, and J.D. Baxter, *The nuclear hormone receptor gene superfamily*, Annu. Rev. Med. 46 (1995), pp. 443–453.

[10] D.L. Bain, A.F. Heneghan, K.D. Connaghan-Jones, and M.T. Miura, *Nuclear receptor structure: Implications for function*, Ann. Rev. Physiol. 69 (2007), pp. 201–220.

[11] J.P. Renaud and D. Moras, *Structural studies on nuclear receptors*, Cell Mol. Life Sci. 57 (2000), pp. 1748–1769.

[12] J.R. Tata, *Signaling through nuclear receptors*, Nat. Rev. Mol. Cell Biol. 3 (2002), pp. 702–710.

[13] M. Robinson-Rechavi, H.E. Garcia, and V. Laudet, *The nuclear receptor superfamily*, J. Cell Sci. 116 (2003), pp. 585–586.

[14] A. Aranda and A. Pascual, *Nuclear hormone receptors and gene expression*, Physiolog. Rev. 81 (2001), pp. 1269–1304.

[15] B.W. O'Malley, S.Y. Tsai, M. Bagchi, N.L. Weigel, W.T. Schrader, and M.J. Tsai, *Molecular mechanism of action of a steroid hormone receptor*, Rec. Prog. Hormone Res. 47 (1991), pp. 1–26.

[16] R. White and M.G. Parker, *Molecular mechanisms of steroid hormone action*, Endocr. Relat. Cancer 5 (1998), pp. 1–14.

[17] W.B. Pratt, K.A. Hutchison, and L.C. Scherrer, *Steroid receptor folding by heat-shock proteins and composition of the receptor heterocomplex*, Trends Endocrinol. Metab. 3 (1992), pp. 326–333.

[18] M. Eckey, U. Moehren, and A. Baniahmad, *Gene silencing by the thyroid hormone receptor*, Mol. Cell Endocrinol. 213 (2003), pp. 13–22.

[19] H. Giersiefen, R. Hilgenfeld, and A. Hillisch, *Modern methods of drug discovery: An introduction*, in *Modern Methods of Drug Discovery*, A. Hillisch and R. Hilgenfeld, eds., Birkhäuser Verlag, Switzerland, 2003, pp. 1–18.

[20] A.J.M. Carpy and N. Marchand-Geneste, *Structural e-bioinformatics and drug design*, SAR QSAR Env. Res. 17 (2006), pp. 1–10.

[21] M.S. Madhusudhan, M.A. Marti-Renom, N. Eswar, B. John, U. Pieper, R. Karchin, S. Min-Yi, and A. Sali, *Comparative protein structure modelling*, in *The Proteomics Protocols Handbook*, J.M. Walker, ed., Humana Press, Totowa, NJ, 2005, pp. 831–860.

[22] A. Godzik, *Fold recognition methods*, in *Structural Bioinformatics*, P. Bourne and H. Weissig, eds., Wiley-Liss, Hoboken, NJ, 2003, pp. 525–546.

[23] L. Lo Conte, S.E. Brenner, T.J. Hubbard, C. Chothia, and A.G. Murzin, *SCOP database in 2002: Refinements accommodate structural genomics*, Nucleic Acids Res. 30 (2002), pp. 264–267.

[24] L. Holm and C. Sander, *Protein folds and families: Sequence and structure alignments*, Nucleic Acids Res. 27 (1999), pp. 244–247.

[25] C.A. Orengo, J.E. Bray, D.W. Buchan, A. Harrison, D. Lee, F.M. Pearl, I. Sillitoe, A.E. Todd, and J.M. Thornton, *The CATH protein family database: A resource for structural and functional annotation of genomes*, Proteomics 2 (2002), pp. 11–21.

[26] F.M. Pearl, D. Lee, J.E. Bray, D.W. Buchan, A.J. Shepherd, and C.A. Orengo, *The CATH extended protein-family database: Providing structural annotations for genome sequences*, Protein Sci. 11 (2002), pp. 233–244.

[27] G.J. Barton, *Protein sequence alignment and database scanning*, in *Protein Structure Prediction: A Practical Approach*, M.J. Sternberg, ed., IRL Press at Oxford University Press, Oxford, 1996, pp. 31–63.

[28] J.D. Thompson, D.G. Higgins, and T.J. Gibson, *ClustalW: Improving the sensitivity of progressive multiple sequence alignment through sequence weighting, position-specific gap penalties and weight matrix choice*, Nucleic Acids Res. 22 (1994), pp. 4673–4680.

[29] R.W.W. Hooft, G. Vriend, C. Sander, and E.E. Abola, *Errors in protein structures*, Nature 381 (1996), p. 272.

[30] P.A. Bates and M.J. Sternberg, *Model building by comparison at CASP3: Using expert knowledge and computer automation,* Proteins 3 (1999), pp. 47–54.

[31] H.E. Dayringer, A. Tramontano, S.R. Sprang, and R.J. Fletterick, *Interactive program for visualization and modelling of proteins, nucleic acids and small molecules*, J. Mol. Graph. 4 (1986), pp. 82–87.

[32] A. Sali and T.L. Blundell, *Comparative protein modelling by satisfaction of spatial restraints*, J. Mol. Biol. 234 (1993), pp. 779–815.

[33] M.C. Peitsch, T. Schwede, and N. Guex, *Automated protein modelling: The proteome in 3D*, Pharmacogenomics 1 (2000), pp. 257–266.

[34] G. Vriend, *WHAT IF — A molecular modelling and drug design program*, J. Molec. Graphics 8 (1990), pp. 52–56.

[35] K.T. Simons, R. Bonneau, I. Ruczinski, and D. Baker, *Ab initio structure prediction of CASP III targets using ROSETTA*, Proteins 3 (1999), pp. 171–176.

[36] A. Fiser, R.K. Do, and A. Sali, *Modeling of loops in protein structures*, Protein Sci. 9 (2000), pp. 1753–1773.

[37] N. Eswar, B. John, N. Mirkovic, A. Fiser, V.A. Ilyin, U. Pieper, A.C. Stuart, M.A. Marti-Renom, M.S. Madhusudhan, B. Yerkovich, and A. Sali, *Tools for comparative protein structure modeling and analysis*, Nucleic Acids Res. 31 (2003), pp. 3375–3380.

[38] R.A. Laskowski, M.W. MacArthur, D.S. Moss, and J.M. Thornton, *PROCHECK: A program to check the stereochemical quality of protein structures*, J. Appl. Crystallog. 26 (1993), pp. 283–291.

[39] A.L. Morris, M.W. MacArthur, E.G. Hutchinson, and J.M. Thornton, *Stereochemical quality of protein structure coordinates*, Proteins 12 (1992), pp. 345–364.

[40] R.A. Laskowski, J.A. Rullmann, M.W. MacArthur, R. Kaptein, and J.M. Thornton, *AQUA and PROCHECK-NMR: Programs for checking the quality of protein structures solved by NMR*, J. Biomol. NMR 8 (1996), pp. 477–486.

[41] T.J. Oldfield, *SQUID: A program for the analysis and display of data from crystallography and molecular dynamics*, J. Mol. Graph. 10 (1992), pp. 247–252.

[42] R. Lüthy, J.U. Bowie, and D. Eisenberg, *Assessment of protein models with three-dimensional profiles*, Nature 356 (1992), pp. 83–85.

[43] A. Zemla, C. Venclovas, J. Moult, and K. Fidelis, *Processing and evaluation of predictions in CASP4*, Proteins 45 (2001), pp. 13–21.

[44] D. Fischer, A. Elofsson, L. Rychlewski, F. Pazos, A. Valencia, B. Rost, A.R. Ortiz, and R.L. Dunbrack Jr., *CAFASP2: The second critical assessment of fully automated structure prediction methods*, Proteins 45 (2001), pp. 171–183.

[45] V.A. Eyrich, M.A. Marti-Renom, D. Przybylski, M.S. Madhusudhan, A. Fiser, F. Pazos, A. Valencia, A. Sali, and B. Rost, *EVA: Continuous automatic evaluation of protein structure prediction servers*, Bioinformatics 17 (2001), pp. 1242–1243.

[46] C. Chothia and A.M. Lesk, *The relation between the divergence of sequence and structure in proteins*, EMBO J. 5 (1986), pp. 823–836.

[47] M.J. Sippl, *Recognition of errors in three dimensional structures of proteins*, Proteins 17 (1993), pp. 355–362.

[48] G.J. Maalouf, W. Xu, T.F. Smith, and S.C. Mohr, *Homology model for the ligand-binding domain of the human estrogen receptor*, J. Biomol. Struct. Dyn. 15 (1998), pp. 841–851.

[49] D.F. Lewis, M.G. Parker, and R.J. King, *Molecular modelling of the human estrogen receptor and ligand interactions based on site-directed mutagenesis and amino acid sequence homology*, J. Steroid Biochem. Mol. Biol. 52 (1995), pp. 55–65.

[50] H. Loebermann, R. Tokuoka, J. Deisenhofer, and R. Huber, *Human α1-proteinase inhibitor. Crystal structure analysis of two crystal modifications, molecular model and preliminary analysis of the implications for function*, J. Mol. Biol. 177 (1984), pp. 531–557.

[51] J.P. Renaud, N. Rochel, M. Ruff, V. Vivat, P. Chambon, H. Gronemeyer, and D. Moras, *Crystal structure of the RAR-γ ligand-binding domain bound to all-trans retinoic acid*, Nature 378 (1995), pp. 681–689.

[52] J.M. Wurtz, U. Egner, N. Heinrich, D. Moras, and A. Mueller-Fahrnow, *Three-dimensional models of estrogen receptor ligand binding domain complexes, based on related crystal structures and mutational and structure-activity relationship data*, J. Med. Chem. 41 (1998), pp. 1803–1814.

[53] W. Bourguet, M. Ruff, P. Chambon, H. Gronemeyer, and D. Moras, *Crystal structure of the ligand-binding domain of the human nuclear receptor RXR-α*, Nature 375 (1995), pp. 377–382.

[54] B.D. Darimont, R.L. Wagner, J.W. Apriletti, M.R. Stallcup, P.J. Kushner, J.D. Baxter, R.J. Fletterick, and K.R. Yamamoto, *Structure and specificity of nuclear receptor-coactivator interactions*, Genes Dev. 12 (1998), pp. 3343–3356.

[55] T. Mori, S. Sumiya, and H. Yokota, *Electrostatic interactions of androgens and progesterone derivatives with rainbow trout estrogen receptor*, J. Steroid Biochem. Mol. Biol. 75 (2000), pp. 129–137.

[56] A.M. Brzozowski, A.C. Pike, Z. Dauter, R.E. Hubbard, T. Bonn, O. Engström, L. Ohman, G.L. Greene, J.A. Gustafsson, and M. Carlquist, *Molecular basis of agonism and antagonism in the oestrogen receptor*, Nature 389 (1997), pp. 753–758.

[57] D.M. Tanenbaum, Y. Wang, S.P. Williams, and P.B. Sigler, *Crystallographic comparison of the estrogen and progesterone receptor's ligand binding domains*, Proc. Natl. Acad. Sci. USA 95 (1998), pp. 5998–6003.

[58] A.K. Shiau, D. Barstad, P.M. Loria, L. Cheng, P.J. Kushner, D.A. Agard, and G.L. Greene, *The structural basis of estrogen receptor/coactivator recognition and the antagonism of this interaction by tamoxifen*, Cell 95 (1998), pp. 927–937.

[59] R.K. DeLisle, S.J. Yu, A.C. Nair, and W.J. Welsh, *Homology modelling of the estrogen receptor subtypeβ (ER-β) and calculation of ligand binding affinities*, J. Mol. Graph. Model. 20 (2001), pp. 155–167.

[60] A.C. Pike, A.M. Brzozowski, R.E. Hubbard, T. Bonn, A.G. Thorsell, O. Engström, J. Ljunggren, J.A. Gustafsson, and M. Carlquist, *Structure of the ligand-binding domain of oestrogen receptor β in the presence of a partial agonist and a full antagonist*, EMBO J. 18 (1999), pp. 4608–4618.

[61] M.N. Jacobs, M. Dickins, and D.F. Lewis, *Homology modelling of the nuclear receptors: Human oestrogen receptor β (hERβ), the human pregnane-X-receptor (PXR), the Ah receptor (AhR) and the constitutive androstane receptor (CAR) ligand binding domains from the human oestrogen receptor α (hERα) crystal structure, and the human peroxisome proliferator activated receptor α (PPARα) ligand binding domain from the human PPARγ crystal structure*, J. Steroid Biochem. Mol. Biol. 84 (2003), pp. 117–132.

[62] N. Marchand-Geneste, M. Cazaunau, A.J.M. Carpy, M. Laguerre, J.M. Porcher, and J. Devillers, *Homology model of the rainbow trout estrogen receptor (rtERα) and docking of endocrine disrupting chemicals (EDCs)*, SAR QSAR Environ. Res. 17 (2006), pp. 93–105.

[63] M. Gangloff, M. Ruff, S. Eiler, S. Duclaud, J.M. Wurtz, and D. Moras, *Crystal structure of a mutant hERα ligand-binding domain reveals key structural features for the mechanism of partial agonism*, J. Biol. Chem. 276 (2001), pp. 15059–15065.

[64] A.D. Costache, P.K. Pullela, P. Kasha, H. Tomasiewicz, and D.S. Sem, *Homology-modeled ligand-binding domains of zebrafish estrogen receptors α, β1, and β2: From in silico to in vivo studies of estrogen interactions in* Danio rerio *as a model system*, Mol. Endocrinol. 19 (2005), pp. 2979–2990.

[65] A.K. Shiau, D. Barstad, J.T. Radek, M.J. Meyers, K.W. Nettles, B.S. Katzenellenbogen, J.A. Katzenellenbogen, D.A. Agard, and G.L. Greene, *Structural characterization of a subtype-selective ligand reveals a novel mode of estrogen receptor antagonism*, Nat. Struct. Biol. 9 (2002), pp. 359–364.

[66] S. Eiler, M. Gangloff, S. Duclaud, D. Moras, and M. Ruff, *Overexpression, purification, and crystal structure of native ERα LBD*, Protein Expr. Purif. 22 (2001), pp. 165–173.

[67] A. Warnmark, E. Treuter, J.A. Gustafsson, R.E. Hubbard, A.M. Brzozowski, and A.C. Pike, *Interaction of transcriptional intermediary factor 2 nuclear receptor box peptides with the coactivator binding site of estrogen receptor α*, J. Biol. Chem. 277 (2002), pp. 21862–21868.

[68] A.M. Leduc, J.O. Trent, J.L. Wittliff, K.S. Bramlett, S.L. Briggs, N.Y. Chirgadze, Y. Wang, T.P. Burris, and A.F. Spatola, *Helix-stabilized cyclic peptides as selective inhibitors of steroid receptor-coactivator interactions*, Proc. Natl. Acad. Sci. USA 100 (2003), pp. 11273–11278.

[69] E.H. Kong, N. Heldring, J.A. Gustafsson, E. Treuter, R.E. Hubbard, and A.C. Pike, *Delineation of a unique protein-protein interaction site on the surface of the estrogen receptor*, Proc. Natl. Acad. Sci. USA 102 (2005), pp. 3593–3598.

[70] E.S. Manas, Z.B. Xu, R.J. Unwalla, and W.S. Somers, *Understanding the selectivity of genistein for human estrogen receptor-β using X-ray crystallography and computational methods,* Structure 12 (2004), pp. 2197–2207.

[71] J.P. Renaud, S.F. Bischoff, T. Buhl, P. Floersheim, B. Fournier, C. Halleux, J. Kallen, H. Keller, J.M. Schlaeppi, and W. Stark, *Estrogen receptor modulators: Identification and structure-activity relationships of potent ERα-selective tetrahydroisoquinoline ligands*, J. Med. Chem. 46 (2003), pp. 2945–2957.

[72] S. Kim, J.Y. Wu, E.T. Birzin, K. Frisch, W. Chan, L.Y. Pai, Y.T. Yang, R.T. Mosley, P.M. Fitzgerald, N. Sharma, J. Dahllund, A.G. Thorsell, F. DiNinno, S.P. Rohrer, J.M. Schaeffer, and M.L. Hammond, *Estrogen receptor ligands. II. Discovery of benzoxathiins as potent, selective estrogen receptor α modulators*, J. Med. Chem. 47 (2004), pp. 2171–2175.

[73] E.S. Manas, R.J. Unwalla, Z.B. Xu, M.S. Malamas, C.P. Miller, H.A. Harris, C. Hsiao, T. Akopian, W.T. Hum, K. Malakian, S. Wolfrom, A. Bapat, R.A. Bhat, M.L. Stahl, W.S. Somers, and J.C. Alvarez, *Structure-based design of estrogen receptor-β selective ligands*, J. Am. Chem. Soc. 126 (2004), pp. 15106–15119.

[74] Y.L. Wu, X. Yang, Z. Ren, D.P. McDonnell, J.D. Norris, T.M. Willson, and G.L. Greene, *Structural basis for an unexpected mode of SERM-mediated ER antagonism*, Mol. Cell 18 (2005), pp. 413–424.

[75] T.A. Blizzard, F. DiNinno, J.D. Morgan, H.Y. Chen, J.Y. Wu, S. Kim, W. Chan, E.T. Birzin, Y.T. Yang, L.Y. Pai, P.M. Fitzgerald, N. Sharma, Y. Li, Z. Zhang, E.C. Hayes, C.A. DaSilva, W. Tang, S.P. Rohrer, J.M. Schaeffer, and M.L. Hammond, *Estrogen receptor ligands. Part 9: Dihydrobenzoxathiin SERAMs with alkyl substituted pyrrolidine side chains and linkers*, Bioorg. Med. Chem. Lett. 15 (2005), pp. 107–113.

[76] J.P. Renaud, S.F. Bischoff, T. Buhl, P. Floersheim, B. Fournier, M. Geiser, C. Halleux, J. Kallen, H.J. Keller, and P. Ramage, *Selective estrogen receptor modulators with conformationally restricted side chains. Synthesis and structure-activity relationship of ERα -selective tetrahydroisoquinoline ligands*, J. Med. Chem. 48 (2005), pp. 364–379.

[77] Q. Tan, T.A. Blizzard, J.D. Morgan, E.T. Birzin, W. Chan, Y.T. Yang, L.Y. Pai, E.C. Hayes, C.A. DaSilva, S. Warrier, J. Yudkovitz, H.A. Wilkinson, N. Sharma, P.M. Fitzgerald, S. Li, L. Colwell, J.E. Fisher, S. Adamski, A.A. Reszka, D. Kimmel, F. DiNinno, S.P. Rohrer, L.P. Freedman, J.M. Schaeffer, and M.L. Hammond, *Estrogen receptor ligands. Part 10: Chromanes: Old scaffolds for new SERAMs*, Bioorg. Med. Chem. Lett. 15 (2005), pp. 1675–1681.

[78] C.W. Hummel, A.G. Geiser, H.U. Bryant, I.R. Cohen, R.D. Dally, K.C. Fong, S.A. Frank, R. Hinklin, S.A. Jones, G. Lewis, D.J. McCann, D.G. Rudmann, T.A. Shepherd, H. Tian, O.B. Wallace, M. Wang, Y. Wang, and J.A. Dodge, *A selective estrogen receptor modulator designed for the treatment of uterine leiomyoma with unique tissue specificity for uterus and ovaries in rats*, J. Med. Chem. 48 (2005), pp. 6772–6775.

[79] R.W. Hsieh, S.S. Rajan, S.K. Sharma, Y. Guo, E.R. DeSombre, M. Mrksich, and G.L. Greene, *Identification of ligands with bicyclic scaffolds provides insights into mechanisms of estrogen receptor subtype selectivity*, J. Biol. Chem. 281 (2006), pp. 17909–17919.

[80] R.W. Hsieh, *Discovery and characterization of novel estrogen receptor agonist ligands and development of biochips for nuclear receptor drug discovery*, Ph.D. diss., Chicago University, 2006.

[81] B.H. Norman, J.A. Dodge, T.I. Richardson, P.S. Borromeo, C.W. Lugar, S.A. Jones, K. Chen, Y. Wang, G.L. Durst, R.J. Barr, C. Montrose-Rafizadeh, H.E. Osborne, R.M. Amos, S. Guo, A. Boodhoo, and V. Krishnan, *Benzopyrans are selective estrogen receptor β agonists with novel activity in models of benign prostatic hyperplasia*, J. Med. Chem. 49 (2006), pp. 6155–6157.

[82] F.F. Vajdos, L.R. Hoth, K.F. Geoghegan, S.P. Simons, P.K. LeMotte, D.E. Danley, M.J. Ammirati, and J. Pandit, *The 2.0 Å crystal structure of the ERα ligand-binding domain complexed with lasofoxifene*, Protein Sci. 16 (2007), pp. 897–905.

[83] K.W. Nettles, J.B. Bruning, G. Gil, E.E. O'Neill, J. Nowak, Y. Guo, Y. Kim, E.R. DeSombre, R. Dilis, R.N. Hanson, A. Joachimiak, and G.L. Greene, *Structural plasticity in the oestrogen receptor ligand-binding domain*, EMBO Rep. 8 (2007), pp. 563–568.

[84] H.B. Zhou, K.W. Nettles, J.B. Bruning, Y. Kim, A. Joachimiak, S. Sharma, K.E. Carlson, F. Stossi, B.S. Katzenellenbogen, G.L. Greene, and J.A. Katzenellenbogen, *Elemental isomerism: A boron-nitrogen surrogate for a carbon-carbon double bond increases the chemical diversity of estrogen receptor ligands*, Chem. Biol. 14 (2007), pp. 659–669.

[85] T.I. Richardson, J.A. Dodge, G.L. Durst, L.A. Pfeifer, J. Shah, Y. Wang, J.D. Durbin, V. Krishnan, and B.H. Norman, *Benzopyrans as selective estrogen receptor β agonists (SERBAs). Part 3: Synthesis of cyclopentanone and cyclohexanone intermediates for C-ring modification*, Bioorg. Med. Chem. Lett. 17 (2007), pp. 4824–4828.

[86] B.H. Norman, T.I. Richardson, J.A. Dodge, L.A. Pfeifer, G.L. Durst, Y. Wang, J.D. Durbin, V. Krishnan, S.R. Dinn, S. Liu, J.E. Reilly, and K.T. Ryter, *Benzopyrans as selective estrogen receptor β agonists (SERBAS). Part 4: Functionalization of the benzopyran A-ring*, Bioorg. Med. Chem. Lett. 17 (2007), pp. 5082–5085.

[87] K.D. Dykstra, L. Guo, E.T. Birzin, W. Chan, Y.T. Yang, E.C. Hayes, C.A. DaSilva, L.Y. Pai, R.T. Mosley, B. Kraker, P.M. Fitzgerald, F. DiNinno, S.P. Rohrer, J.M. Schaeffer, and M.L. Hammond, *Estrogen receptor ligands. Part 16: 2-Aryl indoles as highly subtype selective ligands for ERα*, Bioorg. Med. Chem. Lett. 17 (2007), pp. 2322–2328.

[88] B.R. Henke, T.G. Consler, N. Go, R.L. Hale, D.R. Hohman, S.A. Jones, A.T. Lu, L.B. Moore, J.T. Moore, L.A. Orband-Miller, R.G. Robinett, J. Shearin, P.K. Spearing, E.L. Stewart, P.S. Turnbull, S.L. Weaver, S.P. Williams, G.B. Wisely, and M.H. Lambert, *A new series of estrogen receptor modulators that display selectivity for estrogen receptor β*, J. Med. Chem. 45 (2002), pp. 5492–5505.

[89] M.S. Malamas, E.S. Manas, R.E. McDevitt, I. Gunawan, Z.B. Xu, M.D. Collini, C.P. Miller, T. Dinh, R.A. Henderson, J.C. Keith Jr., and H.A. Harris, *Design and synthesis of aryl diphenolic azoles as potent and selective estrogen receptor-β ligands*, J. Med. Chem. 47 (2004), pp. 5021–5040.

[90] R.E. Mewshaw, R.J. Edsall Jr., C. Yang, E.S. Manas, Z.B. Xu, R.A. Henderson, J.C. Keith Jr., and H.A. Harris, *ERβ ligands. 3. Exploiting two binding orientations of the 2-phenylnaphthalene scaffold to achieve ERβ selectivity*, J. Med. Chem. 48 (2005), pp. 3953–3979.

[91] R.E. McDevitt, M.S. Malamas, E.S. Manas, R.J. Unwalla, Z.B. Xu, C.P. Miller, and H.A. Harris, *Estrogen receptor ligands: Design and synthesis of new 2-arylindene-1-ones*, Bioorg. Med. Chem. Lett. 15 (2005), pp. 3137–3142.

[92] Y. Wang, N.Y. Chirgadze, S.L. Briggs, S. Khan, E.V. Jensen, and T.P. Burris, *A second binding site for hydroxytamoxifen within the coactivator-binding groove of estrogen receptor β*, Proc. Natl. Acad. Sci. USA 103 (2006), pp. 9908–9911.

[93] R.R. Wilkening, R.W. Ratcliffe, E.C. Tynebor, K.J. Wildonger, A.K. Fried, M.L. Hammond, R.T. Mosley, P.M. Fitzgerald, N. Sharma, B.M. McKeever, S. Nilsson, M. Carlquist, A. Thorsell, L. Locco, R. Katz, K. Frisch, E.T. Birzin, H.A. Wilkinson, S. Mitra, S. Cai, E.C. Hayes, J.M. Schaeffer, and S.P. Rohrer, *The discovery of tetrahydrofluorenones as a new class of estrogen receptor β-subtype selective ligands*, Bioorg. Med. Chem. Lett. 16 (2006), pp. 3489–3494.

[94] R.E. Mewshaw, S.M. Bowen, H.A. Harris, Z.B. Xu, E.S. Manas, and S.T. Cohn, *ERβ ligands. Part 5: Synthesis and structure-activity relationships of a series of 4′-hydroxyphenyl-aryl-carbaldehyde oxime derivatives*, Bioorg. Med. Chem. Lett. 17 (2007), pp. 902–906.

[95] T.I. Richardson, B.H. Norman, C.W. Lugar, S.A. Jones, Y. Wang, J.D. Durbin, V. Krishnan, and J.A. Dodge, *Benzopyrans as selective estrogen receptor β agonists (SERBAs). Part 2: Structure-activity relationship studies on the benzopyran scaffold*, Bioorg. Med. Chem. Lett. 17 (2007), pp. 3570–3574.

[96] T.I. Richardson, J.A. Dodge, Y. Wang, J.D. Durbin, V. Krishnan, and B.H. Norman, *Benzopyrans as selective estrogen receptor beta agonists (SERBAs). Part 5: Combined A- and C-ring structure-activity relationship studies*, Bioorg. Med. Chem. Lett. 17 (2007), pp. 5563–5566.

[97] A.C. Pike, A.M. Brzozowski, J. Walton, R.E. Hubbard, A.G. Thorsell, Y.L. Li, J.A. Gustafsson, and M. Carlquist, *Structural insights into the mode of action of a pure antiestrogen*, Structure 9 (2001), pp. 145–153.

[98] S. McDonald, L. Brive, D.B. Agus, H.I. Scher, and K.R. Ely, *Ligand responsiveness in human prostate cancer: Structural analysis of mutant androgen receptors from LNCaP and CWR22 tumors*, Cancer Res. 60 (2000), pp. 2317–2322.

[99] B.F. Luisi, W.X. Xu, Z. Otwinowski, L.P. Freedman, K.R. Yamamoto, and P.B. Sigler, *Crystallographic analysis of the interaction of the glucocorticoid receptor with DNA*, Nature 352 (1991), pp. 497–505.

[100] S.P. Williams and P.B. Sigler, *Atomic structure of progesterone complexed with its receptor*, Nature 393 (1998), pp. 392–396.

[101] N. Poujol, J.M. Wurtz, B. Tahiri, S. Lumbroso, J.C. Nicolas, D. Moras, and C. Sultan, *Specific recognition of androgens by their nuclear receptor. A structure-function study*, J. Biol. Chem. 275 (2000), pp. 24022–24031.

[102] C.A. Marhefka, B.M. Moore, T.C. Bishop, L. Kirkovsky, A. Mukherjee, J.T. Dalton, and D.D. Miller, *Homology modelling using multiple molecular dynamics simulations and docking studies of the human androgen receptor ligand binding domain bound to testosterone and nonsteroidal ligands*, J. Med. Chem. 44 (2001), pp. 1729–1740.

[103] P.M. Matias, P. Donner, R. Coelho, M. Thomaz, C. Peixoto, S. Macedo, N. Otto, S. Joschko, P. Scholz, A. Wegg, S. Bäsler, M. Schäfer, U. Egner, and M.A. Carrondo, *Structural evidence for ligand specificity in the binding domain of the human androgen receptor. Implications for pathogenic gene mutations*, J. Biol. Chem. 275 (2000), pp. 26164–26171.

[104] B. He, R.T. Gampe Jr., A.J. Kole, A.T. Hnat, T.B. Stanley, G. An, E.L. Stewart, R.I. Kalman, J.T. Minges, and E.M. Wilson, *Structural basis for androgen receptor inter-domain and coactivator interactions suggests a transition in nuclear receptor activation function dominance*, Mol. Cell 16 (2004), pp. 425–438.

[105] P.M. Matias, M.A. Carrondo, R. Coelho, M. Thomaz, X.Y. Zhao, A. Wegg, K. Crusius, U. Egner, and P. Donner, *Structural basis for the glucocorticoid response in a mutant human androgen receptor (AR^{ccr}) derived from an androgen-independent prostate cancer*, J. Med. Chem. 45 (2002), pp. 1439–1446.

[106] E. Estébanez-Perpiñá, J.M. Moore, E. Mar, E. Delgado-Rodrigues, P. Nguyen, J.D. Baxter, B.M. Buehrer, P. Webb, R.J. Fletterick, and R.K. Guy, *The molecular mechanisms of coactivator utilization in ligand-dependent transactivation by the androgen receptor*, J. Biol. Chem. 280 (2005), pp. 8060–8068.

[107] C.E. Bohl, W. Gao, D.D. Miller, C.E. Bell, and J.T. Dalton, *Structural basis for antagonism and resistance of bicalutamide in prostate cancer*, Proc. Natl. Acad. Sci. USA 102 (2005), pp. 6201–6206.

[108] C.E. Bohl, D.D. Miller, J. Chen, C.E. Bell, and J.T. Dalton, *Structural basis for accommodation of nonsteroidal ligands in the androgen receptor*, J. Biol. Chem. 280 (2005), pp. 37747–37754.

[109] K. Pereira de Jesus-Tran, P.L. Côté, L. Cantin, J. Blanchet, F. Labrie, and R. Breton, *Comparison of crystal structures of human androgen receptor ligand-binding domain complexed with various agonists reveals molecular determinants responsible for binding affinity*, Protein Sci. 15 (2006), pp. 987–999.

[110] E.B. Askew, R.T. Gampe, T.B. Stanley, J.L. Faggart, and E.M. Wilson, *Modulation of androgen receptor activation function 2 by testosterone and dihydrotestosterone*, J. Biol. Chem. 282 (2007), pp. 25801–25816.

[111] F. Wang, X.Q. Liu, H. Li, K.N. Liang, J.N. Miner, M. Hong, E.A. Kallel, A. van Oeveren, L. Zhi, and T. Jiang, *Structure of the ligand-binding domain (LBD) of human androgen receptor in complex with a selective modulator LGD2226*, Acta Crystallogr. 62 (2006), pp. 1067–1071.

[112] C.E. Bohl, Z. Wu, D.D. Miller, C.E. Bell, and J.T. Dalton, *Crystal structure of the T877A human androgen receptor ligand-binding domain complexed to cyproterone acetate provides insight for ligand-induced conformational changes and structure-based drug design*, J. Biol. Chem. 282 (2007), pp. 13648–13655.

[113] L. Cantin, F. Faucher, J.F. Couture, K. Pereira de Jesus-Tran, P. Legrand, L.C. Ciobanu, Y. Fréchette, R. Labrecque, S.M. Singh, F. Labrie, and R. Breton, *Structural characterization of the human androgen receptor ligand-binding domain complexed with EM5744, a rationally designed steroidal ligand bearing a bulky chain directed toward helix 12*, J. Biol. Chem. 282 (2007), pp. 30910–30919.

[114] J.S. Sack, K.F. Kish, C. Wang, R.M. Attar, S.E. Kiefer, Y. An, G.Y. Wu, J.E. Scheffler, M.E. Salvati, S.R. Krystek Jr., R. Weinmann, and H.M. Einspahr, *Crystallographic structures of the ligand-binding domains of the androgen receptor and its T877A mutant complexed with the natural agonist dihydrotestosterone*, Proc. Natl. Acad. Sci. USA 98 (2001), pp. 4904–4909.

[115] M.E. Salvati, A. Balog, W. Shan, D.D. Wei, D. Pickering, R.M. Attar, J. Geng, C.A. Rizzo, M.M. Gottardis, R. Weinmann, S.R. Krystek, J. Sack, Y. An, and K. Kish, *Structure based approach to the design of bicyclic-1H-isoindole-1,3(2H)-dione based androgen receptor antagonists*, Bioorg. Med. Chem. Lett. 15 (2005), pp. 271–276.

[116] C. Sun, J.A. Robl, T.C. Wang, Y. Huang, J.E. Kuhns, J.A. Lupisella, B.C. Beehler, R. Golla, P.G. Sleph, R. Seethala, A. Fura, S.R. Krystek, Y. An, M.F. Malley, J.S. Sack, M.E. Salvati, G.J. Grover, J. Ostrowski, and L.G. Hamann, *Discovery of potent, orally-active, and muscle-selective androgen receptor modulators based on an N-aryl-hydroxybicyclohydantoin scaffold*, J. Med. Chem. 49 (2006), pp. 7596–7599.

[117] J. Ostrowski, J.E. Kuhns, J.A. Lupisella, M.C. Manfredi, B.C. Beehler, S.R. Krystek, Y. Bi, C. Sun, R. Seethala, R. Golla, P.G. Sleph, A. Fura, Y. An, K.F. Kish, J.S. Sack, K.A. Mookhtiar, G.J. Grover, and L.G. Hamann, *Pharmacological and X-ray structural characterization of a novel selective androgen receptor modulator: Potent hyperanabolic stimulation of skeletal muscle with hypostimulation of prostate in rats*, Endocrinology 148 (2007), pp. 4–12.

[118] E. Hur, S.J. Pfaff, E.S. Payne, H. Grøn, B.M. Buehrer, and R.J. Fletterick, *Recognition and accommodation at the androgen receptor coactivator binding interface*, Plos Biol. 2 (2004), p. E274.

[119] D.G. Jones, X. Liang, E.L. Stewart, R.A. Noe, L.S. Kallander, K.P. Madauss, S.P. Williams, S.K. Thompson, D.W. Gray, and W.J. Hoekstra, *Discovery of non-steroidal mifepristone mimetics: Pyrazoline-based PR antagonists*, Bioorg. Med. Chem. Lett. 15 (2005), pp. 3203–3206.

[120] K.P. Madauss, J.S. Deng, R.J. Austin, M.H. Lambert, I. McLay, J. Pritchard, S.A. Short, E.L. Stewart, I.J. Uings, and S.P. Williams, *Progesterone receptor ligand binding pocket flexibility: Crystal structures of the norethindrone and mometasone furoate complexes*, J. Med. Chem. 47 (2004), pp. 3381–3387.

[121] Z. Zhang, A.M. Olland, Y. Zhu, J. Cohen, T. Berrodin, S. Chippari, C. Appavu, S. Li, J. Wilhem, R. Chopra, A. Fensome, P. Zhang, J. Wrobel, R.J. Unwalla, C.R. Lyttle, and R.C. Winneker, *Molecular and pharmacological properties of a potent and selective novel nonsteroidal progesterone receptor agonist tanaproget*, J. Biol. Chem. 280 (2005), pp. 28468–28475.

[122] K.P. Madauss, E.T. Grygielko, S.J. Deng, A.C. Sulpizio, T.B. Stanley, C. Wu, S.A. Short, S.K. Thompson, E.L. Stewart, N.J. Laping, S.P. Williams, and J.D. Bray, *A structural and* in vitro *characterization of asoprisnil: A selective progesterone receptor modulator*, Mol. Endocrinol. 21 (2007), pp. 1066–1081.

[123] U. Lind, P. Greenidge, M. Gillner, K.F. Koehler, A. Wright, and J. Carlstedt-Duke, *Functional probing of the human glucocorticoid receptor steroid-interacting surface by site-directed mutagenesis: Gln-642 plays an important role in steroid recognition and binding*, J. Biol. Chem. 275 (2000), pp. 19041–19049.

[124] R. Dey, P. Roychowdhury, and C. Mukherjee, *Homology modelling of the ligand-binding domain of glucocorticoid receptor: Binding site interactions with cortisol and corticosterone*, Protein Eng. 14 (2001), pp. 565–571.

[125] D.F. Lewis, M.S. Ogg, P.S. Goldfarb, and G.G. Gibson, *Molecular modelling of the human glucocorticoid receptor (hGR) ligand-binding domain (LBD) by homology with the human estrogen receptor α (hERα) LBD: Quantitative structure-activity relationships within a series of CYP3A4 inducers where induction is mediated via hGR involvement*, J. Steroid Biochem. Mol. Biol. 82 (2002), pp. 195–199.

[126] C. Honer, K. Nam, C. Fink, P. Marshall, G. Ksander, R.E. Chatelain, W. Cornell, R. Steele, R. Schweitzer, and C. Schumacher, *Glucocorticoid receptor antagonism by cyproterone acetate and RU486*, Mol. Pharmacol. 63 (2003), pp. 1012–1020.

[127] R.K. Bledsoe, V.G. Montana, T.B. Stanley, C.J. Delves, C.J. Apolito, D.D. Mckee, T.G. Consler, D.J. Parks, E.L. Stewart, T.M. Willson, M.H. Lambert, J.T. Moore, K.H. Pearce, and H.E. Xu, *Crystal structure of the glucocorticoid receptor ligand binding domain reveals a novel mode of receptor dimerization and coactivator recognition*, Cell 110 (2002), pp. 93–105.

[128] M.R. Yudt, C.M. Jewell, R.J. Bienstock, and J.A. Cidlowski, *Molecular origins for the dominant negative function of human glucocorticoid receptor β*, Mol. Cell Biol. 23 (2003), pp. 4319–4330.

[129] J. von Langen, K.H. Fritzemeier, S. Diekmann, and A. Hillisch, *Molecular basis of the interaction specificity between the human glucocorticoid receptor and its endogenous steroid ligand cortisol*, ChembioChem. 6 (2005), pp. 1110–1118.

[130] B. Kauppi, C. Jakob, M. Färnegårdh, J. Yang, H. Ahola, M. Alarcon, K. Calles, O. Engström, J. Harlan, S. Muchmore, A.K. Ramqvist, S. Thorell, L. Ohman, J. Greer, J.A. Gustafsson, J. Carlstedt-Duke, and M. Carlquist, *The three-dimensional structures of antagonistic and agonistic forms of the glucocorticoid receptor ligand-binding domain: RU-486 induces a transconformation that leads to active antagonism*, J. Biol. Chem. 278 (2003), pp. 22748–22754.

[131] J. Fagart, J.M. Wurtz, A. Souque, C. Hellal-Levy, D. Moras, and M.E. Rafestin-Oblin, *Antagonism in the human mineralocorticoid receptor*, EMBO J. 17 (1998), pp. 3317–3325.

[132] M. Hendlich, P. Lackner, S. Weickus, H. Floeckner, R. Froschauer, K. Gottsbacher, G. Casari, and M.J. Sippl, *Identification of the native protein folds amongst a large number of incorrect models: The calculation of low energy conformation from potentials of mean force*, J. Mol. Biol. 216 (1990), pp. 167–180.

[133] G. Auzou, J. Fagart, A. Souque, C. Hellal-Lévy, J.M. Wurtz, D. Moras, and M.E. Rafestin-Oblin, *A single amino acid mutation of ala-773 in the mineralocorticoid receptor confers agonist properties to 11β-substituted spirolactones*, Mol. Pharmacol. 58 (2000), pp. 684–691.

[134] R. Dey and P. Roychowdhury, *Homology modelling of the ligand binding domain of mineralocorticoid receptor: Close structural kinship with glucocorticoid receptor ligand binding domain and their similar binding mode with DOC (de-oxy corticosterone)*, J. Biomol. Struct. Dyn. 20 (2002), pp. 21–29.

[135] J. Fagart, J. Huyet, G.M. Pinon, M. Rochel, C. Mayer, and M.E. Rafestin-Oblin, *Crystal structure of a mutant mineralocorticoid receptor responsible for hypertension*, Nat. Struct. Mol. Biol. 12 (2005), pp. 554–555.

[136] R.K. Bledsoe, K.P. Madauss, J.A. Holt, C.J. Apolito, M.H. Lambert, K.H. Pearce, T.B. Stanley, E.L. Stewart, R.P. Trump, T.M. Willson, and S.P. Williams, *A ligand-mediated hydrogen bond network required for the activation of the mineralocorticoid receptor*, J. Biol. Chem. 280 (2005), pp. 31283–31293.

[137] Y. Li, K. Suino, J. Daugherty, and H.E. Xu, *Structural and biochemical mechanisms for the specificity of hormone binding and coactivator assembly by mineralocorticoid receptor*, Mol. Cell 19 (2005), pp. 367–380.

[138] J. Huyet, G.M. Pinon, M.R. Fay, J. Fagart, and M.E. Rafestin-Oblin, *Structural basis of spirolactone recognition by the mineralocorticoid receptor*, Mol. Pharmacol. 72 (2007), pp. 563–571.

[139] M. Schapira, B.M. Raaka, S. Das, L. Fan, M. Totrov, Z. Zhou, S.R. Wilson, R. Abagyan, and H.H. Samuels, *Discovery of diverse thyroid hormone receptor antagonists by high-throughput docking*, Proc. Natl. Acad. Sci. USA 100 (2003), pp. 7354–7359.

[140] L. Ye, Y.L. Li, K. Mellström, C. Mellin, L.G. Bladh, K. Koehler, N. Garg, A.M. Garcia Collazo, C. Litten, B. Husman, K. Persson, J. Ljunggren, G. Grover, P.G. Sleph, R. George, and J. Malm, *Thyroid receptor ligands. 1. Agonist ligands selective for the thyroid receptor β1*, J. Med. Chem. 46 (2003), pp. 1580–1588.

[141] A.S. Nascimento, S.M. Dias, F.M. Nunes, R. Aparicio, A.L. Ambrosio, L. Bleicher, A.C. Figueira, M.A. Santos, M. de Oliveira Neto, H. Fischer, M. Togashi, A.F. Craievich, R.C. Garratt, J.D. Baxter, P. Webb, and I. Polikarpov, *Structural rearrangements in the thyroid hormone receptor hinge domain and their putative role in the receptor function*, J. Mol. Biol. 360 (2006), pp. 586–598.

[142] R.L. Dow, S.R. Schneider, E.S. Paight, R.F. Hank, P. Chiang, P. Cornelius, E. Lee, W.P. Newsome, A.G. Swick, J. Spitzer, D.M. Hargrove, T.A. Patterson, J. Pandit, B.A. Chrunyk, P.K. LeMotte, D.E. Danley, M.H. Rosner, M.J. Ammirati, S.P. Simons, G.K. Schulte, B.F. Tate, and P. DaSilva-Jardine, *Discovery of a novel series of 6-azauracil-based thyroid hormone receptor ligands: Potent, TRβ subtype-selective thyromimetics*, Bioorg. Med. Chem. Lett. 13 (2003), pp. 379–382.

[143] B.R. Huber, M. Desclozeaux, B.L. West, S.T. Cunha Lima, H.T. Nguyen, J.D. Baxter, H.A. Ingraham, and R.J. Fletterick, *Thyroid hormone receptor-β mutations conferring hormone resistance and reduced corepressor release exhibit decreased stability in the N-terminal ligand-binding domain*, Mol. Endocrinol. 17 (2003), pp. 107–116.

[144] B.R. Huber, B. Sandler, B.L. West, S.T. Cunha Lima, H.T. Nguyen, J.W. Apriletti, J.D. Baxter, and R.J. Fletterick, *Two resistance to thyroid hormone mutants with impaired hormone binding*, Mol. Endocrinol. 17 (2003), pp. 643–652.

[145] S. Borngraeber, M.J. Budny, G. Chiellini, S.T. Cunha Lima, M. Togashi, P. Webb, J.D. Baxter, T.S. Scanlan, and R.J. Fletterick, *Ligand selectivity by seeking hydrophobicity in thyroid hormone receptor*, Proc. Natl. Acad. Sci. USA 100 (2003), pp. 15358–15363.

[146] J.J. Hangeland, A.M. Doweyko, T. Dejneka, T.J. Friends, P. Devasthale, K. Mellström, J. Sandberg, M. Grynfarb, J.S. Sack, H. Einspahr, M. Färnegårdh, B. Husman, J. Ljunggren, K. Koehler, C. Sheppard, J. Malm, and D.E. Ryono, *Thyroid receptor ligands. Part 2: Thyromimetics with improved selectivity for the thyroid hormone receptor β*, Bioorg. Med. Chem. Lett. 14 (2004), pp. 3549–3553.

[147] B. Sandler, P. Webb, J.W. Apriletti, B.R. Huber, M. Togashi, S.T. Cunha Lima, S. Juric, S. Nilsson, R. Wagner, R.J. Fletterick, and J.D. Baxter, *Thyroxine-thyroid hormone receptor interactions*, J. Biol. Chem. 279 (2004), pp. 55801–55808.

[148] K. Koehler, S. Gordon, P. Brandt, B. Carlsson, A. Bäcksbro-Saedi, T. Apelqvist, P. Agback, G.J. Grover, W. Nelson, M. Grynfarb, M. Färnegårdh, S. Rehnmark, and J. Malm, *Thyroid receptor ligands. 6. A high affinity "direct antagonist" selective for the thyroid hormone receptor*, J. Med. Chem. 49 (2006), pp. 6635–6637.

[149] K. Yamamoto, H. Masuno, M. Choi, K. Nakashima, T. Taga, H. Ooizumi, K. Umesono, W. Sicinska, J. VanHooke, H. F. DeLuca, and S. Yamada, *Three-dimensional modeling of and ligand docking to vitamin D receptor ligand binding domain*, Proc. Natl. Acad. Sci. USA 97 (2000), pp. 1467–1472.

[150] P. Rotkiewicz, W. Sicinska, A. Kolinski, and H.F. DeLuca, *Model of three-dimensional structure of vitamin D receptor and its binding mechanism with 1α,25-dihydroxyvitamin D₃*, Proteins 44 (2001), pp. 188–199.

[151] N. Rochel, J.M. Wurtz, A. Mitschler, B. Klaholz, and D. Moras, *The crystal structure of the nuclear receptor for vitamin D bound to its natural ligand*, Mol. Cell. 5 (2000), pp. 173–179.

[152] G. Tocchini-Valentini, N. Rochel, J.M. Wurtz, A. Mitschler, and D. Moras, *Crystal structures of the vitamin D receptor complexed to superagonist 20-epi ligands*, Proc. Natl. Acad. Sci. USA 98 (2001), pp. 5491–5496.

[153] G. Tocchini-Valentini, N. Rochel, J.M. Wurtz, and D. Moras, *Crystal structures of the vitamin D nuclear receptor liganded with the vitamin D side-chain analogues calcipotriol and seocalcitol, receptor agonists of clinical importance. Insights into a structural basis for the switching of calcipotriol to a receptor antagonist by further side chain modification*, J. Med. Chem. 47 (2004), pp. 1956–1961.

[154] G. Eelen, L. Verlinden, N. Rochel, F. Claessens, P. De Clercq, M. Vandewalle, G. Tocchini-Valentini, D. Moras, R. Bouillon, and A. Verstuyf, *Superagonistic action of 14-epi-analogs of 1,25-dihydroxyvitamin D explained by vitamin D receptor-coactivator interaction*, Mol. Pharmacol. 67 (2005), pp. 1566–1573.

[155] S. Hourai, T. Fujishima, A. Kittaka, Y. Suhara, H. Takayama, N. Rochel, and D. Moras, *Probing a water channel near the A-ring of receptor bound 1α,25-dihydroxyvitamin D₃ with selected 2α-substituted analogues*, J. Med. Chem. 49 (2006), pp. 5199–5205.

[156] J.L. Vanhooke, M.M. Benning, C.B. Bauer, J.W. Pike, and H.F. DeLuca, *Molecular structure of the rat vitamin D receptor ligand binding domain complexed with 2-carbon-substituted vitamin D₃ hormone analogues and a LXXLL-containing coactivator peptide*, Biochemistry 43 (2004), pp. 4101–4110.

[157] J.L. Vanhooke, B.P. Tadi, M.M. Benning, L.A. Plum, and H.F. Deluca, *New analogs of 2-methylene-19-nor-(20S)-1,25-dihydroxyvitamin D₃ with conformationally restricted side chains: Evaluation of biological activity and structural determination of VDR bound conformations*, Arch. Biochem. Biophys. 460 (2007), pp. 161–165.

[158] F. Ciesielski, N. Rochel, and D. Moras, *Adaptability of the vitamin D nuclear receptor to the synthetic ligand Gemini: Remodelling the LBP with one side-chain rotation*, J. Steroid Biochem. Mol. Biol. 103 (2007), pp. 235–242.

[159] N. Rochel, S. Hourai, X. Perez-Garcia, A. Rumbo, A. Mourino, and D. Moras, *Crystal structure of the vitamin D nuclear receptor ligand binding domain in complex with a locked side-chain analog of calcitriol*, Arch. Biochem. Biophys. 460 (2007), pp. 172–176.

[160] M. Schapira, B.M. Raaka, H.H. Samuels, and R. Abagyan, *Rational discovery of novel nuclear hormone receptor antagonists*, Proc. Natl. Acad. Sci. USA 97 (2000), pp.1008–1013.

[161] M. Schapira, B.M. Raaka, H. Herbert, and R. Abagyan, In silico *discovery of novel retinoic acid receptor agonist structures*, BMC Struct Biol. 1 (2001), p. 1.

[162] B.W. Lund, F. Piu, N.K. Gauthier, A. Eeg, E. Currier, V. Sherbukhin, M.R. Brann, U. Hacksell, and R. Olsson, *Discovery of a potent, orally available, and isoform-selective retinoic acid β2 receptor agonist*, J. Med. Chem. 48 (2005), pp. 7517–7519.

[163] B.P. Klaholz, J.P. Renaud, A. Mitschler, C. Zusi, P. Chambon, H Gronemeyer, and D. Moras, *Conformational adaptation of agonists to the human nuclear receptor RARγ*, Nat. Struct. Biol. 5 (1998), pp. 199–202.

[164] B.P. Klaholz, A. Mitschler, and D. Moras, *Structural basis for isotype selectivity of the human retinoic acid nuclear receptor*, J. Mol. Biol. 302 (2000), pp. 155–170.

[165] B.P. Klaholz, A. Mitschler, M. Belema, C. Zusi, and D. Moras, *Enantiomer discrimination illustrated by high-resolution crystal structures of the human nuclear receptor hRARγ*, Proc. Natl. Acad. Sci. USA 97 (2000), pp. 6322–6327.

[166] P. Germain, S. Kammerer, E. Pérez, C. Peluso-Iltis, D. Tortolani, F.C. Zusi, J. Starrett, P. Lapointe, J.P. Daris, A. Marinier, A.R. De Lera, N. Rochel, and H. Gronemeyer, *Rational design of RAR-selective ligands revealed by RARβ crystal structure*, EMBO Rep. 5 (2004), pp. 877–882.

[167] D.F.V. Lewis and B.G. Lake, *Molecular modelling of the rat peroxisome proliferator-activated receptor-α (rPPARα) by homology with the human retinoic acid X receptor α (hRXRα) and investigation of ligand binding interactions I: QSARs*, Toxicol. in Vitro 12 (1998), pp. 619–632.

[168] D.F. Lewis, M.N. Jacobs, M. Dickins, and B.G. Lake, *Molecular modelling of the peroxisome proliferator-activated receptor α (PPARα) from human, rat and mouse, based on homology with the human PPARγ crystal structure*, Toxicol. in Vitro 16 (2002), pp. 275–280.

[169] R.T. Nolte, G.B. Wisely, S. Westin, J.E. Cobb, M.H. Lambert, R. Kurokawa, M.G. Rosenfeld, T.M. Willson, C.K. Glass, and M.V. Milburn, *Ligand binding and co-activator assembly of the peroxisome proliferator-activated receptor-γ*, Nature 395 (1998), pp. 137–143.

[170] F.E. Blaney, *Homology modeling and* ab initio *calculations identify a basis for ligand selectivity for the PPARγ nuclear hormone receptor*, Int. J. Quantum Chem. 73 (1999), pp. 97–111.

[171] R.T. Gampe Jr., V.G. Montana, M.H. Lambert, G.B. Wisely, M.V. Milburn, and H.E. Xu, *Structural basis for autorepression of retinoid X receptor by tetramer formation and the AF-2 helix*, Genes Dev. 14 (2000), pp. 2229–2241.

[172] P.F. Egea, A. Mitschler, N. Rochel, M. Ruff, P. Chambon, and D. Moras, *Crystal structure of the human RXRα ligand-binding domain bound to its natural ligand: 9-cis retinoic acid*, EMBO J. 19 (2000), pp. 2592–2601.

[173] P.F. Egea, A. Mitschler, and D. Moras, *Molecular recognition of agonist ligands by RXRs*, Mol. Endocrinol. 16 (2002), pp. 987–997.

[174] P. Cronet, J.F. Petersen, R. Folmer, N. Blomberg, K. Sjöblom, U. Karlsson, E.L. Lindstedt, and K. Bamberg, *Structure of the PPARα and -γ ligand binding domain in complex with AZ 242; ligand selectivity and agonist activation in the PPAR family*, Structure 9 (2001), pp. 699–706.

[175] H.E. Xu, M.H. Lambert, V.G. Montana, K.D. Plunket, L.B. Moore, J.L. Collins, J.A. Oplinger, S.A. Kliewer, R.T. Gampe Jr., D.D. McKee, J.T. Moore, and T.M. Willson, *Structural determinants of ligand binding selectivity between the peroxisome proliferator-activated receptors*, Proc. Natl. Acad. Sci. USA 98 (2001), pp. 13919–13924.

[176] H.E. Xu, T.B. Stanley, V.G. Montana, M.H. Lambert, B.G. Shearer, J.E. Cobb, D.D. McKee, C.M. Galardi, K.D. Plunket, R.T. Nolte, D.J. Parks, J.T. Moore, S.A. Kliewer, T.M. Willson, and J.B. Stimmel, *Structural basis for antagonist-mediated recruitment of nuclear co-repressors by PPARα*, Nature 415 (2002), pp. 813–817.

[177] H.O. Han, S.H. Kim, K.H. Kim, G.C. Hur, H.J. Yim, H.K. Chung, S.H. Woo, K.D. Koo, C.S. Lee, J.S. Koh, and G.T. Kim, *Design and synthesis of oxime ethers of α-acyl-β-phenylpropanoic acids as PPAR dual agonists*, Bioorg. Med. Chem. Lett. 17 (2007), pp. 937–941.

[178] M.L. Sierra, V. Beneton, A.B. Boullay, T. Boyer, A.G. Brewster, F. Donche, M.C. Forest, M.H. Fouchet, F.J. Gellibert, D.A. Grillot, M.H. Lambert, A. Laroze, C. Le Grumelec, J.M. Linget, V.G. Montana, V.L. Nguyen, E. Nicodeme, V. Patel, A. Penfornis, O. Pineau, D. Pohin, F. Potvain, G. Paulain, C.B. Ruault, M. Saunders, J. Toum, H.E. Xu, R.X. Xu, and P.M. Pianetti, *Substituted 2-[(4-aminomethyl)phenoxy]-2-methylpropionic acid PPAR agonists. 1. Discovery of a novel series of potent HDLC raising agents*, J. Med. Chem. 50 (2007), pp. 685–695.

[179] H.E. Xu, M.H. Lambert, V.G. Montana, D.J. Parks, S.G. Blanchard, P.J. Brown, D.D. Sternbach, J.M. Lehmann, G.B. Wisely, T.M. Willson, S.A. Kliewer, and M.V. Milburn, *Molecular recognition of fatty acids by peroxisome proliferator-activated receptors*, Mol. Cell 3 (1999), pp. 397–403.

[180] I. Takada, R.T. Yu, H.E. Xu, M.H. Lambert, V.G. Montana, S.A. Kliewer, R.M. Evans, and K. Umesono, *Alteration of a single amino acid in peroxisome proliferator-activated receptor-α (PPARα) generates a PPARδ phenotype*, Mol. Endocrinol. 14 (2000), pp. 733–740.

[181] S.A. Fyffe, M.S. Alphey, L. Buetow, T.K. Smith, M.A. Ferguson, M.D. Sφrensen, F. Bjorkling, and W.N. Hunter, *Recombinant human PPAR-β/δ ligand-binding domain is locked in an activated conformation by endogenous fatty acids*, J. Mol. Biol. 356 (2006), pp. 1005–1013.

[182] S.A. Fyffe, M.S. Alphey, L. Buetow, T.K. Smith, M.A. Ferguson, M.D. Sφrensen, F. Björkling, and W.N. Hunter, *Reevaluation of the PPAR-β/δ ligand binding domain model reveals why it exhibits the activated form*, Mol. Cell 21 (2006), pp. 1–2.

[183] R. Epple, M. Azimioara, R. Russo, Y. Xie, X. Wang, C. Cow, J. Wityak, D. Karanewsky, B. Bursulaya, A. Kreusch, T. Tuntland, A. Gerken, M. Iskandar, E. Saez, H. Martin Seidel, and S.S. Tian, *3,4,5-Trisubstituted isoxazoles as novel PPARδ agonists. Part 2*, Bioorg. Med. Chem. Lett. 16 (2006), pp. 5488–5492.

[184] J. Uppenberg, C. Svensson, M. Jaki, G. Bertilsson, L. Jendeberg, and A. Berkenstam, *Crystal structure of the ligand binding domain of the human nuclear receptor PPARγ*, J. Biol. Chem. 273 (1998), pp. 31108–31112.

[185] Y. Li, M. Choi, K. Suino, A. Kovach, J. Daugherty, S.A. Kliewer, and H.E. Xu, *Structural and biochemical basis for selective repression of the orphan nuclear receptor liver receptor homolog 1 by small heterodimer partner*, Proc. Natl. Acad. Sci. USA 102 (2005), pp. 9505–9510.

[186] J.L. Oberfield, J.L. Collins, C.P. Holmes, D.M. Goreham, J.P. Cooper, J.E. Cobb, J.M. Lenhard, E.A. Hull-Ryde, C.P. Mohr, S.G. Blanchard, D.J. Parks, L.B. Moore, J.M. Lehmann, K. Plunket, A.B. Miller, M.V. Milburn, S.A. Kliewer, and T.M. Willson, *A peroxisome proliferator-activated receptor γ ligand inhibits adipocyte differentiation*, Proc. Natl. Acad. Sci. USA 96 (1999), pp. 6102–6106.

[187] P. Sauerberg, I. Pettersson, L. Jeppesen, P.S. Bury, J.P. Mogensen, K. Wassermann, C.L. Brand, J. Sturis, H.F. Wöldike, J. Fleckner, A.S. Andersen, S.B. Mortensen, L.A. Svensson, H.B. Rasmussen, S.V. Lehmann, Z. Polivka, K. Sindelar, V. Panajotova, L. Ynddal, and E.M. Wulff, *Novel tricyclic-α-alkyloxyphenylpropionic acids: Dual PPARα/γ agonists with hypolipidemic and antidiabetic activity*, J. Med. Chem. 45 (2002), pp. 789–804.

[188] S. Ebdrup, I. Pettersson, H.B. Rasmussen, H.J. Deussen, A. Frost Jensen, S.B. Mortensen, J. Fleckner, L. Pridal, L. Nygaard, and P. Sauerberg, *Synthesis and biological and structural characterization of the dual-acting peroxisome proliferator-activated receptor α/γ agonist ragaglitazar*, J. Med. Chem. 46 (2003), pp. 1306–1317.

[189] T. Ostberg, S. Svensson, G. Selén, J. Uppenberg, M. Thor, M. Sundbom, M. Sydow-Bäckman, A.L. Gustavsson, and L. Jendeberg, *A new class of peroxisome proliferator-activated receptor agonists with a novel binding epitope shows antidiabetic effects*, J. Biol. Chem. 279 (2004), pp. 41124–41130.

[190] J.B. Bruning, M.J. Chalmers, S. Prasad, S.A. Busby, T.M. Kamenecka, Y. He, K.W. Nettles, and P.R. Griffin, *Partial agonists activate PPARγ using a helix 12 independent mechanism*, Structure 15 (2007), pp. 1258–1271.

[191] G.Q. Shi, J.F. Dropinski, B.M. MacKeever, S. Xu, J.W. Becker, J.P. Berger, K.L. McNaul, A. Elbrecht, G. Zhou, T.W. Doebber, P. Wang, Y.S. Chao, M. Forrest, J.V. Heck, D.E. Moller, and B.A. Jones, *Design and synthesis of α-aryloxyphenylacetic acid derivatives: A novel class of PPAR α/γ dual agonists with potent antihyperglycemic and lipid modulating activity*, J. Med. Chem. 48 (2005), pp. 4457–4468.

[192] N. Mahindroo, C.F. Huang, Y.H. Peng, C.C. Wang, C.C. Liao, T.W. Lien, S.K. Chittimalla, W.J. Huang, C.H. Chai, E. Prakash, C.P. Chen, T.A. Hsu, C.H. Peng, I.L. Lu, L.H. Lee, Y.W. Chang, W.C. Chen, Y.C. Chou, C.T. Chen, C.M. Goparaju, Y.S. Chen, S.J. Lan, M.C. Yu, X. Chen, Y.S. Chao, S.Y. Wu, and H.P. Hsieh, *Novel indole-based peroxisome proliferator-activated receptor agonists: Design, SAR, structural biology, and biological activities*, J. Med. Chem. 48 (2005), pp. 8194–8208.

[193] N. Mahindroo, C.C. Wang, C.C. Liao, C.F. Huang, I.L. Lu, T.W. Lien, Y.H. Peng, W.J. Huang, Y.T. Lin, M.C. Hsu, C.H. Lin, C.H. Tsai, J.T. Hsu, X. Chen, P.C. Lyu, Y.S. Chao, S.Y. Wu, and H.P. Hsieh, *Indol-1-yl acetic acids as peroxisome proliferator-activated receptor agonists: Design, synthesis, structural biology, and molecular docking studies*, J. Med. Chem. 49 (2006), pp. 1212–1216.

[194] E. Burgermeister, A. Schnoebelen, A. Flament, J. Benz, M. Stihle, B. Gsell, A. Rufer, A. Rufer, B. Kuhn, H.P. Märki, J. Mizrahi, E. Sebokova, E. Niesor, and M. Meyer, *A novel partial agonist of peroxisome proliferator-activated receptor-γ (PPARγ) recruits PPARγ-coactivator-1α, prevents triglyceride accumulation, and potentiates insulin signaling in vitro*, Mol. Endocrinol. 20 (2006), pp. 809–830.

[195] I.L. Lu, C.F. Huang, Y.H. Peng, Y.T. Lin, H.P. Hsieh, C.T. Chen, T.W. Lien, H.J. Lee, N. Mahindroo, E. Prakash, A. Yueh, H.Y. Chen, C.M. Goparaju, X. Chen, C.C. Liao, Y.S. Chao, J.T. Hsu, and S.Y. Wu, *Structure-based drug design of a novel family of PPARγ partial agonists: Virtual screening, X-ray crystallography, and in vitro/in vivo biological activities*, J. Med. Chem. 49 (2006), pp. 2703–2712.

[196] B. Kuhn, H. Hilpert, J. Benz, A. Binggeli, U. Grether, R. Humm, H.P. Märki, M. Meyer, and P. Mohr, *Structure-based design of indole propionic acids as novel PPARα/γ co-agonists*, Bioorg. Med. Chem. Lett. 16 (2006), pp. 4016–4020.

[197] C.R. Hopkins, S.V. O'neil, M.C. Laufersweiler, Y. Wang, M. Pokross, M. Mekel, A. Evdokimov, R. Walter, M. Kontoyianni, M.E. Petrey, G. Sabatakos, J.T. Roesgen, E. Richardson, and T.P. Demuth Jr., *Design and synthesis of novel N-sulfonyl-2-indole carboxamides as potent PPAR-γ binding agents with potential application to the treatment of osteoporosis*, Bioorg. Med. Chem. Lett. 16 (2006), pp. 5659–5663.

[198] N. Mahindroo, Y.H. Peng, C.H. Lin, U.K. Tan, E. Prakash, T.W. Lien, I.L. Lu, H.J. Lee, J.T. Hsu, X. Chen, C.C. Liao, P.C. Lyu, Y.S. Chao, S.Y. Wu, and H.P. Hsieh, *Structural basis for the structure-activity relationships of peroxisome proliferator-activated receptor agonists*, J. Med. Chem. 49 (2006), pp. 6421–6424.

[199] G. Pochetti, C. Godio, N. Mitro, D. Caruso, A. Galmozzi, S. Scurati, F. Loiodice, G. Fracchiolla, P. Tortorella, A. Laghezza, A. Lavecchia, E. Novellino, F. Mazza, and M. Crestani, *Insights into the mechanism of partial agonism: Crystal structures of the peroxisome proliferator-activated receptor γ ligand-binding domain in the complex with two enantiomeric ligands*, J. Biol. Chem. 282 (2007), pp. 17314–17324.

[200] A.L. Ambrosio, S.M. Dias, I. Polikarpov, R.B. Zurier, S.H. Burstein, and R.C. Garratt, *Ajulemic acid, a synthetic nonpsychoactive cannabinoid acid, bound to the ligand binding domain of the human peroxisome proliferator-activated receptor γ*, J. Biol. Chem. 282 (2007), pp. 18625–18633.

[201] S. Chen, D. Zhou, C. Yang, and M. Sherman, *Molecular basis for the constitutive activity of estrogen-related receptorα-1*, J. Biol. Chem. 276 (2001), pp. 28465–28470.

[202] K. Nam, P. Marshall, R.M. Wolf, and W. Cornell, *Simulation of the different biological activities of diethylstilbestrol (DES) on estrogen receptor α and estrogen-related receptor γ*, Biopolymers 68 (2003), pp. 130–138.

[203] M. Suetsugi, L. Su, K. Karlsberg, Y.C. Yuan, and S. Chen, *Flavone and isoflavone phytoestrogens are agonists of estrogen-related receptors*, Mol. Cancer Res. 1 (2003), pp. 981–991.

[204] H. Greschik, J.M. Wurtz, S. Sanglier, W. Bourguet, A. van Dorsselaer, D. Moras, and J.P. Renaud, *Structural and functional evidence for ligand-independent transcriptional activation by the estrogen-related receptor 3*, Mol. Cell 9 (2002), pp. 303–313.

[205] L. Xiao, X. Cui, V. Madison, R.E. White, and K.C Cheng, *Insights from a three-dimensional model into ligand binding to constitutive active receptor*, Drug Metab. Dispos. 30 (2002), pp. 951–956.

[206] R.E. Watkins, G.B. Wisely, L.B. Moore, J.L. Collins, M.H. Lambert, S.P. Williams, T.M. Willson, S.A. Kliewer, and M.R. Redinbo, *The human nuclear xenobiotic receptor PXR: Structural determinants of directed promiscuity*, Science 292 (2001), pp. 2329–2333.

[207] I. Dussault, M. Lin, K. Hollister, M. Fan, J. Termini, M.A. Sherman, and B.M. Forman, *A structural model of the constitutive androstane receptor defines novel interactions that mediate ligand-independent activity*, Mol. Cell Biol. 22 (2002), pp. 5270–5280.

[208] C.Y. Wang, C.W. Li, J.D. Chen, and W.J. Welsh, *Structural model reveals key inter-actions in the assembly of the pregnane X receptor/corepressor complex*, Mol. Pharmacol. 69 (2006), pp. 1513–1517.

[209] J. Kallen, J.M. Schlaeppi, F. Bitsch, I. Filipuzzi, A. Schilb, V. Riou, A. Graham, A. Strauss, M. Geiser, and B. Fournier, *Evidence for ligand-independent transcriptional activation of the human estrogen-related receptorα (ERRα): Crystal structure of ERRα ligand binding domain in complex with peroxisome proliferator-activated receptor coactivator-1α*, J. Biol. Chem. 279 (2004), pp. 49330–49337.

[210] J. Kallen, R. Lattmann, R. Beerli, A. Blechschmidt, M.J. Blommers, M. Geiser, J. Ottl, J.M. Schlaeppi, A. Strauss, and B. Fournier, *Crystal structure of human estrogen-related receptor α in complex with a synthetic inverse agonist reveals its novel molecular mechanism*, J. Biol. Chem. 282 (2007), pp. 23231–23239.

[211] H. Greschik, R. Flaig, J.P. Renaud, and D. Moras, *Structural basis for the deactivation of the estrogen-related receptor γ by diethylstilbestrol or 4-hydroxytamoxifen and determinants of selectivity*, J. Biol. Chem. 279 (2004), pp. 33639–33646.

[212] L. Wang, W.J. Zuercher, T.G. Consler, M.H. Lambert, A.B. Miller, L.A. Orband-Miller, D.D. McKee, T.M. Willson, and R.T. Nolte, *X-ray crystal structures of the estrogen-related receptor-γ ligand binding domain in three functional states reveal the molecular basis of small molecule regulation*, J. Biol. Chem. 281 (2006), pp. 37773–37781.

[213] E.Y. Chao, J.L. Collins, S. Gaillard, A.B. Miller, L. Wang, L.A. Orband-Miller, R.T. Nolte, D.P. McDonnell, T.M. Willson, and W.J. Zuercher, *Structure-guided synthesis of tamoxifen analogs with improved selectivity for the orphan ERRγ*, Bioorg. Med. Chem. Lett. 16 (2006), pp. 821–824.

[214] A. Matsushima, Y. Kakuta, T. Teramoto, T. Koshiba, X. Liu, H. Okada, T. Tokunaga, S. Kawabata, M. Kimura, and Y. Shimohigashi, *Structural evidence for endocrine disruptor bisphenol A binding to human nuclear receptor ERRγ*, J. Biochem. (Tokyo) 142 (2007), pp. 517–524.

[215] R.X. Xu, M.H. Lambert, B.B. Wisely, E.N. Warren, E.E. Weinert, G.M. Waitt, J.D. Williams, J.L. Collins, L.B. Moore, T.M. Willson, and J.T. Moore, *A structural basis for constitutive activity in the human CAR/RXRα heterodimer*, Mol. Cell 16 (2004), pp. 919–928.

[216] K. Suino, L. Peng, R. Reynolds, Y. Li, J.Y. Cha, J.J. Repa, S.A. Kliewer, and H.E Xu, *The nuclear xenobiotic receptor CAR: Structural determinants of constitutive activation and heterodimerization*, Mol. Cell 16 (2004), pp. 893–905.

[217] L. Shan, J. Vincent, J.S. Brunzelle, I. Dussault, M. Lin, I. Ianculescu, M.A. Sherman, B.M. Forman, and E.J. Fernandez, *Structure of the murine constitutive androstane receptor complexed to androstenol: A molecular basis for inverse agonism*, Mol. Cell 16 (2004), pp. 907–917.

[218] R.E. Watkins, J.M. Maglich, L.B. Moore, G.B. Wisely, S.M. Noble, P.R. Davis-Searles, M.H. Lambert, S.A. Kliewer, and M.R. Redinbo, *2.1 Å crystal structure of human PXR in complex with the St. John's Wort compound hyperforin*, Biochemistry 42 (2003), pp. 1430–1438.

[219] R.E. Watkins, P.R. Davis-Searles, M.H. Lambert, and M.R. Redinbo, *Coactivator binding promotes the specific interaction between ligand and the pregnane X receptor*, J. Mol. Biol. 331 (2003), pp. 815–828.

[220] J.E. Chrencik, J. Orans, L.B. Moore, Y. Xue, L. Peng, J.L. Collins, G.B. Wisely, M.H. Lambert, S.A. Kliewer, and M.R. Redinbo, *Structural disorder in the complex of human pregnane X receptor and the macrolide antibiotic rifampicin*, Mol. Endocrinol. 19 (2005), pp. 1125–1134.

[221] Y. Xue, E. Chao, W.J. Zuercher, T.M. Willson, J.L. Collins, and M.R. Redinbo, *Crystal structure of the PXR-T1317 complex provides a scaffold to examine the potential for receptor antagonism*, Bioorg. Med. Chem. 15 (2007), pp. 2156–2166.

[222] J.A. Kallen, J.M. Schlaeppi, F. Bitsch, S. Geisse, M. Geiser, I. Delhon, and B. Fournier, *X-ray structure of hRORα LBD at 1.63A: Structural and functional data that cholesterol or a cholesterol derivative is the natural ligand of RORα*, Structure 10 (2002), pp. 1697–1707.

[223] J. Kallen, J.M. Schlaeppi, F. Bitsch, I. Delhon, and B. Fournier, *Crystal structure of the human RORα ligand binding domain in complex with cholesterol sulfate at 2.2 Å*, J. Biol. Chem. 279 (2004), pp. 14033–14038.

[224] C. Stehlin, J.M. Wurtz, A. Steinmetz, E. Greiner, R. Schüle, D. Moras, and J.P. Renaud, *X-ray structure of the orphan nuclear receptor RORβ ligand-binding domain in the active conformation*, EMBO J. 20 (2001), pp. 5822–5831.

[225] C. Stehlin-Gaon, D. Willmann, D. Zeyer, S. Sanglier, A. Van Dorsselaer, J.P. Renaud, D. Moras, and R. Schüle, *All-trans retinoic acid is a ligand for the orphan nuclear receptor RORβ*, Nat. Struct. Biol. 10 (2003), pp. 820–825.

[226] J.M. Harris, P. Lau, S.L. Chen, and G.E. Muscat, *Characterization of the retinoid orphan-related receptor-α coactivator binding interface: A structural basis for ligand-independent transcription*, Mol. Endocrin. 16 (2002), pp. 998–1012.

[227] S. Kurebayashi, T. Nakajima, S.C. Kim, C.Y. Chang, D.P. McDonnell, J.P. Renaud, and A.M. Jetten, *Selective LXXLL peptides antagonize transcriptional activation by the retinoid-related orphan receptor RORγ*, Biochem. Biophys. Res. Commun. 315 (2004), pp. 919–927.

6 The U.S. Food and Drug Administration's (FDA) Endocrine Disruptor Knowledge Base (EDKB)
Lessons Learned in QSAR Modeling and Applications

Hong Fang, Roger Perkins, Leming Shi,
Daniel M. Sheehan, and Weida Tong

CONTENTS

ABSTRACT

Considerable scientific, regulatory, and popular press attention has been devoted to the endocrine disrupting chemicals (EDCs). A larger number of potential estrogenic EDCs are associated with products regulated by the Food and Drug Administration (FDA), including plastics used in food packaging, phytoestrogens, food additives, pharmaceuticals, and cosmetics. Given the huge number of chemicals, many commercially important, and the expense of testing, Structure-Activity Relationship/ Quantitative Structure-Activity Relationship (SAR/QSAR) has been considered to be an important priority setting strategy for subsequent experimentation. At the U.S. FDA's National Center for Toxicological Research (NCTR), we conducted the Endocrine Disruptor Knowledge Base (EDKB) project, of which SAR/QSARs is a major component. We developed predictive models for estrogen and androgen receptor binding. The strengths and weaknesses of various QSAR methods were assessed to select those most appropriate for regulatory priority setting. This chapter, rather than presenting the work and results of the EDKB program in an exhaustive manner, selectively discusses salient concepts, issues, and challenges, endeavoring to achieve a tutorial outcome. In particular, concepts such as designing training sets, living models, use of QSARs in a regulatory context, predictive model validation, QSAR applicability domain, and prediction confidence estimates are among topics the authors have chosen to highlight. The concepts are presented and discussed using EDKB program results to provide qualitative and quantitative illustrations and examples. We believe the experience and lessons learned in the EDKB program will prove valuable to practitioners of QSAR should they endeavor to extend predictive systems to real-world regulatory implementations.

KEYWORDS

Applicability domain
Chance correlation
Comparative molecular field analysis (CoMFA)
Decision Forest
Endocrine Disruptor Knowledge Base (EDKB)
EDKB datasets
EDKB Web database
Four-phase approach
Model validation
Structure-Activity Relationship/Quantitative Structure-Activity Relationship (SAR/QSAR)

6.1 INTRODUCTION

Evidence that certain man-made chemicals had the ability to disrupt the endocrine systems of vertebrates by mimicking endogenous hormones had sparked intense international scientific discussion and debate [1]. A growing national concern resulted in legislation, including reauthorization of the Safe Drinking Water Act (www.epa. gov/safewater/sdwa/index.html) and passage of the 1996 Food Quality Protection

Act, mandating that the Environmental Protection Agency (EPA) develop a screening program for EDCs (www.epa.gov/endo1). Under this requirement, at least 58,000 existing chemicals would be experimentally evaluated for their potential to disrupt activities in the estrogen, androgen, and thyroid hormone systems. Some were associated with products regulated by the FDA, including plastics used in food packaging, phytoestrogens, food additives, pharmaceuticals, and cosmetics [2]. A battery of *in vitro* and short-term *in vivo* screening assays would be used to provide guidance for subsequent longer-term, more definitive *in vivo* tests for toxicity (www.epa.gov/scipoly/oscpendo/).

There was thus a huge incentive to reduce the cost and speed the screening and testing process to identify those chemicals most likely to induce adverse effects. The objective of the EDC priority setting was to rank order a large number of chemicals from most important to least important for more resource-intensive and costly experimental evaluations. Several criteria were identified for priority setting, such as production volume, persistence and fate in the environment, and human exposure levels, besides the magnitude of relevant endpoints. Most of the 58,000 chemicals without assay data were identified for screening, many of which were important industrial chemicals produced in vast quantity. QSARs were considered as the primary source of biological effect information for priority setting.

The FDA's Endocrine Disruptor Knowledge Base (EDKB) program was initiated in the mid-1990s in the wake of enormous concerns about EDCs. The time period coincided with the FDA's National Center for Toxicological Research (NCTR) reorienting its strategic goals to begin to alter the paradigm of toxicological research, taking direct aim at increasing regulatory efficiency by reducing the time, expense, and animal use in the regulation process. The refocused strategies called for development of knowledge bases and *in silico*–based predictive systems. The EDKB program remained active for many years, and the resulting data and knowledge base remain important today. Knowledge bases are computer-based systems that unify germane literature and data and provide a computational searchable tool to explore structure and toxic information.

The EDKB includes SAR, QSAR, chemometric, and consensus models, and an online structure-searchable database to assist in risk assessments and regulatory decisions for exogenous compounds that may disrupt vertebrate endocrine systems. Model training data for SAR/QSAR were obtained from *in vitro* assays conducted at the NCTR/FDA for estrogen and androgen receptor binding affinity. Chemicals for training sets were selected to obtain a broad range of structure diversity and activity. Quantitative models to predict binding affinity were developed using three-dimensional comparative molecular field analysis (CoMFA) as well as classical QSARs. Numerous qualitative SAR class predictors were also developed. A number of different models were integrated in a hierarchical manner for use in priority setting based on likelihood of activity for tens of thousands of untested chemicals.

6.2   ENDOCRINE DISRUPTOR KNOWLEDGE BASE DATABASE

6.2.1   Web Database

In the fall of 1996, a National Science and Technology Council (NSTC, 1996) report on EDCs identified a need for new databases and information systems. The report called for "a compilation of the results of chemicals in various short-term screening tests and *in vivo* assays to assist in the evaluation of their sensitivity, specificity and general predictiveness." Although these assays and tests have been performed many times by different procedures in many laboratories, the experimental results were scattered throughout the literature, making it difficult for researchers to find, compare, and evaluate relevant data. The EDKB database remedies this situation by aggregating in one easily accessible location experimental data relevant to estrogenic, androgenic, and other EDC data. This collection of experimental results for a wide variety of chemicals and species would provide raw data for modeling methodologies such as QSAR, while serving as a collaborative testing ground for the resulting predictive models.

The primary access to the EDKB database is through the Internet (http://edkb. fda.gov/databasedoor.html). EDKB Web database is a client-server application consisting of a Java front-end and an Oracle database serving as the data repository at NCTR. It has a client application that runs on a user's workstation. Java language programs allow researchers to interactively query the database from Web browsers and return the results to their desktops.

Although the EDKB database currently holds data on EDCs, it was designed to be extensible to other research areas, particularly for other toxicological endpoints where toxicity or disease is associated with chemical structure. Ultimately, the software infrastructure in an updated version may evolve to be a general repository for toxicology data, supporting data mining and meta-analysis activities, as well as the development of robust and validated predictive systems. The EDKB is now integrated with the ArrayTrack software system [3]. ArrayTrack was developed by researchers at NCTR's Center for Toxicoinformatics for a wide array of capabilities in data analysis and pathway mapping of microarray data. ArrayTrack has been made freely available to public, private, and academic users.

User statistics are shown in Figure 6.1. The EDKB database has been maintained since 1997, and is among the few databases of chemical structure and associated biological activity that is open to the public [4,5].

6.2.2 Database Contents and Features

The EDKB contains chemical structures and endocrine disruptor activities for more than 3,200 records tested in several assays. Activity data are represented by binding to the estrogen and androgen nuclear receptors, uterotropic weight gain, E-screen (cell proliferation), and reporter gene assay (Table 6.1). All data are linked to their associated citations. Activities across different assays are scaled relative to the endogenous ligand, 17β-estradiol (E_2), such that they can be viewed together in a Graphical Activity Profile (Figure 6.2). Note that in the graphical display, the biological activities for compounds are displayed first by assay type then by compound along the X-axis. The Y-axis gives relative potency on a log base 10 scale. Each bar for a particular compound and assay has maximum and minimum Y intercepts corresponding to the highest and lowest activity, respectively, measured for that compound in the same assay. Boolean query logic is supported for bibliography fields, compounds, species, and endpoints. Query

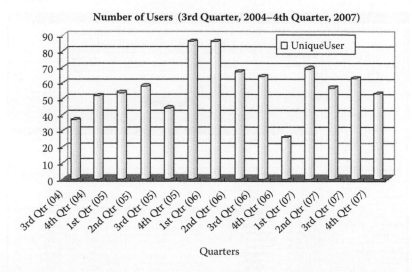

FIGURE 6.1 Endocrine Disruptor Knowledge Base (EDKB) database user statistics.

results are returned in spreadsheet format and Graphic Activity Profile forms (Figure 6.2).

The user can link each chemical to external databases, such as Toxnet, Cactus, ChemACX, Chemfind, ChemIDplus, and NCI DTP, and can search data by assay type or assay type combinations or directly search query result columns in a spreadsheet-like manner. The user can also perform chemical structure or chemical similarity searches available in the upper-left panel. In a similarity search, the 50 most similar chemicals will be reported in a spreadsheet, one compound per row, with multiple columns listing the activity information. The structure of a compound is displayed in the top-left panel when its name is clicked. The user, if interested in a particular compound, can highlight the compound and select Individual Compound from the pull-down list

TABLE 6.1
Summary of Endocrine Disruptor Knowledge Base (EDKB)

Assay Type	Number of Records	Standard Chemical to Be Compared	Endpoint
Estrogen receptor binding	616	Estradiol	RBA[a]
Androgen receptor binding	230	R1881	RBA
Uterotropic	1,707	Estradiol	RP[b]
Cell proliferation	160	Estradiol	RPP[c]
Reporter gene	544	Estradiol	RP

[a] RBA: relative binding affinity.
[b] RP: relative potency.
[c] RPP: relative proliferation potency.

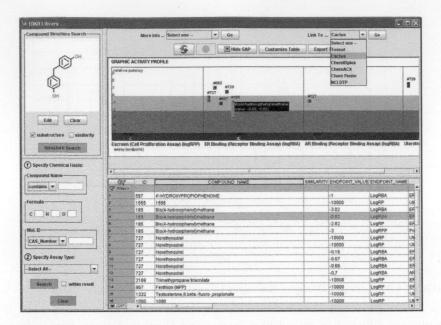

FIGURE 6.2 Endocrine Disruptor Knowledge Base (EDKB) database interface. **(See color insert following page 244.)**

of "More Info…," and a new window will pop up showing the detailed compound information.

Because a broad query could result in a huge and cumbersome output, advanced query capability is available to allow the user to lock down and re-query the database across other data within the realm of the previous query (that is, within the previous output or other tables associated with the previous output).

6.3   EDKB DATASETS

A robust QSAR model to predict activity of a wide variety of chemical structures must start with a training set that contains a sufficiently large number of chemicals with diverse structures that reflect, to some degree, the dataset to be evaluated. Despite decades of studying estrogens, the EDKB program found the existing data to be inadequate to construct sufficient models. Thus, the EDKB developed and validated a rat estrogen receptor (ER) binding assay [6,7] and a human recombinant androgen receptor (AR) binding assay [8] in order to acquire designed training sets for QSAR model development.

The ER competitive binding assay was for many years considered the gold standard. However, many variants were developed, leading to some significant differences in results. The EDKB ER binding assay was rigorously validated to assure quality data requisite for reliable model development, and each experimental value was replicated at least twice. Assays were conducted for 232 chemicals for ER and 202 chemicals for AR (called ER232 and AR202 hereafter) [6,7]. Both ER232 and AR202 contain chemicals that were selected to cover the structural diversity of

Is this chemical likely to be an ER ligand?

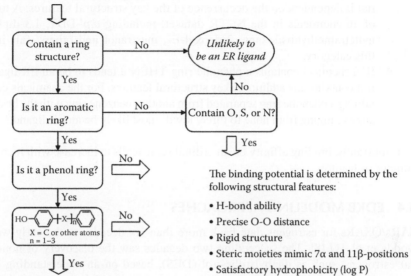

Yes, it is likely to be an ER ligand!

FIGURE 6.3 Flowchart for identification of ER ligands using a set of "IF-THEN" rules: (a) IF a chemical contains no ring structure THEN it is unlikely to be an ER ligand; (b) IF a chemical has a nonaromatic ring structure THEN it is unlikely to be an ER ligand if it does not contain an O, S, N, or other heteroatoms for H-bonding. Otherwise, its binding potential is dependent on the occurrence of the key structural features; (c) IF a chemical has a non-OH aromatic structure THEN its binding potential is dependent on the occurrence of the key structural features; and (d) IF a chemical contains a phenolic ring THEN it tends to be an ER ligand if it contains any additional key structural features. For the chemicals containing a phenolic ring separated from another benzene ring with the bridge atoms ranging from none to three, it will most likely be an ER ligand.

chemicals that bind to both receptors with an activity distribution ranging over six orders of magnitude. They have extensively been used to build and validate a series of SAR/QSAR models [9–16].

SAR evaluations of the 232 diverse chemicals revealed five structural features most important for chemical binding to ER. These findings can be generalized as a set of "IF-THEN" rules for guidance to identify potential ER ligands, depicted in Figure 6.3, and summarized as follows:

1. IF a chemical contains no ring structure THEN it is unlikely to be an ER ligand.
2. IF a chemical has a nonaromatic ring structure THEN it is unlikely to be an ER ligand if it does not contain an O, S, N or other heteroatoms for H-bonding. Otherwise, its binding potential is dependent on the occurrence of the key structural features. Kepone, dihydrotestosteron, norethynodrel, 3α- and 3β-androstanediol are active ER ligands that fall into this category.

3. IF a chemical has a non-OH aromatic structure THEN its binding potential is dependent on the occurrence of the key structural features. A total of 16 chemicals in the NCTR dataset, including o,p'-DDT, 1,3-diphenyltetramethyldisiloxane, 3-deoxyl-E_2, mestranol and others, fall into this category.

4. IF a chemical contains a phenolic ring THEN it tends to be an ER ligand if it contains any additional key structural features. For the chemicals containing a phenolic ring separated from another benzene ring with the bridge atoms ranging from none to three, it will most likely be an ER ligand.

Importantly, binding affinity on an ordinal scale tends to increase with increasing number of features.

6.4 EDKB MODELING APPROACHES

SARs/QSARs for estrogens date back more than six decades to the early work of Dodds et al. [17,18]. The succeeding two decades saw the discovery of nonsteroidal estrogens, such as diethylstilbestrol (DES), based on an understanding of the important structural features governing potency for steroidal estrogens. A number of SAR studies were reported for steroidal estrogens [19] and nonsteroidal estrogens [20]. These were generally focused on identification of structural characteristics for chemicals within similar two-dimensional (2D) structural frameworks, such as E_2 derivatives [19], DES derivatives [21], polychlorinated biphenyls (PCBs) [22], phytoestrogens [23], alkylphenols [24], raloxifenes [25], and others. Other computer-based tools have enabled the development of QSAR models to identify steric and electrostatic features of a molecule in three-dimensional (3D) space for estrogenic activity [21,26–32]. A crystallographic structure of the human ER α subtype (hERα) with a number of ligands, including E_2, DES, raloxifene and 4-OH-tamoxifene, were also reported [33,34]; aligning the four ligands based on the superposition of their ER binding sites demonstrates the common binding characteristics among them and adds the knowledge in the computational modeling process [15].

SAR/QSAR modeling employs statistical approaches to correlate and rationalize variations in the biological activity for a series of chemicals with variations in their molecular structures. The molecular structure is often represented by a set of independent variables commonly known as molecular descriptors. The Endocrine Disruptor Screening and Testing Advisory Committee (EDSTAC) considered SAR/QSAR as an important part of a proposed priority setting process [35] appropriate for use in determining priorities for further testing in biological assays [36–40]. The EDKB explored many SAR/QSAR methods, ranging from the simple structural alerts to the pharmacophore identification, classifiers, and 2D/3D QSARs.

6.4.1 QUANTITATIVE STRUCTURE-ACTIVITY RELATIONSHIPS (QSARS)

The EDKB evaluated three different techniques for QSAR modeling — CoMFA, CODESSA (COmprehensive DEscriptors for Structural and Statistical Analysis), and HQSAR (Hologram QSAR) — for their utility (predictivity, speed, accuracy, and reproducibility) to quantitatively predict ER binding activity [15,29]. Common to the three QSAR methods is the use of the partial least squares (PLS) regression; the differences among these QSAR techniques are primarily in the type of the descriptors used to represent chemical structure. Specifically, CoMFA employs steric and electrostatic field descriptors that encode 3D intermolecular interaction information. CODESSA calculates molecular descriptors on the basis of 2D and 3D structures and quantum-chemical properties, whereas HQSAR uses molecular holograms constructed from counts of substructural molecular fragments. For three relatively small datasets under investigation, the QSAR models generated using CoMFA and HQSAR techniques demonstrated comparable high quality for potential usage to identify ER ligands [29]. CoMFA and HQSAR were further investigated and compared, particularly for their predictivity, by using the NCTR ER232 dataset and two other test sets [15]. CoMFA performed better for the training set as well as for predicting two different test sets.

To develop a CoMFA model, the molecules of interest must first be aligned to maximize superposition of their steric and electrostatic fields. Although a statistically robust CoMFA model is dependent on a number of factors, proper alignment is essential to produce a valid QSAR model. For chemical congeners, the alignment rule is typically defined based on the maximum common substructure among the training set chemicals, which usually leads to a statistically robust CoMFA model [13,15]. The drawback for such models is the unreliability in predicting activities of chemicals whose structures are not similar to the training set, like some previously reported models [21,26–30,41–43]. In contrast, a CoMFA model based on a structurally diverse dataset provides more robust predictions, but alignment can be exceedingly difficult. CoMFA models developed in the EDKB program for ER and AR binding were particularly challenging due to the difficulty in choosing alignment rules, as active ligands have greatly differing structures. Fortunately, crystal structures of four ligands binding to the ER were published [33,34] and are available to guide derivation of rational CoMFA alignment rules. The resultant CoMFA model for ER binding based on the crystal structure-guided alignment met requirements for statistical robustness with conventional r^2 and cross-validated q^2_{LOO} of 0.91 and 0.66, respectively, indicative of both internal consistency and high predictiveness. The CoMFA-calculated versus experimental relative binding affinities (RBAs) (as logs) for the NCTR training set chemicals are plotted in Figure 6.4.

6.4.2 CLASSIFICATION MODELS (OR CLASSIFIERS)

Classification can be either supervised or unsupervised learning techniques. Unless specifically mentioned, classification models discussed in this chapter are associated with the supervised technique that provides categorical prediction (for example, binary active or inactive classification). A number of classification methods were evaluated to categorize chemicals as ER binders or nonbinders. Although the methods differ in a number of ways, they generally produce similar results [14,16,44]. It was found for EDKB models that the nature of the descriptors used, and more specifically the effectiveness with which descriptors encode the structural features

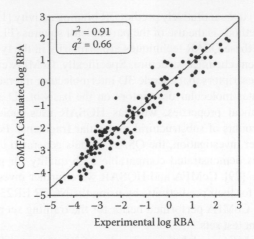

FIGURE 6.4 Plot of comparative molecular field analysis (CoMFA)-calculated log relative binding affinity (log RBA) versus experimental log RBA for the National Center for Toxicological Research (NCTR) dataset. $r^2 = 0.91$; $q^2 = 0.66$.

of the molecules related to the ER binding activity, is far more important than the specific method employed. The selection of biologically relevant descriptors is thus a critical step to develop a robust classification model.

Additionally, the Genetic Algorithm was found to be the preferred method to identify the most biological relevant descriptors from among a large set of descriptors. For example, using the best 10 descriptors selected by the Genetic Function Approximation approach [45,46] from among 153 descriptors, a decision tree (DT) model consisting of 5 meaningful descriptors was constructed:

- The Phenolic Ring Index indicates the presence or absence of the phenolic ring in a chemical, which is considered to be the most important structural feature for ER binding [19].
- The Shadow-XY fraction is a geometric descriptor related to the breadth of a molecule [47]; this is consistent with our observation that substitution of a bulky group at the 7α and 11β positions of E_2 increased the breadth of a chemical and enhanced ER binding [48].
- The log P, Jurs-PNSA-2, and Jurs-RPCS reflect the hydrophobicity (log P) [49] and charged surface area (Jurs-PNSA-2, Jurs-RPCS) [50] that are, in principle, critical for all receptor-mediated activity.

The DT model combining the five primary descriptors is summarized in Figure 6.5. The model identified the Phenolic Ring Index as the most important descriptor for ER binding. If chemicals contained a phenolic moiety but also had log P values larger than 1.49, they were more likely to be ER binders. In contrast, chemicals without a phenolic moiety were less likely to be ER binders unless they had relatively large hydrophobicity and charged surface area, and breadth of the structure.

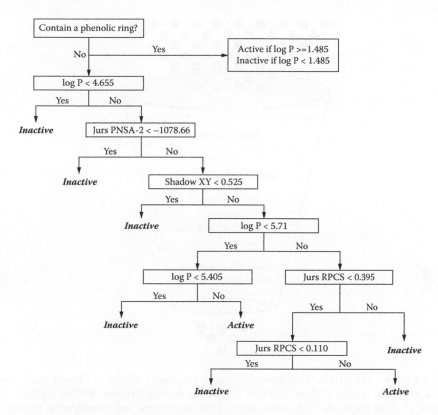

FIGURE 6.5 Decision Tree model. The model displays a series of YES/NO (Y/N) rules to classify chemicals into active (A) and inactive (I) categories based on five descriptors: Phenolic Ring Index, log P, Jurs PNSA-2, Shadow XY, and Jurs RPCS. The squares represent the rules and the circles represent the categorical results.

6.4.3 FDA's NATIONAL CENTER FOR TOXICOLOGICAL RESEARCH (NCTR) "FOUR-PHASE" SYSTEM

The SAR/QSAR approaches described above have strengths and weaknesses, and they all produce a degree of prediction error. All models and particularly those that only provide active/inactive predictions can be optimized to minimize either the overall prediction error or the false negative or positive rate. Decreasing false negatives is achieved at a cost of increasing false positives and vice versa. Because selecting an appropriate single model is problematic, we adopted an approach of rationally combining different QSAR models into a sequential "Four-Phase" scheme according to the strength of each type of model. A progressive phase paradigm is used to screen out chemicals and thus reduce the number of chemicals to be considered in each subsequent phase. The four phases work in a hierarchical manner, incrementally reducing the size of a dataset while increasing precision of prediction during each phase. Within each phase, different models are selected to work complementarily

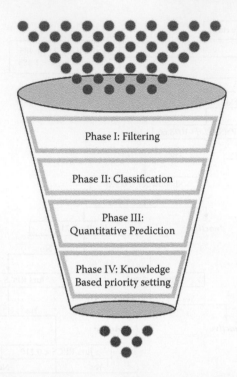

FIGURE 6.6 The NCTR "Four-Phase" approach for priority setting. In Phase I, chemicals with molecular weight <94 or >1,000 or containing no ring structure will be rejected. In Phase II, three approaches (structural alerts, pharmacophores, and classification methods) that include a total of 11 models are used to make a qualitative activity prediction. In Phase III, a three-dimensional (3D) Quantitative Structure-Activity Relationship (QSAR)/comparative molecular field analysis (CoMFA) model is used to make a more accurate quantitative activity prediction. In Phase IV, an expert system is expected to make a decision on priority setting based on a set of rules. Different phases are hierarchical; different methods within each phase are complementary.

in representing key activity-determining structure features in order to minimize the rate of false negatives.

The overall architecture of the NCTR "Four-Phase" system for identification of ER ligands is illustrated in Figure 6.6:

- *Phase I: Filtering* — Two filters, the molecular weight range and ring-structure indicator, were selected to efficiently eliminate chemicals very unlikely to have ER binding activity. The Nishihara dataset [51] was selected to evaluate the performance of these two filters. The Nishihara dataset contained 517 chemicals tested with the yeast two-hybrid assay, of which over 86% were pesticides and industrial chemicals, consistent with the intended application of the NCTR models. The two rejection filters correctly eliminated 98 inactive chemicals from the Nishihara dataset with no false negatives. The data size was reduced some 21%. This suggests that, for real-world applications, the two rejection filters significantly reduce the number of chemicals for further evaluation with a minimum risk of introducing false negatives.

- *Phase II: Active/Inactive Assignment* — This phase categorizes chemicals from Phase I as either active or inactive. The three structural alerts, seven pharmacophore queries, and the DT classification model discussed above were used in parallel to discriminate between active and inactive chemicals. To ensure the lowest possible false negative rate in Phase II, a chemical predicted to be active by any of the 11 models is presumed active and subsequently evaluated in Phase III, and only those predicted to be inactive by all these models are deemed inactive and eliminated for further evaluation. Because structural alert, pharmacophore, and DT methods incorporate and weigh differently the various structural features that endow a chemical with the ability to bind the ER, the combined outputs derived from the three approaches are complementary in minimizing false negatives. All active chemicals in the NCTR, Waller [27], Kuiper [52], and Nishihara [51] datasets were identified by combining the 11 models.
- *Phase III: Quantitative Predictions* — In Phase III, the CoMFA model was used to make a quantitative activity prediction for chemicals categorized as active in Phase II. Chemicals with higher predicted binding affinity are ranked at higher priority for further evaluation in Phase IV. The CoMFA model demonstrated good statistical reliability in both cross-validation and external validation [15].
- *Phase IV: Rule-Based Decision-Making System* — In this final stage of the integrated system, we believe that a set of rules needs to be developed as a knowledge-based or expert system to make a priority setting decision. The system is useful only after incorporating accumulated human knowledge and expertise (that is, rules). This system can make decisions on individual chemicals based on the rules in its knowledge base. Computational chemists, toxicologists, and regulatory reviewers should jointly develop and define the rules. The following are suggestions for such rules:
  - Special attention needs to be placed on the following chemicals, which may need to be reevaluated by assaying or modeling according to the "IF-THEN" scheme depicted in Figure 6.3 [48]:
    - The chemical is predicted to be inactive, but its structure is modified during structural preprocessing.
    - The chemical has a structure that is dissimilar to all those that have been used to train and test the models.
    - The chemical is active in Phase II but inactive in Phase III.
  - Information on the level of human exposure and production, environmental fate, and other public health–related parameters can be used independently or can be jointly incorporated for priority setting.

The NCTR "Four-Phase" system has been validated by a number of existing datasets, including the E-SCREEN assay data [53], the yeast two-hybrid reporter gene assay data [51], and other datasets [24,27,54–57].

6.5 EDKB MODEL VALIDATION STRATEGY

The goodness-of-fit of a model based on the training set can be assessed using various statistical measures. Concordance, specificity, and sensitivity [14] are commonly used quality metrics of a classification model, and a quantitative regression (or nonlinear) model is often assessed using r^2 (the correlation coefficient) [26]. A qualitative or quantitative model is generally deemed statistically significant if concordance or $r^2 \geq 0.9$, respectively, a result not difficult to obtain within the training set if the dataset is of adequate size and quality. However, a QSAR's model purpose is not predicting what is known, but what is unknown. Accurate prediction of the training set might be a necessary condition for accurate prediction of unknown chemicals, but it is not a sufficient condition. Validation of the model's accuracy in predicting untested chemical activity is essential, for which several commonly used approaches are described below.

6.5.1 CROSS-VALIDATION

Cross-validation is widely employed in QSAR modeling to obtain an estimate of prediction accuracy outside the training set. However, modelers should be mindful that cross-validation tends to give overly optimistic estimates of prediction accuracy because, in fact, the entire training set is used for the estimate. In cross-validation a fraction of chemicals in the training set are excluded, and then their activity is predicted by a new model developed from the remaining chemicals. The process has to be repeated until all chemicals in the training set are left out at least once. When each chemical is left out one at a time, and the process repeated for each chemical, this is known as leave-one-out (LOO) cross-validation. A quantitative model with a value of cross-validated $q_{LOO}^2 > 0.5$ is normally considered to possess at least a marginally statistically significant predictive ability. If the training set is divided into N groups with approximately equal numbers of chemicals, the process is called leave-N-out (LNO) cross-validation. Generally, it becomes increasingly difficult to attain $q2 > 0.5$ with increasing N.

The EDKB CoMFA model for ER binding showed a $q_{LOO}^2 = 0.66$, demonstrating a high predictive capability within the training set [15]. In addition to LOO, a more thorough LNO validation was conducted for this CoMFA model. Unlike q_{LOO}^2 that is a single value from the process, q_{LNO}^2 varies for a selected N group in each run because of the random nature of selection of chemicals in the process. Doing LNO 100 times for each random N groups of chemicals ($N = 2$–10, 13, 20, and 50) gave $q_{LNO}^2 > 0.5$ even for the worst case of randomly leaving out 50% ($N = 2$) of the training chemicals, lending confidence that the CoMFA model is reliable for the prediction of structurally diverse chemicals (Figure 6.7).

For the ER classification models, in addition to using LNO, another validation procedure was used to assess the model's predictive capability. In this "out-of-bag" procedure, the dataset was randomly divided into two portions, two-thirds for training and one-third for testing. The classification models were constructed using the training portion and were subsequently used to predict the test portion. The random process was repeated 2,000 times. The validation results for the DT classification

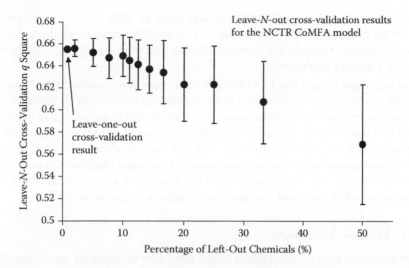

FIGURE 6.7 Leave-*N*-out cross-validation results for CoMFA.

model are shown in Figure 6.8. In the training step, a narrowed distribution was observed in a concordance range of 89% to 99%, indicating that all 2,000 models show comparable results in measuring overall accuracy. In contrast, a much wider distribution of concordance was found in prediction for 2,000 test sets. The difference

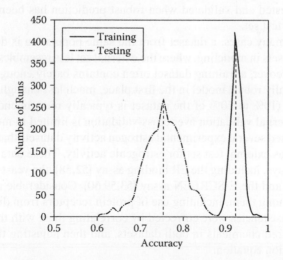

FIGURE 6.8 An extensive cross-validation procedure to validate the Decision Tree classification model. In this method, the National Center for Toxicological Research (NCTR) dataset was divided into two groups, two-thirds for training and one-third for testing. The process was repeated 2,000 times. Concordance was calculated based on the misclassifications divided by the number of training chemicals for the training models and the misclassifications divided by the testing chemicals for prediction.

between the best and worst prediction was as large as 30%, or three times greater than for the training set. The results indicate that the differences in widths and peaks of the concordance distribution between training and testing might be another way to assess a model's predictivity.

It is important to note that LOO and LNO and out-of-bag methods test the stability of the model through perturbation of the correlation coefficients by consecutively omitting chemicals. The methods assess only the internal extrapolation in the training set and have limited indication in predicting untested chemicals, especially for the chemicals with structural disparity with the training chemicals. More specifically, because cross-validation incorporates all of the same chemicals in both training and predicting steps, its estimates are biased and tend to overstate accuracy that will be observed in groups of chemicals not explicitly in the training set.

6.5.2 EXTERNAL VALIDATION

When additional data are available, a model should be validated by predicting other chemicals not used in the training set but whose activities are known (an external test set), a process called external validation. The major differences between cross-validation and external validation are (1) chemicals selected in the latter case are random in a sense, which provides a more rigorous evaluation of the model's predictive capability for untested chemicals than does cross-validation; (2) the external test is not biased by the test set chemicals that are excluded from training; and (3) the external set can be obtained from different populations, serving a better validation means. We feel strongly that the confidence in a model's predictive capability can and should be tested and validated when robust prediction has been demonstrated with an external test set.

However, in many cases, a dataset from the same population is divided into the training and test sets in modeling, where the external set less resembles the real-world application. Moreover, a training dataset often contains barely enough chemicals to create a statistically robust model in the first place, much less enough to set aside an external test set (10% to 20% of the dataset is typically recommended). Thus, the advantage of external validation over cross-validation is limited in most situations.

The EDKB used several experimental estrogen activity datasets that were reported in the literature as external test set for estrogenic activity. These data were obtained from *in vitro* assays, including the ER binding assay [52,58], the yeast-based reporter gene assay [54], and the E-SCREEN assay [53,59,60]. Considerable activity disparity was found among them, including use of protein receptors from different species [61]. Some inconsistencies were corrected by correlating them with the NCTR dataset based on shared chemicals in both datasets, and then adjusting the test set data with the correlation equation.

Another important consideration in selecting test set chemicals is that they are pertinent to the intended use of the model. Because the EDKB models were being developed to predict the activity of environmental chemicals, a dataset reported by Nishihara et al. [51] was also selected as a test set. After structure preprocessing, 463 chemicals remained for a test set, of which 62 chemicals were measured to be active at 10% of $10^{-7}M\ E_2$ [51]. The majority of the chemicals were inactive, which was similar to the real-world situation, not training set distribution where inactive chemicals were expected to dominate.

It is important to note that the yeast two-hybrid assay used for the Nishihara dataset differed from the NCTR binding assay. The ER competitive binding assay measured the binding affinity of a chemical for ER, and the yeast two-hybrid assay measured ER binding–dependent transcriptional and translational activity. Thus, the two assays differed in their sensitivity to distinguish active from inactive chemicals, particularly for weak estrogens and antiestrogens [61]. When assay results for 80 chemicals common to both the Nishihara and NCTR datasets were compared, the assays disagreed for 12 chemicals. Specifically, of 30 active chemicals in the Nishihara dataset, one chemical was inactive in the NCTR dataset; and of 50 inactive chemicals in the Nishihara dataset, 11 chemicals were active in the NCTR dataset. These observations showed that using a model based on EDKB ER binding training set to predict the experimental results from the yeast two-hybrid assay (the Nishihara dataset) would likely result in about 15.0% (12/80) disparity with a 3.3% (1/30) false negative rate and a 22% (11/50) false positive rate. The result amplifies the need for caution in applying models based on one experimental construct to a markedly different experimental construct [14].

6.5.3 Assessment of Chance Correlation

Testing whether a fitted QSAR model is, in fact, a chance correlation is highly recommended. Testing becomes increasingly imperative for smaller training datasets, with increasing numbers of descriptors, with increasing noise in biological data, and with an increasing skew of numbers of chemicals across activity categories. All of these conditions increase the omnipresent risk of obtaining a chance correlation lacking predictive value.

The randomization test is an effective means of evaluating the potential that a correlation that appears good is, in fact, a chance result, and thus useless for prediction. This requires generation of many pseudo datasets (for example, a few thousand) where the activity class is randomly scrambled across all chemicals in the training set. Next, N-fold cross-validation is done for each pseudo dataset. The null distribution — that is, the distribution of prediction accuracy for all pseudo datasets — can then be compared with the distribution of multiple N-fold cross-validation results derived from the real dataset. The degree of chance correlation in the predictive model can be estimated from the overlap of the two distributions. Figure 6.9 shows the results of a test for chance correlation of Decision Forest (DF) classification models to predict liver carcinogenicity based on four different datasets. Although high cross-validation results were obtained for the models for all four datasets (Table 6.2), there was a significant overlap between the null and real distribution for each dataset (Figure 6.9), indicative of an expected high degree of chance correlations for these models.

In another example, a comparison of the null distribution for 2,000 pseudo datasets with real distribution based on 2,000 runs of tenfold cross-validation for ER binding for the EDKB ER232 training dataset is shown in Figure 6.10. The distribution of prediction accuracy of the real datasets centers is about 82%, and

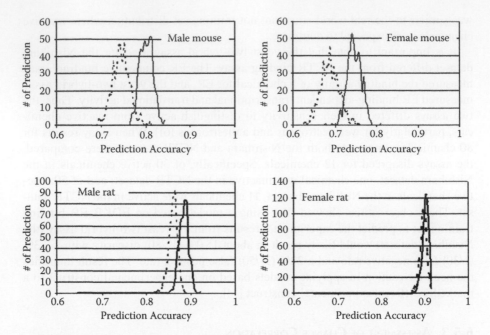

FIGURE 6.9 Assessment of the chance correlation in Decision Forest (DF) for four datasets listed in Table 6.2. For each graph, the null distribution (-------) is generated from the results of tenfold cross-validation on 2,000 pseudo datasets, and the real distribution (———) is derived from 2,000 runs of tenfold cross-validation for the original dataset.

the pseudo datasets are about 50%. The distribution for the real dataset is also much narrower than for the pseudo datasets. Thus, cross-validation shows consistent and high prediction accuracy for the real datasets, whereas the pseudo dataset results vary widely, indicative of the expected large variability of signal/noise ratio. Importantly, there is no overlap between two distributions, indicating that a statistically and biologically relevant DF model could be developed from the real dataset.

TABLE 6.2
Decision Forest Tenfold Cross-Validation Results for Four Liver Carcinogenicity Datasets Obtained from Testing on Two Species, Rat and Mouse, for Both Sexes

Datasets	Number of Compounds	Number of Carcinogens	Number of Noncarcinogens	Cross-Validation Accuracy (%)
Female mouse	247	60	187	74.6
Male mouse	241	48	193	80.1
Male rat	230	28	202	88.5
Female rat	237	21	216	89.8

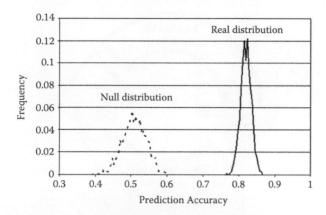

FIGURE 6.10 Assessment of the chance correlation in Decision Forest (DF) for ER232, an estrogenic dataset that contains 232 chemicals tested in an estrogen receptor binding assay. The same assessment described in Figure 6.8 was used in this test.

6.5.4 LIVING MODEL CONCEPT

No matter how rigorous the validation procedure for a model, it is important to realize that the model may give incorrect predictions for some chemicals because the entire chemistry space of active chemicals is unknown. Thus, any QSAR model could be considered as a *living* model that could be improved when new data become available, and the model can aid in deciding what new data are needed.

Such a living model concept views modeling as a recursive process that expands the chemistry space of the training set by alternately incorporating new data in the model and then using the model to choose new chemicals for assay [62,63]. As depicted in Figure 6.11, the process starts with an initial set of chemicals for QSAR modeling [26,28,29]. Next, the preliminary QSAR models are used prospectively to define and rationalize a set of chemicals that may further improve the model's robustness and predictive capability. These new chemicals are assayed, and the data are then used to first challenge and then to refine the QSAR model.

Several benefits accrue from collaborative integration of the experimental and modeling efforts. Immediate feedback can be given to the experimentalists so that suspected assay problems can be rapidly investigated. Also, as the models evolve, the modelers can select the chemicals for subsequent testing, based on considerations of structural diversity and activity range. Each new assay data point directly from the laboratory becomes a challenge to the evolving model; the result is either further confirmation of its validity, identification of a limitation, or an outlier prediction. Failure of the model also provides important information, such as identification of the need for new data based on a rational understanding of the dependence of activity on structure. Regardless of the cause of model failure, a research hypothesis is spawned in each iteration, which should lead to new data and an improved training set, and an improvement to the living model. Despite the obvious benefits, QSAR history has few precedents outside the pharmaceutical industry for sustaining joint modeling and experimental efforts for multiple iterations.

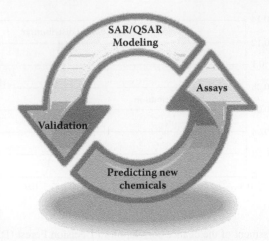

FIGURE 6.11 Depiction of the recursive process used by National Center for Toxicological Research (NCTR) to develop Quantitative Structure-Activity Relationship (QSAR) models for predicting estrogen receptor (ER) binding. The process starts with an initial set of chemicals from the literature for QSAR modeling. Next, the preliminary QSAR models are used prospectively to define a set of chemicals that will further improve the model's robustness and predictive capability. The new chemicals are assayed, and these data are then used to challenge and refine the QSAR models. Validation of the model is critical. The process emphasizes the living model concept.

6.6 APPLICABILITY DOMAIN

As demonstrated in the previous section, QSAR models need to be validated either by cross-validation or external test sets. Assessment of prediction accuracy compared with random chance is another important aspect of validation. Generally speaking, training data of small number, with skewed activity distributions and low signal to noise are most vulnerable to chance solutions, especially with a larger number of independent variables (for example, chemical structure descriptors).

Unfortunately, in both model validation and chance correlation assessment, it is not typical and often not possible to specify accuracy and prediction confidence for specific unknown chemicals with structures not well represented in the training set. It is also not typical and often not possible to specify accuracy and prediction confidence for specific unknown chemicals with structures requiring that the model extrapolate beyond the chemistry space determined by the training set. Neither cross-validation nor randomization testing assess the ability of the model to extrapolate beyond the chemistry space. The chemistry space where predictions for unknown chemicals can be ascribed as reliable is known as the model's applicability domain and is an important concept discussed in this section.

"Applicability domain" is a term often used to define an abstract response and chemical structure space wherein a model's predictions will have known reliability. In practice, applicability domain means different things to different people, and a specific definition is feasible only for a specific method or model. Related to applicability domain is the chemistry space determined by the training set, and this should be assessed in conjunction with prediction accuracy to validate the limitations of a model. Because a single simple definition of applicability domain is not feasible, and its conceptual purpose is in model quality, applicability domain might best be viewed for measures of confidence in each prediction when the overall quality of a model is acceptable.

Tong et al. [12] describe an approach to assess a QSAR's applicability domain with two variables: (1) prediction confidence (that is, certainty for an individual chemical's prediction) and (2) domain extrapolation (that is, the prediction accuracy for a chemical that is outside the chemistry space defined by the training chemicals). The approach is demonstrated using the DF consensus modeling method [12,44]. DF uses the consensus prediction of multiple, comparable, and heterogeneous decision trees. The critical, implicit assumption in consensus modeling is that multiple models will effectively identify and encode more aspects of the SAR relationship than will a single model. DF attempts to minimize overfitting by maximizing the difference among individual trees in order to cancel some random noise when combining the trees. Although applicability domain extrapolation can theoretically be defined for any QSAR method, the DF method has distinct advantages in specifying the applicability domain in terms of prediction confidence and domain extrapolation, and it is thus used here to illustrate the approach.

Prediction confidence is a measure of the certainty of prediction for a specific chemical. Prediction confidence is probabilistically calculated in DF for each unknown chemical by averaging the predictions over all trees that are combined to form the model. Figure 6.12 gives an example illustrating how prediction accuracy and prediction confidence are related. Prediction accuracy is plotted versus prediction confidence for both DF and DT for a problem with 2,000 runs of tenfold cross-validation for the EDKB ER232. A strong trend of increasing accuracy with

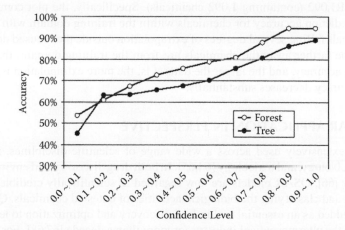

FIGURE 6.12 Decision Forest prediction accuracy versus confidence level for ER232 based on 2,000 runs of tenfold cross-validation.

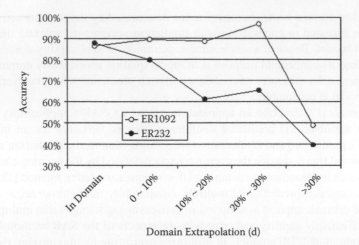

FIGURE 6.13 Decision Forest prediction accuracy versus domains extrapolation for ER232 and ER1092 based on 2,000 runs of tenfold cross-validation. Domain extrapolation (*d*) for a chemical is defined as a percentage away from the training domain, and the prediction accuracy for the domain *d* is calculated by dividing correct predictions by total number of chemicals in this domain.

increasing confidence is readily apparent for both DF and DT, as is the substantially higher accuracy for DF across the entire range of confidence levels.

Domain extrapolation is the extent of extrapolation beyond the training domain for a specific chemical. For DF, it is probabilistically calculated as the average Euclidian distance that an unknown chemical's descriptors in tree paths are outside of the range of those same descriptors across all chemicals in the training set that determines the applicability domain. Figure 6.13 shows the results of evaluation of DF domain extrapolation for two EDKB estrogen receptor binding datasets, ER232, and a larger ER1,092 (containing 1,092 chemicals). Specifically, the plot compares the overall prediction accuracy for chemicals within the training domain with accuracy for chemicals falling several degrees of extrapolation outside the focused domain. In general, the farther away the chemicals are from the training domain, the less the prediction accuracy, and the larger the dataset is, the more extrapolation is tolerated before accuracy decreases substantially.

6.7 QSAR APPLICATION IN PERSPECTIVE

QSAR is extensively used across a wide range of scientific disciplines, including chemistry, biology, and toxicology [64,65]. In both drug discovery and environmental toxicology [66], QSAR models are now regarded as scientifically credible tools for predicting and classifying the biological activities of untested chemicals. QSAR has been imbedded as an essential tool in lead discovery and optimization to lead development in the pharmaceutical industry for more than a decade [67,68]. For example, QSAR is often used early in the drug discovery process as a screening and enrichment tool to eliminate from further development those chemicals lacking drug-like properties [69], or those chemicals predicted to elicit a toxic response. This paradigm portends the spread of QSAR beyond the pharmaceutical industry to human and environmental regulatory authorities for use in toxicology [15,70–75].

The incorporation of QSAR and related computer-based techniques in the drug discovery paradigm is a simple function of economics: if there is an economic benefit, it is used. This is not so in the regulatory paradigm that has minimally changed regarding how efficacy, toxicity, and risk-benefit are evaluated. The regulatory paradigm is anchored in law and precedent, raising the bar for incorporation of QSAR techniques. Decisions are largely adjudicated in animal studies and clinical trials with confidence determined statistically. The results are taken as definitive, by law and precedence, despite known limitations of experimentation and statistics, some of which are low sensitivity, insufficient statistical power to detect sensitive populations, and shortcomings of cross-species extrapolations.

Wider acceptance of QSARs in regulation could result in a constellation of benefits and savings to both the private and public sectors. One way to gain greater acceptance would be the ability to be more definitive about a model's accuracy and limitations for a specific chemical, as opposed to a statistical accuracy across a pool of chemicals such as a blinded, external test set. If a model achieves high accuracy (based on, for example, sensitivity and specificity) for a certain test set, is it justifiable to assert that the same accuracy can be assumed for the next unknown chemical? Defining the model's applicability domain in terms of prediction confidence and domain extrapolation takes a long stride toward being able to ascribe a definitive accuracy (probabilistic) to a specific untested chemical's prediction [44].

As a last thought, caution must be taken against the temptation to overinterpret the SAR/QSAR results. The inherent limitations of SAR/QSAR have to be kept in mind when it becomes the major component and application of a project. For example,

- It is important to point out that any QSAR model will produce some degree of error. This is partially due to the inherent limitation to predict a biological activity solely based on chemical structure. One can argue from the principles of chemistry that the molecular structure of a chemical is a determinate of its physicochemical properties and, ultimately, its biological activity and influence on organisms. Because molecular structure and physicochemical properties are associated with the chemical, the relationship between structure and physicochemical properties should be apparent and, therefore, accessible using QSARs. In contrast, the biological activity of a chemical is an induced response influenced by numerous factors dictated by the level of biological complexity of the system under investigation. The relationship between structure and activity is thus more implicit and thereby poses a more challenging problem in QSAR applications.

- In principle, a chemical can be represented in three distinct but also related structural representations: 2D substructures, 3D pharmacophores, and physicochemical properties. If a biological mechanism is mainly related to the chemical structure, which is probably the case for receptor binding, informative QSARs are possible using structural features. It is often found, however, that even for a simple mechanism such as ER binding, some features may well represent binding dependencies for one structural class, and other features will better represent binding dependencies for a different structural class [48]. In such cases, caution must be taken in interpreting QSAR results for the chemical classes that are not well represented in the training set.

- Despite widespread use of and success with QSARs, developers and users of them should always be mindful that predictions from any model are intrinsically no better than the experimental data employed for modeling. Any limitations of the assay used to generate the training data apply equally to the model's predictions. Errors tend to increase toward the limits of sensitivity of experimental assays, and these errors will be conveyed to a QSAR model. The increased experimental error in close proximity to the sensitivity limit will be transferred to the QSAR model, which, in turn, will increase false predictions for chemicals in the lower activity region.

ACKNOWLEDGMENTS

The authors gratefully acknowledge funding from the FDA's Office of Women's Health that enabled the NCTR Endocrine Disruptor Knowledge Base project to begin and make early rapid progress. The American Chemistry Council (ACC), through a Cooperative Research and Development Agreement (CRADA), supported the estrogen and androgen receptor competitive binding assays and 3D QSAR model development. Finally, through an Interagency Agreement (IAG), the Environmental Protection Agency (EPA) provided support for predictive model development and validation.

REFERENCES

[1] R.J. Kavlock, G.P. Daston, C. DeRosa, P. Fenner-Crisp, L.E. Gray, S. Kaattari, G. Lucier, M. Luster, M.J. Mac, C. Maczka, R. Miller, J. Moore, R. Rolland, G. Scott, D.M. Sheehan, T. Sinks, and H.A. Tilson, *Research needs for the risk assessment of health and environmental effects of endocrine disruptors: A report of the U.S. EPA-sponsored workshop*, Environ. Health Perspect. 104 Suppl 4 (1996), pp. 715–740.

[2] W. Tong, R. Perkins, H. Fang, H. Hong, Q. Xie, S.W. Branham, D.M. Sheehan, and J.F. Anson, *Development of Quantitative Structure-Activity Relationships (QSARs) and their use for priority setting in the testing strategy of endocrine disruptors*, Regul. Res. Perspect. 1 (2002), pp. 1–16.

[3] W. Tong, X. Cao, S. Harris, H. Sun, H. Fang, J. Fuscoe, A. Harris, H. Hong, Q. Xie, R. Perkins, L. Shi, and D. Casciano, *ArrayTrack — Supporting toxicogenomic research at the U.S. Food and Drug Administration National Center for Toxicological Research*, Environ. Health Perspect. 111 (2003), pp. 1819–1826.

[4] A.M. Richard and C.R. Williams, *Distributed structure-searchable toxicity (DSSTox) public database network: A proposal*, Mutat. Res. 499 (2002), pp. 27–52.

[5] A.M. Richard, L.S. Gold, and M.C. Nicklaus, *Chemical structure indexing of toxicity data on the Internet: Moving toward a flat world*, Curr. Opin. Drug Discov. Devel. 9 (2006), pp. 314–325.

[6] R. Blair, H. Fang, W.S. Branham, B. Hass, S.L. Dial, C.L. Moland, W. Tong, L. Shi, R. Perkins, and D.M. Sheehan, *Estrogen receptor relative binding affinities of 188 natural and xenochemicals: Structural diversity of ligands*, Toxicol. Sci. 54 (2000), pp. 138–153.

[7] W.S. Branham, S.L. Dial, C.L. Moland, B.S. Hass, R.M. Blair, H. Fang, L. Shi, W. Tong, R.G. Perkins, and D.M. Sheehan, *Phytoestrogens and mycoestrogens bind to the rat uterine estrogen receptor*, J. Nutr. 132 (2002), pp. 658–664.

[8] H. Fang, W. Tong, W.S. Branham, C.L. Moland, S.L. Dial, H. Hong, Q. Xie, R. Perkins, W. Owens, and D.M. Sheehan, *Study of 202 natural, synthetic, and environmental chemicals for binding to the androgen receptor*, Chem. Res. Toxicol. 16 (2003), pp. 1338–1358.

[9] R. Serafimova, M. Todorov, D. Nedelcheva, T. Pavlov, Y. Akahori, M. Nakai, and O. Mekenyan, *QSAR and mechanistic interpretation of estrogen receptor binding*, SAR QSAR Environ. Res. 18 (2007), pp. 389–421.

[10] J. Devillers, N. Marchand-Geneste, A. Carpy, and J.M. Porcher, *SAR and QSAR modeling of endocrine disruptors*, SAR QSAR Environ. Res. 17 (2006), pp. 393–412.

[11] T. Ghafourian and M.T. Cronin, *The impact of variable selection on the modelling of oestrogenicity*, SAR QSAR Environ. Res. 16 (2005), pp. 171–190.

[12] W. Tong, Q. Xie, H. Hong, L. Shi, H. Fang, and R. Perkins, *Assessment of prediction confidence and domain extrapolation of two structure-activity relationship models for predicting estrogen receptor binding activity*, Environ. Health Perspect. 112 (2004), pp. 1249–1254.

[13] H. Hong, H. Fang, Q. Xie, R. Perkins, D.M. Sheehan, and W. Tong, *Comparative molecular field analysis (CoMFA) model using a large diverse set of natural, synthetic and environmental chemicals for binding to the androgen receptor*, SAR QSAR Environ. Res. 14 (2003), pp. 373–388.

[14] H. Hong, W. Tong, H. Fang, L.M. Shi, Q. Xie, J. Wu, R. Perkins, J. Walker, W. Branham, and D. Sheehan, *Prediction of estrogen receptor binding for 58,000 chemicals using an integrated system of a tree-based model with structural alerts*, Environ. Health Perspect. 110 (2002), pp. 29–36.

[15] L.M. Shi, H. Fang, W. Tong, J. Wu, R. Perkins, R. Blair, W. Branham, and D. Sheehan, *QSAR models using a large diverse set of estrogens*, J. Chem. Inf. Comput. Sci. 41 (2001), pp. 186–195.

[16] L.M. Shi, W. Tong, H. Fang, R. Perkins, J. Wu, M. Tu, R. Blair, W. Branham, J. Walker, C. Waller, and D. Sheehan, *An integrated "Four-Phase" approach for setting endocrine disruption screening priorities — Phase I and II predictions of estrogen receptor binding affinity*, SAR QSAR Environ. Res. 13 (2002), pp. 69–88.

[17] E. Dodds, L. Goldberg, and W. Lawson, *Oestrogenic activity of alkylated stilboestrols*, Nature 142 (1938), p. 34.

[18] E. Dodds and W. Lawson, *Molecular structure in relation to oestrogenic activity. Compounds without a phenanthrene nucleus*, Proc. Royal Soc. 125 (1938), pp. 222–232.

[19] G.M. Anstead, K.E. Carlson, and J.A. Katzenellenbogen, *The estradiol pharmacophore: Ligand structure-estrogen receptor binding affinity relationships and a model for the receptor binding site*, Steroids 62 (1997), pp. 268–303.

[20] J. Dodge and C. Jones, *Non-steroidal estrogens,* in Oettel, M.a.S., E., ed., Springer-Verlag, Heidelberg, 1999.

[21] B.R. Sadler, S.J. Cho, K.S. Ishaq, K. Chae, and K.S. Korach, *Three-dimensional quantitative structure-activity relationship study of nonsteroidal estrogen receptor ligands using the comparative molecular field analysis/cross-validated r^2-guided region selection approach*, J. Med. Chem. 41 (1998), pp. 2261–2267.

[22] K. Korach, P. Sarver, K. Chae, J. McLachlan, and J. McKinney, *Estrogen receptor-binding activity of polychlorinated hydroxybiphenyls: Conformationally restricted structural probes*, Mol. Pharmacol. 33 (1988), pp. 120–126.

[23] R.J. Miksicek, *Interaction of naturally occurring nonsteroidal estrogens with expressed recombinant human estrogen receptor*, J. Steroid Biochem. Mol. Biol. 49 (1994), pp. 153–160.

[24] E.J. Routledge and J.P. Sumpter, *Structural features of alkylphenolic chemicals associated with estrogenic activity*, J. Biol. Chem. 272 (1997), pp. 3280–3288.

[25] T.A. Grese, S. Cho, D.R. Finley, A.G. Godfrey, C.D. Jones, C.W. Lugar, 3rd, M.J. Martin, K. Matsumoto, L.D. Pennington, M.A. Winter, M.D. Adrian, H.W. Cole, D.E. Magee, D.L. Phillips, E.R. Rowley, L.L. Short, A.L. Glasebrook, and H.U. Bryant, *Structure-activity relationships of selective estrogen receptor modulators: Modifications to the 2-arylbenzothiophene core of raloxifene*, J. Med. Chem. 40 (1997), pp. 146–167.

[26] W. Tong, R. Perkins, R. Strelitz, E.R. Collantes, S. Keenan, W.J. Welsh, W.S. Branham, and D.M. Sheehan, *Quantitative structure-activity relationships (QSARs) for estrogen binding to the estrogen receptor: Predictions across species*, Environ. Health Perspect. 105 (1997), pp. 1116–1124.

[27] C.L. Waller, T.I. Oprea, K. Chae, H.K. Park, K.S. Korach, S.C. Laws, T.E. Wiese, W.R. Kelce, and L.E. Gray, Jr., *Ligand-based identification of environmental estrogens*, Chem. Res. Toxicol. 9 (1996), pp. 1240–1248.

[28] W. Tong, R. Perkins, L. Xing, W.J. Welsh, and D.M. Sheehan, *QSAR models for binding of estrogenic compounds to estrogen receptor alpha and beta subtypes*, Endocrinology 138 (1997), pp. 4022–4025.

[29] W. Tong, D.R. Lowis, R. Perkins, Y. Chen, W.J. Welsh, D.W. Goddette, T.W. Heritage, and D.M. Sheehan, *Evaluation of quantitative structure-activity relationship methods for large-scale prediction of chemicals binding to the estrogen receptor*, J. Chem. Inf. Comput. Sci. 38 (1998), pp. 669–677.

[30] T.E. Wiese, L.A. Polin, E. Palomino, and S.C. Brooks, *Induction of the estrogen specific mitogenic response of MCF-7 cells by selected analogues of estradiol-17 beta: A 3D QSAR study*, J. Med. Chem. 40 (1997), pp. 3659–3669.

[31] T.G. Gantchev, H. Ali, and J.E. van Lier, *Quantitative structure-activity relationships/comparative molecular field analysis (QSAR/CoMFA) for receptor-binding properties of halogenated estradiol derivatives*, J. Med. Chem. 37 (1994), pp. 4164–4176.

[32] C.L. Waller, D.L. Minor, and J.D. McKinney, *Using three-dimensional quantitative structure-activity relationships to examine estrogen receptor binding affinities of polychlorinated hydroxybiphenyls*, Environ. Health Perspect. 103 (1995), pp. 702–707.

[33] A.M. Brzozowski, A.C. Pike, Z. Dauter, R.E. Hubbard, T. Bonn, O. Engstrom, L. Ohman, G.L. Greene, J.A. Gustafsson, and M. Carlquist, *Molecular basis of agonism and antagonism in the oestrogen receptor*, Nature 389 (1997), pp. 753–758.

[34] A.K. Shiau, D. Barstad, P.M. Loria, L. Cheng, P.J. Kushner, D.A. Agard, and G.L. Greene, *The structural basis of estrogen receptor/coactivator recognition and the antagonism of this interaction by tamoxifen*, Cell 95 (1998), pp. 927–937.

[35] EDSTAC, Available at www.epa.gov/opptintr/opptendo/finalrpt.htm. Final Report.

[36] J.D. Walker and R.H. Brink, *New cost-effective, computerized approaches to selecting chemicals for priority testing consideration,* in *Aquatic Toxicology and Environmental Fate,* G.W. Suter, II and M.A. Lewis, eds., American Society for Testing and Materials (ASTM), Philadelphia, 1989, pp. 507–536.

[37] J.D. Walker, *Chemical selection by the TSCA Interagency Testing Committee: Use of computerized substructure searching to identify chemical groups for health effects, chemical fate and ecological effects testing*, Sci. Total Environ. 109–110 (1991), pp. 691–700.

[38] J.D. Walker, *Estimation methods used by the TSCA Interagency Testing Committee to prioritize chemicals for testing: Exposure and biological effects scoring and structure activity relationships,* Toxicol. Model. 1 (1995), pp. 123–141.

[39] J.D. Walker, D.A. Gray, and M. Pepling, *Past and future strategies for sorting and ranking chemicals: Applications to the 1998 Drinking Water Contaminants List Chemicals,* in *Identifying Future Drinking Water Contaminants,* National Academy Press, Washington, DC, 1999, pp. 51–102.

[40] J.D. Walker and D.A. Gray, *The Substructure-Based Computerized Chemical Selection Expert System (SuCCSES): Providing chemical right-to-know (CRTK) information on potential actions of structurally-related chemical classes on the environment and human health,* in *Handbook on Quantitative Structure Activity Relationships (QSARs) for Predicting Effects of Chemicals on Environmental-Human Health Interactions,* J.D. Walker, ed., SETAC Press, Pensacola, FL, 2001.

[41] S. Bradbury, O. Mekenyan, and G. Ankley, *Quantitative structure-activity relationships for polychlorinated hydroxybiphenyl estrogen receptor binding affinity — An assessment of conformer flexibility,* Environ. Toxicol. Chem. 15 (1996), pp. 1945–1954.

[42] L. Xing, W.J. Welsh, W. Tong, R. Perkins, and D.M. Sheehan, *Comparison of estrogen receptor alpha and beta subtypes based on comparative molecular field analysis (CoMFA),* SAR QSAR Environ. Res. 10 (1999), pp. 215–237.

[43] W. Zheng and A. Tropsha, *A novel variable selection QSAR approach based on the K-nearest neighbor principle,* J. Chem. Inf. Comput. Sci. 40 (2000), pp. 185–194.

[44] W. Tong, H. Hong, H. Fang, Q. Xie, and R. Perkins, *Decision Forest: Combining the predictions of multiple independent Decision Tree model,* J. Chem. Inf. Comput. Sci. 43 (2003), pp. 525–531.

[45] D.E. Clark and D.R. Westhead, *Evolutionary algorithms in computer-aided molecular design,* J. Comput. Aided Mol. Des. 10 (1996), pp. 337–358.

[46] S. Forrest. *Genetic Algorithms-Principles of natural selection applied to computation,* Science 261 (1993), pp. 872–878.

[47] R.H. Rohrbaugh and P.C. Jurs, *Descriptions of molecular shape applied in studies of structure/activity and structure/property relationships,* Anal. Chim. Acta 199 (1987), pp. 99–109.

[48] H. Fang, W. Tong, L. Shi, R. Blair, R. Perkins, W.S. Branham, S.L. Dial, C.L. Moland, and D.M. Sheehan, *Structure activity relationship for a large diverse set of natural, synthetic and environmental chemicals,* Chem. Res. Toxicol. 14 (2001), pp. 280–294.

[49] J.E. Leffler and E. Grunwald, *Rates and Equilibrium Constants of Organic Reaction,* John Wiley & Sons, New York, 1963.

[50] D.T. Stanton and P.C. Jurs, *Development and use of charge partial surface area structural descriptors in computer-aided quantitative structure-property relationship studies,* Anal. Chem. 62 (1990), pp. 2323–2329.

[51] T. Nishihara, J. Nishikawa, T. Kanayama, F. Dakeyama, K. Saito, M. Imagawa, S. Takatori, Y. Kitagawa, S. Hori, and H. Utsumi, *Estrogenic activities of 517 chemicals by yeast two-hybrid assay,* J. Health Sci. 46 (2000), pp. 282–298.

[52] G.G. Kuiper, J.G. Lemmen, B. Carlsson, J.C. Corton, S.H. Safe, P.T. van der Saag, B. van der Burg, and J.A. Gustafsson, *Interaction of estrogenic chemicals and phytoestrogens with estrogen receptor beta,* Endocrinology 139 (1998), pp. 4252–4263.

[53] A.M. Soto, C. Sonnenschein, K.L. Chung, M.F. Fernandez, N. Olea, and F.O. Serrano, *The E-SCREEN assay as a tool to identify estrogens: An update on estrogenic environmental pollutants,* Environ. Health Perspect. 103 Suppl 7 (1995), pp. 113–122.

[54] K.W. Gaido, L.S. Leonard, S. Lovell, J.C. Gould, D. Babai, C.J. Portier, and D.P. McDonnell, *Evaluation of chemicals with endocrine modulating activity in a yeast-based steroid hormone receptor gene transcription assay,* Toxicol. Appl. Pharmacol. 143 (1997), pp. 205–212.

[55] N.G. Coldham, M. Dave, S. Sivapathasundaram, D.P. McDonnell, C. Connor, and M.J. Sauer, *Evaluation of a recombinant yeast cell estrogen screening assay*, Environ. Health Perspect. 105 (1997), pp. 734–742.

[56] E.J. Routledge, J. Parker, J. Odum, J. Ashby, and J.P. Sumpter, *Some alkyl hydroxy benzoate preservatives (parabens) are estrogenic*, Toxicol. Appl. Pharmacol. 153 (1998), pp. 12–19.

[57] C.A. Harris, P. Henttu, M.G. Parker, and J.P. Sumpter, *The estrogenic activity of phthalate esters* in vitro, Environ. Health Perspect. 105 (1997), pp. 802–811.

[58] G.G. Kuiper, B. Carlsson, K. Grandien, E. Enmark, J. Haggblad, S. Nilsson, and J.A. Gustafsson, *Comparison of the ligand binding specificity and transcript tissue distribution of estrogen receptors alpha and beta*, Endocrin. 138 (1997), pp. 863–870.

[59] A.M. Soto, K.L. Chung, and C. Sonnenschein, *The pesticides endosulfan, toxaphene, and dieldrin have estrogenic effects on human estrogen-sensitive cells*, Environ. Health Perspect. 102 (1994), pp. 380–383.

[60] A. Soto, C. Michaelson, N. Prechtl, B. Weill, and C. Sonnenschein, In-vitro *endocrine disruptor screening,* in *Environmental Toxicology and Assessment,* D. Henshel, ed., American Society for Testing and Materials, West Conshohocken, PA, 1999.

[61] H. Fang, W. Tong, R. Perkins, A. Soto, N. Prechtl, and D.M. Sheehan, *Quantitative comparison of* in vitro *assays for estrogenic activity*, Environ. Health Perspect. 108 (2000), pp. 723–729.

[62] Cambridge, *Cambridge Healthtech Institutes Third Annual Conference on Lab-on-a-Chip and Microarrays*, Pharmacogenomics 1 (2001), pp. 73–77.

[63] R. Perkins, J. Anson, R. Blair, W.S. Branham, S. Dial, H. Fang, B.S. Hass, C. Moland, L. Shi, W. Tong, W. Welsh, J.D. Walker, and D. Sheehan, *The Endocrine Disruptor Knowledge Base (EDKB), a prototype toxicological knowledge base for endocrine disrupting chemicals*, in *Handbook on Quantitative Structure Activity Relationships (QSARs) for Predicting Chemical Endocrine Disruption Potentials,* J.D. Walker, ed., SETAC Press, Pensacola, FL, 2001.

[64] C. Hansch and A. Leo, *Exploring QSAR — Fundamentals and Applications in Chemistry and Biology,* The American Chemical Society, Washington, DC, 1995.

[65] C. Hansch, B.R. Telzer, and L. Zhang, *Comparative QSAR in toxicology: Examples from teratology and cancer chemotherapy of aniline mustards*, Crit. Rev. Toxicol. 25 (1995), pp. 67–89.

[66] S. Bradbury, *Quantitative structure-activity relationship and ecological risk assessment: An overview of predictive aquatic toxicology research*, Toxicol. Lett. 79 (1995), pp. 229–237.

[67] A.J. Hopfinger and J.S. Tokarski, *Practical applications of computer-aided drug design,* in *Practical Applications of Computer-Aided Design,* P.S. Charifson, ed., Marcel-Dekker, New York, 1997, pp. 105–164.

[68] H. Kubinyi, G. Folkers, and Y.C. Martin, *3D QSAR in drug design — Recent advances*, Perspect. Drug Disc. Design 12 (1998), pp. R 5–R 7.

[69] C.A. Lipinski, F. Lombardo, B.W. Dominy, and P.J. Feeney, *Experimental and computational approaches to estimate solubility and permeability in drug discovery and development settings*, Adv. Drug Delivery Rev. 23 (1997), pp. 3–25.

[70] S.P. Bradbury, *Predicting modes of toxic action from chemical structure: An overview*, SAR QSAR Environ. Res. 2 (1994), pp. 89–104.

[71] C.L. Russom, S.P. Bradbury, and A.R. Carlson, *Use of knowledge bases and QSARs to estimate the relative ecological risk of agrichemicals: A problem formulation exercise*, SAR QSAR Environ. Res. 4 (1995), pp. 83–95.

[72] R. Benigni and A.M. Richard, *Quantitative structure-based modeling applied to characterization and prediction of chemical toxicity*, Methods 14 (1998), pp. 264–276.

[73] T.W. Schultz and J.R. Seward, *Health-effects related structure-toxicity relationships: A paradigm for the first decade of the new millennium*, Sci. Total Environ. 249 (2000), pp. 73–84.

[74] C. Hansch, D. Hoekman, A. Leo, L. Zhang, and P. Li, *The expanding role of quantitative structure-activity relationships (QSAR) in toxicology*, Toxicol. Lett. 79 (1995), pp. 45–53.

[75] W. Tong, R. Perkins, J. Wu, L. Shi, M. Tu, H. Fang, R. Blair, W. Branham, and D.M. Sheehan, An integrated computational approach for prioritizing potential estrogens, Kobe, Japan, 1999.

7 A Structure-Activity Relationship (SAR) Analysis for the Identification of Environmental Estrogens
The Categorical-SAR (cat-SAR) Approach

*Albert R. Cunningham, Daniel M. Consoer,
Seena A. Iype, and Suzanne L. Cunningham*

CONTENTS

ABSTRACT

The number of environmental chemicals found to have some level of endocrine activity has led to concern about the possible adverse effects these compounds may have on human and environmental health. We describe herein the development of a new set of structure-activity relationship (SAR) models for estrogenic activity using the categorical-SAR (cat-SAR) expert system. The cat-SAR models are built through a comparison of structural features found among categorized compounds in the model's learning set (for example, estrogenic and not estrogenic compounds). Several cat-SAR models were developed from *in vitro* data for 122 compounds tested for estrogenicity in the E-SCREEN assay. By leave-one-out validation, of the models for the endpoint of relative proliferative potency, the concordance between experimental and predicted results ranged from 63% to 85%, and of the models based on the endpoint of relative proliferative effect, the concordance ranged from 82% to 90%. Evaluation of two-dimensional fragments derived during the modeling process suggested the models are mechanistically sound. The models also compared similarly to previous E-SCREEN models developed by using the MCASE methodology. Based on the results described herein, the cat-SAR method would be a useful approach in screening compounds for estrogen activity as well as for investigating their mechanism of action.

KEYWORDS

Cancer
Categorical structure-activity relationship (cat-SAR)
Environmental estrogens
Structure-activity relationship

7.1   INTRODUCTION

7.1.1   ENVIRONMENTAL ESTROGENS

Of the many classes of chemicals that humans and the environment are exposed to, some of the most problematic and potentially serious are those associated with endocrine system disruption. An endocrine disruptor can be defined as "an exogenous agent that interferes with the production, release, transport, metabolism, binding, action, or elimination of natural hormones in the body responsible for the maintenance of homeostasis and the regulation of developmental processes" [1]. Compounds that mimic the activity of 17β-estradiol (that is, estrogen agonist) or have the ability to bind to and interfere with the estrogen receptor in an antagonistic manner can alter the normal functioning of the estrogen signaling pathway. Endocrine disrupting chemicals with estrogenic or antiestrogenic activity are commonly referred to as xenoestrogens or environmental estrogens. Likewise, chemicals with estrogenic activity found in plants are referred to as phytoestrogens.

The consequence of exposure to one or more estrogenic compounds has been shown to induce a number of toxicological and pharmacological responses including cancer [2–4], developmental abnormalities [2,3, 5–7], altered sexual differentiation [5,6], immune disturbances [7], as well as no observable adverse effects or even beneficial responses [8]. Moreover, it is evident that the timing of exposure (for example, fetal versus adult) is of extreme importance in determining the overall effect from estrogenic compounds [9,10]. For example, in 2007, Cohn et al. found that the age of exposure to dichloro-diphenyl-trichloroethane (DDT) is an important risk factor for breast cancer, noting a fivefold increased risk for breast cancer in women who were exposed mostly under the age of 20 to the insecticide [11].

Even though a significant number of environmental chemicals have been found to possess estrogenic activity, they are mostly weaker estrogens than 17β-estradiol. It has been argued, given the typically weak nature of environmental estrogens along with their typical low concentrations, that the risk for adverse health effects from exposure is questionable [12]. On the other hand, studies have shown that exposure to even low concentrations of estrogenic compounds can induce significant biological responses [13]. For instance, biological effects have been observed in rats treated with the weak estrogen equol and ones fed a soy-based diet (that is, a diet that contains phytoestrogens) [14].

Another group of estrogenically active agents are of medicinal interest and value. These are the selective estrogen receptor modulators (SERMs) that are being used and investigated as both breast cancer chemotherapeutics and as hormone replacement therapies. The widely used tamoxifen and to a lesser extent raloxifene are two such examples. Additionally, interest is focusing on phytoestrogens as chemopreventative agents [8] as well as "alternative" approaches for postmenopausal hormone replacement therapies [15,16]. The phytoestrogen hormone replacement therapies are gaining popularity, given the recent observation that women using pharmaceutical-based estrogen replacement therapy have a higher risk of breast cancer. Although early studies did not reach statistical significance [17], more recent work [18] and a review of the data [19] indicate a positive association between hormone replacement therapy and cancer.

With the obvious usefulness of SERMs, medicinal chemistry has added a great deal of understanding to the phenomena of estrogenicity and some of the beneficial and adverse health effects associated with these compounds. Although useful, the investigation of SERMs does not cover the entire range of toxicological effects or the structural diversity associated with environmental endocrine active agents [20].

The U.S. Environmental Protection Agency (EPA) was mandated under the 1996 Food Quality Protection Act by the U.S. Congress to develop a screening and testing strategy to determine whether exogenous substances may have an effect in humans similar to those of natural hormones [21]. The EPA is considering 87,000 chemicals as potentially requiring analysis for endocrine activity [22], and a goal of the EPA is to pursue computational methods for their analysis [23]. Structure-activity relationship (SAR) modeling has gained acceptance in the regulatory community for investigating ecological [24] and human health [25] effects.

7.1.2 Modeling Estrogenic Activity

Briefly, Waller and others have used quantitative structure-activity relationship (QSAR) comparative molecular field analysis (CoMFA) to study the relative binding affinities (RBAs) of compounds for the estrogen receptor [26–29]. Due to the alignment requirements of CoMFA, these analyses relied on congeneric series of compounds for the training sets, and these models are capable of predicting the activity of compounds that fit this model space.

The National Center for Toxicological Research (NCTR) published a set of rat uterine cytosol RBA data [30]. Shi et al. [31] produced predictive CoMFA and holographic quantitative structure-activity relationship (HQSAR) models, and they also demonstrated the use of structural alerts for estrogen activity in a logical tree-based method to prioritize upward of 58,000 compounds that are of environmental concern [32]. The MCASE expert system was also successfully applied in the development of models based on the estrogen-like action of chemicals as measured in the E-SCREEN assay by us [33] and by Klopman and Chakravarti for estrogen RBA [34].

We report herein the development of a new set of qualitative SAR models from the same E-SCREEN data as we previously modeled with MCASE [33] using the categorical SAR (cat-SAR) expert system. The E-SCREEN assay measures estrogen-like proliferation of human MCF-7 breast cancer cells [35,36]. Given the wide spectrum of biological assays for estrogenicity, the E-SCREEN assay falls about in the middle of the range of biological complexity associated with estrogens (that is, above *in vitro* receptor binding and below *in vivo* whole animal assays). This assay is well characterized, and Soto and colleagues report the estrogenic response of chemicals using two unique parameters. The relative proliferative potency (RPP) is the ratio between the least amount of 17β-estradiol needed to produce maximum proliferation and the least amount of the test chemical needed to produce a comparable effect [37]. That is, RPP compares the estrogenic potency of a compound to the potency of the standard estrogen 17β-estradiol. On the other hand, because many estrogenic compounds, no matter how high the dose, will never produce cell proliferation at the rate of 17β-estradiol, the relative proliferative effect (RPE) measures this effect. The RPE is reported as 100 times the ratio of the greatest cell yield obtained with a test chemical and that obtained by 17β-estradiol [37].

7.1.3 Overview of cat-SAR Expert System

The cat-SAR approach is a computational SAR or *in silico* toxicity prediction "expert system" as classified by Dearden [38]. In a previous analysis of human respiratory sensitizers, the cat-SAR program was able to achieve an overall concordance between experimental and predicted values 92% with sensitivities between 89% and 94% and specificities between 87% and 95% [39] (see Section 7.2 for explanation of validation methods).

The approach we have taken in developing the cat-SAR program diverges from existing commercial SAR expert systems and is more in tune with QSAR techniques wherein there is a high degree of user flexibility in both learning set development and model parameterization. For instance, the user is presented with a number of selectable and adjustable modeling attributes including selection of fragment sizes, fragment types, and rules for identifying important fragments for the final model. Selection of compounds included in the learning set and control of model attributes provide the user with the ability to rigorously explore the relationships between chemical structure and biological activity. Ultimately, this rationale negates any *a priori* requirements that a given set of data must fit the attributes of a predefined and often proprietary modeling process.

Basically, cat-SAR models are built through a comparison of structural features found among categorized compounds in the model's learning set. Generically, these categories are toxicologically active and inactive compounds. Essentially, the cat-SAR approach is transparent in the development of the learning set, the identification of fragments, and the determination of significant or important ones. Moreover, the approach allows user intervention and model optimization throughout the modeling process. This method includes the ability to examine the entire fragment base and to explore and optimize the fragments that have perceived biological relevance.

Moreover, because cat-SAR analyzes categorical data and two-dimensional fragments rather than intact chemicals, the program can examine noncongeneric datasets that are divided into categories of activity (that is, rather than degrees of potency). Thus, unlike HQSAR and CoMFA approaches that require continuous-type data, cat-SAR works by identifying molecular attributes associated with biological activity by comparing attributes of active (for example, estrogenic) to inactive (for example, nonestrogenic) compounds. The models and subsequent predictions based on this dichotomy can then be used to examine structural features associated with estrogenicity and predict the likelihood of estrogenic activity of unknown compounds, respectively.

Overall, the cat-SAR modeling approach discussed herein for environmental estrogens, with its high degree of predictivity and mechanistically interpretable models, can provide a computational method singly or in combination with other techniques to prioritize compounds for further testing and for regulatory classification. These methods could therefore reduce the cost, time, and use of animals associated with meeting the mandate to assess environmental compounds for endocrine disrupting ability.

7.2 MATERIALS AND METHODS

7.2.1 The E-SCREEN Learning Sets

A set of compounds assayed for estrogenic activity in the E-SCREEN assay were chosen for this study. Estrogenic activity was reported as RPP and RPE values. The RPP and RPE learning sets of 122 chemicals each were created from publications of Soto and colleagues [35–37], and each set contained the same chemicals. The RPP learning set consisted of 50 active (that is, estrogenic) and 72 inactive (that is, nonestrogenic) chemicals. The RPE learning set consisted of 73 active and 49 inactive chemicals. The estrogenic potency values obviously differed between the RPP and RPE models. The overall designation of compounds as estrogenic or nonestrogenic also differed between the two. Twenty-three chemicals designated as inactive in the RPP set were listed as active in the RPE model (that is, see compounds 2,2′,3,3′,5,

5'-hexachlorobiphenyl through 6-bromonaphthol-2, Table 7.2). These compounds all had very low RPE values. We note that although the original authors of the studies chose to call these compounds nonestrogens, we regarded them specifically as active in the RPE assay because activity (although minimal) was observed.

7.2.2   THE CATEGORICAL-SAR (CAT-SAR) EXPERT SYSTEM

7.2.2.1   Learning Set Development

The cat-SAR models are built through a comparison of structural features found among two designated categories of compounds in the model's learning set. As mentioned, for these analyses, the categories were estrogenic and nonestrogenic. The cat-SAR learning set consists of the chemical name, its structure as a .MOL2 file, and its categorical designation (for example, one or zero for active and inactive). Typically, organic salts are included as the freebase; simple mixtures and technical grade preparations may be included as the major or active component; metals, metallorganic compounds, polymers, and mixtures of unknown composition are not included.

7.2.2.2   *In Silico* Chemical Fragmentation and the Compound-Fragment Data Matrix

Using the Tripos Sybyl HQSAR module [40], each chemical was fragmented *in silico* into all possible fragments meeting user-specified criteria. HQSAR allows the user to select attributes for fragment determination including atom counts (i.e., the size of the fragments), bond types, atomic connections (i.e., the arrangement of atoms in the fragment), explicit hydrogen atoms, chirality, and hydrogen bond donor and acceptor groups. Fragments can be linear, branched, or cyclic moieties. Models developed herein contained fragments between three and seven atoms in size and considered atoms, bond types, and atomic connections.

Upon completion of the fragmentation routine, a Sybyl HQSAR add-on procedure produces a compound-fragment data matrix as a text file. In the matrix, the rows are intact chemicals and columns are the molecular fragments. Thus, for each chemical a tabulation of all its fragments is recorded across the table rows, and for each fragment all chemicals that contain it are tabulated down the columns.

The HQSAR module is not used for statistical analysis or model development. Rather, the compound-fragment matrix is analyzed with the cat-SAR programs we developed in order to identify structural features associated with categorized active and inactive compounds. The cat-SAR programs, E-SCREEN database, and the compound-fragments matrix are available through the corresponding author.

7.2.2.3   Identifying "Important" Fragments of Activity and Inactivity

A measure of each fragment's association with biological activity is next determined. To ascertain an association between each fragment and biological activity (or inactivity), a set of rules is parameterized to choose "important" active and inactive fragments. It should be noted that the cat-SAR program uses a weight-of-evidence approach to select "important" fragments, rather than statistical analysis to select "significant" ones.

The first selection rule is the number of times a fragment is identified in the learning set. For this exercise, it was set at three compounds in the learning set (that is, ~2.5% of compounds). We surmise this is a reasonable choice considering that if a fragment is found in only one or two compounds in the learning set, no matter how large the set, it may be a chance occurrence. It should be noted, however, that these minimally occurring fragments may be outliers or important but underrepresented descriptors of activity. On the other hand, because the E-SCREEN learning sets are composed of 122 relatively structurally diverse compounds, if we required fragments to be found in more than three compounds, we would expect to miss important features based on the diverse nature of the learning set.

The second rule considers the frequency of active or inactive compounds that contribute to each fragment. We derived models based on three frequency set-point rules ($freq \geq 0.65$, 0.75, and 0.85). We reasoned that even if a particular fragment is associated with activity, there may yet be other reasons for the compounds that it is derived from to be classified as inactive (for example, other fragments or chemicophysical properties), and thus, it would not be expected to be found in 100% of the active compounds. The likewise is true for inactive fragments. Thus, if we considered only those fragments found exclusively in active or inactive compounds, we would rarify the fragments pool to an unreasonable level and risk losing valuable information. On the other hand, we expected that fragments found to be present approximately equally in the active and inactive fragment sets would not be associated with biological activity. Such fragments may serve as structural scaffolds holding the biologically active features and are not directly related to activity or inactivity.

As mentioned, the frequency of active or inactive compounds was set to initial set-points of $freq \geq 0.65$, 0.75, and 0.85 for both active and inactive fragments. However, because the RPP and RPE learning sets did not have an equal number of active and inactive compounds contained in them, the initial $freq$ set-points were adjusted to standardize the frequency of active and inactive fragments required to identify important ones. Basically, this adjustment compensated for the unbalanced nature of the learning set. The RPP model contained 50 estrogens out of 122 compounds (i.e., 41% active and 59% inactive), and the RPE model had 73 estrogens out of 122 compounds (i.e., 59% active and 41% inactive). Therefore, for fragments to be selected as important for the RPP model, the frequency for active fragments was increased by 0.09, and for the inactive ones, it was decreased by the same amount. Thus, the $freq \geq 0.65$ model was adjusted to $freq_{ACT} \geq 0.74$ and $freq_{IN} \geq 0.56$, the $freq \geq 0.75$ model was adjusted to $freq_{ACT} \geq 0.84$ and $freq_{IN} \geq 0.66$, and the $freq \geq 0.85$ model was adjusted to $freq_{ACT} \geq 0.94$ and $freq_{IN} \geq 0.76$ (see Table 7.1).

7.2.2.4 Predicting Activity

The resulting list of important fragments can then be used for mechanistic analysis, or to predict the activity of an unknown compound. In the latter circumstance, the model determines which, if any, fragments from the model's learning set the test compound contains. If none are present, no prediction of activity is made for the compound (that is, no default prediction). If one or more fragments are present, the number of active and inactive compounds containing each fragment is determined.

TABLE 7.1

Validation Summary for the Relative Proliferative Potency (RPP) and Relative Proliferative Effect (RPE) cat-SAR Models

	Fragments			Self-Fit			Leave-One-Out			Multiple Leave-Many-Out		
Model	Model[c]	Active[d]	Inactive[e]	Sensitivity[f]	Specificity[g]	Concordance[h]	Sensitivity[f]	Specificity[g]	Concordance[h]	Sensitivity[f]	Specificity[g]	Concordance[h]
RPE												
Freq 65 ss[a]	1,465	1,051	414	0.96(70/73)	0.83(40/48)	0.91(110/121)	0.84(61/73)	0.80(37/46)	0.82(98/119)	0.84(5.62/6.67)	0.81(3.55/4.4)	0.82(9.17/11.2)
Freq 65 conc[b]							0.85(62/73)	0.78(36/46)	0.82(98/119)			
Freq 75 ss[a]	1,276	902	374	0.99(72/73)	0.91(40/44)	0.96(112/117)	0.84(61/73)	0.84(37/44)	0.84(98/117)	0.83(5.52/6.62)	0.87(3.45/3.98)	0.85(8.96/10.6)
Freq 75 conc[b]							0.96(70/73)	0.77(34/44)	0.89(104/117)			
Freq 85 ss[a]	1,060	760	300	0.98(63/64)	0.97(33/34)	0.98(96/98)	0.90(56/62)	0.88(28/32)	0.89(84/94)	0.85(5/5.88)	0.89(2.78/3.13)	0.87(7.83/9)
Freq 85 conc[b]							0.98(61/62)	0.75(24/32)	0.90(85/94)			
RPP												
Freq 65 ss[a]	1,028	679	349	0.80(40/50)	0.85(52/61)	0.83(92/111)	0.60(30/50)	0.66(40/61)	0.63(70/111)	0.65(2.99/4.57)	0.74(4.41/5.92)	0.71(7.40/10.49)
Freq 65 conc[b]							0.60(30/50)	0.66(40/61)	0.63(70/111)			
Freq 75 ss[a]	816	530	286	0.95(40/42)	0.90(46/51)	0.92(86/93)	0.83(35/42)	0.81(44/54)	0.82(79/96)	0.67(2.77/4.12)	0.74(3.78/5.08)	0.71(6.55/9.20)
Freq 75 conc[b]							0.83(35/42)	0.81(44/54)	0.82(79/96)			
Freq 85 ss[a]	588	304	284	1.00(40/40)	0.91(39/43)	0.95(79/83)	0.86(36/42)	0.85(40/47)	0.85(76/89)	0.80(3.17/3.94)	0.77(3.3/4.28)	0.79(6.47/8.22)
Freq 85 conc[b]							0.86(36/42)	0.85(40/47)	0.85(76/89)			

a Validation results based on model with closest sensitivity and specificity.

b Validation results based on model with highest overall concordance.

c Number of fragments specified rules of the model.

d Number of fragments meeting specified rules to be considered as active.

e Number of fragments meeting specified rules to be considered as inactive.

f Number of correct positive predictions/total number of positives.

g Number of correct negative predictions/total number of negatives.

h Observed correct predictions: number of correct predictions/total number of predictions.

The probability of activity or inactivity is then calculated based on the total number of active and inactive compounds containing the fragments.

The probability of activity of a test chemical is calculated from the average probability of the active and inactive fragments contained in it and naturally weighted to the number of active and inactive compounds that go into deriving each fragment. For example, if a compound contains two fragments, one being found in 9/10 active compounds in the learning set (i.e., 90% active) and the other being found in 3/3 inactive compounds (i.e., 100% inactive), the test compound will be predicted to have a 69% chance of activity (i.e., 9/10 actives + 0/3 actives = 9/13 actives or 69% active).

As described, a cat-SAR prediction of activity or inactivity is based on two separate fragment sets (the active fragments and the inactive ones), and the predicted activity of a chemical is based on the average probability of all the active and inactive compounds contributing to its structure. Therefore, to classify compounds back to an active or inactive category (rather than a probability of activity), the program identifies an optimal cut-off point that best separates the prediction of active and inactive compounds based on the probabilistic values of activity derived from a model validation analysis (described below and see Table 7.2) [41]. Depending on the application of the model, the cut-off point that separates active from inactive categorization, for example, can be adjusted, wherein a model with the best overall concordance can be selected (i.e., a most predictive model), one with equal sensitivity and specificity (i.e., a balanced model that does not overly predict active compounds at the cost of wrongly predicting inactive ones and vice versa), or one with high sensitivity (i.e., a risk averse model).

7.2.3 MODEL VALIDATION

A self-fit (that is, leave-none-out [LNO]) and two validation routines (that is, leave-one-out [LOO] and multiple leave-many-out [LMO]) were conducted for each model (see Table 7.1). For the LNO self-fit, a model was developed from the complete learning set of 122 compounds, and that model was used to predict the activity of each compound in the learning set. For the LOO validation, each chemical, one at a time, was removed from the model's total fragment set, and the $n - 1$ model was derived. The activity of the removed chemical was then predicted using the $n - 1$ model. Predicted versus experimental values for each chemical were then compared, and the model's overall concordance, sensitivity, and specificity were determined, where

$$Concordance = \frac{Correct\ predictions}{Total\ predictions}$$

and

$$Sensitivity = \frac{Correct\ positive\ predictions}{Total\ positive\ predictions}$$

and

$$Specificity = \frac{Correct\ negative\ predictions}{Total\ negative\ predictions}$$

TABLE 7.2
Experimental Results and cat-SAR Predictions for Relative Proliferative Potency (RPP) and Relative Proliferative Effect (RPE)

Chemical	RPP Exp	RPP cat-SAR Predictions[a]			RPE Exp	RPP cat-SAR Predictions[a]		
		Freq 0.65	Freq 0.75	Freq 0.85		Freq 0.65	Freq 0.75	Freq 0.85
1,2-Dichloropropane	—	0.56(−)	NP	NP	—	NP	NP	NP
1-Naphthol	—	0.61(−)	NP	NP	—	NP	NP	NP
2,3,7,8-TCDD	—	0.62(−)	0.62(−)	0.00(−)	—	0.75(−)	0.00(−)	0.00(−)
2,4-DB acid	—	0.60(−)	0.64(−)	0.00(−)	—	0.71(−)	0.45(−)	NP
2,4-Dichloro-phenoxyacetic acid	—	0.60(−)	0.64(−)	0.00(−)	—	0.73(−)	0.33(−)	NP
2-Naphthol	—	0.61(−)	NP	NP	—	NP	NP	NP
4-Butooxyphenol	—	0.64(−)	0.67(−)	NP	—	0.67(−)	0.28(−)	NP
4-Hexyloxyphenol	—	0.64(+)	0.67(−)	0.77(−)	—	0.68(−)	0.38(−)	NP
5,6,7,8-Tetrahy-dronaphthol-2	—	0.73(+)	0.75(+)	0.88(+)	—	0.82(+)	0.98(+)	1.00(+)
Alachlor	—	0.63(−)	0.63(−)	0.00(−)	—	0.66(−)	0.10(−)	0.00(−)
Atrazine	—	0.01(−)	0.01(−)	0.00(−)	—	0.01(−)	0.01(−)	0.01(−)
Bendiocarb	—	0.62(−)	0.62(−)	0.00(−)	—	0.63(−)	0.15(−)	0.00(−)
Butylate	—	0.26(−)	0.27(−)	0.00(−)	—	0.32(−)	0.32(−)	0.04(−)
Butylated hydroxytoluene	—	0.68(+)	0.71(−)	0.89(+)	—	0.77(+)	0.75(−)	1.00(+)
Carbaryl	—	0.54(−)	0.00(−)	0.00(−)	—	0.17(−)	0.00(−)	0.00(−)
Carbofuran	—	0.67(+)	0.69(−)	0.72(−)	—	0.70(−)	0.40(−)	0.44(−)
Chlordimeform	—	0.65(+)	0.66(−)	0.00(−)	—	0.74(−)	0.05(−)	0.05(−)
Chlorothalonil	—	0.00(−)	0.00(−)	0.00(−)	—	0.69(−)	0.00(−)	0.00(−)
Chlorpyrifos	—	0.48(−)	0.00(−)	0.00(−)	—	0.68(−)	0.02(−)	0.00(−)
Cyanazine	—	0.11(−)	0.11(−)	0.00(−)	—	0.14(−)	0.07(−)	0.07(−)
Dacthal	—	0.00(−)	0.00(−)	0.00(−)	—	0.60(−)	0.29(−)	0.00(−)
Diamyl phthalate	—	0.61(−)	0.70(−)	NP	—	0.41(−)	0.33(−)	NP
Diazinon	—	0.57(−)	0.54(−)	0.33(−)	—	0.47(−)	0.24(−)	0.00(−)
Dibutyl phthalate	—	0.60(−)	0.68(−)	NP	—	0.38(−)	0.33(−)	NP
Dimethyl isophthalate	—	NP	NP	NP	—	0.38(−)	0.32(−)	NP
Dimethyl terephthalate	—	NP	NP	NP	—	0.38(−)	0.32(−)	NP
Dinonyl phthalate	—	0.63(−)	0.73(+)	0.77(−)	—	0.45(−)	0.42(−)	NP
Dinoseb	—	0.67(+)	0.70(−)	0.75(−)	—	0.74(−)	0.61(−)	0.52(−)
Hexachlorobenzene	—	NP	NP	NP	—	0.77(+)	NP	NP
Hexazinone	—	0.45(−)	0.18(−)	0.00(−)	—	0.48(−)	0.04(−)	0.04(−)
Kelthane	—	0.77(+)	0.77(+)	1.00(+)	—	0.83(+)	0.99(+)	1.00(+)
Lindane	—	0.64(−)	0.70(−)	NP	—	0.87(+)	0.89(+)	NP
Malathion	—	0.35(−)	0.37(−)	0.00(−)	—	0.43(−)	0.10(−)	0.10(−)
Maneb or zineb	—	0.04(−)	0.04(−)	0.00(−)	—	0.04(−)	0.04(−)	0.04(−)
Metolachlor	—	0.61(−)	0.62(−)	0.00(−)	—	0.63(−)	0.09(−)	0.01(−)
Methoprene	—	0.69(+)	0.72(+)	0.81(−)	—	0.86(+)	0.95(+)	0.99(−)
Mirex	—	0.69(+)	0.76(+)	0.82(−)	—	0.94(+)	0.95(+)	1.00(+)

TABLE 7.2 (CONTINUED)
Experimental Results and cat-SAR Predictions for Relative Proliferative Potency (RPP) and Relative Proliferative Effect (RPE)

		RPP cat-SAR Predictions[a]				RPP cat-SAR Predictions[a]		
Chemical	RPP Exp	Freq 0.65	Freq 0.75	Freq 0.85	RPE Exp	Freq 0.65	Freq 0.75	Freq 0.85
Octachlorostyrene	—	0.69(+)	0.69(−)	0.69(−)	—	0.70(−)	0.72(−)	0.00(−)
Parathion	—	0.62(−)	0.63(−)	0.00(−)	—	0.72(−)	0.00(−)	0.00(−)
Phenol	—	0.65(+)	0.67(−)	NP	—	0.77(+)	NP	NP
Picloram	—	0.00(−)	0.00(−)	0.00(−)	—	0.58(−)	0.19(−)	0.00(−)
Propazine	—	0.01(−)	0.01(−)	0.00(−)	—	0.01(−)	0.01(−)	0.01(−)
Rotenone	—	0.69(+)	0.72(−)	0.81(−)	—	0.69(−)	0.63(−)	0.87(−)
Simazine	—	0.01(−)	0.01(−)	0.00(−)	—	0.01(−)	0.01(−)	0.01(−)
Styrene	—	0.86(+)	0.86(+)	0.86(+)	—	0.54(−)	0.86(+)	NP
Tetrachloroethylene	—	NP	NP	NP	—	0.86(+)	0.86(+)	NP
Thiram	—	0.04(−)	0.00(−)	0.00(−)	—	0.04(−)	0.04(−)	0.04(−)
Trifluralin	—	0.59(−)	0.60(−)	0.00(−)	—	0.47(−)	0.03(−)	0.03(−)
Ziram	—	0.04(−)	0.00(−)	0.00(−)	—	0.04(−)	0.04(−)	0.04(−)
2,2′,3,3′,5,5′-Hexachlorobiphenyl	—	0.75(+)	0.75(+)	NP	1	0.97(+)	1.00(+)	1.00(+)
2,3,3′,4,5-Pentachloro-biphenyl	—	0.59(−)	NP	NP	1	0.97(+)	1.00(+)	1.00(+)
3,5-Dichloro-4-hydroxybiphenyl	—	0.64(−)	0.69(−)	1.00(+)	1.5	0.90(+)	1.00(+)	1.00(+)
4-Monochlorobiphenyl	—	0.58(−)	NP	NP	2.1	0.96(+)	1.00(+)	1.00(+)
2,3′,5-Trichloro-biphenyl	—	NP	NP	NP	2.2	0.98(+)	1.00(+)	1.00(+)
3,5-Dichlorobiphenyl	—	NP	NP	NP	2.7	0.96(+)	1.00(+)	1.00(+)
2,3,5,6-Tetrachloro-biphenyl	—	NP	NP	NP	3.1	0.97(+)	1.00(+)	1.00(+)
2,6-Dichlorobiphenyl	—	NP	NP	NP	3.4	0.96(+)	1.00(+)	1.00(+)
Decachlorobiphenyl	—	0.62(−)	0.75(+)	NP	3.5	0.97(+)	1.00(+)	1.00(+)
2,5-Dichlorobiphenyl	—	NP	NP	NP	3.7	0.97(+)	1.00(+)	1.00(+)
Chlordene	—	0.65(+)	0.70(−)	0.76(−)	4	0.95(+)	0.97(+)	1.00(+)
Gibberellic acid	—	0.79(+)	0.83(+)	0.93(+)	4	0.90(+)	0.93(+)	1.00(+)
2,3,4,5,6-Pentachloro-biphenyl	—	0.59(−)	NP	NP	4.4	0.97(+)	1.00(+)	1.00(+)
2-Monochlorobiphenyl	—	NP	NP	NP	4.4	0.97(+)	1.00(+)	1.00(+)
2,3,4,4′-Tetrachloro-biphenyl	—	0.59(−)	NP	NP	4.7	0.97(+)	1.00(+)	1.00(+)
2′,3′,4′,5,5-Pentachloro-2-hydroxybiphenyl	—	0.64(−)	0.65(−)	0.00(−)	4.8	0.93(+)	0.99(+)	1.00(+)
4-Ethylphenol	—	0.68(+)	0.69(−)	0.81(−)	5	0.76(+)	0.86(+)	NP

(continued)

TABLE 7.2 (CONTINUED)
Experimental Results and cat-SAR Predictions for Relative Proliferative Potency (RPP) and Relative Proliferative Effect (RPE)

Chemical	RPP Exp	RPP cat-SAR Predictions[a] Freq 0.65	Freq 0.75	Freq 0.85	RPE Exp	RPP cat-SAR Predictions[a] Freq 0.65	Freq 0.75	Freq 0.85
Chlordane	—	0.66(+)	0.71(−)	0.81(−)	5	0.94(+)	0.97(+)	1.00(+)
3,5-Dichloro-2-hydroxybiphenyl	—	0.61(−)	0.64(−)	0.00(−)	5.4	0.90(+)	0.99(+)	0.99(−)
2,3,6-Trichlorobiphenyl	—	NP	NP	NP	5.8	0.97(+)	1.00(+)	1.00(+)
Heptachlor	—	0.65(+)	0.70(−)	0.76(−)	8	0.95(+)	0.97(+)	1.00(+)
4-Propylphenol	—	0.69(+)	0.70(−)	0.79(−)	17	0.76(+)	0.86(+)	NP
6-Bromonaphthol-2	—	0.72(+)	0.79(+)	1.00(+)	38	0.81(+)	1.00(+)	1.00(+)
t-Butylhydroxyanisol	0.00006	0.64(+)	0.66(−)	0.52(−)	30	0.67(−)	0.40(−)	0.00(−)
2′,5′-Dichloro-2-hydroxybiphenyl	0.0001	0.63(−)	NP	NP	13	0.94(+)	1.00(+)	1.00(+)
2′,3′,4′,5′-Tetrachloro-3-hydroxybiphenyl	0.0001	0.55(−)	NP	NP	35.3	0.93(+)	1.00(+)	1.00(+)
2,3,4,5-Tetrachloro-biphenyl	0.0001	0.25(−)	NP	NP	39.2	0.97(+)	1.00(+)	1.00(+)
1-Hydroxychlordene	0.0001	0.60(−)	0.64(−)	0.53(−)	40	0.95(+)	0.97(+)	1.00(+)
Toxaphene	0.0001	0.64(−)	0.75(+)	0.82(+)	51.9	0.92(+)	0.96(+)	1.00(+)
Dieldrin	0.0001	0.65(+)	0.75(+)	0.83(+)	54.89	0.94(+)	0.97(+)	1.00(+)
Methoxychlor	0.0001	0.68(+)	0.73(+)	0.83(+)	57	0.74(−)	0.60(−)	NP
2,2′,3,3′,6,6′-Hexachlorobiphenyl	0.0001	0.25(−)	NP	NP	61.6	0.97(+)	1.00(+)	1.00(+)
2,2′,4,5-Tetrachlorobiphenyl	0.0001	0.25(−)	NP	NP	61.6	0.97(+)	1.00(+)	1.00(+)
2′,5′-Dichloro-3-hydroxybiphenyl	0.0001	0.63(−)	NP	NP	69.9	0.93(+)	1.00(+)	1.00(+)
p,p′-DDT	0.0001	0.70(+)	0.73(+)	0.83(+)	71	0.78(+)	0.85(−)	NP
2,4,4′,6-Tetrachlorobiphenyl	0.0001	0.26(−)	NP	NP	75.7	0.97(+)	1.00(+)	1.00(+)
2,3,4-Trichlorobiphenyl	0.0001	0.26(−)	NP	NP	77	0.97(+)	1.00(+)	1.00(+)
Endosulfan	0.0001	0.64(−)	0.74(+)	0.84(+)	81.25	0.95(+)	0.97(+)	1.00(+)
Kepone	0.0001	0.65(+)	0.76(+)	0.84(+)	84	0.91(+)	0.95(+)	1.00(+)
o,p′-DDD	0.0001	0.70(+)	0.73(+)	0.83(+)	84	0.78(+)	0.85(−)	NP
o,p′-DDT	0.0001	0.70(+)	0.73(+)	0.83(+)	86.14	0.78(+)	0.85(−)	NP
4-*tert*-Butylphenol	0.0003	0.67(+)	0.73(+)	0.85(+)	71	0.77(+)	0.87(+)	NP
4-*sec*-Butylphenol	0.0003	0.68(+)	0.74(+)	0.87(+)	76	0.78(+)	0.88(+)	NP
Bisphenol A	0.0003	0.69(+)	0.75(+)	0.84(+)	82	0.78(+)	0.87(+)	NP
4,4′-Dihydroxyiphenyl	0.0003	0.64(+)	0.86(+)	0.86(+)	84	0.92(+)	1.00(+)	1.00(+)
4-Hydroxybiphenyl	0.0003	0.64(+)	0.86(+)	0.86(+)	87	0.92(+)	1.00(+)	1.00(+)

TABLE 7.2 (CONTINUED)
Experimental Results and cat-SAR Predictions for Relative Proliferative Potency (RPP) and Relative Proliferative Effect (RPE)

Chemical	RPP Exp	RPP cat-SAR Predictions[a] Freq 0.65	Freq 0.75	Freq 0.85	RPE Exp	RPP cat-SAR Predictions[a] Freq 0.65	Freq 0.75	Freq 0.85
Butylbenzylphthalate	0.0003	0.47(–)	0.46(–)	0.00(–)	90	0.31(–)	0.23(–)	0.17(–)
4-*iso*-Pentylphenol	0.0003	0.59(–)	0.33(–)	0.32(–)	93	0.63(–)	0.34(–)	0.22(–)
4-*tert*-Pentylphenol	0.0003	0.68(+)	0.74(+)	0.87(+)	105	0.78(+)	0.88(+)	NP
Tamoxifen	0.001	0.65(+)	0.69(–)	0.84(+)	11	0.62(–)	0.54(–)	0.62(–)
2,2′,5-Trichloro-4-hydroxybiphenyl	0.001	0.60(–)	0.33(–)	0.33(–)	37.8	0.93(+)	1.00(+)	1.00(+)
2′,5′-Dichloro-4-hydroxybiphenyl	0.001	0.64(+)	0.86(+)	0.86(+)	71.2	0.93(+)	1.00(+)	1.00(+)
2′,3′,4′,5′-Tetrachloro-4-hydroxybiphenyl	0.001	0.56(–)	0.86(+)	0.86(+)	92	0.93(+)	1.00(+)	1.00(+)
Coumestrol	0.001	0.62(–)	0.72(+)	0.86(+)	93	0.84(+)	0.89(+)	1.00(+)
Bisphenol A dimethacrylate	0.003	0.67(+)	0.75(+)	0.84(+)	84	0.75(–)	0.70(–)	NP
4-Nonylphenol	0.003	0.67(+)	0.73(+)	0.91(+)	100	0.78(+)	0.90(+)	1.00(+)
2′,4′,6′-Trichloro-4-hydroxybiphenyl	0.01	0.60(–)	0.86(+)	0.86(+)	99.8	0.92(+)	1.00(+)	1.00(+)
4-Octylphenol	0.03	0.67(+)	0.73(+)	0.91(+)	100	0.78(+)	0.90(+)	1.00(+)
5-Octylphenol	0.03	0.69(+)	0.76(+)	0.90(+)	100	0.79(+)	0.92(+)	1.00(+)
16-Hydroxyestrone	0.1	0.73(+)	0.80(+)	0.91(+)	0	0.86(+)	0.94(+)	1.00(+)
Pseudo diethylstilbestrol	0.1	0.68(+)	0.83(+)	0.97(+)	100	0.70(–)	0.97(+)	1.00(+)
Equilenin	1	0.75(+)	0.81(+)	0.92(+)	82	0.88(+)	0.95(+)	1.00(+)
Zearalenone	1	0.58(–)	0.81(+)	0.89(+)	88	0.57(–)	0.54(–)	0.65(–)
Zearalenol	1	0.57(–)	0.78(+)	0.88(+)	93	0.57(–)	0.51(–)	0.53(–)
Estrone	1	0.73(+)	0.80(+)	0.91(+)	95	0.86(+)	0.94(+)	1.00(+)
Allenolic acid	1	0.62(–)	0.65(–)	0.00(–)	105	0.74(–)	0.75(–)	1.00(+)
Estriol	10	0.74(+)	0.81(+)	0.90(+)	95	0.85(+)	0.93(+)	1.00(+)
Indenestrol	10	0.68(+)	0.74(+)	0.97(+)	100	0.71(–)	0.97(+)	1.00(+)
Ethynylestradiol	100	0.73(+)	0.80(+)	0.90(+)	92	0.86(+)	0.94(+)	1.00(+)
17β-Estradiol	100	0.74(+)	0.81(+)	0.90(+)	100	0.85(+)	0.93(+)	1.00(+)
11β-Chloromethyl-estradiol	1000	0.74(+)	0.81(+)	0.90(+)	110	0.85(+)	0.93(+)	1.00(+)
Moxestrole	1000	0.74(+)	0.81(+)	0.90(+)	110	0.84(+)	0.93(+)	1.00(+)
Diethylstilbestrol	1000	0.68(+)	0.83(+)	0.97(+)	112	0.70(–)	0.97(+)	1.00(+)

Note: NP: No prediction by cat-SAR.

[a] Values reported as probability of activity and final determination (+) for active and (–) for inactive.

For the LMO validation, randomly selected sets of 10% of the chemicals were removed from the model's total fragment set, and the $n - 10\%$ model was derived. The activity of each of the removed chemicals was then predicted using the $n - 10\%$ model. Predicted versus experimental values for the chemicals in the left-out sets were then compared, and the $n - 10\%$ model's concordance, sensitivity, and specificity were determined. This was repeated 100 times to compute the model's average concordance, sensitivity, and specificity.

7.2.4 MODEL COMPARISON: THE CHEMICAL DIVERSITY APPROACH

To compare the MCASE and cat-SAR modeling approaches and to analyze the potential of chemicals demonstrating estrogenic activity in the E-SCREEN assay to induce other toxicological phenomena, including cancer and developmental toxicity, we used the "Chemical Diversity Approach." This is a method based upon comparisons of the SAR-predicted toxicological profiles of a group of 10,000 chemicals chosen to represent a random assortment of all chemicals and chemical features [42]. These chemicals were derived from chemical structure libraries and from a random sample of chemical structures from the National Cancer Institute Repository of potential cancer chemotherapeutic agents. The various toxicological properties of these chemicals are predicted using validated SAR models, including the models for RPE and RPP. The prevalence of chemicals predicted to possess two toxicological properties simultaneously is then quantified and compared to the expected prevalence. If the two effects are assumed to be independent of one another (that is, null hypothesis), the observed and expected values should be nearly equal. A significantly greater observed than expected prevalence suggests a similarity in mechanism among the toxicological effects that are being studied. Likewise, a significantly lower observed than expected prevalence suggests a possible antagonism between the phenomena under investigation. The applicability of the methodology to the study of diverse toxicological phenomena has been demonstrated by successfully estimating the number of potential *Salmonella* mutagens in the environment [43], and the inhibition of gap junctional intercellular communication is related to rodent carcinogenesis through cellular and systemic toxicity but not genotoxicity [42].

7.3 RESULTS AND DISCUSSION

7.3.1 PREDICTIVE PERFORMANCE AND MODEL SELECTION

Together, six cat-SAR models were derived from the E-SCREEN dataset, three each for RPE and RPP data. Each of the models was based on fragments that considered atoms, bonds, and connections and were identified in three or more compounds in the learning sets. The models differed in the set-points for the frequency of active and inactive compounds each fragment was derived from. The three frequency set-points used were $freq \geq 0.65$, 0.75, and 0.85. The model validation results are shown in Table 7.1 and include summaries for models with near-equal sensitivity and specificity (labeled "ss" in Table 7.1) and with highest overall concordance between predicted and experimental values (labeled "conc" in Table 7.1).

Generally, considering both models with highest concordance or with balanced sensitivity and specificity, comparable RPE models outperformed the RPP ones (Table 7.1). For example, the *Freq 85* RPE model with highest concordance was 90% and that of the RPP one was 85% (Table 7.1) This is consistent with our previous models developed with MCASE where the RPE model outperformed the RPP with concordance values of 88% and 72%, respectively [33]. Moreover, the cat-SAR RPE models also consistently were able to make predictions on more compounds than comparable RPP ones (Table 7.1).

The LNO self-fit analysis for the RPE balanced sensitivity and specificity model yielded concordance values between 91% and 98%, sensitivity between 96% and 99%, and specificity between 83% and 97% (Table 7.1). The RPP balanced model yielded concordance values between 83% and 95%, sensitivity between 80% and 100%, and specificity between 85% and 91% (Table 7.1).

The LOO validation for the RPE balanced sensitivity and specificity model yielded concordance values between 82% and 89%, sensitivity between 84% and 90%, and specificity between 80% and 88% (Table 7.1). The RPP balanced model yielded concordance values between 63% and 85%, sensitivity between 60% and 86%, and specificity between 66% and 85% (Table 7.1). Moreover, the best model, based on the highest achieved concordance was the RPE *freq* ≥ 0.85 model that achieved a concordance of 90%, a sensitivity of 98%, and a specificity of 75%.

The LMO validation for the RPE balanced sensitivity and specificity model yielded concordance values between 82% and 87%, sensitivity between 83% and 85%, and specificity between 81% and 89% (Table 7.1). The RPP balanced model yielded concordance values between 71% and 79%, sensitivity between 65% and 80%, and specificity between 74% and 77% (Table 7.1).

Overall, the RPE model clearly outperformed the RPP one and as expected, the higher frequency set-point models (*freq* ≥ 0.85) were able to make substantially more accurate predictions versus the lower one (*freq* ≥ 0.65) but at the cost of making less predictions. We selected the RPE *freq* ≥ 0.75 with balanced sensitivity and specificity model for further analyses. This model was balanced with regard to sensitivity and specificity (that is, it did not make predictions of activity and inactivity at the cost of the other), had a relatively high rate of accurate predictions (that is, LOO concordance, sensitivity, and specificity of all of 84%), and was able to make LOO prediction for 96% (117 out of the 122) (Table 7.1) of the compounds in the learning set (Table 7.1).

7.3.2 DIFFERENCE BETWEEN RPP AND RPE MODELS

As mentioned, 23 compounds (compounds 2,2',3,3',5,5'-hexachlorobiphenyl through 6-bromonaphthol-2, Table 7.2) had disparate activity between the RPP and RPE learning sets. These were weakly active RPE compounds and inactive RPP ones. In our previous MCASE model, 22 of the 23 compounds (96%) were accurately predicted as active in the RPE model, but the RPP model was able to predict only 11 of the 23 as inactive (48%). This same trend was apparent in the $freq \geq 0.75$ cat-SAR models, wherein for the RPE model all 23 compounds (100%) were accurately predicted as active, but the RPP model made predictions on only 11, 7 of which (64%) were classified correctly as inactive compounds (Table 7.2).

The predictivity of a model has been used as an acceptable measure for assessing the "meaningfulness" of a model [44], and we observed that good predictivity is related to mechanistically sound models [41,45]. Therefore, again we consider the RPE model, which includes the very weak estrogens as a more informative model than that based on the RPP dataset. This finding is significant with respect to applying the model to environmental estrogens and phytoestrogens, many of which are exceedingly weak compared to 17β-estradiol.

7.3.3 Examples of cat-SAR Predictions

The structural components that make up cat-SAR predictions of estrogenic activity are consistent with previous analyses describing the interactions of the closely related estrogen receptors (α and β) and their ligands. Briefly, effects of 17β-estradiol and its mimics are mediated by intracellular estrogen receptors [46], and binding of 17β-estradiol to the receptor involves all four rings and both terminal hydroxyl groups. The phenolic OH (located on the A ring) contributes approximately 1.9 kcal/mol of binding energy as a hydrogen bond donor, the 17β-hydroxyl group contributes about 0.6 kcal/mol as a hydrogen bond acceptor, and the aromatic ring contributes approximately 1.5 kcal/mol through polar interactions. Binding to the estrogen receptor is generally inhibited by the addition of polar substituents and by larger hydrophobic substitutions in a number of positions [46].

We used the LOO cross-validation results for the RPE model based on atoms, bonds, and connection. Fragments had to be derived from at least three compounds, and the frequency set-point was $freq \geq 0.75$, which equated to active fragments having to be found in at least 84% of compounds that contained the fragment ($freq_{ACT} \geq 0.84$) and inactive ones having to be found in at least 66% of compounds that contained it ($freq_{IN} \geq 0.66$). To visualize fragments and compounds, the aromaticity of both were normalized by Sybyl Unity [47] (see Figure 7.1 through Figure 7.3). Moreover, because the structures of the fragments are independent from the intact chemicals (from which they are derived or for which they are used to predict), the conformation of the rendered fragments in Figure 7.1 through Figure 7.4 may appear different from the structures of the whole molecules they overlap (see Figure 7.1 through Figure 7.4).

The cat-SAR prediction for 17β-estradiol is shown in Figure 7.1. The fragments contributing to the prediction of activity for 17β-estradiol were derived predominately from active compounds (that is, $freq_{ACT} \geq 0.84$), and there were no fragments associated with inactivity. The fragments can be divided into four general sets that covered 17β-estradiol rings A through D. Estradiol Fragment Sets A and D specifically covered the aromatic portion of 17β-estradiol, and Fragment Set A was specific for the phenolic A ring (Figure 7.1). Fragment Set C covered the interior lipophilic

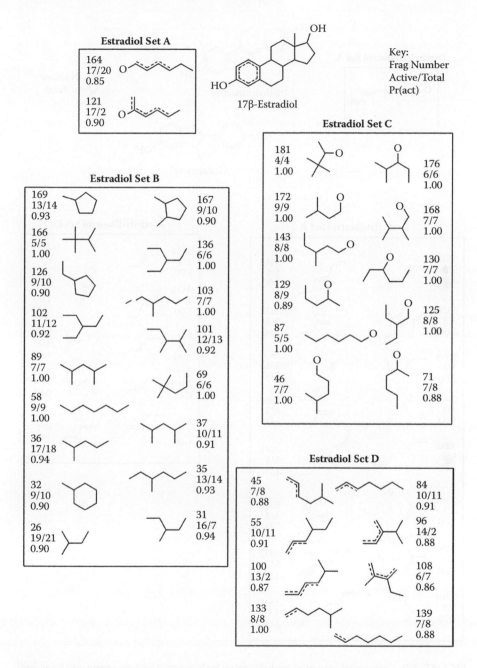

FIGURE 7.1 Categorical structure-activity relationship (cat-SAR) fragment analyses and predictions of the estrogenic compound 17β-estradiol.

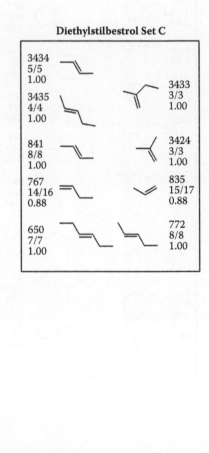

Diethylstilbestrol Set A

3361
3/3
1.00

Diethylstilbestrol

Indenestrol

Key:
Frag Number
Active/Total
Pr(act)

Diethylstilbestrol Set B

3415
5/5
1.00

3418
4/4
1.00

3406
3/3
1.00

3414
3/3
1.00

3398
5/5
1.00

3400
4/4
1.00

3394
3/3
1.00

3397
3/3
1.00

3386
5/5
1.00

3388
4/4
1.00

3382
3/3
1.00

3385
3/3
1.00

3376
3/3
1.00

3377
5/5
1.00

3368
3/3
1.00

3369
5/5
1.00

Diethylstilbestrol Set C

3434
5/5
1.00

3433
3/3
1.00

3435
4/4
1.00

841
8/8
1.00

3424
3/3
1.00

767
14/16
0.88

835
15/17
0.88

650
7/7
1.00

772
8/8
1.00

FIGURE 7.2 Categorical structure-activity relationship (cat-SAR) fragment analyses and predictions of the estrogenic compounds diethylstilbestrol and its metabolite indenestrol.

and aliphatic B, C, and D rings, and Estradiol Fragment Set C specifically covered the 17β hydroxy moiety of the D ring (Figure 7.1).

The cat-SAR prediction for the synthetic estrogen diethylstilbestrol and its metabolite indenestrol are shown in Figure 7.2. Like the fragments used for the prediction of activity for 17β-estradiol, none of the fragments used to predict the activity of diethylstilbestrol and indenestrol were predominately associated with inactivity. The hydroxylated rings of diethylstilbestrol and indenestrol are roughly equivalent to the A to D rings of 17β-estradiol. As seen in Figure 7.2, Diethylstilbestrol Fragment Set A covered the *para*-hydroxylated phenyl ring of both compounds, Fragment Set B mostly covered part of the aromatic ring and the alkene bond linking the two phenolic moieties of the molecule, and Fragment Set C was specific for the alkene section of the molecule (Figure 7.2).

The cat-SAR prediction of the phytoestrogen coumestrol is shown in Figure 7.3. Similar to the previously described predictions, Coumestrol Fragment Sets A and B covered the two phenolic moieties on the molecule, and Fragment Set C covered the alkyl interior part of the molecule. Again, the chemicals that contributed to these fragments were predominately active compounds ($freq_{ACT} \geq 0.84$). However, Coumestrol Fragment Set D covered the three oxygen atoms in the interior of coumestrol, and the chemicals that contributed to these fragments were predominately inactive ($freq_{IN} \geq 0.66$).

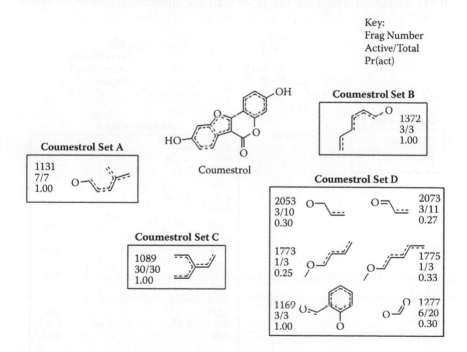

FIGURE 7.3 Categorical structure-activity relationship (cat-SAR) fragment analyses and predictions of the estrogenic compound coumestrol.

These results are similar to an observation we made earlier that the oxygen-rich areas of phytoestrogens impart a level of water solubility to the interior of the molecule resulting in compounds of this type having different a biological activity [48,49].

We speculated that these differences in chemical features of estrogenic compounds could induce different biological responses [48–50]. Because then, the estrogen receptor alpha (ERα) ligand binding domain was crystallized, and its atomic coordinates resolved with those of bound estradiol and raloxifene [51], genistein [52], and 4-hydroxytamoxifen and diethylstilbestrol [53]. It was noted that the lipophilic cavity is nearly twice the size of estradiol, which may explain in part the receptor's promiscuity [54]. Most importantly, it was observed by these authors that estrogen antagonists induce a different conformational change in the AF-2 region compared to that for the natural ligand. Together, these analyses demonstrated the utility of SAR analyses to not only generate predictive models that are explainable by current knowledge but also their ability to generate and investigate hypotheses regarding the mechanistic action of toxicants.

Finally, malathion, an organophosphorous ester widely used as an insecticide, was evaluated to provide an example of a negative cat-SAR prediction, and is shown in Figure 7.4. Malathion Fragment Set A consisted of one fragment that made up part of malathion's ester moiety and linked it to the dimethoxyphosphinothioylthio moiety. Chemicals that contributed to this fragment were all active compounds ($freq_{ACT} \geq 0.84$). Malathion Fragment Set B covered the dimethoxyphosphinothioylthio

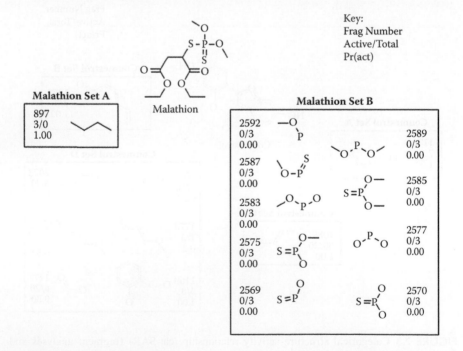

FIGURE 7.4 Categorical structure-activity relationship (cat-SAR) fragment analyses and predictions of the nonestrogenic compound malathion.

portion of the molecule and resulted in the overall prediction of inactivity for malathion because the chemicals that contributed to these fragments were all inactive compounds ($freq_{IN} \geq 0.66$).

Regarding the importance of the phenolic A ring in 17β-estradiol and other estrogenic compounds, our previous study with MCASE did not specifically identify it, only rather a biophore that indicated a *para*-substitute feature on an aromatic ring [33]. Interestingly, MCASE yields five biophores [33] versus ~1,000 important model fragments cat-SAR produced (Table 7.1). Moreover, as seen in the above examples, several different cat-SAR fragments for the phenolic A ring were identified. Together, this suggests that cat-SAR produced a more complex model for estrogenic activity, and it is also producing a more mechanistically interpretable model. As mentioned, the cat-SAR prediction for coumestrol consisted of important fragments that covered both the phenolic part of the molecule and its more water-soluble interior, but the MCASE RPE model misclassified coumestrol as inactive [33].

7.3.4 COMPARISON OF E-SCREEN MODELS TO OTHER SAR MODELS

Comparisons between the E-SCREEN cat-SAR models and other SAR models were conducted to assess the likelihood that these models might be related and have common underlying biological mechanisms of action using the CDA method. For these analyses, we considered cat-SAR models for the RPE and RPE E-SCREEN assay, rat carcinogenesis based on data from the Carcinogenic Potency Database (CPDB) [55], a model of mammary carcinogens versus nonmammary carcinogens, also based on data from the CPDB, *Salmonella* mutagens based on data from the National Toxicology Program [56], and human developmental toxicity based on data from Ghanooni et al. [57]. Additionally, the cat-SAR RPE and RPP models were compared to the previously published MCASE RPE and RPP models [33] to assess whether the two modeling programs were consistent.

To assess the overall applicability of the cat-SAR program to analyze environmental estrogens, several basic CDA analyses were performed. First, there was a significant overlap (112.3%, $p < 0.0001$) observed between the *Salmonella* mutagenesis and the rat carcinogenesis model (Analysis 1, Table 7.3). This result is consistent with the electrophilic theory of carcinogenesis [58] as well as with our previous CDA analyses of mutagens and carcinogens with the MCASE program [43].

Next, the CDA analysis between the cat-SAR RPE model and the MCASE RPE model (69.2%, $p < 0.0001$) (Analysis 2, Table 7.3) and between the cat-SAR RPP model and the MCASE RPP model showed a high degree of similarity (58.0%, $p < 0.0001$) (Analysis 3, Table 7.3). As such, the comparisons between the cat-SAR and MCASE models shows that the two modeling programs produce generally consistent results.

Next, comparison between the cat-SAR RPE and cat-SAR RPP models also showed a high degree of similarity (147.4.0%, $p < 0.0001$) (Analysis 4, Table 7.3). This suggests that while the RPE and RPP methods measure different estrogenic endpoints (that is, proliferative potency and proliferative effect relative), these endpoints are highly related.

Considering just cat-SAR analyses, comparison of the RPE model to one for *Salmonella* mutagenesis showed a significantly less than expected overlap (−43.6%,

TABLE 7.3
Mechanistic Relationship Analyses between cat-SAR E-SCREEN Models, MCASE E-SCREEN Models, Mutagens, Carcinogens, and Breast Carcinogens

Analysis	Observed[a]	Expected[b]	Δ[c]	100Δ/Expected[d]	p-Value[e]
Salmonella vs.					
1. Rat CPDB	1,524	718	806	112.3	<0.0001
RPE (cat-SAR) vs.					
2. RPE (MCASE)	494	292	202	69.2	<0.0001
RPP (cat-SAR) vs.					
3. RPP (MCASE)	678	429	249	58.0	<0.0001
RPE vs.					
4. RPP	1,331	538	793	147.4	<0.0001
RPE vs.					
5. *Salmonella*	395	700	−305	−43.6	<0.0001
6. Rat CPDB	389	629	−240	−27.6	<0.0001
7. Developmental toxicity	817	678	139	20.5	0.0002

[a] Number of compounds simultaneously identified.
[b] Number of compounds expected by chance.
[c] Difference of observed from expected.
[d] Percent difference from expected.
[e] Difference of two means test.

$p < 0.0001$) (Analysis 5, Table 7.3). Similarly, when comparing the RPE and rat carcinogenesis models, a significantly less than expected overlap was also observed (-27.6%, $P<0.0001$) (Analysis 6, Table 7.3). It should be noted that in our previous CDA analysis of *Salmonella* mutagenesis, rat carcinogenesis, and estrogenicity, no relationships were observed [33]. Although the reason for this difference between no relationships observed with MCASE and negative ones observed with cat-SAR is not clear at this time, both sets of analyses are consistent to the point that no positive relationship was suggested between estrogenicity as measured by the E-SCREEN RPE assay, mutagenicity, and rat carcinogenesis. Finally, as also seen with MCASE, there was a significant overlap between estrogenicity and developmental toxicity in humans (20.5%, $p = 0.0002$).

7.4   CONCLUSIONS

Overall, the cat-SAR validation results (Table 7.1), examples of cat-SAR predictions (Table 7.2 and Figure 7.1 through Figure 7.4), and the CDA analyses (Table 7.3) all provide evidence that the cat-SAR method is an applicable method for SAR analysis of environmental estrogens. The present analysis using the cat-SAR program again demonstrated the effectiveness of fragment-based SAR approaches to develop mechanistically sound and predictive models for categorizing compounds as either having estrogenic activity or not. As seen in Table 7.1, because the cat-SAR program is controllable regarding model sensitivity and specificity versus number of compounds it is capable of making predictions for, it has the potential to be especially useful for screening and prioritizing suspected environmental estrogens for subsequent testing. This is especially important for screening methods, wherein the rate of false negative predictions should be kept low so as not to miss toxicologically active compounds. Moreover, as we mentioned previously, because no computational screening mechanism is perfect, it seems prudent that both cat-SAR and MCASE could contribute as part of a battery of computational tools aimed at prioritizing suspected environmental estrogens for further testing.

ACKNOWLEDGMENTS

We gratefully acknowledge support for this work from the Congressionally Directed Medical Research Program for Breast Cancer Idea Award DAMD17-01-0376 and by NIH Grant Number P20RR018733 from the National Center for Research Resources.

REFERENCES

[1] R.J. Kavlock, G.P. Daston, C. DeRosa, P. Fenner-Crisp, L.E. Gray, S. Kaatari, G. Lucier, M. Luster, M.J. Mac, C. Maczka, R. Miller, J. Moore, R. Rolland, G. Scott, D.M. Sheehan, T. Sinks, and H.A. Tilson, *Research needs for the risk assessment of health and environmental effects of endocrine disruptors: A report of the U.S. EPA-sponsored workshop*, Environ. Health Perspect. Suppl. 104 (Suppl 4) (1996), pp. 715–740.

[2] M. Marselos and L. Tomatiz, *Diethylstilbestrol: I, Pharmacology, toxicology and carcinogenicity in humans*, Eur. J. Cancer 28A (1992), pp. 1182–1189.

[3] M. Marselos and L. Tomatiz, *Diethylstilbestrol: II, Pharmacology, toxicology and carcinogenicity in experimental animals*, Eur. J. Cancer 29A (1993), pp. 149–155.

[4] IARC, *Monographs on the Evaluation of the Carcinogenic Risk of Chemicals to Humans, Sex Hormones (II)*, International Agency for Research on Cancer, Lyon, 1979.

[5] F.S. vom Saal, M.M. Montano, and M.H. Wang, *Sexual differentiation in mammals*, in *Chemically-Induced Alterations in Sexual Development: The Wildlife/Human Connection*, T. Colborn and C. Clement, eds., Princeton Scientific, Princeton, NJ, 1992, pp. 17–84.

[6] J.L.E. Gray, *Chemical-induced alterations of sexual differentiation: A review of effects in humans and rodents*, in *Chemically-Induced Alterations in Sexual Development: The Wildlife/Human Connection*, T. Colborn and C. Clement, eds., Princeton Scientific, Princeton, NJ, 1992, pp. 203–230.

[7] B.B. Blair, *Immunologic studies of women exposed in utero to diethylstilbestrol*, in *Chemically-Induced Alterations in Sexual Development: The Wildlife/Human Connection*, T. Colborn and C. Clement, eds., Princeton Scientific, Princeton, NJ, 1992, pp. 289–294.

[8] H. Adlercreutz, *Phytoestrogens: Epidemiology and a possible role in cancer protection*, Environ. Health Perspect. 103 (Suppl 7) (1993), pp. 103–112.

[9] F.S. vom Saal, S.C. Nagel, P. Palanza, M. Boechler, S. Parmigiani, and W.V. Welshons, *Estrogenic pesticides: Binding relative to estradiol in MCF-7 cells and effects of exposure during fetal life on subsequent territorial behaviour in male mice*, Toxicol. Lett. 77 (1995), pp. 343–350.

[10] P. Palanza, K.L. Howdeshell, S. Parmigiani, and F.S. vom Saal, *Exposure to a low dose of bisphenol A during fetal life or in adulthood alters maternal behavior in mice*, Environ. Health Perspect. Suppl. 110 (Suppl 3) (2002), pp. 415–422.

[11] B.A. Cohn, M.S. Wolff, P.M. Cirillo, and R.I. Sholtz, *DDT and breast cancer in young women: New data on the significance of age at exposure*, Environ. Health Perspect. 115 (2007), pp. 1406–1414.

[12] S.H. Safe, *Environmental and dietary estrogens and human health: Is there a problem?*, Environ. Health Perspect. 103 (1995), pp. 346–351.

[13] W.V. Welshons, K.A. Thayer, B.M. Judy, J.A. Taylor, E.M. Curran, and F.S. vom Saal, *Large effects from small exposures. I. Mechanisms for endocrine-disrupting chemicals with estrogenic activity*, Environ. Health Perspect. 111 (2003), pp. 994–1006.

[14] V. Selvaraj, M.A. Zakroczymski, A. Naaz, M. Mukai, Y.H. Ju, D.R. Doerge, J.A. Katzenellenbogen, W.G. Helferich, and P.S. Cooke, *Estrogenicity of the isoflavone metabolite equol on reproductive and non-reproductive organs in mice*, Biol. Reprod. 71 (2004), pp. 966–972.

[15] T.B. Clarkson, M.S. Anthony, J.K. Williams, E.K. Honore, and J.M. Cline, *The potential of soybean phytoestrogens for postmenopausal hormone replacement therapy*, Proc. Soc. Exp. Biol. Med. 217 (1998), pp. 365–368.

[16] B.H. Arjmandi, *The role of phytoestrogens in the prevention and treatment of osteoporosis in ovarian hormone deficiency*, J. Am. Coll. Nutr. 20 (2001), pp. 398s–402s.

[17] B. Ettinger, G.D. Friedman, T. Bush, and C.P. Quesenberry, *Reduced mortality associated with long-term postmenopausal therapy*, Obstet. Gynecol. 87 (1996), pp. 6–12.

[18] V. Beral and M.W S. Collaborators, *Breast cancer and hormone-replacement therapy in the Million Women Study*, Lancet 362 (2003), pp. 414–415.

[19] H.D. Nelson, L.L. Humphrey, P. Nygren, S.M. Teutsch, and J.D. Allan, *Postmenopausal hormone replacement therapy: Scientific review*, J. Am. Med. Assoc. 288 (2002), pp. 872–881.

[20] J.A. Katzenellenbogen, *The structural pervasiveness of estrogenic activity*, Environ. Health Perspect. Suppl. 103(Suppl 7) (1995), pp. 99–101.

[21] Environmental Protection Agency (EPA), *Endocrine Disruptor Screening and Testing Advisory Committee (EDSTAC) Final Report* (1998). Available at www.epa.gov/endo/pubs/edspoverview/finalrpt.htm.

[22] Environmental Protection Agency (EPA), *Priority-setting in the Endocrine Disruptor Screening Program (EDSP) — Background* (2002). Available at www.epa.gov/endo/pubs/prioritysetting/index.htm.

[23] G. Timm, *Progress in implementing the U.S. Endocrine Disruptor Screening Program* (2002). Available at http://caat.jhsph.edu/programs/workshops/testsmart/endocrine 2002/proceedings/timm.htm.

[24] M.T.D. Cronin, J.D. Walker, J.S. Jaworska, M.H.I. Comber, C.D. Watts, and A.P. Worth, *Use of quantitative structure-activity relationships in international decision-making frameworks to predict ecological effects and environmental fate of chemical substances*, Environ. Health Perspect. 111 (2003), pp. 1376–1390.

[25] M.T.D. Cronin, J.S. Jaworska, J.D. Walker, M.H.I. Comber, C.D. Watts, and A.P. Worth, *Use of quantitative structure-activity relationships in international decision-making frameworks to predict health effects of chemical substances*, Environ. Health Perspect. 111 (2003), pp. 1376–1390.

[26] C.L. Waller, T.I. Oprea, K. Chae, H.-K. Park, K.S. Korach, S.C. Laws, T.E. Wiese, W.R. Kelce, and J.L.E. Gray, *Ligand-based identification of environmental estrogens*, Chem. Res. Toxicol. 9 (1996), pp. 1240–1248.

[27] W. Tong, R. Perkins, R. Strelitz, E.R. Collantes, S. Keenan, W.J. Welsh, W.S. Branham, and D.M. Sheehan, *Quantitative structure-activity relationships (QSARs) for estrogen binding to the estrogen receptor: Predictions across species*, Environ. Health Perspect. 105 (1997), pp. 1116–1124.

[28] W. Tong, R. Perkins, L. Xing, W.J. Welsh, and D.M. Sheehan, *QSAR models for binding of estrogenic compounds to estrogen receptor alpha and beta subtypes*, Endocrinology 138 (1997), pp. 4022–4025.

[29] W. Tong, D.R. Lowis, R. Perkins, Y. Chen, W.J. Welsh, D.W. Goddette, T.W. Heritage, and D.M. Sheehan, *Evaluation of quantitative structure-activity relationship methods for large-scale prediction of chemicals binding to the estrogen receptor*, J. Chem. Inf. Comput. Sci. 38 (1998), pp. 669–677.

[30] R.M. Blair, H. Fang, W.S. Branham, B.S. Hass, S.L. Dial, C.L. Moland, W. Tong, L. Shi, R. Perkins, and D.M. Sheehan, *The estrogen receptor relative binding affinities of 188 natural and xenochemicals: Structural diversity of ligands*, Toxicol. Sci. 54 (2000), pp. 138–153.

[31] L.M. Shi, H. Fang, W. Tong, J. Wu, R. Perkins, R.M. Blair, W.S. Branham, S.L. Dial, C.L. Moland, and D.M. Sheehan, *QSAR models using a large diverse set of estrogens*, J. Chem. Inf. Comput. Sci. 41 (2001), pp. 186–195.

[32] H. Hong, W. Tong, H. Fang, L. Shi, W. Xie, J. Wu, R. Perkins, J.D. Walker, W. Branham, and D.M. Sheehan, *Prediction of estrogen receptor binding for 58,000 chemicals using an integrated system of a tree-based model with structural alerts*, Environ. Health Perspect. 110 (2002), pp. 29–36.

[33] A.R. Cunningham, S.L. Cunningham, and H.R. Rosenkranz, *Structure activity approach to the identification of environmental estrogens: The MCASE approach*, SAR QSAR Environ. Res. 15 (2004), pp. 55–67.

[34] G. Klopman and S.K. Chakravarti, *Structure-activity relationship study of a diverse set of estrogen receptor ligands (I) using MultiCASE expert system*, Chemosphere 51 (2003), pp. 445–459.

[35] A.M. Soto, T.-M. Lin, H. Justicia, R.M. Silvia, and C. Sonnenschein, *An "in culture" bioassay to assess the estrogenicity of xenobiotics (E-SCREEN)*, in *Chemically-Induced Alterations in Sexual Development: The Wildlife/Human Connection*, T. Colborn and C. Clement, eds., Princeton Scientific, Princeton, NJ, 1992, pp. 295–309.

[36] A.M. Soto, C. Sonnenschein, K.L. Chung, M.F. Fernandez, N. Olea, and F.O. Serrano, *The E-SCREEN assay as a tool to identify estrogens: An update on estrogenic environmental pollutants.*, Environ. Health Perspect. 103 (Suppl 7) (1995), pp. 113–122.

[37] C. Sonnenschein, A.M. Soto, M.F. Fernandez, N. Olea, M.F. Olea-Serrano, and M.D. Ruiz-Lopez, *Development of a marker of estrogenic exposure in human serum*, Clin. Chem. 41 (1995), pp. 1888–1895.

[38] J.C. Dearden, *In silico prediction of drug toxicity*, J. Comput. Aided Mol. Des. 17 (2003), pp. 119–127.

[39] A.R. Cunningham, S.L. Cunningham, D.M. Consoer, S.T. Moss, and M.H. Karol, *Development of an information-intensive structure-activity relationship model and its application to human respiratory chemical sensitizers*, SAR QSAR Environ. Res. 16 (2005), pp. 273–285.

[40] D.R. Lowis, *HQSAR: A new, highly predictive QSAR technique* (1997). Available at www.tripos.com/data/SYBYL/HQSAR_Application_Note_072605.pdf.

[41] A.R. Cunningham, H.S. Rosenkranz, and G. Klopman, *Identification of structural features and associated mechanisms of action for carcinogens in rats*, Mutat. Res. 405 (1998), pp. 9–28.

[42] N. Pollack, A.R. Cunningham, G. Klopman, and H.S. Rosenkranz, *Chemical diversity approach for evaluating mechanistic relatedness among toxicological phenomena*, SAR QSAR Environ. Res. 10 (1999), pp. 533–543.

[43] H.S. Rosenkranz and A.R. Cunningham, *Prevalence of mutagens in the environment: Experimental data vs. simulation*, Mutat. Res. 484 (2001), pp. 49–51.

[44] L.M. Shi, Y. Fan, T.G. Myers, P.M. O'Connor, K.D. Paull, S.H. Friend, and J.N. Weinstein, *Mining the NCI anticancer drug discovery database: Genetic function approximation for the QSAR of anticancer ellipticine analogues*, J. Chem. Inf. Comput. Sci. 38 (1998), pp. 189–199.

[45] A.R. Cunningham, H.S. Rosenkranz, Y.P. Zhang, and G. Klopman, *Identification of "genotoxic" and "non-genotoxic" alerts for cancer in mice: The carcinogenic potency database*, Mutat. Res. 398 (1998), pp. 1–17.

[46] G.M. Anstead, K.E. Carlson, and J.A. Katzenellenbogen, *The estradiol pharmacophore: Ligand structure-estrogen receptor binding affinity relationships and a model for the receptor binding site*, Steroids 62 (1997), pp. 268–303.

[47] Tripos, *Unity* (2008). Available at www.tripos.com/data/SYBYL/Unity_072505.pdf.

[48] A.R. Cunningham, G. Klopman, and H.S. Rosenkranz, *A dichotomy in the lipophilicity of natural estrogens/xenoestrogens and phytoestrogens*, Environ. Health Perspect. Suppl. 105 (Suppl 3) (1997), pp. 665–668.

[49] A.R. Cunningham, H.S. Rosenkranz, and G. Klopman, *Structural analysis of a group of phytoestrogens for the presence of a 2-D geometric descriptor associated with non-genotoxic carcinogens and some estrogens*, Proc. Soc. Exp. Biol. Med. 217 (1998), pp. 288–292.

[50] H.S. Rosenkranz, A. Cunningham, and G. Klopman, *Identification of a 2-D geometric descriptor associated with non-genotoxic carcinogens and some estrogens and antiestrogens*, Mutagenesis 11 (1996), pp. 95–100.

[51] A.M. Brzozowski, A.C. Pike, Z. Dauter, R.E. Hubbard, T. Bonn, O. Engstrom, L. Ohman, G.L. Greene, J.A. Gustafsson, and M. Carlquist, *Molecular basis of agonism and antagonism in the oestrogen receptor*, Nature 389 (1997), pp. 753–758.

[52] A.C. Pike, A.M. Brzozowski, R.E. Hubbard, T. Bonn, A.-G. Thorsell, O. Engstrom, J. Ljunggren, J.-A. Gustafsson, and M. Carlquist, *Structure of the ligand-binding domain of oestrogen receptor β in the presence of a partial agonist and a full antagonist*, EMBO J. 18 (1999), pp. 4608–4618.

[53] A.K. Shiau, D. Barstad, P.M. Loria, L. Cheng, P.J. Kushner, D.A. Agard, and G.L. Greene, *The structural basis of estrogen receptor/coactivator recognition and the antagonism of this interaction by tamoxifen*, Cell 95 (1998), pp. 927–937.

[54] A.K. Hihi and W. Wahli, *Structure and function of the estrogen receptor*, in *Estrogens and Antiestrogens I*, M. Oettel and E. Schillinger, eds., Springer, Berlin, 1999, pp. 111–126.

[55] L.S. Gold, *Carcinogenic Potency Database* (2007), Available at http://potency.berkeley.edu.

[56] National Toxicity Program (NTP), *Results, status, and publication information on all NTP chemicals* (2004).

[57] M. Ghanooni, D.R. Mattison, Y.P. Zhang, O.T. Macina, H.S. Rosenkranz, and G. Klopman, *Structural determinants associated with risk of human developmental toxicity*, Am. J. Obstet. Gynecol. 176 (1997), pp. 799–806.

[58] J.A. Miller and E.C. Miller, *Ultimate chemical carcinogens as reactive mutagenic electrophiles*, in *Origins of Human Cancer*, H.H. Hiatt, J.D. Watson, and J.A. Winsten, eds., Cold Spring Harbor Laboratory Press, Cold Spring Harbor, NY, 1977, pp. 605–627.

8 Kohonen and Counterpropagation Neural Networks Employed for Modeling Endocrine Disruptors

Marjana Novič and Marjan Vračko

CONTENTS

ABSTRACT

A methodology is presented for modeling and classifying endocrine disruptors based on Kohonen and counterpropagation neural networks. Three different datasets were considered. The first dataset consists of 106 substances extracted from the list of 553 chemicals that were inspected by the European Union Commission for the scientific evidence of their endocrine disruption activity. For this dataset, we present the classification model designed for a preliminary assessment of potential endocrine disruptors, which would help the assessors to make the priority list for a large amount of chemicals that have to be tested with more expensive *in vitro* and *in vivo* methods. The second dataset consists of 132 compounds of known chemical structures, which were tested for their binding affinities to the mice estrogen receptor. We compared the counterpropagation neural network models for the prediction of relative binding affinity with two other multivariate modeling methods (partial least square regression, and error-back-propagation neural network. The results were assessed with the aim to get insight into the mechanisms involved in the binding of estrogenic compounds to the receptor. The third dataset encompasses 60 diverse chemicals tested for the binding affinity to human estrogen receptors α and β (ER-α and ER-β). To obtain the structure-activity relationship, the three-dimensional (3D) structures of ligands and receptors were taken into consideration. Structural features of ligands having the strongest influence to the binding affinities were investigated.

KEYWORDS

Binding affinity
Counterpropagation neural network
Endocrine disruptors
Estrogen receptor
List of chemicals of the European Commission

8.1 INTRODUCTION

Over decades, the epidemiological and scientific evidence has raised awareness that some chemicals present in our environment have adverse impacts on the reproductive systems of human and wildlife species. For example, epidemiological studies showed the increase of cancer in human reproductive organs (men and women), there is evidence of a decrease of sperm count in men, or of reduced fertility. In wildlife species, several abnormalities have been observed, such as an increase in birth abnormalities, abnormalities in mating behavior, or feminization/masculinization. Although all of these changes can be caused by different factors, one suspects the common origin — the disruption of the endocrine system. The most sensitive targets for endocrine disruptors is human and animal reproductive systems. The OECD (Organization for Economic Cooperation and Development) adopted the definition of reproductive endocrine disruptors as follows [1]:

> A sex hormone–disruptor is an exogenous substance that causes adverse health effects related to the reproductive function of an intact organism or its progeny, consequent to changes in endocrine function.
>
> A potential sex hormone-disruptor is a substance that possesses properties that might be expected to lead to endocrine disruption of the reproductive processes in an intact organism.

The mechanisms of hormone disruptor–related biological activity are complicated and manifold. Basically, these chemicals may act as agonists, when they bind to particular physiological receptors in the cells of sex organs, or as antagonists, when they block or reduce the binding of natural hormones. In both cases, a potential hormone-disruptor expresses any kind of structural similarity to natural hormones. Further modes of action are modification of postreceptor pathways within the cells causing alteration of the functioning of the organism as a whole. Such examples include interference in the hypothalamic–pituitary–gonadal axis, or in the neurotransmitters in the central nervous system. Alternatively, a chemical may be directly involved in activation or inhibition of enzymes, causing changes in the production of hormones or changes of the carrier proteins in the blood, thus disrupting the delivery of hormones or feedback control.

The regulators, which control the chemical market and consumption of chemicals, are faced with a very difficult problem. Some chemicals represent a potential danger due to their endocrine disruption activity. But, it is not possible to design a single test that could definitively clarify if a chemical acts as an endocrine disruptor or not. None of the tests reported in the OECD Test Guidelines was specifically designed to directly detect endocrine disruption. However, various modifications of the tests used to detect different adverse effects can be applied to detect effects related to endocrine disruption. For example, in the mammalian acute toxicity test, a single dose can induce effects in reproductive organs. These effects are mostly related to estrogen agonistic or antagonistic activity of chemicals [2–4]. A second example represents the mammalian subchronic toxicity tests [5–8] that could be modified in order to consider the endpoints of relevance to endocrine disruption activity, similar to the mammalian long-term toxicity test [9–11] or the mammalian teratogenicity test [12]. Due to the nature of these tests, they can be performed for only a limited number of compounds. Regulators express an urgent need for new *in vitro* or *in silico* approaches to assess a large number of chemicals. OECD proposes using SAR (Structure-Activity Relationship) analysis at the screening level (for example, in the elucidation of mechanisms of activity and to set testing priorities). Furthermore, the QSAR (Quantitative Structure-Activity Relationship) methods are applied when specific targets are considered (for example, the binding of compounds on a specific sex hormone receptor). Many authors have developed QSAR models to predict the endocrine activity of chemicals. The estrogen receptors are the main targets of these studies. Kovalishyn et al. [13] applied the volume learning algorithm artificial neural network and partial least squares (PLS) to model relative potency to estrogen receptor. Liu et al. [14] applied different classification methods on a set of 232 chemicals treating the estrogen binding activity. Vedani et al. [15] studied six proteins as possible targets for 693 chemicals. Devillers et al. [16] reported the SAR study performed on a set of 11,416 chemicals for which 13 different endocrine activities were considered. Harju et al. [17] studied a set of 26 brominated flame retardants for their endocrine disruption activity. The *in vitro* assays included interactions with androgen, progesterone, estrogen, and dioxin (aryl hydrocarbon) receptors, the competition with thyroxin, and inhibition of estradiol sulfation. Metabolic rates were also considered. Ghafourian and Cronin [18] studied estrogenic receptor binding affinity using a three-dimensional approach (CoMFA, comparative molecular field analysis). They compared various linear methods like stepwise regression, PLS, and recursive partitioning to nonlinear methods such as counterpropagation neural network and support vector machine. Tamura et al. [19] studied the androgen receptor activity of a list of chemicals, SPEED 98 issued from the Ministry of Environment, Japan. They used CoMFA and studied agonist and antagonist activity. Saliner et al. used a two-descriptor decision tree to classify the compounds as active or inactive endocrine disruptors [20]. Two descriptors that measured the activation of estrogenic genes were used. The model has been extensively commented on in terms of OECD principles for validation of QSAR models [21]. Marchand-Geneste and coworkers performed a QSAR study on rainbow trout estrogen receptor as a homolog to human receptor [22]. Asikainen et al. applied decision trees, learning vector quantization, and k-nearest neighbor methods to classify active and nonactive estrogen compounds. A set of 311 compounds was used to build the model based on DRAGON descriptors [23]. Akahori et al. used a two-step method (discriminant analysis and multilinear regression) to study the binding of chemicals to the estrogen receptor [24]. Zhao et al. [25] compared three modeling techniques — multiple linear regression, radial basis function neural network, and support vector machine — to study a set of 146 natural, synthetic, and environmental chemicals and their relative binding affinity to androgen receptor. Tong et al. [26] proposed an ensemble of decision trees to classify the chemicals according to their ability to bind to the estrogen receptor.

In this chapter, we present three concepts of QSAR modeling related to endocrine disruption activity elaborated in several case studies during the last few years [27–30]. As mentioned above, endocrine disruption is a complex biological process that involves many different mechanisms, and to get a mechanistic insight into the entire process, a series of different tests must be performed for each chemical. This is a costly and time-consuming process that, in addition, requires a lot of testing on animals, raising ethical concerns. But, the number of chemicals put on the market is increasing rapidly and some of them may act as endocrine disruptors. Because all chemicals cannot be experimentally tested, there is a need for alternative, computational, or *in silico* methods for the prediction of their biological activity [31,32].

Our first proposal was to construct a robust model that gave information about the possibility of a compound to be an endocrine disruptor. The European Union Commission reported about the candidate list of 553 man-made substances that are potential endocrine disruptors and 9 synthetic/natural hormones [33]. We applied a counterpropagation neural network with an architecture that is suitable for classification. More details about the dataset and modeling technique are given in Section 8.2.

Our second proposal was to study the endocrine disruption on a model system, in our case, mice. On a set of 132 compounds, we studied the binding activity to a mice estrogen receptor. We compared the counterpropagation neural network models for the prediction of relative binding affinity with two other multivariate modeling methods — PLS and error-back-propagation neural network.

Our third proposal represented the study of binding affinity to human ER-α and ER-β. The dataset consisted of 60 compounds. To obtain the SAR, the 3D structures of ligands and receptors were taken into consideration. We compared two approaches in 3D structure determination — host independent conformations obtained by minimal energy optimization, and host dependent conformation originating from the ligands docked in the host protein. Structural features of ligands having the strongest influence to the binding affinities were investigated.

8.2 SELF-ORGANIZING MAPS AND COUNTERPROPAGATION NEURAL NETWORK

A self-organizing map (SOM) or Kohonen artificial neural network is a basic type of neural network. Its architecture represents a network of neurons organized in a rectangular or hexagonal two-dimensional (2D) lattice. The training, which runs iteratively, is a mapping from multidimensional descriptor space into a 2D lattice in a way that similar objects are located close to each other. The result of training is a network where objects populate neurons in a way that a neuron can be empty or populated with one or more objects. It is to emphasize that SOM are unsupervised models, which means that only input variables (descriptors) are involved in the training. It is obvious that such a network shows similarity relationships among objects (that is, one can indicate clusters or outliers within the dataset). Further application of SOM is the division of data into a training and a test set. The test set must be within the applicability domain of the model or, in other words, both sets must equivocally cover the entire information space. To get an even partition of data, one usually divides the entire network into subparcels, and from each subparcel some objects are assigned to the training set and others to the test set. Another application of SOM is the selection of descriptors. In this procedure, a SOM is trained with the transposed matrix (that is, with the data matrix where indices indicating objects or descriptors are exchanged). A result of such training is a network where neurons are occupied with descriptors and for further modeling only one or two descriptors from each neuron are selected. An additional criterion, from which descriptors are selected, is the Euclidian distance between descriptors and the neuron. The Euclidian distance is expressed in Equation 8.1 where $X^T_{j,s}$ and $W_{j,i}$ indicate descriptors and neuron weights, respectively. Usually, the descriptors with the shortest and the largest distances are selected.

$$d_{s,i} = \sqrt{\sum_{j=1}^{N_{mol}} (X^T_{j,s} - W_{j,i})^2} \tag{8.1}$$

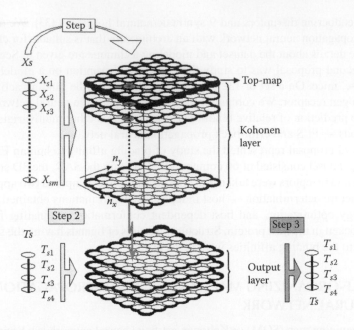

FIGURE 8.1 Counterpropagation artificial neural network architecture. The circles represent the weights of individual neurons in the Kohonen and output layers, as well as the components of the molecular structure representation vector and target vector shown vertically on the left side of the plot. Step 1, mapping of the molecule X_s into the Kohonen layer; Step 2, correction of the weights in both Kohonen and output layers; Step 3, prediction of the four-dimensional target T_s. The position of the molecule after Step 1 ($n_x = 4$, $n_y = 1$) is visualized by a tiny dashed vertical line and gray circles within the Kohonen layer and in the top-map. (From A. Roncagnioli et al., *J. Chem. Inf. Comput. Sci.*, 44, 304, 2004. With permission.)

A counterpropagation artificial neural network (CP ANN) is a generalization of the SOM. The architecture of CP ANN is shown in Figure 8.1. It consists of two layers: the input layer is the same as in SOM, and the output layer is associated with output values (properties). The training runs in two steps. In the first, unsupervised step, the input layer is constructed as described above — objects are organized considering the descriptors and similarity relationships among them. In the second, the supervised step, the positions of objects are projected to output layers where the weights are adjusted to their property values. In the prediction phase (when a new object is presented to the model, it is first situated into the input layer on the best-fitting neuron), the position of the neuron is projected to the output layer, which gives the predicted property value. The distance between the selected neuron and the new object can now be calculated. It is a measure of how well a new object fits in the applicability domain of the model. If the outputs (properties) are expressed as real numbers (for example, biological activity), the dimension of the output layer is one. If the outputs are classes, the output layer is multidimensional or, more precisely, the dimension of the output layer is equal to the number of classes (p). A corresponding target is a p-dimensional vector with the elements zero or one. The value one on jth place indicates that the object belongs to the jth class. During the training, all p layers are constructed independently; after the training, each weight has value between zero and one. For the final prediction of classes, the response surface values must again be transformed into discrete values, zeros and ones. The threshold value between 0.01 and 0.99 must be determined for each of the p classes. Below the threshold, all predictions are negative and denoted by a zero; the predictions above the threshold are positive and denoted by one. The threshold is determined according to the number of correct/wrong class predictions or by taking into account additional criteria. Different cases can occur, first, one value is above the threshold and other values are very small. In this case, the object is unambiguously classified to the class with highest value. Second, two vector components are similar. In this case, the object is classified between two classes. It is our decision to accept or reject such a classification. Third, three or all vector components are about the same. Here a model cannot make any decision. We know *a priori* that the model cannot classify the object, and this is valuable information.

8.3   CASE STUDY 1

8.3.1   DATA

The initial pool of substances, which was taken from the report of the EU Commission, was a set of 553 man-made chemicals suspected to act as endocrine disruptors [33]. The substances were classified by the EU Commission according to the data from the literature on several effects related to the endocrine disruption potency.

For further study, a subset of 146 compounds was selected (see table 2 in [27]) in which two additional criteria were investigated: high production volume compounds (that is, the compounds that production or import exceeds 1,000 tonnes per year on the EU market, and high persistent compounds). Compounds were classified into three categories. From the subset of 146 compounds, 39 were additionally excluded because their structures were not uniquely defined. For compounds like polymers, salts, mixtures of isomers, or mixtures, we cannot obtain the structural descriptors necessary for the modeling. The remaining 107 structures were designed and then optimized with the MOPAC program (MOPAC 93 for Windows IBM-PC compatible), using AM1 or PM3 semiempirical method. A small group of tin (Sn) compounds were optimized within PM3 approximation, which has a proper parameterization for the Sn atom.

The endocrine disruptor (ED) categories associated with the 106 compounds from the dataset were defined as follows:

- *Category 1: Endocrine disruptor*   At least one study was found providing evidence of endocrine disruption in an intact organism. Not a formal weight of evidence approach.
- *Category 2: Potential endocrine disruptor — In vitro* data indicating the potential for endocrine disruption in intact organisms. Also includes effects *in vivo* that may, or may not, be ED mediated. May include structural analyses and metabolic considerations.

- *Category 3: Nonendocrine disruptor* — No scientific basis for inclusion in list of endocrine disruptors.
  - 3A: No certain evidence for non-ED: No data available on wildlife-relevant or mammal-relevant endocrine effects.
  - 3B: Some evidence for non-ED: Some data are available, but the evidence is insufficient for identification.
  - 3C: Certain evidence for non-ED: Data available indicate no scientific basis for inclusion into the list of active ED chemicals.

For the purposes of our study, categories 3A and 3B were merged into one class, and 3C was taken as the fourth class ($p = 4$). We decided to split the third category into two classes because the uncertain evidence of 3A and 3B is not strong enough for such an important decision. On the other hand, it is essential for the modeling that the training dataset contains compounds that are nonactive.

8.3.2 Modeling Strategy

To obtain the descriptors, the following working scheme was adopted:

- 3D structure optimization (AM1 and PM3 semiempirical methods for minimization of total molecular energy) to obtain atom coordinates.
- From CODESSA (COmprehensive DEscriptors for Structural and Statistical Analysis), five classes of structural descriptors were obtained: constitutional, geometrical, topological, electrostatic, and quantum-chemical [34].
- In addition to the descriptors calculated by CODESSA from molecular 3D coordinates, the octanol/water partition coefficient log P was added to the descriptor pool [35]. The experimental values for a great part of the chemicals were taken from the literature [36,37], a few additional values were added from Hansch's book [38], and others were estimated with the KowWin program [39].
- All descriptors were autoscaled (that is, normalized with mean = 0 and standard deviation = 1).

The CP ANN with architecture suitable for classification was applied as the modeling method (see Section 8.2).

8.3.3 Results and Discussion

8.3.3.1 Training-Test Set Division

For splitting the data into training set and test set, we applied SOM as described in Section 8.2 [42,43]. The dimension of SOM was 5×5, which enables the mapping of objects onto 25 positions. A part of a representative object from each position in the Kohonen map was chosen for the training set, taking into account the original proportion among different classes and the predefined 2:1 ratio between training and test objects. Seventy-one compounds were assigned to the training set, and 35 compounds were assigned to the test set.

8.3.3.2 Descriptor Selection

For selecting the descriptors, we applied Kohonen mapping using "the transposed matrix method." We used a SOM with $5 \times 5 = 25$ neurons producing a map with 25 positions. All 266 descriptors were placed onto these 25 positions (neurons), meaning that each neuron was occupied on average by 11 descriptors. Only two descriptors from each neuron were chosen for final representation of molecular structure, one with the smallest and one with the largest distance from the excited neuron. The distribution of objects (descriptors) in the 5×5 top-map and the distances of all objects on one neuron were examined. The network trained for 300 epochs was chosen for the final descriptor selection procedure because of the most even distribution of objects and small differences between the maximal and minimal distances calculated at each neuron. The reduced set contained 50 descriptors, two from each neuron: the most similar one and most different one regarding the distance from the particular neuron (Equation 8.1). These descriptors are further referred to as Reduced Descriptors 1 (RD1). Analysis showed that there were eight neurons at which the clusters of descriptors were homogenously distributed with calculated differences between the maximal and minimal distances below 2.4 ($1.0 < d_{max} - d_{min} < 2.4$). To get a larger reduction of the descriptor set, only one descriptor from each of these neurons was taken. From the remaining 17 neurons, two descriptors were selected, yielding $8 + 34 = 42$ descriptors. These are further referred to as Reduced Descriptors 2 (RD2). Three different sets of descriptors were prepared: the nonreduced set of 266 descriptors, the reduced set of 50 descriptors (RD1), and the reduced set of 42 descriptors, chosen on the basis of the above-described selection procedure. The reduced sets of descriptors were distributed among the four types of descriptors as follows: constitutional 34 (all), 4 (RD1), 4 (RD2); topological 36 (all), 4 (RD1), 4 (RD2), geometrical 12 (all), 2 (RD1), 1 (RD2), electrostatic 57 (all), 10 (RD1), 8 (RD2), and quantum-chemical 127 (all), 30 (RD1), 25 (RD2). In addition, the octanol/water partition coefficient log P was added to each set of descriptors. According to the three different descriptor selections described above, the structures of all 106 molecules were encoded, and three datasets were obtained:

- Dataset 1 (DS1) → 106 molecular structures represented by 267 descriptors
- Dataset 2 (DS2) → 106 molecular structures represented by 51 descriptors
- Dataset 3 (DS3) → 106 molecular structures represented by 43 descriptors

8.3.3.3 Construction and Validation of Models

Datasets DS1, DS2, and DS3 were divided into a training set (71 molecules) and a test set (35 molecules). As described, the responses in the output layer, which consists of four levels, are real numbers between zero and one ($Out = (out_1, out_2, out_3, out_4)$, $0.00 \le Out_j \le 1.00$). The crucial point is to determine the threshold value (T^+) above which the prediction for a jth class is positive (confirmative). T^+ enables the transformation of the model output values to discrete class predictions — one for a confirmative and zero for a rejecting answer. There are four classes, so we need four threshold values for each of the constructed models (T_j^+, $j = 1$, 4). Below the threshold, all predictions are rejected and denoted by a zero, which means that the compound does not belong to the jth class, but the predictions above the threshold are positive and denoted by one (the compound belongs to the jth class). Figure 8.2 shows examples of the determination of T_j^+ for one of models.

One considers two extreme situations. If the T_j^+ is close to zero, the predictions of the jth class for most of the molecules from the training set will be confirmed ($Out_j > 0$). So the molecules that are really in the jth class will be correctly predicted, but the predictions for those that are not will be wrong, or false positive. On the other hand, if the T_j^+ is close to one, the majority of predictions will be rejecting for class j. Now the predictions of molecules that are actually in the jth class will be wrong, or false negative. One way to define the threshold is to select the point where the cumulative error of training set (sum of false positive and false negative predictions) is the lowest. The resulting models were validated by checking the class predictions for 35 test molecules. The misclassification tables obtained by comparison of actual and predicted classes of test compounds are shown in Figure 8.3.

The resulting predictions from all models were inspected to choose the optimal model. The prediction performance of 12 models demonstrated in Figure 8.3 includes the examples with the largest number of correct predictions (sum of the diagonal elements). However, to choose the optimal model, we have to consider additional criteria, such as the lowest number of false negative predictions. From a regulatory point of view, false negatives represent more severe errors than false positives, because in this case a harmful compound is classified as a nontoxic one. An important indicator about the quality of a model is the sum of predictions that are wrong for more than one category. For example, the element at position (1,4) in model (a) of Figure 8.3 is equal to 1. It stands for a prediction of one molecule as being the class E, but in fact it is N (nonactive), which is three categories lower. Taking into account all listed possible criteria, it was found that the lowest total number of false predictions was 11 (Figure 8.3, model (a) DS1; 9×9; 100 epochs), the lowest number of false negative predictions was 2 (Figure 8.3, models (c) DS1; 12×12; 100 epochs, and (i) DS3; 9×9; 100 epochs), and the model with zero predictions erroneous for more than one class was model (k) DS3; 12×12; 100 epochs. It is not a straightforward decision which model to propose to be the best among the four listed above. At first glance, 69% of correct prediction in model (a) is the best result. It suggests that the descriptors' space covered by the training compounds contains significant information to obtain the model, which makes the relationship between structure and toxicity class general enough to reasonably predict classes of test compounds. However, if the QSAR model is used to make the priority list of compounds that have to be tested by more assured *in vivo* methods, model (k) is better, because it makes the range-list of tested chemicals from most to least harmful less erroneous (a mistake is never larger than for one class). It is to emphasize that the predictions of all of the above-described models refer to the test set compounds, which did not participate in the training procedure. Considering model (a) from Figure 8.3, the information about the SAR contained

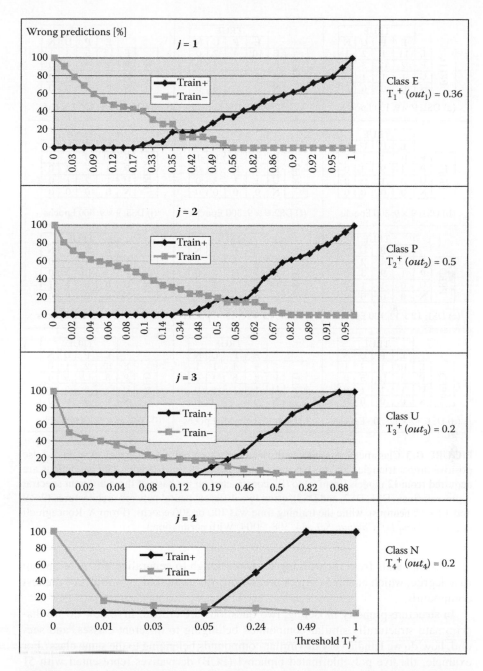

FIGURE 8.2 The thresholds T_j^+ determined for the class predictions in the model from the counterpropagation neural network of 9×9 neurons, trained for 100 epochs. The diamonds and squares stand for positive (confirmative) and negative (rejecting) predictions, respectively. (From A. Roncagnioli et al., *J. Chem. Inf. Comput. Sci.*, 44, 307, 2004. With permission.)

PRED.	TRUE			
	E	P	U	N
E	11	1	2	1
P	2	12	3	0
U	1	1	1	0
N	0	0	0	0

(a) DS1; 9 × 9; 100 Epochs

PRED.	TRUE			
	E	P	U	N
E	10	4	2	0
P	3	9	3	0
U	1	1	1	1
N	0	0	0	0

(e) DS2; 9 × 9; 100 Epochs

PRED.	TRUE			
	E	P	U	N
E	12	5	3	1
P	2	9	3	0
U	0	0	0	0
N	0	0	0	0

(i) DS3; 9 × 9; 100 Epochs

PRED.	TRUE			
	E	P	U	N
E	10	4	1	1
P	4	9	3	0
U	0	1	2	0
N	0	0	0	0

(b) DS1; 9 × 9; 300 Epochs

PRED.	TRUE			
	E	P	U	N
E	9	2	0	1
P	4	11	3	0
U	1	1	3	0
N	0	0	0	0

(f) DS2; 9 × 9; 300 Epochs

PRED.	TRUE			
	E	P	U	N
E	10	5	1	1
P	3	7	4	0
U	1	2	1	0
N	0	0	0	0

(j) DS3; 9 × 9; 300 Epochs

PRED.	TRUE			
	E	P	U	N
E	13	4	2	1
P	1	9	3	0
U	0	1	1	0
N	0	0	0	0

(c) DS1; 12 × 12; 100 Epochs

PRED.	TRUE			
	E	P	U	N
E	10	3	2	1
P	3	11	2	0
U	1	0	2	0
N	0	0	0	0

(g) DS2; 12 × 12; 100 Epochs

PRED.	TRUE			
	E	P	U	N
E	10	4	0	0
P	4	10	5	0
U	0	0	1	1
N	0	0	0	0

(k) DS3; 12 × 12; 100 Epochs

PRED.	TRUE			
	E	P	U	N
E	10	1	2	1
P	3	12	3	0
U	1	1	1	0
N	0	0	0	0

(d) DS1; 12 × 12; 300 Epochs

PRED.	TRUE			
	E	P	U	N
E	9	4	1	1
P	4	8	4	0
U	1	1	1	0
N	0	0	0	0

(h) DS2; 12 × 12; 300 Epochs

PRED.	TRUE			
	E	P	U	N
E	10	2	1	0
P	3	12	5	0
U	1	0	0	1
N	0	0	0	0

(l) DS3; 12 × 12; 300 Epochs

FIGURE 8.3 Classification tables with the number of correct (diagonal elements), false positive (upper triangle), and false negative predictions (lower triangle). The predictions are acquired from 12 models [from (a) to (l)], constructed on the basis of three different spectral representations (DS1, DS2, and DS3), using two different neural network architectures (9 × 9 and 12 × 12 neurons), while the training time was 100 or 300 epochs. (From A. Roncagnioli et al., *J. Chem. Inf. Comput. Sci.*, 44, 308, 2004. With permission.)

in 71 compounds from the training set was enough to generalize this relationship to a degree, which enabled correct predictions for more than two-thirds of the test compounds.

In structure-property modeling, two questions are crucial: first, how do we discriminate structurally similar compounds belonging to different classes, and second, how do we handle very dissimilar compounds belonging to the same class? For example, the five polychlorinated biphenyl (PCB) derivatives represented with 51 descriptors (DS2) are structurally so similar that they occupy the same neuron. Two compounds from the test set are typed in bold; the other three are from the test set, yet they occupy the same neuron. The model was not able to discriminate between these compounds, which were at the end of the training located on the same neuron in the network. They belong to three different classes, which causes a conflicting situation in the network. An additional criterion for selection of the final model is the low number of conflicts.

Unfortunately, the number of compounds declared as nontoxic (class N) in this study was low. It is not unusual that the experimental data of very toxic compounds is available in a larger extent than for less harmful compounds. This is a consequence of circumstances that, for many reasons, from economical to ecological ones, make it more urgent to obtain the toxicity tests of the compounds that are likely to be harmful (pesticides, herbicides, phytoestrogens, drugs). With a dataset of more even distribution of compounds between categories, the prediction results obtained by the QSAR models proposed in this paper would certainly be improved.

8.4 CASE STUDY 2

8.4.1 DATA

A dataset of 132 compounds with their relative binding affinities (RBAs) to rat uterus estrogen receptor known was collected from the literature and from the Web page of the U.S. Environmental Protection Agency (EPA) [44,45]. The available sets were pruned for modeling purposes according to the previously defined criteria. The CAS numbers, log RBAs, and ID numbers of the molecules were published elsewhere [28].

8.4.2 MODELING STRATEGY

8.4.2.1 Descriptor Generation

The descriptors were calculated from 3D molecular structures. Structures were optimized within AM1 approximation using MOPAC program package [46]. For calculation of descriptors, the program CODESSA was applied [34]. Calculated were 280 descriptors classified as constitutional, topological, geometrical, electrostatic, and quantum-chemical. Additionally, the octanol/water partition coefficient (log P) was added, with values obtained from the experimental database or estimated with the KowWin program [36–39]. All descriptors were autoscaled (that is, normalized with mean = 0 and standard deviation = 1).

8.4.2.2 Modeling Methods

Results of three methods — PLS regression, CP ANN, and Error Back Propagation ANN (BP ANN) — were compared. Initially, the ANN models were built using all the computed descriptors and validated with the leave-one-out procedure. Successively, genetic algorithms (GAs) were used to select the relevant descriptors.

PLS projection to latent structures is one of the most used regression techniques in QSAR. Mathematically, it is a linear projection of independent variables (descriptors) and dependent variables (property) onto a single space of latent variables. Then a linear inner relationship is modeled between the projections of the dependent and independent variables. PLS-R appears as a widely applied and accurate predictive modeling method [47], although due to its linear character, it may appear as an oversimplification of the actual relationship between independent and dependent variables.

CP ANN [40–42] was described in Section 8.2, where it was used to build a classification model. In this second example, the aim is to build a model for prediction of one property — the relative binding affinity RBA. The supervised part of the CP ANN has the same structure as described above; in contrary to the previous case, the output layer consists only in one response surface.

The multilayer feedforward organization of the units is the most used neural network architecture [48] for the applications in chemistry. It consists of computational units organized into three kinds of layers: input, hidden, and output layers. The units (neurons) in each layer all receive the same information — an output vector ($\mathbf{X}$) from the previous layer in turn sends its output vector as input to the neurons in the successive layer. The output of an individual neuron is calculated as a sigmoid function of the input signals. In the jth layer, the output (sf) is calculated as follows:

$$sf_j = \frac{1}{1 + \exp(-\sum_{i=1}^{m} w_{ji}x_i)} \tag{8.2}$$

The units of the input layer receive their input in the form of a data file, and the units of the output layer produce the output signal, which is the overall result of the network. This multilayer architecture is often used in conjunction with the back-propagation weight update rule, according to which a supervised form of learning is implemented. The error back-propagation algorithm (BP) is essentially an iterative weight update on the basis of a steep descent criterion, so as to minimize the root-mean square error (RMS) between the desired and the actual target of the network.

8.4.2.3 Genetic Algorithms (GAs)

For the reduction of descriptor number and the selection of relevant descriptors, a GA was applied. It is an advanced optimization technique that mimics the selection processes occurring in nature, Darwinian evolution: crossover, mutation, and selection [49–51]. The main idea is the survival of individuals having the highest fitness score; under specified conditions they are the most likely to prevail in the next generation. Moreover, crossover and random mutations introduce genetic variety to the offspring, so that an increasingly wide solution space is spanned during the computing. In this study, GAs were used in combination with ANN techniques, to select relevant descriptors.

In particular, computational GA consists of four steps: In a first step, an initial population of chromosomes has to be generated. The chromosomes are represented as binary strings of bits (zeros and ones) and indicate whether or not a descriptor is taken in a modeling experiment. In this work, 50 chromosomes were randomly generated as the initial pool of each population used. The code "1" in the ith position of the chromosome indicates the inclusion of the ith variable to the modeling procedure. In a second step, the performance of each chromosome must be evaluated. With the indicated variables, a model is generated and the fitness score is computed as the predictive ability over the test set expressed in terms of the r_{test}, the correlation coefficient between the actual and the predicted target property. The chromosomes are ranked according to a decreasing value of the fitness score. In a third step, the best chromosome is "protected" and copied without further modification to the next generation, and the remaining chromosomes of the child offspring are created by crossover (that is, mutually exchanging a selected part in pairs of randomly selected chromosomes). In the last step, a random mutation is introduced to modify a single position of a chromosome by changing the value of one of its bits.

All four steps are repeated until a stopping criterion is satisfied; in this work the criterion was a maximal number of generations.

8.4.3 RESULTS AND DISCUSSION

First we applied the PLS-regression technique to model the relationship between the log RBA and 281 molecular descriptors. The predictive ability of the model was assessed by the leave-one-out cross-validation (LOO-CV). Using all 281 input variables as descriptors, initially only one significant component was extracted from the dataset on the basis of the Q^2 value resulting in a model with an R_X^2 and R_Y^2 of 0.28 and 0.36, respectively, and with $Q^2 = 0.31$. Here, Q^2 is the fraction of the total variation of the dependent variable that can be predicted by a component according to the LOO-CV, and   and R_Y^2 are fractions of the sums of squares (SS) of all the independent and dependent variables explained by the chosen component [52]. Despite the insignificant second component, the use of a three-component model increased the Q^2 to 0.41, with R_X^2 and R_Y^2 rising correspondingly to 0.59 and 0.55, respectively. In the PLS-R method, the importance of individual variable was evaluated with the "variable importance in projection" (VIP) index. It accounts for the contribution of each individual descriptor for explaining the variable, summed over the model dimension, and has been computed according to the formula reported in the SIMCA-P User's Guide © Umertics AB [52]. The descriptors with a VIP value larger than 1 are the most relevant to model the variation in the dependent variable, so at first all the variables whose contribution was less than 1 were ruled out. The following variables were on the top of the list: HOMO-1 energy, information content (order 1), principal moment of inertia A, bonding information content (order 2), total molecular 2-center exchange energy, information content (order 2), PPSA-1 partial positive surface area [Semi-MO PC], moment of inertia C, complementary information content (order 0), complementary information content (order 1), WPSA-1 weighted PPSA (PPSA1*TMSA/1000) [Zefirov's PC], maximum total interaction for a C-C bond, count of H-donor sites [Semi-MO PC], count of H-donor sites [Zefirov's PC], and log P. Successively, the other variables were iteratively excluded until an optimal model (in terms of Q^2) was obtained. At last, 68 descriptors (reported in [29] together with their individual VIP score) were retained to build the optimal PLS model. The resulting three-component model showed an improved modeling ($R_X^2 = 0.74$ and $R_Y^2 = 0.64$) and predictive ($Q^2 = 0.62$) ability. Three factors seem to significantly influence the value of the relative binding affinity for the investigated set of compounds. The high positive coefficient for log P and the corresponding negative contribution from the absolute number of oxygen atoms suggests that the polarity of the sample is involved in modulating the binding of the endocrine disruptor to the receptor site: specifically, less polar molecules, characterized by a high octanol/water partition coefficient and by a small number of oxygen atoms, are supposed to be more favored with respect to highly polar ligands. Furthermore, the highly negative coefficients corresponding to the moments of inertia along an orthogonal axis C and the first principal axis A indicate that small and flexible molecules are privileged over bulky ones. Last, a significant positive correlation is observed between the dependent variable and the final heat of formation of the molecule, the energy of the HOMO-1, and the maximum total interaction for a C-C bond. This issue suggests that probably the binding of the ligand to the receptor site involves a certain degree of electron transfer. High HOMO-1 energy (the second highest energy of valence electron) indicates that less-bound valence electrons enhance the ligand–protein complex formation. On the other hand, the van der Waals interactions, which are present and important for the binding mechanism, are accounted for by the polarity terms described before. Several CP ANN were tested with LOO CV and some of the models performed better than the PLS models, some are even better in cross-validated prediction. The best CP ANN performed with a R_Y^2 value of 0.88 and Q^2 = 0.62. To look for a chemical interpretation of the result obtained by the use of CP ANN, both the top-map and the weight maps were examined. The top-map is a 2D representation of the samples projected onto the neurons organized in a quadratic neighborhood called Kohonen layer and can account for the "quality" of the projection into the Kohonen layer. Through inspection of the weight maps (that is, the distribution of the values of each input and of the output along the Kohonen layer) and the output layer, one can find correlation between each descriptor and the dependent variable. Careful inspection of all the weights confirmed the considerations reported above on the PLS section. Most of the correlations observed in the interpretation of the PLS-R coefficients are also indicated in the CP ANN model. In particular, the high positive correlation between the log P value and the relative binding affinity was found, as shown in Figure 8.4 (the contours around high values on both plots coincide). Contrary, a negative correlation was found between the dependent variable (output layer) and the layer associated with the number of oxygen atoms,

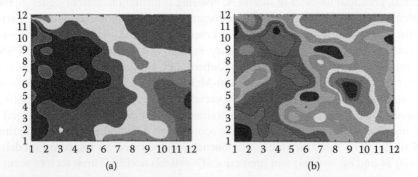

(a)   (b)

FIGURE 8.4 Weight maps of CP ANN model corresponding to a variable log P (a) and the output layer — that is, the response surface of log RBA (b). The colors indicate the distribution of log P and log RBA values from low (blue) to high (red). (Adapted from F. Marini et al., *J. Chem. Inf. Model.*, 45, 1514, 2005. With permission.) **(See color insert following page 244.)**

which seems to confirm the hypothesis for the mechanism of action obtained from the PLS-R model. Also, the correlation between HOMO energy terms and relative binding affinity is large, which indicates that ligands with high RBA values have high HOMO energy.

GAs were employed to reduce the number of variables to be included in the CP ANN model. Three independent GAs run from different random origins were performed, considering a population of 50 chromosomes evolved for 300 generations and looking for the combination of variables that led to better predictive ability, as evaluated by the test set defined in the training/test splitting procedure [43]. The samples to be included in each set were chosen according to a Kohonen-based intelligent selection, by mapping the 132 samples onto a 12×12 SOM and selecting at least one sample from each of the occupied units, having care that the whole experimental range was spanned. Additionally, samples with a high LOO-CV prediction error were added to the training set in order to include the information of their unique structure–property relationship into the model, so that they could be modeled as well as possible. The root mean squared errors of test compounds (RMS test) were 0.494, 0.525, and 0.529 in the three independent GA runs, resulting in 50, 44, and 31 selected variables, respectively. Close inspection of the variables involved in the best models shows that most of the relevant PLS variables are also present in the GA-selected set of descriptors. This could result in additional confirmation of the proposed interpretation of our results. Furthermore, an additional variable-selection experiment was conducted based on the SOM technique for descriptor selection. The projection of 132-dimensional sample space to a 5×5 2D Kohonen network enabled the selection of 50 descriptors (the nearest and the farthest variables from the centroid of each unit were included in the set). The modeling and predicting ability of the resulting counterpropagation network appeared to be slightly lower ($R^2 = 0.85$ and $Q^2 = 0.55$) than that of the corresponding 281-descriptor model or reduced representation models obtained with GA variable selection.

The third modeling technique was BP ANN. As for the other multivariate methods, in a first stage, a model containing all the descriptors was built, having care to choose the dimensionality of the problem (particularly the number of hidden neurons) in a way as to not risk overfitting. In this respect, the final number of adjustable parameters (the connection weights) was kept significantly lower than the number of independent data available. Different network architectures were tested. The optimal BP ANN model constructed with the complete descriptor set resulted in a 281-18-1 architecture (5,095 connection weights including bias nodes) trained for 225 epochs with learning rate $\eta = 0.10$ and momentum $\mu = 0.15$. A GA was used to reduce the number of descriptors. Ten independent GA runs were performed, considering a 50-chromosome population, which was evolving for 300 generations, using the predictive ability over the test set as the fitness criterion. The test set was chosen on the basis of SOM as described in Section 8.2. The best 15 chromosomes of each independent GA run were then considered, and their predictive ability was finally evaluated using LOO-CV. The percentages of occurrences of each descriptor in the best 150 models (15 best chromosomes of each of the 10 independent runs), weighted by the fitness value scored by each of the models were calculated and are available for all 281 descriptors (available from the authors). The most frequently chosen descriptors were log P, number of C atoms, relative number of C atoms, number of H atoms, relative number of single bonds, relative number of double bonds, Randic index (orders 0 to 3), and Kier & Hall index (order 0), to list only those with more than 50% of occurrence on average. The best overall regression model was found to be a 50-10-1 network (521 weighted connections including bias nodes) trained for 2,500 epochs with learning rate $\eta = 0.15$ and momentum $\mu = 0.15$. This optimal BP ANN model performed better than the PLS-R or CP-ANN models, with $R^2 = 0.92$ and $Q^2 = 0.71$.

Inspection of the 49 selected descriptors used in the optimal BP-ANN model shows some overlapping features, if confronted with the best models resulting from the use of the other modeling techniques considered in this work. In particular, we can observe the selection of the log P as a relevant variable present in all the reduced sets of descriptors used for modeling the binding affinity of the examined samples. Moreover, the selection of the XY and YZ shadows and of the moment of inertia A confirms a possible influence of the shape and flexibility of the molecule on the binding affinity, and the inclusion of the maximum and minimum atomic charge and of other terms accounting for the electronic distribution of the inspected molecules is indicative of a correlation between the polarity of the molecules and the value of the dependent variable. Log P is the most frequently selected descriptor, being present on average in more than 80% of the best models. Moreover, terms accounting for each of the contributions to the relative binding affinities suggested by previous models (PLS-R and CP ANN) are present among these ten variables: charge, polarity and van der Waals effects, shape, and orbital electronic population.

8.4.4 DISCUSSION OF CASE 2

Different modeling methods (PLS-R, CP ANN, BP ANN) were successfully employed to model RBA with different degrees of prediction ability. For a reliable comparison, the model validation procedure must be strict and comparable for all investigated modeling methods. The most objective and independent on the chosen modeling method of relatively small datasets is the complete LOO CV method [53]; Hua Gao et al. [54] showed that it may also be applied for larger datasets, however, not in the procedure of variable selection, but for final comparison of different models. This was also our modeling strategy. The LOO CV was applied to assess the predictive ability of all models obtained by the three investigated modeling methods. For the variable selection process, a GA was used, considering the prediction of the test as the fitness function.

It is obvious that for different modeling methods, different variable selection/ reduction procedures are appropriate. In the case of QSAR models based on a large number of descriptors used as molecular structure representation vectors, it is not feasible to expect a unique optimal set of selected descriptors for all modeling methods. This is especially true when the descriptors are grouped into several clusters, each being descriptive for certain structural features or particularities. Having obtained the optimal sets of descriptors for each of the three applied modeling methods, we are able to compare them and find clusters of descriptors overlapping in the three optimal sets.

The best modeling results were obtained with the BP ANN model; however, it was not the only aim of the presented research to obtain the best model. A comparison of chosen descriptors for the best models in all three different modeling methods accounts for the essence of information about the structure–property relationship. We mapped the three sets of variables, selected as individual optimal descriptors for the three confronted methods, onto the maps from Kohonen neural network (Kohonen top-map) and compared them with the map of the original, nonreduced set of 218 descriptors. The resulting 2D distributions of descriptors show a reasonable overlap. As an example, if we examine the selection of log P as one of the descriptors for the reduced set in different modeling methods, it was present in the reduced sets obtained with PLS-R and BP ANN modeling methods but not with the CP ANN method. Inspection of Kohonen top-maps of a nonreduced set of descriptors shows log P at the same position (at identical neuron) as the variable "complementary information content — order," which was selected in the CP ANN reduced set of variables. If two descriptors occupy the same neuron (position in the Kohonen top-map), they must be similar enough to be indistinguishable in the mapping procedure.

8.5 CASE STUDY 3

In this study, the modeling methodologies of the receptor-dependent and the receptor-independent approach were compared. In both cases, the 3D structures of the estrogen binders were determined once optimized *in vacuo* and once optimized together with the receptor.

8.5.1 DATA

The studied dataset ("Kuiper dataset") consists of 60 molecules tested for human ER-α and ER-β binding affinity reported in the literature [66]. The affinity was expressed as log RBA, where RBA is a logarithm of relative binding affinity toward natural ligand 17ß-estradiol (E2), for human ER α and for ER-β. The library contains both environmental estrogenic chemicals (PCB, DDT, and derivatives, methoxychlor, etc.) and phytoestrogens (genistein, coumestrol, zearalenone, among others [29,66]. The second dataset ("Harris dataset") contains almost the same compounds [67]. It follows from the original publications that the experimental log RBA values of common compounds are not identical. However, a correlation between the log RBA of common compounds in the two datasets exists. It is defined with two equations, for log RBA-α and log RBA-β, Equation 8.3 and Equation 8.4, respectively:

$$\log \text{RBA}-\alpha(\text{Harris-new}) = 1.0141 \times \log \text{RBA} -\alpha(\text{Harris-original}) - 0.3731 \quad (8.3)$$

$$\log \text{RBA}-\beta(\text{Harris-new}) = 0.9448 \times \log \text{RBA}-\beta(\text{Harris-original}) - 0.1244 \quad (8.4)$$

The compounds from "Harris dataset" [67], which were not present in the "Kuiper dataset" [66], were used as an external validation set (see Table 8.1).

8.5.2 STRUCTURES AND DESCRIPTORS

The geometrical structures were obtained with a two-step optimization. First, the conformational space of structures was analyzed with the Merck Molecular Force Field

TABLE 8.1
Predictions for Log RBA-α and Log RBA-β and Distances (d) to the Neuron from Which the Individual Prediction Was Obtained, for the External Test Compounds

Name	Experimental Log RBA-α Harris New Eq. 8.3	PMαA (All Variables) Log RBA-α		d	PMαC (Reduced Var.) Log RBA-α		d	Experimental Log RBA-β Harris New Eq. 8.4	PMβA (All Variables) Log RBA-β		d	PMβC (Reduced Var.) Log RBA-β		d
		RI[a]	RD[b]	RI[a]	RI[a]	RD[b]	RI[a]		RI[a]	RD[b]	RI[a]	RI[a]	RD[b]	RI[a]
16α-Iodo-estadiol	1.13	2.00	−0.81	2.79	0.16	0.38	0.41	0.22	2.00	−0.41	2.79	0.58	0.98	0.64
Ent-estadiol	−0.25	2.0	1.80	2.75	−0.14	1.93	0.29	0.79	2.00	1.33	2.75	0.58	0.98	0.65
Estriol	0.24	2.00	0.88	2.79	−0.06	0.38	0.55	0.96	2.00	0.55	2.79	0.63	0.98	1.08
17α-Ethinylestadiol	1.58	2.19	0.50	3.08	0.42	−9.93	1.21	1.19	2.17	1.33	3.08	0.63	0.74	0.83
17α-Iodovinyl-11β-methoxyestradiol(E)	0.80	2.19	−2.10	2.97	0.72	−9.96	0.62	1.19	2.17	−2.19	2.97	−0.63[c]	0.74	0.83
17α-Iodovinyl-11β-methoxyestradiol(Z)	1.37	−3.21[c]	0.02	3.05	0.72	−2.90	1.27	1.49	−3.34[c]	−2.19	3.05	0.58	0.74	0.83
EM-800	−1.56	1.84[c]	0.75	3.70	4.43[c]	−2.69	3.13	−0.45	1.20[c]	−2.86	3.70	−0.81	0.77	1.19
GW-5638	0.27	−4.19[c]	−2.54	2.91	−1.12	−0.04	0.93	0.88	−4.21[c]	−2.86	2.91	−0.56	−4.97	0.50
ICI-182780	1.14	1.84	0.75	4.41	2.01	−2.69	2.19	1.40	1.20	−2.86	4.41	−1.43[c]	0.77	2.47
Lasofoxifene	0.65	0.60	−2.60	3.59	2.63[c]	1.19	1.79	1.08	0.48	−2.86	3.59	−0.81[c]	0.77	1.86
Levormeloxifene	−0.18	0.60	−2.60	2.87	2.63[c]	1.19	1.07	0.13	0.48	−2.86	2.87	−0.81	0.77	0.90

Source: E. Boriani et al., *Mol. Divers.*, 11, 168, 2007. (With permission.)

Note: Predictive models PmαA, PMαC, PmβA, and PMβC were tested.

[a] RI, receptor-independent approach.
[b] RD, receptor-dependent approach.
[c] Outliers.

(MMFF) method, which implements Monte Carlo Multiple Minimum (MCMM) search. The method is a part of the Chem3D package. The second step of optimization of the geometries toward the lowest energies was performed for the gas phase using AM1 parameterization using the MOPAC program package [46]. Based on the results from MOPAC, CODESSA software was used to calculate quantum chemical, constitutional, topological, geometrical, and electrochemical descriptors [34]. Log P, which was obtained from an experimental value database [36] or estimated with the KowWin [39] program, was added to the pool of CODESSA descriptors. If different values of log P were found, the medium value between the experimental and calculated values was accepted in case the difference was not bigger than 1 logarithmic unit. The final number of descriptors was 279.

For the receptor-dependent approach, 3D geometry parameters of molecules were obtained in a docking procedure, facilitated by the availability of crystal structure for estrogen receptors α and β. Virtual docking was used to determine active conformation for ligands. Docking methodologies use knowledge of the 3D structure of a receptor in an attempt to optimize the bound ligand or a series of molecules into the active site.

The docking procedure was performed using LigandFit (Accelrys Software Inc. [68]), a shape-based method for docking ligands into protein-active sites. It employs a Monte Carlo conformational search for generating ligand poses consistent with the active site shape. Accuracy was assessed by analyzing the resulting poses of native ligands and examining the correlation between scoring functions and activity values on the entire dataset. Several steps are needed to perform virtual docking. First, the structures of ER-α and ER-β were downloaded from the Research Collaboratory for Structural Bioinformatics (RCSB) Web site [69]; hydrogens were added and minimized using Dreiding 2.21 (smart minimizer, high convergence, max 5,000 iterations for protein and 2,500 for native ligands). During the docking procedure, conserved water molecules were not included. Second, the binding site has to be defined. We used a Site Search tool for the docked ligand, a shape-based procedure that maps the binding site using a grid around the native ligand. Third, the Monte Carlo method was employed to position the conformations into the site according to the Principal Moment of Inertia (conformational search and alignment). In the last step, five empirical scoring functions (LigScore1, LigScore2, PLP1, PLP2, Jain) and one knowledge-based potential mean force (PMF) scoring function were used for ligand scoring. JAIN for Estrogen Receptor α and LIGSCORE1 for Estrogen Receptor β showed the highest correlation with experimental binding affinity values and were used to rank and choose among the output conformations for each single compound.

8.5.3 RESULTS AND DISCUSSION

The Kohonen mapping technique was applied to divide the 60-compound dataset into the test and training datasets. Technical parameters of the model were selected as follows: dimension network was 6×6; number of training epochs was 60, and learning rate 0.5. In the selection procedure, we considered as an additional criterion a distance between an object and neuron. The Euclidean distance between the object represented by a set of descriptors and the neuron is defined in Equation 8.1. With this procedure [28,43], the dataset was divided into a 48-compound training set and a 12-compound test set.

GA, basically the same as described in the previous section, was used in order to select relevant descriptors for the CP ANN models. The whole procedure of variable selection depends on the choice of the evaluation score, which in the present study may be any measure of the predictive ability of the obtained CP ANN model. As the fitness score was computed as the predictive ability over the test compounds expressed in terms of the RMS_{test}, the root mean square error between the actual and the predicted property (target) for test set compounds, which were previously eliminated from the dataset. The architecture and adjustable parameters of the CP ANN used for the evaluation score in the GA procedure were chosen on the basis of performance of the LOO-CV procedure applied on nonreduced sets of descriptors.

Several CP ANN trainings were performed to define the network architecture, varying: number of neurons, number of epochs, and maximal correction factor (learning rate parameter). These parameters were adjusted during the optimization procedure. The root mean square error and correlation of the experimental log RBA values with the predictions of the training compounds (data from the training set), with the LOO-CV predictions, and with the predictions of the test compounds (RMS_{model}, RMS_{loo}, and RMS_{test}, and R_{model}, R_{loo}, R_{test}, respectively, see Table 8.2) were the estimates considered during the model optimization.

TABLE 8.2
Performance Values Summary for Receptor Dependent and Receptor Independent Approach

Performances		Training — Test Sets			External Set		
Approach	Model	Error Rate (%)	R	RMS	Error Rate (%)	R	RMS
Receptor independent approach	All variables alpha	7	0.62	0.61	0	−0.11	2.43
	Reduced variables alpha	0	0.95	0.17	0	−0.49	2.20
	All variables beta	7	0.66	0.41	0	−0.21	2.34
	Reduced variables beta	0	0.98	0.24	0	0.13	1.33
Receptor dependent approach	All variables alpha	5	0.50	1.61	0	−0.09	2.12
	Reduced variables alpha	0	0.96	0.36	18	−0.39	5.14
	All variables beta	12	0.44	1.79	0	0.12	2.83
	Reduced variables beta	2	0.73	0.85	0	−0.05	1.86

Source: M. Spreafico et al., *Mol. Divers.*, 11, 179, 2007. (With permission.)

Notes: Error rate, % of misclassified compounds (classification models); R, correlation coefficient (predictive models); RMS, root mean squared error of prediction (predictive models).

In the initial step, the best neural network architecture was chosen on the basis of RMS_{loo} and was used for further selecting the descriptors with the GA. The chosen ANN used in the GA procedure was 10×10 neurons; 1,000 epochs; and 0.1 maximal correction factor. Once the number of generations in the GA optimization process with the minimal RMS_{test} value were obtained with a chosen random seed number, the GA was repeated five times, each time with a different random seed number. Two different analyses of the GA results were tested to explore the reduced sets of selected variables and corresponding models, first, to choose the most frequent variables obtained by a statistical analysis of a variety of five different GA runs and, second, to choose the variables from the best chromosome (that is, the run with the lowest RMS_{test}). For the first analysis, we took the best ten chromosomes from the last population of each random seed. Thus, obtaining 50 chromosomes, we analyzed the descriptors with the highest occurrence. The most frequently present descriptors (present in at least 20 models) were kept and used to build a reduced set for new CP ANN. In the second analysis, the best chromosome was used to select descriptors. It turned out that this method outperforms the first.

In both cases, the developed model was first used for classification, to rule out the inactive compounds. These models were labeled by CMαC, CMαD, CMβC, and CMβD, see Figure 8.5. For every classification model, a cutoff value was defined to discriminate active–inactive compounds as described above for the nonreduced sets

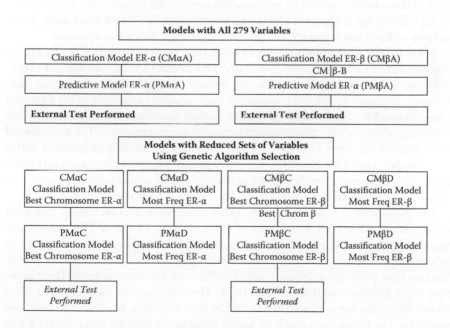

FIGURE 8.5 The processing of the research analysis to perform the final model. (From E. Boriani et al., *Mol. Divers.*, 11, 153, 2007. With permission.)

of variables. The cutoff value –5.0 was obtained for the models with reduced sets of variables as a satisfactory threshold to classify all the compounds correctly. The predictive models with reduced variables for α and β receptors were PMαC and PMβC for the "best chromosome" selection, and PMαD and PMβD for the "most frequent variables" selection.

Different CP ANN models were developed in order to find a way to predict ER binding affinity for a large variety of molecules that can act as endocrine disruptors or mimic endocrine behavior. The developed methodology was encompassed in the following steps: (1) training of CP ANN models with the complete set of variables (279 descriptors), (2) classification of active/inactive compounds, (3) testing the CP ANN models for the predictive ability of active compounds only, (4) optimization of models with variable selection using GA, and (5) validation of the best models with the external set of compounds. All steps and consecutive models are described in the scheme shown in Figure 8.5.

8.5.3.1   Models with All Variables

The initial modeling step was the classification of active versus inactive inhibitors of α and β receptors with CP ANN–based models (CMαA and CMβA, respectively) trained with all the variables calculated by CODESSA. The CP ANN architecture of 10×10 neurons; 1,000 epochs; and 0.1 maximal correction factor was used. The cutoff value to discriminate between active and inactive compounds was determined on the basis of the minimal number of misclassified compounds obtained for the LOO-CV on the training set. It appears to be at –5.0 in both models, CMαA and CmβA, in the receptor-independent approach, and –6.0 in the receptor-dependent approach.

The same CP ANNs (trained with all molecules) were also used for predicting the binding affinities of active compounds only. This means that the core network coefficients (weights) were obtained by training with all the molecules, including the inactive ones. Because the inactive compounds were classified according to the threshold value obtained for the CMαA and CMβA models described above, the predictions of binding affinities were now obtained for active compounds only. The second-level application of the same CP ANN for the prediction of binding affinities of active compounds is called the predictive CP ANN model. The predictive models (PMαA and PMβA) were tested for the predictive ability of training and test compounds. The performance of the models using all the descriptors is shown in Table 8.2.

The classification results of the receptor-independent approach show that after setting the cutoff to –5.0, in the case of ER-α dataset, 15 out of 17 inactive compounds were correctly classified. Two inactive compounds were misclassified as false positive, and two active compounds were classified as inactive (false negative). The two false positive compounds were biochanin A and formononetin. False negatives were 5-androstenedion and o,p'-DDT. The classification model for the ER-β dataset shows the same four errors as for ER-α — two false positives and two false negatives. The predictive models of active compounds with all variables for ER-α and ER-β (PMαA and PMβA) are reported in Table 8.2. A large RMS error in the prediction of test compounds for ER-α and ER-β (1.20 and 1.22, respectively) was ascribed to the detected outliers, which are three in both models (4-OH-tamoxifen and apigenin from the test set and genistein from the training set).

The outlier 4-OH-tamoxifen has experimental log RBA 2.41 and 2.37 for α and β receptors, respectively. This molecule occupied the neuron with the coordinates (1,2), and the analog compound tamoxifen from the training set (experimental log RBA 0.6 and 0.48 for α and β receptors, respectively) was located at the neighboring neuron with the coordinates (2,1). Unfortunately, the molecule from the training set (inactive methoxychlor) located at position (3,1) had such an influence on the neuron (1,2) that the predicted log RBA value of the test molecule (4-OH-tamoxifen) for both α and β receptors was much too low.

Apigenin from the test set and genistein from the training set are both outliers; their structures differ only in the position of the single ring, for genistein in ortho position and for apigenin in meta position. This influences the ligand binding affinity more than assumed by our model, which does not recognize significant differences in descriptors representing the two structures. They are both positioned at the neuron (3,10). Additional error was introduced by positioning an inactive compound from the training set (biochanin A) to the same neuron (3,10), which was already reported as a false positive error during the classification.

The classification in the receptor-dependent approach, a threshold RBA value of −6.0, was chosen to obtain minimal classification errors in the training set to discriminate active from inactive compounds. Three molecules (two from the training set and one from the test set) are not correctly classified. In particular, one slightly active compound is considered inactive and two inactive compounds are predicted slightly active. As inactive, we classified the o,p'-DDT, although its experimental value showed weak activity. Actually, it has the lowest possible value for active compounds. This compound is similar to other inactive DDT derivatives. For this reason it is placed in the cluster of inactive DDTs in the CP ANN model, and its weak activity is not recognized. Methoxychlor and formononetin, inactive compounds, are classified as active with a low RBA value (below −4), so this error could be corrected with raising the threshold from −6.0 to −4.0, but then two the active compounds from the test set would be classified as inactive. All the molecules from the external set are correctly classified as active, but quantitative predictions are not satisfactory for all the compounds. Compound EM-800 is not correctly predicted, and that can be explained by the size and characteristics of this molecule (it is much bigger compared to the other molecules in the training set). Also, ent-estradiol is not properly predicted, but its nature is more similar to natural ligands. Deleting these two outliers from the library, the classification results improved significantly. See Table 8.2 for detailed prediction results with all variables.

In addition to these large errors discussed, the predictions of the other compounds in the training and in the test sets are not satisfactory. The external validation set is reported in Section 8.5.3.5. We assumed on the basis of the results obtained from the models with all of the variables that a selection of influential variables, which would stress the structural difference of compounds mentioned above, would improve the models' predictive ability. Consequently, a variable reduction procedure was applied to select influential descriptors out of the complete set of 279 descriptors. The results are reported in the following sections.

8.5.3.2 Analysis of the GA Model ER-α

In the receptor-independent approach, the chromosome with lowest RMS was chosen from the five GA runs with different random seed numbers; a reduced set of 38 selected variables was built. The resulting classification model was 100% correct. With the cutoff of –5 for active/nonactive compounds division, no classification error appeared. The results of the model predictive ability tested for 43 active compounds (34 of the training set and 9 of the test set) are shown in Table 8.2. There were no outliers detected, and the RMS error in the best model PMαC fell to 0.44 (0.17) for test (training and test) compounds. The largest error in prediction was observed for the compound 4-OH-tamoxifen. In the Predictive Model (PMαC), this compound is positioned at the neuron (1,9), the same as the analog training compound tamoxifen, which is considerably less active. The structural descriptors of the two compounds were still too similar to be differentiated within the PMαC model. For this reason, the prediction for log RBA α of OH-tamoxifen was too low (0.55 instead of 2.41).

In case of receptor-dependent modeling, 48 variables selected after GA reduction of variables were used to build the model for ERr-α. The results are shown in Table 8.2. The selected variables for ER-α are representative of almost all the descriptor classes and are numerous (48 variables). Some are scoring functions from docking; others are counters of particular atoms (H, C, N, Cl). Topological, electrostatic, and quantum chemical descriptors are also present. By choosing a threshold at the value of –6.0, the same as with nonreduced descriptors sets, no classification errors are made on training and test sets, and predictions of active compounds of these two sets are very good, with the RMS error below 0.5 on average for all the compounds. For the external set, two errors in classifications are detected: two active compounds are classified as inactive. These compounds are 17α-ethinyl estradiol and 17α-iodiovinyl-11β-methoxyestradiol (E). Predictions are not always satisfactory for the external set, while for training and test sets they are very good. The compounds with less accurate predictions are 17α-ethinyl estradiol, 17α-iodovinyl-11β-methoxyestradiol (E), 17α-iodovinyl-11β-methoxyestradiol (Z), ICI-182780. For the rest of the compounds, very good predictions are obtained (RMS = 1.13).

8.5.3.3 Analysis of the GA Model ER-β

In the case of ER-β in the receptor-independent approach, the five different GA runs with different seed numbers resulted in the best chromosome containing 29 descriptors. Again, the procedure for classification was repeated with the same threshold (–5.0), and all the compounds were correctly classified. The predictions for active compounds (34 in the training and 9 in the test set) are reported in Table 8.2. The error in the prediction for 4-OH-tamoxifen in PMβC model is smaller than the error in the PMαC model. A detailed analysis of the CP ANN model shows that OH-tamoxifen is positioned at neuron (9,2), which is not occupied by any of the training compounds. The analog training compound tamoxifen is positioned in a close neighborhood — position (10,2). For this reason, the prediction of the 4-OH-tamoxifen is more precise because it is not obtained from the neuron influenced only by the tamoxifen activity, but also by the compounds from the neighboring neurons. Experimental and predicted log RBA-β of OH-tamoxifen is 2.37 and 1.33, respectively. The tamoxifen and OH-tamoxifen show a large difference in ER binding affinity, despite a small difference in chemical structure. This causes problems because there is a sharp change in the structure–property relationships that our models try to encompass. The mechanism of binding of tamoxifen and OH-tamoxifen is described in several papers, and the common hypothesis and crystal structure evidence is that in the positioning of the OH-tamoxifen in the binding pocket, the possibility to make an additional H bond increases the ligand binding affinity of OH-tamoxifen compared to tamoxifen [70–72]. A similar GA reduction of variables was performed in the receptor-dependent modeling approach. Eighteen variables were selected to build the model for ERr-β (see Table 8.2).

Choosing a threshold at the value of −6.0, one classification error was detected on training set data, but the compounds from the test and external sets were all classified correctly. The misclassified compound is, as before, compound 42, o,p'-DDT, which is classified as inactive while it is slightly active. This compound is similar from a chemical point of view to other DDT derivatives that are all active, and this may lead to the error in classification and prediction, as already discussed for the ER-α model.

The results in classification are therefore very good in training, test, and external sets. Predictions are not as satisfactory as results in classification, especially for the external set, where differences among compounds are not adequately recognized. The selected variables are less numerous than for ER-α (18 selected variables for β versus 48 for α). No docking scoring function was selected, while groups of descriptors like constitutional, electrostatic, and quantum chemical descriptors, together with log P value, were chosen.

8.5.3.4 Variables Analysis

The initial set of variables (278 CODESSA descriptors and log P) was reduced as described above, using GA to optimize the performance of the CP ANN trained with selected variables. To investigate further the reasons why the chosen variables influence the quality of the model, we compared the resulting sets of selected variables for ER-α and ER-β models. Common variables for α and β receptor models in the first approach with the receptor-independent ligand conformations are as follows: relative number of N atoms, Kier Shape index-2, minimum and average electrophilic reactivity index for C atom, descriptors related to H-donors and H-bonding surface area, IIDCA H-donors charged surface area, HBCA H-bonding charged surface area, and principal moment of inertia.

There is also a series of variables chosen specifically for α and β models, but they appear close to each other in the Kohonen map, which means that they are similar and they are not receptor-subtype specific. The variables specific for the ER-α model are the following: descriptors for the polar interaction between molecules and functional group portions (in agreement with the reports [70–72]), such as

CPSA variables, PPSA-1, partial positive surface area (sum of surface area on positive parts of molecule) and PPSA-2, and total charge weighted CPSA. Additionally, the LUMO+1 energy, the second-lowest unoccupied molecular level, is specific for the α receptor.

It is interesting that among them there are the following variables: number of S atoms and relative number of S atoms, and HOMO energy. In the map describing the distribution of variables, the variables "No. of S atoms" and "relative No. of S atoms," are located in the area where there is no similar descriptor selected for alpha receptor; those two descriptors seem to influence the beta model only. The descriptors characterizing polar interactions may be useful to discriminate between structurally different chemical compounds that bind to the ER-α and ER-β with specific interaction not only dependent on the binding pocket residues and bonds but also on the interactions around the pocket that are able to modify the size of the cavity and the affinity of the ligand for the receptor [70,71].

Several studies referred to the shape, dimension, and polar interaction as parameters to define the selectiveness of receptors for the alpha or beta subtype. From our study, we can conclude that one cannot obtain a common, well-selective model for ER-α and ER-β binding affinities. It is better to have two separate models and apply them sequentially for the determination of ER-α and ER-β binding affinities of unknown compounds. Obviously, not just a few common influential descriptors, but different descriptors are important for describing structure–property relationships of different receptor types.

In the second approach with the descriptor-dependent ligand conformations, a similar analysis of selected variables was performed. A comparison of selected variables for ER-α and ER-β with the receptor-dependent and receptor-independent approaches sheds light on the nature of ligand–protein interactions. Certain variables were selected in both cases, and for these variables their importance in determining RBA for ER can be argued. For instance, the number of nitrogen atoms may play an important role in the binding affinity toward both proteins, ER-α and ER-β, while the presence of sulfur atoms exhibits selective importance only in the ER-β binding mechanism. This conclusion arises from the analysis of selected variables, but one has to be aware of the fact that the selection procedure is related to the structural domain defined by the training/test compounds. Thus, the generalization of such rules has to be conditioned by the structure applicability domain of constructed models. In addition, a stimulating contribution of nitrogen or sulfur atoms to the binding may also be restricted to certain positions in the molecules, which are inherent in the dataset and must be identified by additional exploration of structures involved in the study.

The descriptor "min electroph. react. index for a C atom" was selected in both approaches for both receptors, indicating that this information is relevant but independent of the receptor isoform. The size and mass of the molecule accounted by the moment of inertia were found to be influential in the case of ER-α, and the atomic charge seems to be important for ER-β.

Evaluating the two approaches described here, we observe that the performance in general is better for the receptor independent approach, indicating that information from the docking procedure in this case does not significantly improve the quality of models. Analyzing more in detail the performances of models built with reduced sets of variables, we find that predictions for ER-α are more accurate with the receptor-based approach, but in the case of ER-β, the receptor-based approach is not able to adequately describe RBA for the studied compounds. Results on the external test set are not always satisfactory because of the diversity of compounds belonging to this set compared with the training-test sets; among nonapplicable compounds, one could find the compound containing iodine or very big synthetic compounds like EM-800 thus not adequately predicted.

A more accurate modeling of the structure and the dynamics of the receptor bound to the ligands may improve the quality of results for the receptor-dependent approach. Molecular dynamics and thermodynamic calculations have been proven to accurately predict binding affinities for various ligands binding to ER [73]. These methods are usually computationally expensive and therefore not applicable to large systems or to large and diverse sets of compounds. However, the single-step perturbation approach [74] was recently demonstrated to be accurate in prediction and significantly reduces computational time when compared to traditional free energy perturbation studies and, therefore, it represents a promising methodology in the study of receptors with different subtypes, or to diverse sets of ligands, like the object of the present study.

8.5.3.5 External Test Set

In order to validate the obtained models the external set [67] was first presented to the classification models CMαA and CMβA. All molecules from the external test set were correctly predicted as active compounds. The same is true for CMαC and CMβC classification models with reduced sets of variables. Then the external set was used to validate the best predictive models obtained with all the variables and those with selected variables for α and β receptors, respectively: PMαA and PMβA and PMαC and PMβC (see Figure 8.5 for model definitions). More than half of the 11 compounds tested were well predicted by all the models. Outliers are compounds outside the structural domain of the training set, such as IC-182780, levormeloxifene, lasofoxifene, and EM-800, which are very big molecules. They are not comparable with the training compounds because of larger dimension or because of the presence of new atoms, as for example the iodine atom in 16α-Iodo-estradiol, 7α iodovinyl-11-β-methoxyestradiol (E) and its isomer (Z) (Table 8.1). In case of the models with all the variables (PMαA and PMβA), three outliers were detected: 17α iodovinyl-11-β-methoxyestradiol (E), EM-800, and GW-5638. Analysis of the positioning of test compounds in the CP ANN model gives us some explanation about the outliers. 17α iodovinyl-11-β-methoxyestradiol (E) is positioned at neuron (2,2) in the CP ANN network of dimension 10 × 10 neurons. No training compounds are in this neuron; the nearest compounds from the training set are in position (2,1) tamoxifen and (1,1) raloxifene. It has been discussed already in Section 8.5.3.1 that the source of the low prediction at position (2,2) is the inactive compound methoxychlor from the training set, positioned at (3,1), which influenced the neighboring positions. The second outlier, EM-800 is positioned at (1,1), the same position as raloxifene in the training test. Although EM-800 appears to be structurally the most similar to raloxifene among all training compounds, the difference is still large, which is reflected in a substantially different binding affinity (the differences are 3.40

and 1.65 for ER-α and ER-β, respectively). This is an indication that we do not have the compounds similar to 17α iodovinyl-11-β- methoxyestradiol (E) in our training set. The third outlier, GW-5638, is positioned at (3,2), and the nearest compounds from the training set are in position (2,1) tamoxifen, in position (3,1) methoxychlor, and in position (4,1) ipriflavone. All are inactive and, consequently, the influence of these inactive compounds to the nearest neighboring neurons is high, and the error in prediction of GW-5638 is big. Although the inactive compounds have been removed according to the threshold defined in the classification model, information about their structure and binding affinities is retained in the neural network model (within the neurons in the Kohonen and output layer). When validating the models for prediction of log RBA-α with reduced sets of variables after GA selection (PMαC), we obtained three outliers: EM-800, lasofoxifene, and levormeloxifene.

ID 24 is positioned at (1,10) in the CP ANN model, like in the case for the model done with all the variables, on the same neuron as raloxifene. Again, the considerable structural difference is the source of large prediction error. Fortunately, an inherent CP ANN feature can be associated with the difference in chemical structure between the training and predicted compounds. This feature is the distance between the selected (most similar) neuron (W) and the compound, which is represented as a vector of descriptors (X), see Equation 8.1. If the distance is large, the prediction for the particular compound is considered less reliable than if the distance is small. For EM-800, the distance found was 3.13, which is considerably larger than for the raloxifene from the training set (0.06). A specific parameter, the reliability factor [43] can be calculated to describe the consistency of predictions of the external test study.

The compounds lasofoxifene and levormeloxifene are positioned at (1,9), at the same position as tamoxifen. Again the distances (1.79 and 1.07) confirm that these two predictions are not reliable; the distance of the training compound tamoxifen is 0.05. The external set was also used for validation of the models with reduced structure representations for prediction of log RBA-β (CMβC and PMβC). There were no misclassified compounds, but there are three outliers in the predictive model: 17-α-iodovinyl-11βmethoxyestradiol (E), ICI-182780, and lasofoxifene. The first is positioned on the neuron (6,10), where we can find the compound OH-PCB6 from the training set. The distance of 17-α-iodovinyl-11β-methoxyestradiol (E) to the neuron (6,10) is 0.83, while that of the training compound OH-PCB6 to the same neuron is 0.03. The compound ICI-182780 is on the neuron (10,10) with the distance 2.47, where we can find the training compound endosulfan with the distance 0.01. Lasofoxifene and levormeloxifene are on the neuron (10,2), with the distances 1.86 and 0.90, respectively, while the training compound tamoxifen occupied this neuron with the distance 0.06. Levormeloxifene with a smaller distance is better predicted (see Table 8.1).

All the predictions for the external test compounds obtained by the models PmαA, PMαC, PmβA, and PMβC, together with the distances to the neuron from which the individual prediction was obtained, are collected in Table 8.1. As can be seen from Table 8.1, the large distances do not always coincide with large prediction errors. One could anticipate that in this case (large distance, small prediction error, see PMβC for ICI-182780 in Table 8.1, row 9), the difference in chemical structure, although large if compared with the training compounds positioned on the same or neighboring neurons, is not essential for the ER binding mechanism.

8.6 CONCLUSIONS

In this chapter, we present CP ANN as a powerful technique for structure-property modeling. The questions important in modeling, such as how to build predictive and classification models, how to select the relevant descriptors, and how to select the training and test set, have been addressed. In comparison to other neural network methods, the CP ANN has a relatively simple architecture and learning algorithm. This enables users to better interpret results. This is especially important when the models are used for regulatory purposes. Extended comments of CP ANN modeling procedure in terms of OECD Principles for validation of QSAR models [22] are reported in Vračko et al. [75]. The combination of an unsupervised and a supervised learning algorithm makes the CP ANN a suitable tool for modeling even in cases where the relationship between descriptors and property is weak. We looked at the problem of endocrine disruption from three different aspects. First, we constructed a robust classification model. Compounds and their classes, which are defined according to evidence on endocrine disruption potency, were taken from documents published by the European Commission. Models were tested with a test set, which was selected from data using the SOM technique. The class predictive power of constructed models shows that the method is promising, despite the weakly defined endpoint. The second case was "a standard QSAR modeling problem," where we modeled the RBAs to rat uterus estrogen receptor for a set of 132 compounds.

We report the comparison between three different modeling methods — linear model based on partial least square regression (PLS-R model), CP ANN model, and multilayer feedforward artificial neural networks trained by the back-propagation of errors algorithm (BP ANN model), where the last has shown the best predictive ability. The results were compared with similar studies from the literature [55–65]. Although the direct comparison between different reports was not possible because of different technical details of modeling, our modeling results were comparable with the most successful models reported (BP ANN with reduced set of selected descriptors, $R^2 = 0.92$ and $Q^2 = 0.71$). The interpretation of selected descriptors can be used to obtain a mechanistic explanation of the model, and this question is addressed by principle 5 of the OECD Principles. In our case study, we applied three methods for descriptor selection. Interestingly, the logarithm of the partition coefficient log P resulted as a relevant variable present in all the reduced sets of descriptors in the two modeling methods, PLS-R and BP ANN, while in the CP ANN modeling method, log P was not selected. However, another descriptor ("complementary information content — order 2") was present, which, as far as the studied compounds are concerned, has been shown to provide the same kind of information because it was located at the same position as log P in the Kohonen map of descriptors. The influence of shape and flexibility of the molecule on the binding affinity was hypothesized on the basis of descriptors like XY and YZ shadows or the moment of inertia chosen in the variable reduction procedure. The selection of the descriptors accounting for the electronic distribution of the inspected molecules (such as max and min atomic charge) indicates a correlation between the polarity of the molecules and the modeled property.

Third, we extended our consideration of endocrine disruption in such a way that we introduced information about receptors in the model. The CP ANN models were built for the prediction of relative binding affinity to ER-α and ER-β and for classification (active/nonactive). Information about receptors was included via molecular geometries. Once the molecular structures were optimized *in vacuum*, they were then optimized in their environment within a receptor. In the following steps, the descriptors were calculated and selected with GA. The variables selection underlines the importance of some constitutional descriptors such as the number of N atoms common for both types of receptors, while the number of S atoms and relative number of S atoms are specific for the β receptor. The CP ANN method in combination with GA descriptor search could explore in depth the differences between the two subtypes of receptors and underline the meaning of some descriptors chosen specifically for the ER-α or ER-β model. Analyzing the source of large prediction errors, we observed several times that the inactive compounds had a negative influence on the predictive power, because the same trained networks were used in two consecutive steps for both classification and prediction models. Good results were observed for the training and the test set compounds, but for the external set, the obtained predictions were not satisfactory. This lack of performance on the external set is probably due to the inadequate applicability domain, which means that the compounds from the training set do not sufficiently match the chemical structure of the compounds from the external set. To introduce information about receptors, the conformations of the ligands were obtained, applying virtual docking methodologies to determine active poses of native ligands inside the receptor. Docking methodologies utilize knowledge of the 3D structure of a receptor (crystal structures for estrogen receptor α and β are available) in an attempt to optimize the ligand or a series of molecules bound into the active site. The aim of the comparative study was to improve the structure–activity correlation obtained in the receptor-independent approach in which the conformations of ligands were obtained by minimal energy optimization. We anticipated that new information obtained from the receptor–ligand complex derived from the docking study would contribute to a better model. Opposite our expectations, the results of the comparative study were not in favor of this hypothesis. Obviously, the refinement of molecular conformation in this modeling approach does not contribute significantly to the quality of the models; the error introduced by inaccurate conformation seems to be compensated for by the optimization of model parameters and variable selection procedure.

ACKNOWLEDGMENTS

The authors acknowledge the financial support of the Ministry of Higher Education, Science, and Technology of Slovenia for financing the research through the project grant P1-017, as well as the European Union Fifth and Sixth framework program schemes for the financial support through the program Marie Curie Host Training Site no. HPMT-CT-2001-00240, CASCADE project, (contract number: FOOD-CT-2004-506319), and CEASAR project (SSPI-022674). We thank Professor A.

Katritzky (University of Florida) and Professor M. Karelson (University of Tartu) for the use of CODESSA, and Professor R. Todeschini (University Milano Bicocca) for DRAGON.

REFERENCES

[1] OECD Environment Health and Safety Publications Series on Testing and Assessment No. 21, Appraisal on test methods for sex hormone disrupting chemicals, March 2001, p. 24.

[2] OECD 401, At their November meeting in 2001, the OECD Joint Meeting of the Chemicals Committee and Working Party on Chemicals, Pesticides and Biotechnolgy, agreed that the LD50 for acute oral toxicity, known as OECD Test Guideline 401 and heavily criticized by the animal welfare protectorate shall be abolished and deleted from the OECD manual of internationally accepted Test Guidelines.

[3] OECD 420, Acute Oral Toxicity — Fixed Dose Procedure, December 2001.

[4] OECD 423, Acute Oral Toxicity — Acute Toxic Class Method, December 2001.

[5] OECD 407, Repeated Dose 28-Day Oral Toxicity Study in Rodents, October 2008.

[6] OECD 408, Repeated Dose 90-Day Oral Toxicity Study in Rodents, September 1998.

[7] OECD 409, Repeated Dose 90-Day Oral Toxicity Study in Non-Rodents, September 1998.

[8] OECD 412, Repeated Dose Inhalation Toxicity: 28-Day or 14-Day Study, May 1981.

[9] OECD 451, Carcinogenicity Studies, May 1981.

[10] OECD 452, Chronic Toxicity Studies, May 1981.

[11] OECD 453, Combined Chronic Toxicity/Carcinogenicity Studies, May 1981.

[12] OECD 414, Teratogenicity, January 2001.

[13] V.V. Kovalishyn, V. Kholodovych, I.V. Tetko, and W.J. Welsh, *Volume learning algorithm significantly improved PLS model for predicting the estrogenic activity of xenoestrogens*, J. Mol. Graph. Model. 26 (2007), pp. 591–594.

[14] H. Liu, E. Papa, J.D. Walker, and P. Gramatica, *In silico screening of estrogen-like chemicals based on different nonlinear classification models*, J. Mol. Graph. Model. 26 (2007), pp. 135–144.

[15] A. Vedani, M.A. Lill, and M. Dobler, *Predicting the toxic potential of drugs and chemicals* in silico, ALTEX-Alternativen zu Tierexperimenten 24 (2007), pp. 63–66.

[16] J. Devillers, N. Marchand-Geneste, J.C. Doré, J.M. Porcher, and V. Poroikov, *Endocrine disruption analysis of 11,416 chemicals from chemometrical tools?*, SAR QSAR Environ. Res. 18 (2007), pp. 181–193.

[17] M. Harju, T. Hamers, J.H. Kamstra, E. Sonneveld, J.P. Boon, M. Tysklind, and P.L. Andersson, *Quantitative structure-activity relationship modeling on in vitro endocrine effects and metabolic stability involving 26 selected brominated flame retardants*, Environ. Toxicol. Chem. 26 (2007), pp. 816–826.

[18] T. Ghafourian and M.T.D. Cronin, *The effect of variable selection on the non-linear modeling of estrogen receptor binding*, QSAR Comb. Sci. 25 (2006), pp. 824–835.

[19] H. Tamura, Y. Ishimoto, T. Fujikawa, H. Aoyama, H. Yoshikawa, and M. Akamatsu, *Structural basis for androgen receptor agonists and antagonists: Interaction of SPEED 98 — Listed chemicals and related compounds with the androgen receptor based on an in vitro reporter gene assay and 3-D QSAR*, Bioorg. Med. Chem. 14 (2006), pp. 7160–7174.

[20] A.G. Saliner, T.I. Netzeva, and A.P. Worth, *Prediction of estrogenicity: Validation of classification model*, SAR QSAR Environ. Res. 17 (2006), pp. 195–223.

[21] OECD, Guidance Document on the Validation of (Quantitative) Structure Activity Relationship ((Q)SAR) Models, ENV/JM/MONO(2007)2.

[22] N. Marhand-Geneste, M. Cazaunau, A.J.M. Carpy, M. Laguerre, J.M. Porcher, and J. Devillers, *Homology model of the rainbow trout estrogen receptor (rtER α) and docking of endocrine disrupting chemicals (EDCs)*, SAR QSAR Environ. Res. 17 (2006), pp. 93–105.

[23] A. Asikainen, M. Kolehmainen, J. Ruuskanen, and K. Tuppurainen, *Structure-based classification of active and inactive estrogenic compounds by decision tree, LVQ and kNN methods*, Chemosphere 62 (2006), pp. 658–673.

[24] Y. Akahori, M. Nakai, Y. Yakabe, A. Takatsuki, M. Mizutani, M. Matsuo, and Y. Shimohigashi, *Two-step models to predict binding affinity of chemicals to the human estrogen receptor α by three-dimensional quantitative structure-activity relationship (3D-QSAR) using receptor-ligand docking simulation*, SAR QSAR Environ. Res. 16 (2005), pp. 323–337.

[25] C.Y. Zhao, R.S. Zhang, H.X. Zhang, C.X. Xue, H.X. Liu, M.C. Liu, Z.D. Zu, and B.T. Fan, *QSAR study of natural, synthetic and environmental endocrine disrupting compounds for binding to the androgen receptor*, SAR QSAR Environ. Res. 16 (2005), pp. 349–367.

[26] W.D. Tong, W. Xie, H.S. Hong, L.M. Shi, H. Fang, and R. Perkins, *Assessment of prediction confidence and domain extrapolation of two structure-activity relationship models for predicting estrogen receptor binding activity*, Environ. Health Perspect. 112 (2004), pp. 1249–1254.

[27] A. Roncaglioni, M. Novič, M. Vračko, and E. Benfenati, *Classification of potential endocrine disrupters on the basis of molecular structure using a nonlinear modeling method*, J. Chem. Inf. Comput. Sci. 44 (2004), pp. 300–309.

[28] F. Marini, A. Roncaglioni, and M. Novič, *Variable selection and interpretation in structure-affinity correlation modeling of estrogen receptor binders*, J. Chem. Inf. Model. 45 (2005), pp. 1507–1519.

[29] E. Boriani, M. Spreafico, E. Benfenati, and M. Novič, *Structural features of diverse ligands influencing binding affinities to estrogen α and estrogen β receptors. Part I: Molecular descriptors calculated from minimal energy conformation of isolated ligands*, Mol. Divers. 11 (2007), pp. 153–169.

[30] M. Spreafico, E. Boriani, E. Benfenati, and M. Novič, *Structural features of diverse ligands influencing binding affinities to estrogen α and estrogen β receptors. Part II. Molecular descriptors calculated from conformation of the ligands in the complex resulting from previous docking study*, Mol. Divers. 11 (2007), pp.171–181.

[31] T.W. Schultz, M.T.D. Cronin, and T. Netzeva, *The present status of QSAR in toxicology*, J. Mol. Struc.-Theoch. 622 (2003), pp. 23–38.

[32] S.P. Bradbury, C.L. Russom, G.T. Ankley, T.W. Schultz, and J.D. Walker, *Overview of data and conceptual approaches for derivation of quantitative structure-activity relationships for ecotoxicological effects of organic chemicals*, Environ. Toxicol. Chem. 22 (2003), pp. 1789–1798.

[33] European Commission, *Communication from the Commission to the Council and the European Parliament on the Implementation of the Community Strategy for Endocrine Disrupters — A Range of Substances Suspected of Interfering with the Hormone Systems of Humans and Wildlife*, COM 262 (2001) Final, Brussels, 14 June 2001. Available at http://europa.eu.int/comm/environment/docum/01262_en.htm#bkh.

[34] A.R. Katritzky, V.S. Lobanov, and M. Karelson, *CODESSA 2.0, Comprehensive Descriptors for Structural and Statistical Analysis*, Copyright©1994–1996 University of Florida.

[35] C. Hansch, D. Hoekman, A. Leo, D. Weininger, and C.D. Selassie, *Comparative QSAR at the interface between chemistry and biology*, Chem. Rev. 102 (2002), pp. 783–812.

[36] Physical Properties Database — PHYSPROP. Available at http://www.syrres.com/esc/physprop.htm.

[37] B. Brunstrom, J. Axelsson, and K. Halldin, *Effects of endocrine modulators on sex differentiation in birds*, Ecotoxicology 12 (2003), pp. 287–295.

[38] L. Hansch and A. Leo, *Exploring QSAR Fundamentals and Applications in Chemistry and Biology*, American Chemical Society, Washington, DC, 1995.

[39] KowWin program v1.66, online demo. Available at http://www.syrres.com/Esc/kowin.htm.

[40] R. Hecht-Nielsen, *Counterpropagation networks*, Appl. Optics 26 (1987), pp. 4979–4984.

[41] J. Dayhof, *Neural Network Architectures, An Introduction*, Van Nostrand Reinhold, New York, 1990.

[42] J. Zupan, M. Novič, and I. Ruisanchez, *Kohonen and counterpropagation artificial neural networks in analytical chemistry*, Chemom. Intell. Lab. Syst., 38 (1997), pp. 1–23.

[43] M. Novič and J. Zupan, *Investigation of infrared spectra-structure correlation using Kohonen and counterpropagation neural network*, J. Chem. Inf. Comput. Sci. 35 (1995), pp. 454–466.

[44] ICPS-WHO, International Programme on Chemical Safety, Global Assessment of the State-of-the-Science of Endocrine Disruptors, T. Damstra, S. Barlow, A. Bergman, R. Kavlock, G. Van Der Kraak, eds., WHO, 2002, pp. 89–105.

[45] Database available at http://edkb.fda.gov/databasedoor.html.

[46] M.J.S. Dewar, E.G Zoebisch, E.F. Healy, and J.J.P. Stewart, *The development and use of quantum-mechanical molecular-models .76. Am1 — A new general-purpose quantum-mechanical molecular-model*, J. Am. Chem. Soc. 107 (1985), pp. 3902–3909.

[47] C. Osborne, *Statistical calibration — A review*, Internat. Statist. Rev. 59 (1991), pp. 309–336.

[48] D.E. Rumelhart, G.E. Hinton, and R.J. Williams, *Learning internal representations by error propagation*, in *Microstructures of Cognition, 1*, D.E. Rumalhart and J.L. McClelland, eds., MIT Press, Cambridge, 1986. pp. 31–8362.

[49] L. Davis, *Handbook of Genetic Algorithms*, Van Nostrand Reinhold, New York, 1991.

[50] B. Hibbert, *Genetic algorithm in chemistry*, Chemom. Inell. Lab. Syst. 19 (1993), pp. 277–293.

[51] J. Zupan and M. Novič, *Optimisation of structure representation for QSAR studies*, Anal. Chim. Acta 388 (1999), pp. 243–250.

[52] SIMCA-P User's Guide © Umertics. Available at www.umetrics.com/default.asp/pagename/software_simcap/c/3.

[53] H.A. Martens and P. Dardenne, *Validation and verification of regression of small data sets*, Chemom. Intell. Lab. Syst. 44 (1998), pp. 99–121.

[54] H. Gao, M.S. Lajiness, and J. Van Drie, *Enhancement of binary QSAR analysis by a GA-based variable selection method*, J. Mol. Graph. Model. 20 (2002), pp. 259–268.

[55] H.X. Hong, W.D. Tong, H. Fang, L.M. Shi, Q. Xie, J. Wu, R. Perkins, J.D. Walker, W. Branham, and D.M. Sheehan, *Prediction of estrogen receptor binding for 58,000 chemicals using an integrated system of a tree-based model with structural alerts*, Environ. Health Persp. 110 (2002), pp. 29–36.

[56] T.W. Schultz, G.D. Sinks, and M.T.D. Cronin, *Structure-activity relationships for gene activation oestrogenicity; Evaluation of a diverse set of aromatic chemicals*, Environ. Toxicol. 17 (2002), pp. 14–23.

[57] P. de Voogt and B. van Hattum, *Critical factors in exposure modeling of endocrine active substances*, Pure Appl. Chem. 75 (2003), pp. 1933–1948.

[58] W.D. Tong, H. Fang, H.X. Hong, Q. Xie, R. Perkins, J. Anson, and D.M. Sheehan, *Regulatory application of SAR/QSAR for priority setting of endocrine disruptors: A perspective*, Pure Appl. Chem. 75 (2003), pp. 2375–2388.

[59] L.M. Shi, H. Fang., W. Tong, J. Wu, R. Perkins, R.M. Blair, W.S. Branham, S.L. Dial, C.L. Moland, and D.M. Sheehan, *QSAR models using a large diverse set of estrogens*, J. Chem. Inf. Comput. Sci. 41 (2001), pp. 186–195.

[60] A. Asikainen, J. Ruuskanen, and K. Tuppurainen, *Consensus kNN QSAR: A versatile method for predicting the estrogenic activity of organic compounds* in silico. *A comparative study with five estrogen receptors and a large, diverse set of ligands*, Environ. Sci. Technol. 38 (2004), pp. 6724–6729.

[61] T. Ghafourian and M.T.D. Cronin, *The impact of variable selection on the modeling of oestrogenicity*, SAR QSAR Environ. Res., 16 (2005), pp. 171–190.

[62] G. Klopman and K.S. Chakravarti, *Structure–activity relationship study of a diverse set of estrogen receptor ligands (I) using MultiCASE expert system*, Chemosphere 51 (2003), pp. 445–459.

[63] C.L. Waller, *A comparative QSAR study using CoMFA, HQSAR, and FRED/SKEYS paradigms for estrogen receptor binding affinities of structurally diverse compounds*, J. Chem. Inf. Comput. Sci. 44 (2004), pp. 758–765.

[64] P. Wolohan and D.E. Reichert, *CoMFA and docking study of novel estrogen receptor subtype selective ligands*, J. Comput. Aid. Mol. Des. 17 (2003), pp. 313–328.

[65] F.A. Pasha, H.K. Srivastava, and P.P. Singh, *QSAR study of estrogens with the help of PM3-based descriptors*, Int. J. Quantum Chem. 104 (2005), pp. 87–100.

[66] G.G.J. Kuiper, J.G. Lemmen, B. Carlsson, J.C. Corton, S.H. Safe, P.T. van der Saag, P. van der Burg, and J.A. Gustafsson, *Interaction of estrogenic chemicals and phytoestrogens with estrogen receptor β*, Endocrinology 139 (1998), pp. 4252–4263.

[67] H.A. Harris, A.R. Bapat, D.S. Gonder, and D.E. Frail, *The ligand binding profiles of estrogen receptors alpha and beta are species dependent*, Steroids 67 (2002), pp. 379–384.

[68] C.M. Venkatachalam, X. Jiang, T. Oldfield, and M. Waldman, *LigandFit: A novel method for the shape-directed rapid docking of ligands to protein active sites*, J. Mol. Graph. Model. 21 (2003), pp. 289–307.

[69] H.M. Berman, J. Westbrook, Z. Feng, G. Gilliland, T.N. Bhat, H. Weissig, I.N. Shindyalov, and P.E. Bourne, *The Protein data bank*, Nucleic Acids Res. 28 (2000), pp. 235–242. Available at www.pdb.org/.

[70] J. Sun, J. Baudry, J.A. Katzenellenbogen, and B.S. Katzenellenbogen, *Molecular basis for the subtype discrimination of the estrogen receptor β selective ligand Diarylpropionitrile*, Mol. Endocrinol. 17 (2003), pp. 247–258.

[71] J. Sun, Y.R. Huang, W.R. Harrington, S. Sheng, J.A. Katzenellenbogen, and B.S. Katzenellenbogen, *Antagonists selective for estrogen receptor α*, Endocrinology 143 (2002), pp. 941–947.

[72] J.A. Katzenellenbogen, R. Muthyala, and B.S. Katzenellenbogen, *Nature of the ligand-binding pocket of estrogen receptor α and β: The search for subtype-selective ligands and implications for the prediction of estrogenic activity*, Pure Appl. Chem. 75 (2003), pp. 2397–2403.

[73] B.C. Oostenbrink, J.W. Pitera, M.M.H. van Lipzig, J.H.N. Meerman, W.F. van Gunsteren, *Simulations of the estrogen receptor ligand-binding domain: Affinity of natural ligands and xenoestrogens*, J. Med. Chem. 43 (2000), pp. 4594–4605.

[74] B.C. Oostenbrink and W.F. van Gunsteren, *Free energies of binding of polychlorinated biphenyls to the estrogen receptor from a single simulation*, Proteins 54(2004), pp. 237–246.

[75] M. Vračko, V. Bandelj, P. Barbieri, E. Benfenati, Q. Chaudhry, M.T.D. Cronin, J. Devillers, A. Gallegos, G. Gini, P. Gramatica, C. Helma, P. Mazzatorta, D. Neagu, T. Netzeva, M. Pavan, G. Patlewicz, M. Randić, I. Tsakovska, and A.P. Worth, *Validation of counter propagation neural network models for predictive toxicology according to the OECD principles: A case study*, SAR QSAR Environ. Res. 17 (2006), pp. 265–284.

9 Quantitative Spectrometric Data-Activity Relationships (QSDAR) Models of Endocrine Disruptor Binding Activities

Richard D. Beger, Dan A. Buzatu, and Jon G. Wilkes

CONTENTS

ABSTRACT

This chapter covers a strategy for model building known as Spectrometric Data-Activity Relationships (SDARs). SDAR models are based on the correlation between the spectral and activity leg of the triangular structure-spectra-activity relationship, whereas traditional three-dimensional Quantitative Structure-Activity

Relationship (3D-QSAR) models are based on the structure-activity relationship. This chapter covers one-dimensional (1D) comparative spectral analysis (CoSA) and two-dimensional (2D) comparative structural connectivity spectral analysis (CoSCoSA) modeling methods for endocrine disruptors. A qualitative 1D CoSA model segregated 108 compounds into 20 strong, 15 medium, and 73 weak relative binding affinity (RBA) classifications to the estrogen receptor, based on ^{13}C NMR (nuclear magnetic resonance) data alone and gave a leave-one-out (LOO) cross-validation of 75.0%. A CoSA model, based on a composite of ^{13}C NMR and electron ionization mass spectra (EI MS) data, gave a LOO cross-validation of 82.4%. Most of the compounds classified by the SDAR models incorrectly were between those exhibiting weak and medium RBAs. A quantitative 2D CoSCoSA model of 130 diverse compounds binding to the estrogen receptor employed 16 bins selected from the ^{13}C-^{13}C COSY spectral data and had an r^2 of 0.827 and a leave-13-out cross-validation average (q_{13}^2) of 0.78. Another CoSCoSA model of estrogen receptor binding that used 15 bins plus one additional distance-related 3D constraint had an r^2 of 0.833 and an average q_{13}^2 of 0.78. CoSCoSA modeling is based on chemical shifts and electrostostatic properties like that used in QSAR models along with added structural atom-to-atom distance information in order to produce a more powerful modeling method. Our results show that CoSA and CoSCoSA models can be used to build computationally rugged, objective, and predictive models of endocrine disruptor activity.

KEYWORDS

Comparative spectral analysis (CoSA)
Comparative structural connectivity spectral analysis (CoSCoSA)
Estrogen receptor

9.1 INTRODUCTION

The Endocrine Disruptor Screening Program (EDSP) was developed by the U.S. Environmental Protection Agency (EPA) in response to a Congressional mandate in the Federal Food, Drug, and Cosmetic Act (FFDCA). The aim of the EDSP is to determine whether certain substances may have an effect in humans that is similar to an effect produced by naturally occurring estrogen. An endocrine disruptor has been defined as an "exogenous agent that interferes with the production, release, transport, metabolism, binding, action or elimination of natural hormones in the body responsible for the maintenance of homeostasis and the regulation of developmental processes" [1]. The National Center for Toxicological Research (NCTR) of the U.S. Food and Drug Administration (FDA) developed the Endocrine Disruptor Knowledge Base (EDKB) to house information on estrogen and androgen binding activities [2–4], and QSAR models of binding to the estrogen and androgen receptors [5,6]. The EDKB was intended to be a public resource by aiding regulatory scientists and fostering computational models that would reduce the need for animal testing.

It was recently reported that male and female pearl dace fish exposed to 17α-ethynylestradiol (EE2) experienced sweeping biochemical changes including edema of the ovaries, inhibited development of testicular tissue, and kidney damage [7]. In a 7-year study, when minnows were exposed to low concentrations of EE2, males exhibited feminization and altered gonadal development, and females had altered oncogenesis that ultimately led to near extinction [8]. These data on EE2, along with the mounting evidence that fish living in waters from municipal wastewater treatment plants (MWTPs) exposed to chemicals have altered reproductive functions, has revived interest in studying endocrine disruptors by *in silico* methods [9]. In addition, each year, 36 million cattle in the United States are fed estrogens in order to make them grow bigger, faster. These cattle excrete large quantities of the estrogens in urine [10], which constitutes another environmental source of synthetic compounds with a strong estrogenic capacity.

Currently, Quantitative Structure-Activity Relationship (QSAR) and Structure-Activity Relationship (SAR) models are used for drug discovery and to form ADME (absorption, distribution, metabolism, and excretion) models [11–14]. Three-dimensional Quantitative Structure-Activity Relationship (3D-QSAR) models are typically based on physical fields obtained by superimposing each compound as a whole on a 3D-grid; SAR modeling simplifies the process by breaking large compounds into secondary structural motifs but does not attempt to provide a comprehensive physical basis for modeling the biological activity [11,15–20]. The calculations used in building QSAR models depend on physical constants that vary in significance between structures of different types and may also exhibit nonlinear relationships for models based on a single structural type. Further, the selection of the most appropriate 3D structure for each molecule requires a number of assumptions. By trial and error, one can build a model for a finite number of molecules, adjusting conformational alignments to determine which orientations give the model the best relationship to known biological activities from a training set. However, the same process is much more prone to error when applied to structures for which the biological activity is unknown. The necessary conformational and other assumptions can give results of uncertain validity when applied for predictive purposes to structures of unknown activity.

This chapter covers an alternative strategy for modeling known as Spectrometric Data-Activity Relationships (SDARs). SDAR models are based on correlations along alternate sides of the triangular structure-spectra-activity relationship: whereas the aforementioned 3D-QSAR models use the structure-activity relationship directly, SDAR models use a combination of the structure-spectrum and spectrum-activity legs. 3D-QSAR models are based partly on electrostatics and partly on molecular geometry [15–19]. 3D-QSAR results show that receptor binding of a compound can be modeled successfully, but only when the model is carefully constructed using somewhat subjective measures of the 3D conformation.

In contrast, molecular spectra result from fundamental physical phenomena that are unbiased and yet provide the kind of information used by QSAR. Mass spectra are generated by fragmentation patterns that reflect the composition and geometry of the original molecules. Ultraviolet (UV), infrared (IR), and nuclear magnetic resonance (NMR) spectra are characteristic of the electronic and geometric configurations of molecules and are explained quite well by the principles of quantum mechanics. For model building purposes, like those described here, an advantage of NMR is that spectral features (chemical shifts) can be assigned to specific atoms within the molecules that produce them, whereas this is not generally possible for UV, IR, or mass spectra (MS).

The ^{13}C NMR spectrum of a compound contains a pattern of frequencies that correspond directly to the quantum mechanical properties of an carbon nuclear magnetic dipole in a magnetic field [21]. The diamagnetic term of an NMR chemical shift tensor is directly related to the electrostatic potential at its nucleus, and the paramagnetic term in the NMR chemical shift tensor is dependent upon the orbital configuration [21]. For ^{13}C NMR spectra, the differences in magnitude between the diamagnetic and paramagnetic terms are very large, so the spectral regions for different carbon orbital configurations are generally well separated from each other. Typically, sp^3 hybridized carbon atoms have ^{13}C NMR chemical shifts in the 0 to 100 ppm range, with the more upfield shifts having a positive electrostatic potential (e.g., methyl groups) and the downfield shifts having a more negative electrostatic potential (e.g., ester bonds). Likewise, sp^2 hybridized carbon atoms usually have ^{13}C NMR chemical shifts in the 100 to 220 ppm range, with the more upfield shifts having a positive electrostatic potential (e.g., benzyl groups) and the downfield shifts having a more negative electrostatic potential (e.g., carbonyl groups). Superimposed on these basic bonding relationships are effects arising from the electrostatics of substituents. The effect of substituents on the ^{13}C NMR chemical shifts can be felt from as far as five bonds away or at even greater bond numbers directly through space, depending on how the molecule has folded back upon itself. Thus, the final disposition of a molecule's NMR spectral features depends on structural, atomic connectivity, and conformational and electrostatic components that, according to SAR theory, also determine the biological activity of the molecule. It follows that patterns can be discovered by a model that correlates a finite training set of NMR spectra with the known biological, chemical, or physical characteristics of the compounds in the set. Once the patterns are validated, they can then be used for predicting the same biological, chemical, or physical activity for other compounds not used in the model building process. This is the conceptual foundation for SDAR and QSDAR.

We used several biological endpoints to identify the relationship between a molecule's biological activity and its experimental 1D ^{13}C NMR spectral data [22–27]. This modeling technique has been referred to as comparative spectral analysis (CoSA) [24–28]. QSDAR CoSA modeling is based on the general triangular relationship between a molecule, its spectra, and its biological or chemical activity. First, in QSDAR modeling, the structures of a set of compounds are generated and their ^{13}C NMR spectra are predicted using *in silico* methods or are found in a database. The ^{13}C NMR spectra are saved as a set of ordered pairs: chemical shift frequencies in ppm and the area under the chemical shift peak. The area under a chemical shift peak is first normalized to an integer. For historical reasons, we assigned an area of 100 to represent a single atom; a doubly degenerate frequency (two ^{13}C NMR chemical shifts at the same frequency) has an area of 200, and so forth, as the number of essentially equivalent atoms increases. Normalization is done so that (1) all the spectra have a similar signal-to-noise ratio and (2) line width variations due to differences in NMR instrumental field strengths, shimming, coupling to protons, temperature, pH, or solvent are eliminated. Bin intensities define the number of chemical shift peaks within a ppm range that are considered equivalent. The bin width used to input the ^{13}C NMR spectra can be optimized by variations from between 0.5 and 10.0 ppm and by determining which bin width produces the best cross-validated and predictive models for a particular training set and pattern definition type.

9.2   SDAR CoSA MODELS OF RELATIVE ESTROGEN RECEPTOR MODELS BINDING USING 1D ^{13}C NMR DATA

We developed two SDAR classification models based on composites of ^{13}C NMR and electron ionization mass spectra (EI MS) data for 108 compounds whose relative binding affinities (RBAs) to the estrogen receptor are published [2,22,29,30]. The RBA to the estrogen receptor is defined as 100 times the ratio of the molar concentrations of 17β-estradiol and the competing compound required to decrease the receptor-bound compound by 50%. Thus, 17β-estradiol had an RBA of 100 and log RBA of 2.0.

The ^{13}C NMR and EI MS data were used as spectrometric digital fingerprints to reflect the electronic and structural characteristics of the compounds. Most of the ^{13}C NMR spectrometric and EI MS data were obtained through public spectral databases [31–34] or predicted using ACD/Labs CNMR prediction software Version 6 (ACD/Labs, Toronto, Canada). In cases where the NMR or MS spectrum was not available from existing databases, the compound was obtained and the spectrum was generated in the laboratory. Both SDAR models segregated the 108 compounds into 20 strong, 15 medium, and 73 weak RBA categories. Strong binders to the estrogen receptor were classified as those with a log RBA over –0.30, weak binders were classified as those with a log RBA less than or equal to –2.70, and medium binders were those with RBAs between –2.70 and –0.30. The classification boundaries were set by trial and error and principal component cluster analysis. The ^{13}C NMR data were binned into 1 ppm wide bins, autoscaled, and Fisher-weighted before discriminate function (DF) analyses were applied to form classification models for a compound's estrogen RBA.

Figure 9.1a shows the SDAR discriminant function (DF) model based on ^{13}C NMR data alone that gave a leave-one-out (LOO) cross-validation of 75.0%. The SDAR model is based on 22 PCs that accounted for 89.8% of the total variance in the NMR data. Figure 9.1b shows the SDAR model based on a composite of ^{13}C NMR and EI MS data that gave a LOO cross-validation of 82.4%. This SDAR model used 21 PCs that accounted for 82.4% of the total variance in the data. Most of the misidentifications in each SDAR model arose from confusing medium and weak classifications, because there were fewer specific spectrometric characteristics to support the distinction of the two classes. This is precisely the behavior expected from a valid modeling process. That is, valid models do not make distinctions based on weak or insignificant differences.

SDAR models were built using real or *in silico*-predicted ^{13}C 1D NMR chemical shifts with or without the addition of composite MS spectra, four models in all, and were used on an external set of five compounds to test the resulting effects on predictive quality. Predictions were made for 2-ethylphenol, 3-deoxyestradiol, dimethylstilbesterol, 3-methylestriol, and 4,4'-dihydroxystilbene. SDAR model predictions as strong binders for 3-deoxyestradiol and dimethylstilbesterol were correct in all four

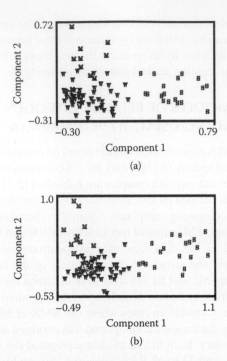

FIGURE 9.1 (a) The discriminant function using ^{13}C NMR data in the SDAR model. (b) The discriminant function using ^{13}C nuclear magnetic resonance (NMR) and electron ionization (EI) mass spectral data in the Spectrometric Data-Activity Relationship (SDAR) model. The X-axis is the first principal component and the Y-axis is the second principal component. Compounds shown with an *S* have a strong classification, compounds shown with an *M* have a medium classification, and compounds shown with a *W* have a weak classification. (From R.D. Beger, J. Freeman, J. Lay Jr., J. Wilkes, and D.W. Miller, *Toxicol. Appl. Pharmacol.*, 169, 17, 2000. Reproduced with permission.)

models. The predictions for 2-ethylphenol were correct as a low relative binder in all four SDAR models. The predictions for three of four models incorrectly predicted 3-methylestriol as a strong binder. The model using only predicted ^{13}C NMR data was the only SDAR that correctly predicted 3-methylestriol as a medium binder. The RBAs for the medium binding 4,4′-dihydroxystilbene were wrong all four times with one strong and three weak predictions. The prediction of compounds in the medium binding classification is less reliable because it can arise from nonspecific binding to the estrogen receptor — that is, binding not related to the underlying structure-activity relationship and not modeled by either SAR or SDAR modeling

9.3 COMBINING STRUCTURAL INFORMATION WITH SPECTRAL INFORMATION

One limitation with 1D CoSA models is that they lack direct 3D structural information. The most obvious way to combine structural and spectral information is to establish one particular molecule as a best or normal representative of those causing

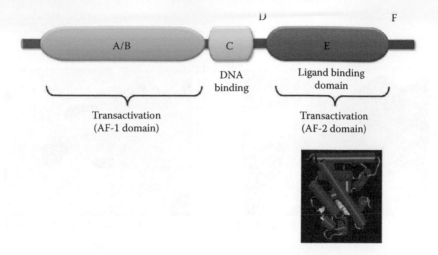

FIGURE 5.1 The structural and functional organization of nuclear receptors.

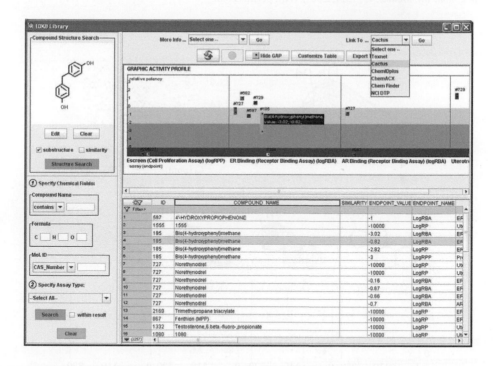

FIGURE 6.2 Endocrine Disruptor Knowledge Base (EDKB) database interface.

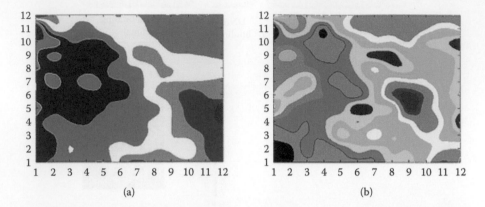

FIGURE 8.4 Weight maps of CP ANN model corresponding to a variable log P (a) and the output layer — that is, the response surface of log RBA (b). The colors indicate the distribution of log P and log RBA values from low (blue) to high (red). (Adapted from F. Marini et al., *J. Chem. Inf. Model.*, 45, 1514, 2005. With permission.)

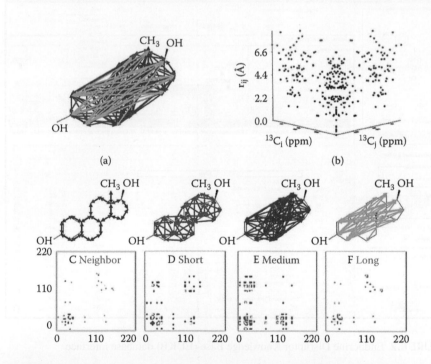

FIGURE 9.2 (a) 3β-estradiol atom-to atom connectivities. (b) The three-dimensional (3D) spectral connectivity matrix of 3β-estradiol. (c) Two-dimensional (2D) nearest neighbor spectral connectivities. (d) 2D Short-distance spectral connectivities. (e) 2D Medium-distance spectral-connectivities. (f) 2D Long-distance spectral connectivities.

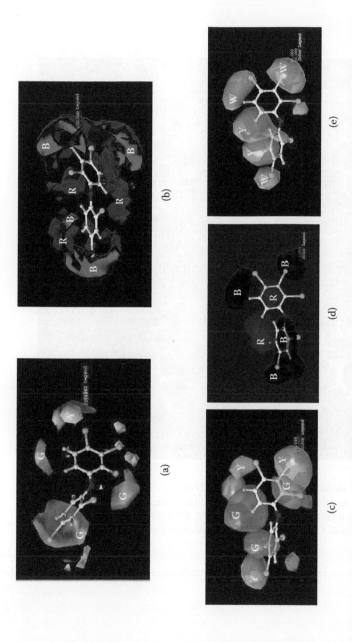

FIGURE 11.4 Comparative molecular field analysis (CoMFA) and comparative molecular similarity analysis (CoMSIA) contour maps. 2,2′,3,4,4′-bromodiphenylether is shown only as reference structure. (a) CoMFA steric contour map. Green contours (G on black and white figure) indicate regions where a rela-ively bulky substitution would increase the binding affinity, whereas yellow contours (Y) indicate areas where a bulkier sub-stituent would decrease the binding affinity. (b) CoMFA electrostatic contour map. Red contours (R) indicate regions where a negative-charged substitution will increase affinity, whereas blue contours (B) show areas where a negative-charged substitution would decrease affinity. (c) CoMSIA steric contour map and (d) CoMSIA electrostatic contour map: same graphic convention as for CoMFA. Greater values of the binding affinity are correlated with more bulk near green or less bulk near yellow, and similarly, more positive charge near blue and more negative charge near red. (e) CoMSIA hydrophobic contour map. Yellow (Y) contours indicate region where hydrophobic group will increase affinity, whereas white contours (W) show areas where a hydrophilic group favors affinity. (Reproduced with permission, from Y. Wang, H. Liu, C. Zhao, H. Liu, Z. Cai, and J. Jiang, *Environ. Sci. Technol.*, 35, 4961, 2005.)

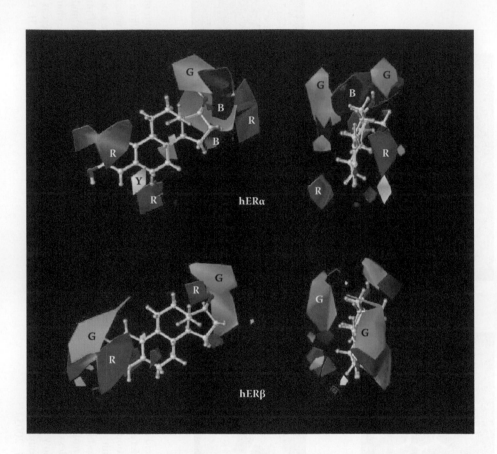

FIGURE 11.5 Comparative molecular field analysis (CoMFA) contour plots for human ERα and ERβ. E$_2$ is shown only as reference structure. Color code is the same as in Figure 11.4. (Reproduced with permission from T. Zhu, G.Z. Han, J.Y. Shim, Y. Wen, and X.R. Jiang, *Endocrinology*, 147, 4132, 2006. Copyright 2006, The Endocrine Society.)

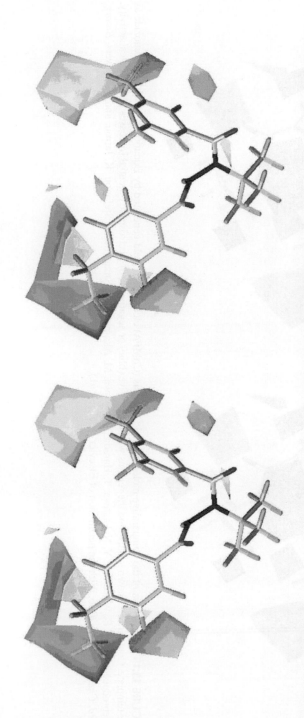

FIGURE 13.14 Stereoviews of the comparative molecular field analysis (CoMFA) steric field generated by Equation 13.9 with tebufenozide as the template. Contours are shown to surround the regions where increased steric bulk increases (green) or decreases (yellow) the biological activity. (Reproduced from C.E. Wheelock, Y. Nakagawa, T. Harada, N. Oikawa, M. Akamatsu, G. Smagghe, D. Stefanou, K. Iatrou, and L. Swevers, *Bioorg. Med. Chem.* 14, 1143, 2006. With the permission of Elsevier Science Ltd.)

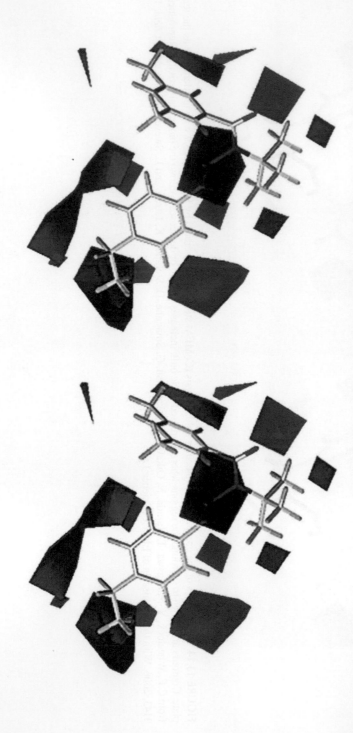

FIGURE 13.15 Stereoviews of the comparative molecular field analysis (CoMFA) electrostatic field generated by Equation 13.9 with tebufenozide as the template. Contours are shown to surround the regions where a positive (blue) or negative (red) electrostatic potential increases the biological activity. (Reproduced from C.E. Wheelock, Y. Nakagawa, T. Harada, N. Oikawa, M. Akamatsu, G. Smagghe, D. Stefanou, K. Iatrou, and L. Swevers, *Bioorg. Med. Chem.* 14, 1143, 2006, With the permission of Elsevier Science Ltd.)

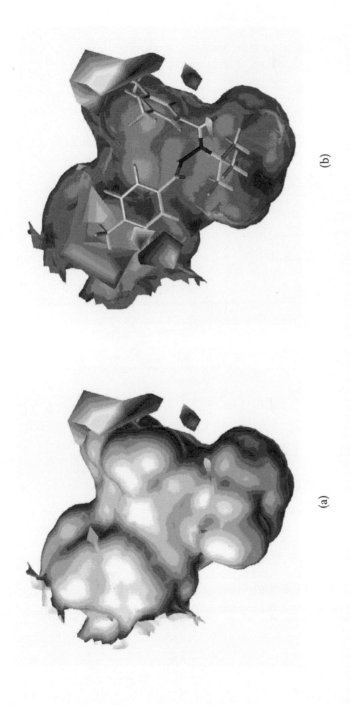

(a)

(b)

FIGURE 13.16 (a) Surface of the EcR ligand binding cavity of *Bombyx mori* for *N,N'*-dibenzoyl-*N*-*t*-butylhydrazine-type ecdysone agonists. (b) Superimposition of comparative molecular field analysis (CoMFA) steric boundary over the *B. mori* EcR cavity, with tebufenozide as the template; a part of the cavity surface was cut down to see the inside. (Reproduced from C.E. Wheelock, Y. Nakagawa, T. Harada, N. Oikawa, M. Akamatsu, G. Smagghe, D. Stefanou, K. Iatrou, and L. Swevers, Bioorg. Med. Chem. 14, 1143, 2006. With the permission of Elsevier Science Ltd.)

a particular biological effect. Each carbon atom in this compound's backbone is numbered, and all other compounds to be modeled must use the same backbone numbering system. Then, each compound's pattern is defined by the ordered pair (carbon number, chemical shift) rather than by the previously described system (chemical shift bin number, occupancy number). This means that the chemical shifts have been assigned to the carbon atoms that produced them. The pattern as defined is correlated with the biological activity of each molecule. The resulting model combines structural information with the assigned simulated ^{13}C NMR chemical shifts. We named this type of 1D SDAR method "comparative structurally assigned spectra analysis" (CoSASA) to distinguish it from the substantially different, unassigned methods previously described as CoSA [24,25]. One supposes that the ability to include spatial relationships in SDAR modeling should improve the quality of the results. In fact, when used on the same spectral training set, we observed inferior results by CoSASA compared to CoSA [24,25]. This unexpected result challenged our understanding of the factors affecting model quality. In addition to inferior results, a limitation of CoSASA is that it can only be used to predict compounds with the same carbon backbone as the representative training molecule.

Another way to combine structural and spectral information is to express geometrical information in spectral space. In the same way that 2D, 3D, and 4D NMR experiments use additional spectral dimensions to reduce spectral overlap [35–40], we conceptualized an analogous way of combining spectral and structural information into a single multidimensional spectral-connectivity matrix that could be generated without the need to run NMR experiments. We hypothesized that the spectra-connectivity relationship defined in such a matrix would produce improved QSDAR models. We also realized that the matrix definition would not limit compounds to those sharing the same backbone template. Compounds in the same model could have quite dissimilar structures; they could differ in the number and connectivity of carbon atoms, as well as the number and identity of constituents or atoms. Below we describe how to define a multidimensional QSDAR data structure matrix.

A molecule's 3D-spectral-connectivity data matrix can be built by displaying all the possible carbon-to-carbon spatial relationships with their assigned carbon NMR chemical shifts on two dimensions and the distance between the two atoms in a third, orthogonal dimension. Even for atoms not directly connected in a molecule, this discussion uses the term "connectivity" to indicate their spatial relationship. Figure 9.2a and Figure 9.2b show all the atom-to-atom connectivities and the 3D-spectral-connectivity data matrix for 3β-estradiol, respectively. The X-axis represents the chemical shift of carbon i, the Y-axis represents the chemical shift of carbon j. The Z-axis represents the through-space distance between carbon i and carbon j (r_{ij}). By representing molecules with this spectral-connectivity matrix, the subjective superposition of molecules on a template is avoided. Other parameterization (for example, selecting dielectric constants or energy minimization to develop typical 3D-QSAR models) is either avoided or minimized. For flexible molecules, some atom-to-atom distances can vary, so representations of multiple conformations in the 3D-spectral-connectivity data matrix format can be accommodated. A method to accomplish this will be explained later.

The 3D-spectral-connectivity data matrix shown in Figure 9.2b is symmetrical about the X-Y diagonal. That is, for every connection between atom i and atom j, the

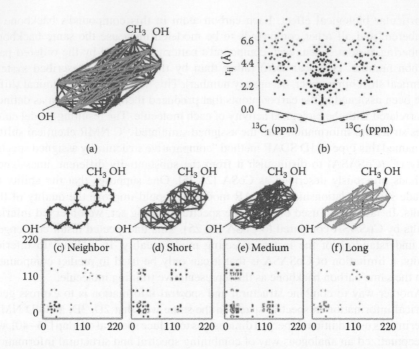

FIGURE 9.2 (a) 3β-estradiol atom-to atom connectivities. (b) The three-dimensional (3D) spectral connectivity matrix of 3β-estradiol. (c) Two-dimensional (2D) nearest neighbor spectral connectivities. (d) 2D Short-distance spectral connectivities. (e) 2D Medium-distance spectral-connectivities. (f) 2D Long-distance spectral connectivities. **(See color insert following page 244.)**

identical relationship is represented across the diagonal at the connection between atom j and atom i. Along the X-Y diagonal at $r_{ij} = 0$ is the 1D ^{13}C NMR spectrum of 3β-estradiol. At $r_{ij} \approx 1.4$ Å are the nearest neighbor atom-to-atom connections, and at $r_{ij} > 1.4$ Å are all the other distance-related atom-to-atom connections.

The information in a 3D-spectral-connectivity data matrix overdetermines a compound's structure, so the information in the matrix that is actually used for a model can be reduced to simplify and accelerate computations. One way to simplify the 3D matrix is to cut it, somewhat arbitrarily, into a set of four 2D spectral planes. The first 2D plane is the nearest neighbor through-bond connectivity plane. Summing or projecting a range of actual through-space atom-to-atom distances with specific distance intervals included in a particular 2D plane set by the modeler constructs the three other 2D planes. This effectively compresses and greatly simplifies atom-to-atom distance information along the Z-axis. In addition to the nearest neighbor spectral-connectivity plane, there is a plane for short atom-to-atom spectral-connections (2.0 Å < r_{ij} < 3.6 Å, perhaps two bond lengths but also through space), another for medium-range atom-to-atom connections (3.6 Å < r_{ij} < 6.0 Å), and a fourth for long distance atom-to-atom connections ($r_{ij} > 6.0$ Å). Figure 9.2c through Figure 9.2f show the four 2D planes and all the atom-to-atom connections in each 2D plane for the specific 2D-spectral-connectivity plane of the 3D-spectral-connectivity matrix of 3β-estradiol. Similarities between the pattern of 2D spectral data associated with the biological activity of the training set compounds and the spectral data for the test compound are detected and used to determine whether the compound is predicted to exhibit the biological activity. We call this procedure comparative structural connectivity spectra analysis (CoSCoSA), to distinguish it from CoSA and CoSASA [41–43].

In 3D-QSDAR, 2D ^{13}C-^{13}C distance spectra associated with short, medium, and long atom-to-atom spectral-connectivity patterns would be difficult to obtain from experimental NMR spectra but easily determined using predicted spectra and the compounds' structures. Because the spectrum is predicted based on a compound's structure, the atom producing each chemical shift is known. It is then possible to calculate atom-to-atom distances (r_{ij}) for each pair of atoms and to associate the r_{ij} values with the corresponding spectral features. The shorter distance relationships in 3D-QSDAR are analogous to 2D ^1H-^1H nuclear Overhauser effect spectroscopy (NOESY) NMR experiments where correlations through space are found for neighboring protons that are generally less than 5 Å apart. As with correlation spectroscopy (COSY), NOESY spectra are expressed as a matrix with off-diagonal cross-peaks [36]. The volumes of the cross-peaks in a NOESY experiment are dependent on the inverse of the distance between the protons, the mixing time of the experiment, and the number of different NOE spin diffusion pathways available for dipolar magnetization transfer. The NOESY peak intensity falls off quickly because it is divided by $r_{ij}^{(-1/6)}$. Thus, NOESY experiments could be conceived but cannot be practically executed for medium- and long-distance interatomic distances. In contrast, because it can be based on simulated spectra, the corresponding 3D-QSDAR matrix is definable and its utility for *in silico* modeling is practical.

9.4 2D QSDAR MODELING OF 130 COMPOUNDS BINDING AFFINITY TO THE ESTROGEN RECEPTOR

Simulated ^{13}C NMR spectra were determined using the ACD/Labs CNMR predictor software (ACD, Toronto, Canada). With this prediction program, NMR spectra are predicted by a substructure similarity technique called hierarchically ordered spherical description of environment (HOSE) [44], which correlates similar substructures with similar NMR chemical shifts. Occasionally, a compound is in the spectral database that is used for the HOSE calculation, which results in their predicted spectra being equivalent to experimental NMR spectra. Because many of our QSDAR models were based only on simulated ^{13}C NMR spectra, QSDAR modeling was driven completely *in silico*.

Table 9.1 shows the log of RBA for 130 structurally diverse compounds used to train 2D CoSCoSA models [27]. Binding data were produced at the National Center for Toxicological Research (NCTR) using a competitive estrogen receptor (ER) assay with radiolabeled estradiol ([$^3H]E_2$) in rat uterine cytosol obtained from ovariectomized uteri of Sprague-Dawley rats [2,3,5]. This dataset spanned seven orders in magnitude ranging from a log (RBA) value of −4 for a compound with weak binding

TABLE 9.1
Compounds and Training Set Predictions of Estrogen Receptor Binding

Compound Name	Number	Exp. Log (RBA)	16 Bin CoSCoSA Log (RBA)	15 Bin + L<7.5Å CoSCoSA Log (RBA)
Diethylstibestrol	1	2.60	1.44	1.43
meso-Hexestrol	2	2.48	2.86	2.70
Ethinyl estradiol	3	2.28	1.44	1.43
4-Hydroxyestradiol	4	2.24	2.45	2.39
4-Hydroxytamoxifen	5	2.24	0.58	0.58
17β-Estradiol	6	2.00	2.36	1.85
α-Zearalenol	7	1.63	0.51	0.51
ICI182780	8	1.57	1.08	1.12
Dienestrol	9	1.57	1.44	1.43
α-Zearalanol	10	1.48	0.51	0.51
2-Hydroxyestradiol	11	1.47	1.26	1.32
Diethylstilbestrol monomethyl ether	12	1.31	1.44	1.43
3,3′-Dihydroxyhestrol	13	1.19	0.75	0.60
Droloxifene	14	1.18	1.63	1.63
Dimethylstibestrol	15	1.16	0.04	0.15
ICI164384	16	1.16	1.08	1.12
Moxestrol	17	1.14	1.44	1.43
17-Deoxyestradiol	18	1.14	0.62	0.79
2,6-Dimethylhexestrol	19	1.11	0.64	0.62
Estriol	20	0.99	0.62	0.79
Monomethyl ether hexestrol	21	0.97	0.51	0.93
Estrone	22	0.86	0.62	0.79
p-meso-Phenol	23	0.60	1.35	1.26
17α-Estradiol	24	0.49	1.17	0.79
Dihydroxymethoxychlorolefin	25	0.42	−0.10	−0.10
Mestranol	26	0.35	1.44	1.43
Zearalanone	27	0.32	0.51	0.51
Tamoxifen citrate	28	0.21	0.58	0.58
Toremifene citrate	29	0.14	0.58	0.58
α,α-Dimethylbethyl allenolic acid	30	−0.02	−0.04	−0.02
Coumestrol	31	−0.05	0.43	0.05
4-Ethyl-7-OH-(*p*-methoxyphenol)-dihydro-1-benzopyran-2-one	32	−0.05	0.11	0.15
Nafoxidine	33	−0.14	0.58	0.58
Comiphene citrate	34	−0.14	−0.47	−0.47
1,3,5-Estratrien-3,6α-17β-triol	35	−0.15	−0.60	−0.61
β-Zearalanol	36	−0.19	0.51	0.51
3-Hydroxy-estra-1,3,5-trien-16-one	37	−0.29	−0.08	0.25
3-Deoxyestradiol	38	−0.30	−1.47	−1.55
3,6,4′-Trihydroxyflavone	39	−0.35	−0.33	−0.35
Genistein	40	−0.36	−2.66	−2.61

TABLE 9.1 (CONTINUED)
Compounds and Training Set Predictions of Estrogen Receptor Binding

Compound Name	Number	Exp. Log (RBA)	16 Bin CoSCoSA Log (RBA)	15 Bin + $L_{<7.5Å}$ CoSCoSA Log (RBA)
4,4'-Dihroxystilbene	41	−0.55	−0.56	−0.51
Dihydroxymethoxychlor (HPTE)	42	−0.60	−1.47	−2.21
Monohydroxymethoxychlorolefin	43	−0.63	−0.10	−0.10
2,3,4,5-TetraCl-4'-biphenylol	44	−0.64	−1.61	−1.56
Norethynodrel	45	−0.67	−2.66	−2.61
2,2',4,4'-Tetrahydroxybenzil	46	−0.68	−0.80	−0.81
β-Zearalenol	47	−0.69	0.51	0.51
4,6-Dihydroxyflavone	48	−0.82	−2.07	−1.95
Equol	49	−0.82	−0.66	0.05
Monohydroxymethoxychlor	50	−0.89	−2.07	−1.95
3β-Androstanediol	51	−0.92	−2.66	−2.61
Bisphenol B	52	−1.07	−2.66	−2.61
Phloretin	53	−1.16	−0.80	−0.81
Diethylstilbestrol dimethyl ether	54	−1.25	−0.66	−0.68
2',4,4'-Trihydroxychalcone	55	−1.26	−1.73	−1.71
2,5-Dichloro-4'-biphenylol	56	−1.44	−1.61	−1.56
4,4'-[1,2-Ethanediyl)bisphenol	57	−1.44	−2.66	−2.61
17β-Estradiol-16β-OH-16-methyl-3-ether	58	−1.48	−1.34	−1.34
Aurin	59	−1.50	−0.56	−0.51
Nordihydroguariareticacid	60	−1.51	−2.66	−2.61
4-Nonylphenol	61	−1.53	−1.61	−1.56
Apigenin	62	−1.55	−2.07	−1.95
Kaempferol	63	−1.61	−2.66	−2.61
Daidzein	64	−1.65	−1.61	−1.56
3-Methylestriol	65	−1.65	−1.34	−1.34
4-Dodecylphenol	66	−1.73	−2.66	−2.61
2-Ethylhexyl-4-hydroxybenzoate	67	−1.74	−2.66	−2.61
4-tert-Octylphenol	68	−1.82	−2.66	−2.61
Phenolphthalein	69	−1.87	−1.47	−1.30
Kepone	70	−1.89	−2.66	−2.61
Heptyl-4-hydroxybenzoate	71	−2.09	−2.66	−2.61
Bisphenol A	72	−2.11	−2.66	−2.61
Naringenin	73	−2.13	−2.66	−2.61
4-Chloro-4'-biphenylol	74	−2.18	−2.66	−2.61
3-Deoxyestrone	75	2.20	−1.47	−1.55
4-Octylphenol	76	−2.31	−2.66	−2.61
Fisetin	77	−2.35	−2.14	−2.09
3',4',7-Trihydroxyisoflavone	78	−2.35	−2.66	−2.61
Biochanin A	79	−2.37	−2.66	−2.61

(continued)

TABLE 9.1 (CONTINUED)
Compounds and Training Set Predictions of Estrogen Receptor Binding

Compound Name	Number	Exp. Log (RBA)	16 Bin CoSCoSA Log (RBA)	15 Bin + L$_{<7.5Å}$ CoSCoSA Log (RBA)
4-Hydroxychalcone	80	−2.43	−2.66	−2.61
4′-Hydroxychalcone	81	−2.43	−2.66	−2.61
2,2′-Methylenebis[4-chlorophenol)	82	−2.45	−2.07	−1.95
4,4′-Dihydroxybenzophenone	83	−2.46	−2.66	−2.61
Benzyl-4-hydroxybenzoate	84	−2.54	−2.66	−2.61
2,4-Dihyroxybenzophenone	85	−2.61	−2.66	−2.61
4′-Hydroxyflavanone	86	−2.65	−3.15	−2.96
3α-Androstanediol	87	−2.67	−2.66	−2.61
4-Phenethylphenol	88	−2.69	−2.66	−2.61
Prunetin	89	−2.74	−2.66	−2.61
Doisynoestrol	90	−2.74	−2.66	−2.61
Myricetin	91	−2.75	−2.66	−2.61
2-Chloro-4-biphenylol	92	−2.77	−3.21	−2.61
Triphenylethylene	93	−2.78	−2.66	−2.61
3′-Hydroxyflavanone	94	−2.78	−3.43	−3.31
Chalcone	95	−2.82	−2.66	−2.61
o,p′,-DDT	96	−2.85	−2.66	−2.61
4-Heptyloxyphenol	97	−2.88	−2.66	−2.61
Dihydrotestosterone	98	−2.89	−2.66	−2.61
Formononetin	99	−2.98	−2.66	−2.61
bis-[4-Hydroxyphenyl)methane	100	−3.02	−2.66	−2.61
p-Phenylphenol	101	−3.04	−2.66	−2.61
6-Hydroxyflavanone	102	−3.05	−2.14	−2.09
4,4′-Sulfonyldiphenol	103	−3.07	−1.47	−1.30
Butyl-4-hydroxybenzoate	104	−3.07	−2.66	−2.61
Diphenolic acid	105	−3.13	−2.66	−2.61
1,3-Diphenyltetramethyldisiloxane	106	−3.16	−2.66	−2.61
Propyl-4-hydroxybenzoate	107	−3.22	−2.66	−2.61
Ethyl-4-hydrobenzoate	108	−3.22	−2.66	−2.61
Phenol red	109	−3.25	−2.66	−2.61
3,3′,5,5′-TetraCl-4,4′-biphenyldiol	110	−3.25	−2.66	−2.61
4-tert-Amylphenol	111	−3.26	−2.66	−3.53
Baicalein	112	−3.35	−2.66	−2.61
Morin	113	−3.35	−2.66	−2.61
4-sec-Butylphenol	114	−3.37	−2.07	−1.95
4-Chloro-3-methylphenol	115	−3.38	−2.66	−3.53
6-Hydroxyflavone	116	−3.41	−2.66	−2.61
4-Benzyloxyphenol	117	−3.44	−2.66	−2.61
3-Phenylphenol	118	−3.44	−2.14	−2.09
Methyl-4-hydrobenzoate	119	−3.44	−2.66	−3.53
2-sec-Butylphenol	120	−3.54	−3.20	−3.04

TABLE 9.1 (CONTINUED)
Compounds and Training Set Predictions of Estrogen Receptor Binding

Compound Name	Number	Exp. Log (RBA)	16 Bin CoSCoSA Log (RBA)	15 Bin + $L_{<7.5Å}$ CoSCoSA Log (RBA)
2,4′-Dichlorobiphenyl	121	−3.61	−2.66	−2.61
4-*tert*-Butylphenol	122	−3.61	−3.75	−3.53
2-Chloro-4-methylphenol	123	−3.66	−2.66	−3.53
Phenolphthalin	124	−3.67	−2.66	−2.61
4-Chloro-2-methylphenol	125	−3.67	−2.66	−3.53
7-Hydroxyflavanone	126	−3.73	−2.66	−2.61
3-Ethylphenol	127	−3.87	−2.90	−3.70
Rutin	128	−4.09	−2.66	−2.61
4-Ethylphenol	129	−4.17	−3.75	−3.53
4-Methylphenol	130	−4.50	−3.75	−3.53

Source: R.D. Beger, K. Holm, K.D. Buzatu, and J.G. Wilkes, *Int. Elec. J. Mol. Des.,* 2, 435, 2003. (Reproduced with permission.)

affinity to the estrogen receptor to a log (RBA) greater than 2. For a particular molecule, the RBA to the estrogen receptor is defined as one hundred times the ratio of the molar concentrations of 17β-estradiol and the concentration of the competing compound required to decrease the amount of receptor-bound 17β-estradiol by 50%. Thus, 17β-estradiol by definition had an RBA of 100 and a $\log_{10}$(RBA) of 2.0.

For each of the 130 compounds, the 2D ^{13}C-^{13}C COSY NMR spectrum was simulated using the ACD/Labs CNMR predictor version 6.0 software (ACD/Labs, Toronto, Canada). The ^{13}C NMR chemical shifts with the identity of their associated nuclei are required to produce the simulated 2D ^{13}C-^{13}C COSY NMR spectral data that were used to develop a model for 130 diverse compounds whose RBAs to the estrogen receptor had been determined. Using the NMR spectral assignments obtained from predicted carbon chemical shifts to identify the nearest neighboring carbon atoms and establish carbon-to-carbon through-bond connectivity, spectral patterns of each compound formed the simulated 2D ^{13}C-^{13}C COSY NMR spectra. The 2D ^{13}C-^{13}C COSY spectra for all 130 compounds were binned into 2.0 ppm by 2.0 ppm square bins.

Forward multiple linear regression (MLR) analysis was used on a selected subset of spectral bins to build quantitative 2D QSDAR models of relative binding affinity to the estrogen receptor. After binning all 130 compounds, only 605 bins from the 7,381 bins had nonzero elements (called "hits") in them. Of the 605 populated bins, only 337 bins had more than one "hit." From the remaining 337 multiply populated bins, a number of bins were selected and used to train MLR models until a model was obtained that had an r^2 greater than 0.8 and an *F*-value greater than 30. This process identified 16 bins

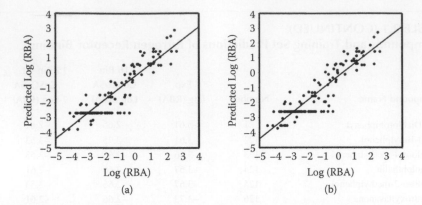

FIGURE 9.3 (a) Plot of the predicted log (RBA) and experimental log (RBA) based on 16 correlation spectroscopy (COSY) bins. (b) Plot of the predicted log (RBA) and experimental log (RBA) based on 15 COSY bins + $L_{<7.5\text{Å}}$ variable. (From R.D. Beger, K. Holm, K.D. Buzatu, and J.G. Wilkes, *Int. Elec. J. Mol. Des.*, 2, 435, 2003. Reproduced with permission.)

that produced a QSDAR model that had an r^2 of 0.82. Bins that had fewer than 2 "hits" were not used in the model building process. The reason for this is that a bin with one "hit" can inappropriately add to the r^2 of a model but cannot improve the LOO cross-validation (q_l^2). The use of a large number of very small, singly populated bins is the reason that Bursi [28] had a high r^2 and very low q_l^2. To more rigorously test the validity of the 2D QSDAR models, two leave-10%-out (L13O, with 13 representing 10% of the 130 test set compounds) cross-validations were performed on each of the models. In these "leave-multiple samples-out" experiments, the compounds omitted were varied, and the results of the two corresponding experiments for omitted compounds were averaged.

Figure 9.3a shows results of a 2D QSDAR MLR model that was based on 16 selected 2D bins from the 2D ^{13}C-^{13}C COSY spectral data. The 16 bin COSY model for the 130 estrogenic compounds had an explained variance r^2 of 0.827, a LOO q_l^2 of 0.78, and an average L13O cross-validated variance (q_{l3}^2) of 0.78 ± 0.01. The 2D QSDAR model was based on COSY bins 28-12 (bin 1), 72-20 (bin 2), 54-28 (bin 3), 50-38 (bin 4), 64-56 (bin 5), 164-104 (bin 6), 152-108 (bin 7), 156-110 (bin 8), 126-112 (bin 9), 140-112 (bin 10), 142-112 (bin 11), 154-112 (bin 12), 154-114 (bin 13), 156-114 (bin 14), 128-116 (bin 15), and 126-120 (bin 16). All 2.0 ppm bins were written using the format a-b, where a and b are the ppm values corresponding to the two "nearest neighbor connected" atoms. The MLR equation for the 16 bin CoSCoSA model is

Predicted $\log_{10}$(RBA) = 0.00999 * bin 1 + 0.03173 * bin 2 + 0.0071 * bin 3

+ 0.01196 * bin 4 + 0.02191 * bin 5 + 0.0093 * bin 6

+ 0.2329 * bin 7 + 0.01324 * bin 8 + 0.00737 * bin 9

+ 0.02558 * bin 10 + 0.0392 * bin 11 + 0.03094 * bin 12

− 0.00545 * bin 13 + 0.00526 * bin 14 + 0.00298 * bin 15

− 0.00768 * bin 16

All bins had more than three "hits" except for bins 152-108 and 140-112, each of which had only two "hits." The correlation matrix for the 16 bins was calculated and only two sets of bins had a correlation between them greater than 0.5. The average correlation between a bin and any of the other 17 bins was 0.04, and many of the average correlations were much lower than that. The lack of a large correlation among bins suggests that the resulting patterns were based on essentially orthogonal data. The bin 28-12 was most often associated with the CH_3 carbon connected to the CH_2 in the ethyl groups in diethylstilbestrol and hexestrol-like compounds. Twelve of the fourteen compounds with a hit in 28-12 had a log_{10} (RBA) greater than −0.05. Compounds that populated bin 154-112 were most often associated with the 3 carbon position connected to the 2 carbon position in the A-ring of 17β-estradiol-like compounds. Nine of the ten compounds with a hit in bin 154-112 had a log_{10} (RBA) greater than −0.05. Fourteen compounds had a "hit" in the bin at 128-116. The 128-116 bin was most often associated with the 2 to 3 and 5 to 6 carbon positions in a phenol ring. Twelve of the fourteen compounds with a hit in bin 128-116 had a log_{10} (RBA) less than 0.60. The 24 compounds that had a hit or multiple hits in the bin 156-114 came from a hydroxylated carbon of a phenol ring (156) connected to its two nearest neighboring unsubstituted carbons (114). Only 5 of the 24 compounds with a hit in bin 156-114 had log_{10} (RBA) less than −1.65. Six compounds had a hit in bin 64-56 derived from the two carbons between the oxygen ester and the nitrodimethyl of tamoxifen-like compounds. Similar spectral-structural associations could be made for the other bins effectively used in the 2D QSDAR models.

Figure 9.3b shows results for the 2D QSDAR model based on the MLR analysis of 15 selected 2D ^{13}C-^{13}C COSY bins plus a distance variable. The distance variable, $L_{<7.5Å}$ was assigned a value of 1 when the maximum distance between nonhydrogen atoms in a compound was less than 7.5 Å (compact) and a value of zero for all other compounds. The $L_{<7.5Å}$ variable replaced the COSY bin at 154-114 (bin 13) in the previous 16 bin CoSCoSA model. The MLR equation for the 15 bin-with-$L_{<7.5Å}$ 2D QSDAR model is

$$\text{Predicted log (RBA)} = 0.00969 * \text{bin } 1 + 0.03122 * \text{bin } 2 + 0.00637 * \text{bin } 3$$
$$+ 0.01066 * \text{bin } 4 + 0.02142 * \text{bin } 5 + 0.00902 * \text{bin } 6$$
$$+ 0.2263 * \text{bin } 7 + 0.01275 * \text{bin } 8 + 0.00732 * \text{bin } 9$$
$$+ 0.02507 * \text{bin } 10 + 0.03934 * \text{bin } 11 + 0.02666 * \text{bin } 12$$
$$+ 0.00526 * \text{bin } 14 + 0.00329 * \text{bin } 15 - 0.00701 * \text{bin } 16$$
$$-0.91773 * L_{<7.5Å}$$

This 15 bin-with-$L_{<7.5Å}$ model had an r^2 of 0.83, a q_l^2 of 0.79, and an average q_{l3}^2 of 0.78 ± 0.01. In this model, the $L_{<7.5Å}$ variable selected nine compounds, all of which had a log (RBA) lower than −3.26. Smaller, compact molecules tended to bind weakly. This bin's significant association with lack of binding is reflected in its large negative β coefficient, −0.91773, in the MLR equation.

In Figure 9.3a, the line of compounds predicted to have a log_{10}(RBA) of −2.60 is a set of compounds that did not have a hit in any of the 16 bins used to formulate the two 2D QSDAR models. There were 56 compounds in the 16 bin 2D QSDAR model that did not have a hit in one of the 16 bins. The 15 bin-with-$L_{<7.5Å}$ model had 52 compounds that did not have a hit in any of the 15 bins or $L_{<7.5Å}$. Removal of compounds with no hits from the 2D QSDAR models did not change the r^2 or q^2 of the model by more than 2%. Three of the compounds in the 16 bin 2D QSDAR model had residuals greater than two standard deviations (3β-androstanediol, genistein, and norethynodrel). In the 16 bin model, only one other compound had a predicted residual greater than two standard deviations and it was 4-hydroxy-tamoxifen. The 15 bin-with-$L_{<7.5Å}$ model had four compounds with no hits that had residuals greater than two standard deviations. They are the same three compounds (plus 4,4′-sulfonyldiphenol) poorly modeled by the 16 bin model. There were two compounds in the 15 bin-with-$L_{<7.5Å}$ model with predicted residuals greater than two standard deviations, 4-hydroxy-tamoxifen and dihydroxymethoxychlor (HPTE). Almost all of the compounds with no hits in the 16 bins had experimental $\log_{10}$ (RBA) lower than –1.0. The 2D QSDAR models did not find a spectral relationship for these weakly binding compounds. Most of the other bins in both 2D QSDAR models were used to establish a relationship between a spectral bin and binding to the estrogen receptor with a $\log_{10}$ (RBA) greater than –2.60.

To further test the ruggedness, the trained 2D QSDAR models of estrogen receptor binding were used to predict the $\log_{10}$(RBAs) of compounds from two published external datasets [29,30]. The $\log_{10}$(RBAs) from those external datasets possessed a greater variability in binding activity. So, by a previously published method, a set of compounds that had their binding activity determined in each of three labs [29,30] was used to normalize the external datasets to the NCTR data [5,45]. The MLR equations shown above were used to predict the $\log_{10}$(RBA) of the compounds in the test set. The published normalized $\log_{10}$(RBA) for 27 compounds from Waller and Kuiper data was used in our external testing of the 2D QSDAR MLR models [5]. However, many of the occupied bins for the new compounds from the external dataset did not fall into the original 605 occupied bins. (The original set of bins comprised only 8.2% of the 2D COSY spectral plane.) We hypothesized that, in the different molecular context of the external datasets, NMR chemical shift information was expressed in adjacent but nonincluded bins. NMR chemical shifts exist along a continuum, and the process of binning them for this type of pattern recognition inherently compromises the pattern when it barely misses a selected bin. To account for this source of confusion with the external data, we tried adding various fractions of "near-miss" signals into each compound's spectrum. With this in mind, we used the CoSCoSA model's MLR equation to predict the normalized log (RBA) of the compounds in the external test set. Compounds from the external test set with bins that were one bin away (one of eight bins surrounding a 2D bin) from the original 605 populated bins were modeled using none, one-quarter, and one-half of that bin's intensity in the nearest neighboring bin used in the original 2D QSDAR model. Table 9.2 shows the predictions for 27 compounds using the 2D QSDAR models. The first 21 compounds show the predictions for Waller's dataset [29] using both the 16 bin and 15 bin-plus-$L_{<7.5Å}$ model of estrogen binding. Only 7 of the 27 compounds

TABLE 9.2
External Test Set Predictions of Estrogen Receptor Binding

Name	Normalized log (RBA)	16-CoSASA	15 + L$_{<7.5Å}$ -CoSCoSA	CoMFA
2-*tert*-Butylphenol	−4.55	−2.66	−3.53	−3.83
3-*tert*-Butylphenol	−4.82	−1.50 ± 0.64	−2.34 ± 0.66	−3.33
2,4,6,-TriCl-4′-biphenylol	−0.16	−1.61	−1.56	−1.60
2-Chloro-4,4′-biphenyldiol	−0.61	−1.61	−1.56	−1.49
2,6-Dichloro-4′-biphenylol	−1.11	−1.61	−1.56	−2.41
2,3,5,6,TetraCl-4,4′-biphenyldiol	−2.18	−1.61	−1.56	−0.82
2,2′,3,3′,6,6′-HexaCl-4-biphenylol	−2.74	−2.14	−2.08	−3.06
2,2′,3,4′,6,6′-HexaCl-4-biphenylol	−2.60	−1.61	−1.56	−2.48
2,2′,3,6,6′-PentaCl-4-biphenylol	−1.97	−1.61	−1.56	−3.07
2,2′5,5′-TetraCl-biphenyl	−2.67	−2.66	−2.61	−2.74
2,2′,4,4′,5,5′-HexaCl-biphenyl	−2.83	−2.66	−2.61	−1.52
2,2′,4,4′,6,6′-HexaCl-biphenyl	−1.87	−2.66	−2.61	−1.83
2,2′,3,3′,5,5′-HexaCl-6′-biphenylol	−2.69	−2.36	−2.29	−3.01
4′-Deoxyindenestrol	−1.37	2.89 ± 0.63	2.96 ± 0.67	−0.53
4′-Deoxyindenestrol	−0.23	2.89 ± 0.63	2.96 ± 0.67	0.11
5′-Deoxyindenestrol	−0.59	−0.61	−0.59	−1.00
5′-Deoxyindenestrol	0.35	−0.61	−0.59	−0.59
Indenestrol A (R)	1.08	3.95 ± 0.64	4.01 ± 0.67	0.29
Indenestrol A (S)	2.39	3.95 ± 0.64	4.01 ± 0.67	0.62
R 5020	−1.81	−2.45 ± 0.18	−2.48 ± 0.17	−0.70
Zearalenone	0.91	0.51	0.51	−0.12
5-Androstenediol	−0.49	−2.66	−2.61	−0.66
16*a*-Bromoestradiol	1.41	−0.11	0.05	0.33
16-Ketoestradiol	−0.38	−0.11	0.05	0.58
17-*epi*-Estriol	0.98	−0.11	0.05	−0.16
2-Hydroxyestrone	−0.19	1.26	1.32	0.36
Raloxifene	1.34	0.17 ± 0.63	0.20 ± 0.66	−0.24

Note: Plus and minus signs reveal the variation seen when using none and one-half of a bin's intensity in neighboring bins used to formulate the comparative structural connectivity spectral analysis (CoSCoSA) model. CoMFA, comparative molecular field analysis.

Source: R.D. Beger, K. Holm, K.D. Buzatu, and J.G. Wilkes, *Int. Elec. J. Mol. Des.,* 2, 435, 2003. (Reproduced with permission.)

from Waller and Kuiper external datasets had binned COSY chemical shifts that were within one bin of those 16 bins used to formulate a 2D QSDAR model. Table 9.2 shows that for seven compounds, we report the predicted the $\log_{10}(RBA)$ using one-quarter intensity in a neighboring bin along with the deviation seen when predicting the $\log_{10}(RBA)$ using none and one-half intensity in the neighboring bin used for a 2D QSDAR model. For Waller's test set and one-quarter of a bin's intensity in neighboring bins, we achieved a q_{pred}^2 of 0.50 for the 16 bin 2D QSDAR model and a q_{pred}^2 0.57 for the 15 bin-plus-$L_{<7.5Å}$ 2D QSDAR model. When using one-half of a bin's intensity in a neighboring bin, we got a q_{pred}^2 of 0.45 for the 16 bin 2D QSDAR model and a q_{pred}^2 0.52 for the 15 bin-plus-$L_{<7.5Å}$ 2D QSDAR model. Using none of a bin's intensity in a neighboring bin, we got a q_{pred}^2 of 0.55 for the 16 bin 2D QSDAR model and a q_{pred}^2 of 0.62 for the 15 bin-plus-$L_{<7.5Å}$ 2D QSDAR model. A CoMFA model had a q_{pred}^2 of 0.70 for Waller's normalized test set [29]. When Indenstrol A (R), Indenestrol A (S), 4'-deoxyindenestrol (R), and 4'-deoxyindenestrol (S) are removed from Waller's test set and none of a bin's intensity from neighboring bins a q_{pred}^2 of 0.59 for the 16 bin model and a q_{pred}^2 0.74 for the 15 bin-plus-$L_{<7.5Å}$ model are achieved. Further inspection of the predictions for Indenstrol A (R), Indenestrol A (S), 4'-deoxyindenestrol (R), and 4'-deoxyindenestrol (S) revealed one chemical shift prediction that was highly suspect (142 ppm). On closer examination of the spectral prediction process, we found that the database structures used as the basis for prediction all had corresponding chemical shifts from 134 to 139 ppm, not 142 ppm. It appears that an error in the spectral prediction process may have contributed to poorer modeling results for these outliers. Undoubtedly, this type of error will be less frequently observed with models based on newer versions of the predictor software which access many more reference compounds.

9.5 FUTURE RESEARCH

9.5.1 PRODUCING A ^{13}C AND ^{15}N HETERONUCLEAR SPECTRAL-CONNECTIVITY DATA MATRIX

Chemicals with potentially useful pharmaceutical value or those that exhibit toxicity, and compound types to be modeled will contain atoms besides carbon, oxygen, and hydrogen. In these cases, other NMR-active atoms besides ^{13}C can be used. ^{15}N is the most prominent atom that is both biologically important and for which accurate NMR prediction software is available. Additional NMR nuclei that could be used for SDAR or QSDAR modeling include ^{17}O, ^{19}F, and ^{31}P. The biological endpoint to model and the availability of an endpoint-characterized training set will determine which nuclei are useful in development of the matrix. A typical 2D spectral connectivity data matrix layout for a ^{13}C-^{15}N heteronuclear CoSCoSA model is characterized by combining four spectral connectivity data matrices [46]. There is one matrix for ^{13}C-^{13}C data, one matrix for ^{15}N-^{15}N data, one matrix for ^{13}C-^{15}N data, and one matrix for ^{15}N-^{13}C data. As before for ^{13}C-^{13}C 3D spectral-connectivity data matrices, symmetry-based data duplicates mean that only half of the four spectral-connectivity data matrices are required to develop a model.

9.5.2 Using Molecular Dynamics of Compounds to Produce a 4D Spectral-Connectivity Data Matrix

A 4D-spectral connectivity data matrix can be defined as the sum of an arbitrary number, say 100, 3D-connectivity matrices. In the simplified version of this 4D-connectivity matrix concept, chemical shifts of atom i and atom j are assumed not to change as the molecule bends or twists with time, but the distance between atom i and atom j would differ as a function of the molecular conformation. For any two atoms, molecular dynamics programs can estimate interatomic distance, a value that may change over some range and for which the percentage of time that the distance is within a certain bin will vary, depending on molecular connections, degrees of freedom, and so forth. This concept applied to CoSCoSA modeling would treat the distance between atoms as a potential variable rather than as a constant. A score of 100 in a 4D spectral-connectivity data matrix will represent unvarying distances between two atoms as seen in bonds and also between more distant atoms if the molecules are very rigid. For two atoms in flexible molecules, there will be a distribution of distance hits along the z-axis varying from some minimum to a maximum, and where the summation over time of these distances will add to 100 for each atom-to-atom pair. The distance distributions can then be used to assign occupation probabilities from zero to some maximum (most probable conformation) among the four distance bins. For all the possible atom pairs, distance distributions will be Gaussian, or skewed-Gaussian functions when there is a single local maximum distance. When there is more than one maximum, more complex distributions will be seen. In the simplified regime of CoSCoSA distance modeling, the occupancy pattern for a particular molecule and a particular through-space distance bin would be 100 if the full range of possible distances are within that 2D bin. If 30% of the time the molecular shape were such that the atomic distances fall outside of that distance range, the atom-to-atom bin occupancy for that distance range would be represented with 70, and the remaining 2D bin(s) would share the other 30 points. The fourth dimension is effectively time, during which the molecule twists through its available conformations, and interatomic distances vary as described above. We will now discuss why this way of representing molecular conformational characteristics confers a significant advantage when building a CoSCoSA model, particularly for cases in which the training set includes very diverse compound types.

Binding of a ligand to a receptor occurs because of a lowering of Gibbs free energy for the combined ligand-receptor system. As in physical and chemical systems, lowered Gibbs free energy of biological systems depends on two factors: favorable changes in enthalpy and entropy changes. Traditional 3D-QSAR approaches are biased to reflect enthalpy changes and ignore entropy contributions. Classical 3D-QSAR models are based on electrostatic and stereospecific patterns in space as they correlate with biological activity. These correlations work because they reflect electrostatic differences and the corresponding changes in enthalpy that occur during binding. But these QSAR models neglect entropy changes. Entropy calculations are routinely neglected because they are often misunderstood, and because it is not easy to conceive a way in which such phenomena can be reflected in models that are built directly using molecular 3D-structural conformations. Modelers typically use only the minimum energy conformer for each compound during model development, and this ignores the changes in molecular conformation that are statistically possible and that certainly occur during the compound's interaction with its enzymatic substrate when binding occurs.

A 4D spectral-connectivity data matrix allows for multiple conformations of a molecule to be considered in modeling and activity prediction. Multiple conformations can be calculated using molecular dynamics principles and used in an "entropy-like" equation to estimate the effect of configurational entropy [47–49]:

$$S \alpha \ 1/N \sum_{i<j} p_{ij} \ In \ p_{ij}$$

where p_{ij}, a probability that must lie between zero and one, is calculated as the percentage of time the distance between atoms i and j lie within the r_{ij} spatial distance bin, with this percentage then divided by 100 to express the occupancy within the probability range required by the equation.

9.5.3 FUTURE APPLICATIONS OF SDAR MODELING

The ability of SDAR models to accurately predict the biological, chemical, or physical activity of molecules has tremendous application in the pharmaceutical, chemical, and materials science industries. LITMUS Molecular Design, LLC, licensed the commercialization rights to SDAR modeling and has been applying SDAR to the creation of new molecular entities that would be predicted to have certain characteristics. The initial task was to develop an antipsychotic drug that was like clozapine in its antipsychotic efficacy, yet without the serious adverse affect of agranulocytosis, a potentially fatal blood disorder. SDAR models of Dopamine D2 and 5-HT2A receptor binding were created as SDAR models for side effects like hERG inhibition and agranulocytosis risk were also created. The latter model was based on epidemiological data, extending SDAR's capabilities into the realm of nonexperimental data, which can be used to predict clinical characteristics for which there are no *in vitro* or *in vivo* models. These SDAR models were used to develop leads that had the desired receptor binding affinities to D2 and 5-HT2A. The application of SDAR to rational drug design has the potential to cut years of time and millions of dollars from the lead candidate creation and selection phases. The antipsychotic program described was accomplished in roughly 3 months. This would normally take 7 years and $250 million to accomplish using traditional methods. Other areas of exploration involve using the model data to identify mechanisms of toxicity through analysis of structural components that were implicated as predictive by the model. Thus, in addition to broad applicability, SDAR has great potential as an investigative tool.

An important step beyond modeling is externally validating the model's predictions with biological assays. We developed QSDAR models for the toxic equivalency factors (TEFs) of the 29 polychlorinated dioxin-like compounds (PCDDs, PCDFs, or PCBs) for which nonzero TEFs have been defined [50]. A separate QSDAR model predicted TEFs of 0.037 and 0.004, respectively, for 1,3,7,8-tetrachlorodibenzo-p-dioxin (TCDD) and 1,2,3,4,7-pentachlorodibenzo-p-dioxin (PeCDD), both of which are among the 390 polychlorinated dioxin-like congeners for which zero value TEFs are assumed [51]. A QSDAR model of relative potency (REP) values estimated the corresponding values as 0.115 and 0.020 for these two dioxins [51]. Results from both models indicated that these two congeners are likely to exhibit significant dioxin-like toxicity. Biological validation of the QSDAR models predictions was required. We decided to use the luciferase gene expression assay based on mouse liver cells to experimentally determine the REPs for 1,3,7,8-TCDD and 1,2,3,4,7-PeCDD binding to the aryl hydrocarbon receptor. The luciferase assay determined that the REP for 1,3,7,8-TCDD was 0.027 and the REP for 01,2,3,4,7-PeCDD was 0.013. The corresponding gene-expression assayed values were in agreement with the QSDAR model predictions for 1,3,7,8-TCDD and 1,2,3,4,7-PeCDD [51].

9.6 CONCLUSIONS

Two accurate SDAR CoSA models of relative binding affinity to the estrogen receptor have been made using 1D ^{13}C NMR spectra and ^{13}C NMR combined with MS data. The LOO accuracy of the SDAR models was 75% and 82%, respectively. These SDAR models were able to predict correctly the classification for 3-deoxyestradiol and dimethylstilbesterol as strong relative binders and 2-ethylphenol as a low relative binder. The predictions for 3-methylestriol and 4,4'-dihydroxystilbene tended to be incorrect, which may be due to the fact that they are medium relative binders. Accurate prediction of compounds in the medium binding classification is difficult because of nonspecific binding to the estrogen receptor. To help address this issue, we combined spectral information with nearest neighbor structural information in the development of 2D QSDAR models. 2D QSDAR models of relative binding affinity for 130 compounds were trained with r^2 of 0.83 and cross-validated with leave-10%-out of 0.78. The 2D QSDAR models were tested with external datasets. The external datasets had the spectra predicted and binned, and many of the bins were not in the training set. To account for this, we used some of a nonselected bin's intensity if it was near one of the bins used in the training set. The predictions ranged from q_{pred}^2 of 0.45 for the 16 bin 2D QSDAR model and a q_{pred}^2 0.62 for the 15 bin-plus-$L_{<7.5\text{Å}}$ model depending on whether the neighboring bin's intensity was used in the prediction. With the development of 2D spectral binning capabilities of the 3D-spectral-connectivity matrix, we can now build models that will include 3D through-bond and through-space structural information, and we believe this information will add to the QSDAR model accuracy and predictive capability.

ACKNOWLEDGMENTS AND DISCLAIMER

We would like to thank ACD/Labs for the "beta test" of ACD/Labs CNMR predictor software. The views presented in this chapter do not necessarily reflect those of the U.S. Food and Drug Administration.

REFERENCES

[1] R.J. Kavlock, G.P. Daston, C. DeRosa, P. Fenner-Crisp, L.E. Gray, S. Kaattari, G. Lucier, M. Luster, M.J. Mac, C. Maczka, R. Miller, J. Moore, R. Rolland, G. Scott, D.M. Sheehan, T. Sinks, and H.A. Tilson, *Research needs for the risk assessment of health and environmental effects of endocrine disruptors: A report of the U.S. EPA-sponsored workshop,* Environ. Health Perspect. 104 Suppl. (1996), pp. 715–740.

[2] R.M. Blair, H. Fang, W.S. Branham, B.S. Hass, S.L. Dial, C.L. Moland, W. Tong, L. Shi, R. Perkins, and D.M. Sheehan, *The estrogen receptor relative binding affinities of 188 natural and xenochemicals: Structural diversity of ligands,* Toxicol. Sci. 54 (2000), pp. 138–153.

[3] W.S. Branham, S.L. Dial, C.L. Moland, B.S. Hass, R.M. Blair, H. Fang, L. Shi, W. Tong, R.G. Perkins, and D.M. Sheehan, *Phytoestrogens and mycoestrogens bind to the rat uterine estrogen receptor,* J. Nutr. 132 (2002), pp. 658–664.

[4] H. Fang, W. Tong, W. Branham, C.L. Moland, S.L. Dial, H. Hong, Q. Xie, R. Perkins, W. Owens, and D.M. Sheehan, *Study of 202 natural, synthetic and environmental chemicals for binding to the androgen receptor,* Chem. Res. Toxicol. 16 (2003), pp.1338–1358.

[5] L.M. Shi, H. Fang, W. Tong, J. Wu, R. Perkins, R. Blair, W. Branham, S.L. Dial, C.L. Moland, and D. Sheehan, *QSAR models using a large diverse set of estrogens,* J. Chem. Inf. Comp. Sci. 41 (2001), pp. 186–195.

[6] H. Hong, H. Fang, Q. Xie, R. Perkins, D.M. Sheehan, and W. Tong, *Comparative Molecular Field Analysis (CoMFA) model using a large diverse set of natural, synthetic and environmental chemicals for binding to the androgen receptor,* SAR QSAR Environ. Res. 14 (2003), pp. 373–388.

[7] V.P. Palace, K.G. Wauter, R.E. Evans, P.J. Blanchfield, K.H. Mills, S.M. Chalanchuk, D. Godard, M.E. McMaster, and G.R. Tetreault, *Biochemical and histopathological effects in pearl dace* (Margarisus margarita) *chronically exposed to a synthetic estrogen in a whole lake experiment,* Environ. Toxicol. Chem. 25 (2006), pp. 1114–1125.

[8] K.A. Kidd, P.J. Blanchfield, K.H. Mills, V.P. Palace, R.E. Evans, J.M. Lazorchak, and R.W. Flick, *Collaspe of a fish population after exposure to a synthetic estrogen,* PNAS 104 (2007), pp. 8897–8901.

[9] J.E. Harries, D. Sheehan, S. Joblin, P. Matthiessen, P. Neall, J.P. Sumpter, T. Taylor, and N. Zaman, *Estrogenic activity in five United Kingdom rivers detected by measurement of vitellogenesis in caged male trout,* Environ. Toxicol. Chem. 16 (1997), pp. 534–542.

[10] J. Raloff. *Hormones: Here's the beef,* Science News 161 (2002), p. 10.

[11] R.D. Cramer, J.D. Bunce, and D.E. Patterson, *Cross-validation, bootstrapping, and partial least squares compared with multiple regression in conventional QSAR studies,* Quant. Struct. Act. Relat. 7 (1988), pp. 18–25.

[12] R.D. Cramer, D.E. Paterson, and J.D. Bunce, *Comparative molecular field analysis (CoMFA). 1. Effect of shape on binding of steroids to carrier proteins,* J. Am. Chem. Soc. 110 (1988), pp. 5959–5967.

[13] G. Klopman, *Artificial intelligence approach to structure-activity studies. Computer automated structure evaluation of biological activity of organic molecules,* J. Am. Chem. Soc. 106 (1984), pp. 7315–7321.

[14] G. Klopman, *MULTICASE1. A hierarchial computer automated structure evaluation program,* Quant. Struct. Act. Rel. 11 (1992), pp. 176–184.

[15] C. Hansch and A. Leo, *Exploring QSAR — Fundamentals and Applications in chemistry and biology,* The American Chemical Society, Washington, DC, 1995.

[16] W. Tong, R. Perkins, L. Xing, W.J. Welsh, and D.M. Sheehan, *QSAR models for binding of estrogenic compounds to estrogen receptor α and β subtypes,* Endocrinology 138 (1997), pp. 4022–4025.

[17] A.R. Katritzky, L. Mu, V.S. Labanov, and M. Karelson, *Correlation of boiling points with molecular structure. 1. A training set of 298 diverse organics and a test set of 9 simple inorganics,* J. Phys. Chem. 100 (1996), pp. 10400–10407.

[18] T. Fujita, J. Iwasa, and C. Hansch, *A new substituent constant, π, derived from partition coefficient,* J. Am. Chem. Soc. 86 (1964), pp. 5175–5180.

[19] S.P. Branbury, *Quantitative structure-activity relationship and ecological risk assessment: An overview of predictive aquatic toxicology research,* Toxicology 25 (1995), pp. 67–89.

[20] A.C. De Dios, J.G. Pearson, and E. Oldfield, *Secondary and tertiary structural effects on protein NMR chemical shifts: An ab initio approach,* Science 260 (1993), pp. 1491–1496.

[21] J.W. Emsley, J. Feeney, and L.H. Sutcliffe, *High Resolution Nuclear Magnetic Resonance,* Volume I, Pergamon Press, Oxford, 1965.

[22] R.D. Beger, J. Freeman, J. Lay Jr., J. Wilkes, and D.W. Miller, *^{13}C NMR and EI mass spectrometric data-activity relationship (SDAR) model of estrogen receptor binding,* Toxicol. Appl. Pharmacol. 169 (2000), pp. 17–25.

[23] L. Shade, R.D. Beger, and J.G. Wilkes, *New computerized method for modeling binding affinities to the aryl hydrocarbon receptor using ^{13}C NMR spectra,* Environ. Toxicol. Chem. 22 (2003), pp. 501–509.

[24] R.D. Beger and J.G. Wilkes, *Developing ^{13}C NMR quantitative spectrometric data-activity relationship (QSDAR) models of steroid binding to the corticosteroid binding globulin,* J. Comput. Aid. Mol. Des. 15 (2001), pp. 659–669.

[25] R.D. Beger and J.G. Wilkes, *^{13}C NMR quantitative spectrometric data-activity relationship (QSDAR) models to the aromatase enzyme,* J. Chem. Inf. Comput. Sci. 41 (2001), pp. 1360–1366.

[26] R.D. Beger and W.G. Wilkes, *Models of polychlorinated dibenzodioxins, dibenzofurans, and biphenyls binding affinity to the aryl hydrocarbon receptor developed using ^{13}C NMR data,* J. Chem. Inf. Comput. Sci. 41 (2001), pp. 1322–1329.

[27] R.D. Beger, K. Holm, K.D. Buzatu, and J.G. Wilkes, *Using simulated 2D ^{13}C-^{13}C NMR spectral data to model a diverse set of estrogens,* Int. Elec. J. Mol. Des. 2 (2003), pp. 435–453.

[28] R. Bursi, T. Dao, T. van Wilk, M. de Gooyer, E. Kellenbach, and P. Verwer, *Comparative spectra analysis (CoSA): Spectra as three-dimensional molecular descriptors for the prediction of biological activities,* J. Chem. Inf. Comput. Sci. 39 (1999), pp. 861–867.

[29] C.L. Waller, T.I. Opera, K. Chae, H.K. Park, K.S. Korach, S.C. Laws, T.E. Wiese, W.R. Kelce, and L.E. Gray Jr., *Ligand-based identification of environmental estrogens,* Chem. Res. Toxicol. 9 (1996), pp. 1240–1248.

[30] G.G.J.M. Kuiper, B. Carlsson, K. Grandien, E. Enmark, J. Haggblad, S. Nilsson, and J.-A. Gustafsson. *Comparison of the ligand binding specificity and transcript tissue distribution of estrogen receptors α and β,* Endocrinology 138 (1997), pp. 63–870.

[31] Integrated Spectral Data Base System for Organic Compounds. Available at www.aist.go.jp/RIODB/SDBS/.

[32] NIST MS database software version 1.6.

[33] *The Aldrich Library of ^{13}C and ^1H FT NMR Spectra* Edition 1, C.J. Pouchert and J. Behnke, eds., vol. 1–3, Aldrich Chemical Company, Milwaukee, WI, 1993.

[34] *Spectral Data for Steroids,* M. Frenkel and K.N. Marsh, eds., Thermodynamics Research Center, College Station, TX, 1994.

[35] A. Bax and S. Grzesiek, *Methodological advances in protein NMR*, Acc. Chem. Res. 26 (1993), pp. 131–138.

[36] W.P. Aue, E. Bartholdi, and R.R. Ernst, *Two-dimensional spectroscopy. Application to nuclear magnetic resonance*, J. Chem. Phys. 64 (1976), pp. 2229–2246.

[37] G. Bodenhausen and D.J. Ruben, *Natural abundance nitrogen-15 NMR by enhanced heteronuclear spectroscopy*, Chem. Phys. Lett. 69 (1980), pp. 185–189.

[38] A. Bax, R.H. Griffey, and B.L. Hawkins, *Sensitivity-enhanced correlation of 15N and 1H chemical shifts in natural-abundance samples via multiple quantum coherence*, J. Am. Chem. Soc. 105 (1983), pp. 7188–7190.

[39] A. Bax and M.F. Summers, *1H and 13C Assignments from sensitivity-enhanced detection of heteronuclear multiple-bond connectivity by 2D multiple quantum NMR*, J. Am. Chem. Soc. 108 (1986), pp. 2093–2094.

[40] A. Kumar, R.R. Ernst, and K. Wuthrich, *A two-dimensional nuclear Overhauser enhancement (2D NOE) experiment for the elucidation of complete proton-proton cross-relaxation networks in biological macromolecules*, Biochem. Biophys. Res. Comms. 95 (1980), pp. 1–6.

[41] R.D. Beger, D. Buzatu, and J.G. Wilkes, *Combining NMR spectral and structural data to form models of polychlorinated dibenzodioxins, dibenzofurans, and biphenyls binding to the AhR*, J. Comp. Aided Molec. Design 16 (2002), pp. 727–740.

[42] R.D. Beger, D. Buzatu, J.G. Wilkes, and J.O. Lay, Jr., *Developing comparative structural connectivity spectra analysis (CoSCoSA) models of steroid binding to the corticosteroid binding globulin*, J. Chem. Inf. Comput. Sci. 42 (2002), pp. 1123–1131.

[43] R.D. Beger and J.G. Wilkes, *Comparative structural connectivity spectra analysis (CoSCoSA) models of steroids binding to the aromatase enzyme*, J. Mol. Recog. 15 (2002), pp. 154–162.

[44] W. Bremser, *HOSE — A novel substructure code*, Anal. Chim. Acta. 103 (1978), pp. 355–365.

[45] H. Fang, W. Tong, R. Perkins, A. Soto, N. Prechtl, and S.D. Sheehan, *Quantitative comparison of in vitro assays for estrogenic assays*, Environ. Health Perspect. 139 (1998), pp. 723–729.

[46] R.D. Beger, D.A. Buzatu, and J.G. Wilkes, *Combining NMR spectral information with associated structural features to form computationally non-intensive, rugged, and objective models of biological activity*, in *Drug Discovery Handbook Volume 1. Pharmaceutical Development and Research Handbook*, S.C. Gad, ed., John Wiley & Sons, New York, 2005, pp. 227–286.

[47] R.L.L. Compadre, R.A. Pearlstein, A.J. Hopfinger, and J.K. Seydel, *A quantitative structure-activity relationship analysis of some 4-aminophenyl sulfone antibacterial agents using linear free energy and molecular modeling methods*, J. Med. Chem. 30 (1987), pp. 900–906.

[48] S.D. Pickett and M.J.E. Sternberg, *Empirical scale of side-chain conformational entropy in protein folding*, J. Mol. Biol. 231 (1993), pp. 825–839.

[49] M. Karplus and J.N. Kushick, *Method for estimating the configurational entropy of macromolecules*, Macromolecules 14 (1981), pp. 325–332.

[50] D. Buzatu, D.R.D. Beger, J.G. Wilkes, and J.O. Lay Jr., *Predicting toxic equivalence factors from ^{13}C NMR spectra for dioxins, furans, and PCBs using multiple linear regression and artificial neural networks*, Environ. Toxicol. Chem. 23 (2004), pp. 24–31.

[51] J.G. Wilkes, B.S. Hass, D. Buzatu, L.M. Pence, J.C. Archer, R.D. Beger, L. Schnackenberg, M.K. Halbert, L. Jennings, and R.L. Kodell, *QSDAR prediction of TEF values for 1,2,3,4,7-PCDD and 1,3,7,8-TCDD is experimentally verified by CALUX-derived binding values*, Toxicol. Sci. 102 (2008), pp. 187–195.

10 Mechanism-Based Modeling of Estrogen Receptor Binding Affinity
A Common Reactivity Pattern (COREPA) Implementation

Ovanes Mekenyan and Rossitsa Serafimova

CONTENTS

ABSTRACT

The capabilities of the probabilistic classification scheme for identification of the common reactivity pattern (COREPA) of biologically similar chemicals to model binding affinity to human estrogen receptor (hER) will be reviewed. The COREPA reactivity patterns based on the distance between nucleophilic sites allow the identification of distinct ER binding interaction mechanisms. Three-dimensional (3D) structural and parametric boundaries of different binding mechanisms are defined and used for building a categorical quantitative structure-activity relationship (QSAR) model. In addition to the traditional molecular orbital (MO) reactivity descriptors, local hydrophobicity parameters are used to assess the interaction of ligands with the hydrophobic pockets of ER. The interaction mechanisms are less pronounced at the lowest activity range where the categories are defined by chemical classes (phenols, phthalates, and so forth). Once the chemicals are grouped by mechanism, then COREPA models are developed to determine the potency of chemicals belonging to each relative binding affinity (RBA) range. Eventually, each category is associated with a specific binding mechanism and activity bin. In the integrated model, the categories are organized in a hierarchical battery of local models associated with various interaction mechanisms and activity bins. To assess the effect of the metabolic activation of a chemical, the categorical scheme is combined with a metabolism model simulating rat liver S-9 mix activation. The details of the 3D categorical model for ER binding affinity will be presented along with illustration for the model performance with and without metabolic activation of chemicals.

KEYWORDS

Common reactivity pattern (COREPA)
Estrogen receptor (ER) binding affinity
Interaction mechanism
Molecular flexibility

10.1 INTRODUCTION

One of the Organization for Economic Cooperation and Development (OECD) principles for valid Quantitative Structure-Activity Relationship (QSAR) is to model well-defined toxic pathways [1]. The diversity of toxic pathways that could disrupt the endocrine system makes the risk assessing of potential endocrine disrupting chemicals (EDCs) a challenging task. The focus of the present chapter is the direct interactions of chemicals with ER — that is, ER binding mediated endocrine disruption. This interaction is considered as a primary event that triggers the ultimate biological responses [2]. In this respect, the structural tolerance of ER allowing a large number of exogenous chemicals to mimic the action of natural hormones is of high concern. The latest legislative and governmental efforts have focused on finding out simple screening tools for identifying those chemicals most likely to bind ER without experimental testing of all chemicals of regulatory concern. The QSAR approach is a powerful *in silico* technique that should be considered for prioritizing chemicals for subsequent empirical assessments. This is supported by the five-level conceptual framework of the OECD [3], according to which the use of *in vitro* and QSARs is foreseen before use of *in vivo* tests.

The high specificity of the interaction between ligands and ER inevitably requires analyzing the structure-activity relationship at three-dimensional (3D QSARs) and four-dimensional (4D QSAR) levels. A number of 3D approaches have been used to develop QSARs for ER binding affinity, such as automated docking models (ADAM) [4], comparative molecular field analysis (CoMFA) [5–7], and a common reactivity pattern (COREPA) approach [8].

The involvement of the 3D molecular structure in the modeling inevitably requires us to take into account the molecular flexibility, or "fourth dimension," of molecular structure [9]. It was shown that the variation in stereoelectronic parameter values for the conformers of same chemical could be as large as the variation in stereoelectronic parameter values for conformers of distinct chemicals [10]. Therefore, the ignorance of the effect of conformational flexibility of ligands on their potential activity could lead to large errors. To address the issue, it is assumed that in complex environments, such as with biological tissues and fluids, chemicals could exist in conformations other than the lowest gas phase energy state. In fact, the lowest-energy conformations might be the least likely to interact with macromolecules [11], and solvation and binding interactions could more than compensate for energy differences among the conformers of a chemical [12–18]. Based on theoretical considerations [19], it was determined that multiple conformational states differ in energy within several kcal/mole of the lowest-energy conformer that makes the transition between them feasible both thermodynamically and kinetically. On the other hand, conformers of an individual chemical that have free energy within approximately 20 kcal/mol from the lowest-energy structure (usually accepted energy range) often exhibited significant variation in electronic descriptor values. For example, conformers of phloretin had a range of 0.51 eV for E_{LUMO}, 0.55eV for E_{HOMO}, 0.65 eV for $E_{HOMO-LUMO}$, and 5.09 D for dipole moment. Moreover, it was found that the lowest-energy conformer of phloretin was not the active one with respect to binding to ER (Scheme 10.1).

The fact that small energy differences between conformers can result in significant variations in electronic structure demonstrated the necessity of including all energetically reasonable conformers when defining reactivity pattern, which is common for biologically similar chemicals [20,21]. This imposed the necessity of developing effective algorithms for investigating conformational space of the molecules. Alternatively, one could produce the wrong screening results if the conformational space of the chemicals is not well represented. To resolve combinatorial difficulties with systematic algorithms for conformer generation, much more effective nondeterministic methods were developed [22]. They, in turn, demonstrated insufficiencies due to irreproducibility of the generated conformers. To overcome this problem, the nondeterministic algorithms have been improved by further saturating conformational space [23].

Another "dimension" of the molecular structure that should be taken into account in QSAR studies is its metabolic transformations that are important in the evaluation of its toxic potential. For example, the determination of metabolic stability can

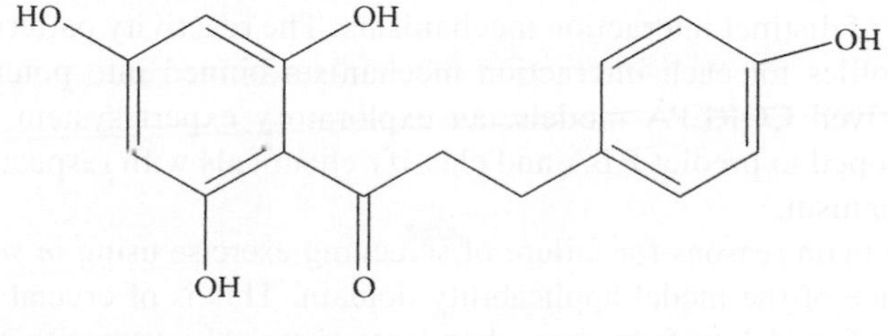

SCHEME 10.1 Phloretin, CAS # 60-82-2.

provide information on the potential for bioaccumulation and whether a chemical is likely to be converted to a more active form (metabolic activation) or a less active form (metabolic detoxification). With regard to endocrine disruptors, one should take into account that metabolism of the xenobiotic may alter its own potency by creating metabolites that are either more or less active than the parent compound. Thus, methoxychlor (bis[p-methoxydichlorophenyl]trichloroethane-MC) has estrogenic activity *in vivo* while it is inactive or virtually inactive *in vitro* [24]. Substances like the di-hydroxylated metabolite of methoxychlor, 2,2-bis(p-hydroxyphenyl)-1, 1,1-trichloroethane (HPTE), are estrogenic and antiandrogenic *in vitro,* and their formation *in vivo* can explain the estrogenicity of methoxychlor.

The COREPA technique [20,21] we used for predicting ER binding affinity is a probabilistic scheme for estimating the COREPA of biologically similar chemicals, accounting for their molecular flexibility. The common reactivity patterns are identified as areas of stereoelectronic structural space populated by the conformers of active chemicals. In our earlier exercise for deriving of reactivity patterns for ER relative binding affinity (RBA), the common reactivity patterns were derived in the structural space defined by the global nucleophilicity, interatomic distances between nucleophilic sites, and local electron donor capability of the nucleophilic sites [20,21].

It is known that a reliable QSAR model should be developed within the same interaction mechanism. Another challenging issue in predicting the estrogenicity of chemicals is the "opposite" task — namely, the possibility of inferring the interaction mechanisms by QSAR methods. The latter became realistic after recent successful attempts to elucidate the nature of the ligand-binding pocket of estrogen receptors α and β [25,26]. Thus, it was found that a significant increase in binding affinity was reached by placing lipophilic substituents on steroidal estrogens, notable at positions 7α and 11β. It was assumed that the intimate contact between the protein and the ligand in the ligand-binding pocket at these predefined positions is favorable for binding. Alternatively, even small substituents at other ligand sites caused dramatic losses of binding affinity. Apparently, the differences in the characteristics of the pocket regions should be closely associated with a diversity of interaction mechanisms.

In our recent exercise for modeling the hER binding affinity, we focused on the categorical approach for modeling where each category is associated with an anticipated interaction mechanism. A training set of 645 chemicals, including 497 steroid and environmental chemicals and 148 chemicals synthesized for medicinal purposes, were used to explore hER-structure interactions. Analysis of reactivity patterns based on the distance between nucleophilic sites resulted in identification of distinct interaction mechanisms. The reactivity patterns provided descriptor profiles for each interaction mechanism binned into potency ranges. Based on derived COREPA models, an exploratory expert system was subsequently developed to predict RBA and classify chemicals with respect to the hER binding mechanism.

One of the main reasons for failure of screening exercise using *in silico* models is the ignorance of the model applicability domain. This is of crucial importance for reliability of model performance when large chemical inventories are screened for potential EDCs. The model domain is also important for expending initial exploratory 3D QSAR models beyond current training sets to increase applicability to more diverse structures in large chemical inventories.

The aim of this study was to review the recent achievements of the Laboratory of Mathematical Chemistry to model hER binding affinity of chemicals and mechanisms of binding interaction by making use of the COREPA modeling technique. The focus of the review will be the model published in Serafimova et al. [27]. In the current review, we will also present the latest exercises of application of this model combined with a simulator for rat liver metabolism. In Section 10.2, the upgraded modeling tools are presented, including the multiparameter formulation of the COREPA method, the procedure for investigating conformational flexibility of chemicals, the new molecular descriptor for assessing local hydrophobicity in order to estimate the receptor binding pocket, and the principles of used model applicability domain. The ultimate screening system is constituted as a hierarchical tree of ER binding models for each of the identified interaction mechanisms and potency bins and was combined with a liver metabolic simulator. The capabilities of the models and decision tree were illustrated by the screening exercise with 387 chemicals tested for hER binding affinity by the Ministry of Economy, Trade and Industry of Japan (METI) [28]. The performance of the model combined with simulator of rat liver metabolism is illustrated for 25 chemicals with documented metabolism and available data for ER binding affinity of parents and metabolites.

10.2 METHODS

10.2.1 TRAINING SET CHEMICALS

The training database used for deriving the hER binding affinity model (645 chemicals) was published recently [27] and could be found as supplementary information on the Laboratory of Mathematical Chemistry Web site (www.oasis-lmc.org/). It combines two datasets: a collection of 148 chemicals provided by Katzenellenbogen and Anstead, from the University of Illinois, Urbana, and the University of Kentucky, Albert B. Chandler, respectively (reported in about 300 papers; see [25,26], as representative of this collection), and part of the METI database (497 chemicals) [27]. The potency values are expressed as RBA, with estradiol having an affinity of 100%.

10.2.2 LIGAND-BINDING POCKET AND INTERACTION MECHANISMS

The ligand-binding pocket is described in detail in Anstead et al. [26], and it was summarized in our preceding publication [27]. Several receptor-binding sites are identified. That not occupied by the receptor and thus available to the ligands and their analogs is called receptor-excluded volume (RExV) of about 700 $Å^3$, which is significantly larger than the volume of the ligands (for example, 245 $Å^3$ for E2) [29]. This suggests that the receptor-binding site can be distorted considerably without loss of RBA. The extensive synthetic work in the area allowed further elucidation of the electronic and hydrophobic interaction sites in the receptor pocket, which is summarized as follows: (1) the hydroxyl group at position 3 donates a hydrogen bond to weakly charged histidine residue (it was also hypothesized that this OH group could also act as H-bond acceptor); (2) the A-ring with small excess negative charge on the faces of π-electron could be engaged in a weak polar interaction with a slightly positively charged receptor residue, situated on the β-face of the steroid; similarly, the H-atoms at its periphery with small excess positive charge could get in such interactions with the negatively charged atoms (oxygen, sulfur) of the receptor; (3) the hydrophobic pocket at the 7α-position has a minimum volume of 17 Å³; (4) the hydrophobic pocket at the 11β-position has a minimum volume of at least 45 Å³; (5) 14 α, 16 α, and 17 α are small flexible subsites; some of them like 16 α could tolerate large nonpolar groups; and (6) the 17 β-hydroxyl group probably acts as an H-bond acceptor from a receptor residue.

Based on the above model for the ligand-binding pocket, it was concluded that the binding affinity is conditioned by electronic and hydrophobic interactions. In addition to the traditional two sites interacting with the nucleophiles on A and D rings, a third receptor site situated at one of the binding pockets (11 β) could interact with the ligand. These three sites are denoted as A, B, and C; moreover, the A site provides the strongest interaction energy, whereas the B and C sites have approximately the same strength [30]. Four electronic interaction mechanisms could be anticipated based on these three sites: A–B, A–C, A–B–C (AD), and A only, as shown in Figure 10.1.

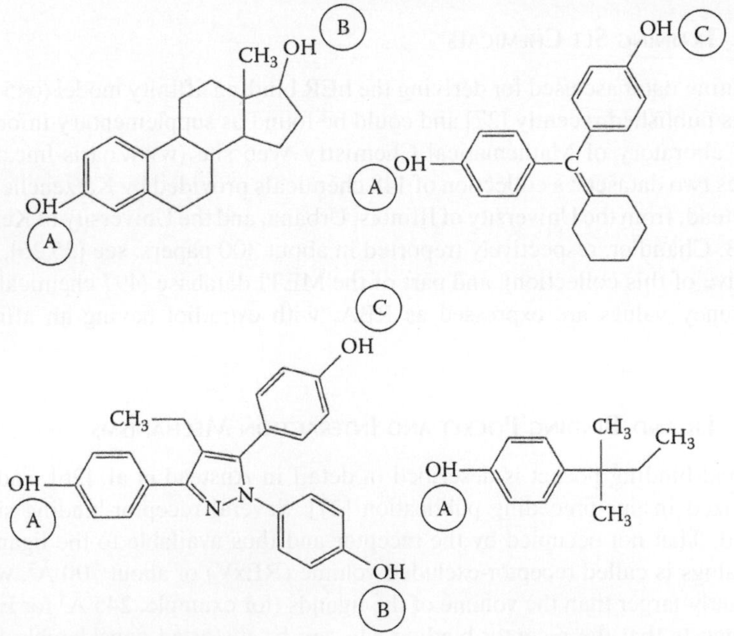

FIGURE 10.1 Estrogen receptor (ER) electronic binding mechanisms.

10.2.3 MOLECULAR MODELING APPROCHES

10.2.3.1 Conformational Analysis by Genetic Algorithm

To avoid the combinatorial complexity of the systematic approaches [31–33] a nondeterministic algorithm for coverage of the conformational space by a limited number of conformers was developed [22]. It is based on a genetic algorithm (GA) aiming to minimize 3D similarity among the generated conformers. The 3D similarity of a pair of conformers is assumed reciprocal to the root-mean-square (RMS) distance between identical atomic sites in an alignment providing its minimum. In contrast to traditional GA, the fitness of a conformer is not quantified individually but as part of the population to which it belongs. The RMS-based function for assessing conformer diversity is applied in combination with another fitness function based on Shannon entropy formulae [34]. The joint application of both functions provides better coverage of conformational space by generated structures. To better reflect the flexibility of saturated polycyclic structures, a new conformer variable (flip of Csp3 pyramids) is introduced in addition to rotation around single bonds and flip of free corners (in saturated cyclic structures). During the conformer generation process, the stereochemistry of the active enantiomer is maintained. Each of the generated conformations is submitted to a geometry optimization procedure by quantum-chemical methods. Usually, MOPAC 93 [35,36] is employed by making use of the AM1 Hamiltonian. Next, the conformers are screened to eliminate those whose heat of formation, ΔH^0_f, is greater from the ΔH^0_f associated with the conformer with absolute energy minimum by user-defined threshold. Usually, the 20 kcal mol^{-1} (or 15 kcal mol^{-1}) threshold is employed. Subsequently, conformational degeneracy, due to molecular symmetry and geometry convergence, is detected within a user-defined torsion angle resolution.

To minimize the effects of nondeterministic character of GA and smoothing parameter h of the continuous (probabilistic) conformer distribution [19] on the reproducibility and diversity of generated conformers, a new procedure for saturation of conformation space was developed [23]. The goal of the saturation is to determine the optimal number of conformers providing stable conformational distributions across selected molecular descriptors that are no longer perturbed by the addition of new conformers. Such conformer distributions are expected to eventually provide reliable COREPA patterns of biologically similar chemicals. The approach for conformer saturation is based on the independent application of the GA for conformer generation. This algorithm is applied consecutively a number of times until conformational space is saturated within a user-defined population density. Details on the saturation algorithm and more illustrations were reported elsewhere [23].

10.2.3.2 Molecular Descriptors — Local Hydrophobicity Index

A variety of mechanistically sound molecular descriptors assessing specific steric and electronic interactions are used in the OASIS software to model the receptor binding interactions [20,21]. Among them, one should mention the energies of frontier orbitals (E_{LUMO}, E_{HOMO} [eV]) assessing the global electrophilicity and nucleophilicity of molecules, respectively; the difference between these energies (Egap [eV]) as a measure of molecular reactivity; electronegativity EN [eV]; dipole moment ([D]); volume polarizability (VolP [a.u./eV]) as an averaged ability of a chemical to change electron density at its atoms during chemical interactions; degree of stretching or compactness (quantified as sum of interatomic steric distances, GW); greatest interatomic distance (L_{max}); planarity (normalized sum of torsion angles in a molecule); van der Waals surface; and solvent accessible surface calculated by making use of the Connoly algorithm [37]. The local specificity of molecular structure should also be used, such as the distances between fragments and atomic sites, d_{ij} [A°], atomic charges (q_i [a.u.]), frontier atomic charges (f^i_{HOMO} and f^i_{LUMO}, [a.u.]), donor and acceptor superdelocalizabilities (S^i_E and S^i_N [a.u./eV]), charged partial surface areas (CPSAs) as introduced by Stanton and Jurs [38]. Among the CPSAs one could distinguish partial positive surface area (PPSA) and partial negative surface area (PNSA). Less specific molecular descriptors have also been used in describing receptor mediated effects, such as water solubility and log Kow, which are important for the nonreactive component of the effect (penetration, diffusion).

The analysis of the ligand-receptor interactions, however, clearly showed that parameters associated with the local hydrophobic interactions should also be included in the list of molecular parameters used for modeling ER binding affinity of molecules. Recently, a set of new parameters was introduced to describe the magnitude and directional character of hydrophobic interactions in addition to previously described electronic components. The new parameters are derived by analyzing the distribution of the hydrophobic "bumps" (in terms of van der Waals volume, surface, and charged surfaces), across the line between nucleophilic sites previously shown to be associated with high-affinity ER binding [20,21].

The following parameters are inferred to assess the dimensionality of a hydrophobic bump (Figure 10.2): the distance from the projection of the outermost bump atom to the most remote nucleophilic site across the line between nucleophilic sites (R_bump [Å]); a similar distance applied to the center of a cylinder encompassing the bump

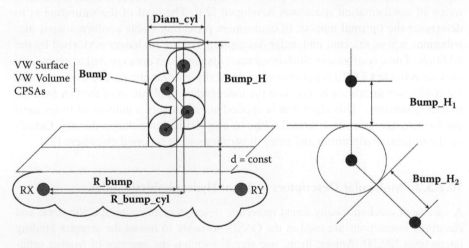

FIGURE 10.2 Local hydrophobic descriptors, named "bump" parameters, and the respective abbreviations.

(R_bump_cyl [Å]); diameter of this cylinder (Diam_cyl [Å]); and height of the cylinder, that is, the bump (Bump_H [Å]). The volumetric parameters of the bumps are formed by atoms above the plane situated at a user-defined distance ($d = 5$Å) from the line between nucleophilic sites. Thus, the van der Waals volumes of the bump Bump_VW_Volume [Å^3]) are calculated by summing their atomic van der Waals volumes; similarly, the surface (Bump_VW_Surface [Å^2]) and charged surface (Bump_CPSAs [Å^2] [38]) of the bump are calculated by summing the respective atomic contributions. The above set of local hydrophobicity parameters are referred to as functional because they are inherently dependent (for example, the volume of the bump is estimated for all distance between two sites the bump is originating from. Another functional parameter is the distance between two atoms combined with their charges; this is already a 3D parameter in terms of the distance between atoms and both charges. Within the COREPA formalism, the probability distributions of functional parameters are calculated as multidimensional distributions. For example, a functional parameter is used in the model for AB interaction type and RBA > 10%. (Figure 10.5); it was named Q_DISTANCE, combining the distance between two atomic sites (nucleophilic atoms in this specific case) and charge of one of these sites.

10.2.3.3 Basic Principles of the COREPA Method

The COREPA is a probabilistic classification scheme identifying parametric criteria that will classify an unknown object into predefined classes of biological similarity using a training set of objects from multiple classes. The COREPA formalism uses a Bayesian probabilistic method to identify common structural characteristics among chemicals that elicit similar biological activities. Instead of comparing and aligning 3D structures of the conformers of biologically similar chemicals, their probabilistic conformational distributions in the molecular descriptor space are analyzed and compared, thus accounting for molecular flexibility of the chemicals. The COREPA is developed through seeking overlap between conformer distributions of biologically similar chemicals in the specific structural space (Figure 10.3).

Thus, the problem of structure alignment traditionally used for similarity assessments is circumvented in COREPA by overlapping and comparing conformational distributions of chemicals across the descriptor axis. For a mathematical formalism of the current algorithm, the reader is encouraged to consult the literature [9,27,39].

10.2.3.4 OASIS Model Applicability Domain

The reliability of the predictions made by OASIS models was evaluated by the recently developed stepwise approach of determining the model applicability domain [40]. Four stages were applied to account for the diversity and complexity of QSAR models, reflecting both their mechanistic rationale (including metabolic activation of chemicals) and transparency. General parametric requirements were imposed in the first stage, specifying the domain for only those chemicals that fall in the range of variation of selected physicochemical parameters of chemicals in the training set. The second stage defined the structural similarity between target chemicals and those correctly predicted by the model. Different molecular features (usually atom-centered

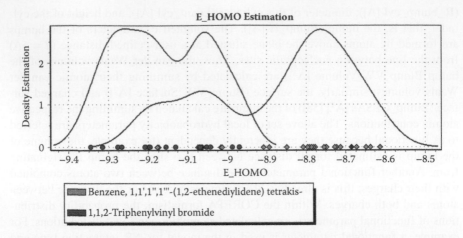

FIGURE 10.3 The conformer distributions of two chemicals: benzene, 1,1′,1″,1‴-(1,2-ethenediylidene) tetrakis- (CAS 632-51-9) and 1,1,2-Triphenylvinyl bromide (CAS 1607-57-4), across E_{HOMO}. The overlap between conformer distributions is used in the COREPA method to evaluate the similarity between chemicals with respect to E_{HOMO}.

fragments) are used to determine this similarity. The training set chemicals for which the QSAR model provides correct predictions (within user-defined accuracy thresholds) were used for extracting atom-centered fragments ("good fragments") that form the model structural domain. If the atom-centered fragments for each atom constituting an external chemical were determined to be elements of this list, then that chemical belongs to the structural domain of the model. The third stage in defining the domain is based on a mechanistic understanding of the modeled phenomenon — that is, the domain of the mechanistic hypothesis. It is defined by specific reactive (alerting) groups, hypothesized to cause the effect and the boundaries of explanatory variables, which determined the parametric requirements for functional groups to elicit their reactivity. Finally, the reliability of simulated metabolism (metabolites, pathways, and maps) was taken into account in assessing the reliability of predictions, if metabolic activation of chemicals was part of the (Q)SAR model. Some of the stages of the proposed approach for defining the model domain could be disabled depending on the type of QSAR model (for example, accounting or not the metabolic activation of chemicals) as well as availability and quality of the experimental data used to derive the model.

10.2.3.5 Simulation of Metabolism

A probabilistic approach for simulating metabolism is developed in the OASIS software. It consists of a list of hierarchically ordered transformations and a substructure-matching engine for their implementation. According to the probabilistic nature of the approach, the hierarchy of transformations is defined by the probabilities of transformations determined in a way to reproduce a database of documented metabolic transformations or the rate of chemical disappearance. The transformation probabilities are not rate constants; however, they are related to them by assessing the feasibility of occurrence of reactions within the duration of the metabolism tests. It is assumed that the transformations are independent and performed sequentially. Each molecular transformation consists of parent submolecular fragments, transformation products, and inhibiting masks serving as reaction inhibitors. If the fragment assigned as a mask is attached to the target subfragment, the execution of the transformation on the parent chemical is prevented. The presence of groups that can promote or inhibit metabolic reactions significantly increases the number of principal transformations in metabolism simulators. Currently, 343 principal transformations are used to model metabolism in the liver (mammalian). The transformations are separated into two major classes: nonrate-determining and rate-determining reactions. The first class includes 41 abiotic and enzyme-controlled reactions that occur at a very high rate as compared to the duration of the tests. Transformations of highly reactive groups and intermediates are included here. Various chemical equilibrium processes like tautomerism are also included in this class of transformations. The second type of reaction includes 302 metabolic (Phase I and II) transformations such as oxidative, redox, reductive, hydrolytic, and synthetic reactions.

In the multipathway formulation of the algorithm, the parent chemical is submitted to the list of transformations, and all transformations meeting the associated substructures are implemented on the parent producing the list of the first level metabolites. Each of the generated metabolites is then submitted to the same list of transformations to produce the second level of metabolites, and so forth. The procedure is repeated for the newly formed chemicals until the product of probabilities of consecutively performed transformations reaches a user-defined threshold. The mathematical formalism is based on the assumption that transformations occur sequentially — that is, the most probable transformation is applied first to the parent chemical and, then, the remainder of nonmetabolized parent molecules undergo the second transformation with lower probability, and so on [41,42]. The reaction probabilities of the metabolic simulator were adjusted to reproduce a database with 332 documented maps for rat (mammalian) liver metabolism [41,43]. The performance of the simulator is assessed by the degree of reproducibility of the training set with documented maps. Similarly, assessments evaluating the reliability of generated metabolites and metabolic maps were introduced [41,42].

10.3   RESULTS AND DISCUSSION

The hypothesis for interaction mechanisms between ligands and ERα, developed by Katzenellenbogen [30], was used as a starting point for the hER binding model we developed. These mechanisms were investigated by making use of the COREPA analysis. Based on the assumption that distances between electrophilic sites in the receptor determine the requirements for the most favorable binding mechanism, the probabilistic conformer distribution of most active chemicals (RBA > 10%) across distances between nucleophilic sites in the molecules was analyzed. The assumption was that the mechanisms will be best elicited at the highest activity bin due to lack of structural or parametric factors hampering these interactions (Figure 10.4).

Three interaction mechanisms were identified on the basis of the observed peaks in the probabilistic distribution depicted in Figure 10.4: from 7.7 to 9.7 Å, from 10.2 to 11.0 Å, and from 11.3 to 14.0 Å. The comparative analysis with the mechanisms

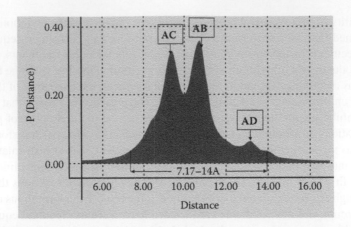

FIGURE 10.4 The probabilistic conformer distribution of most active binders (RBA > 10%) across the distances between nucleophilic sites. The distance ranges corresponding to different binding mechanisms are indicated. The entire range of 7.7–14.0 Å between two O atoms appeared to be a necessary requirement for eliciting the ER binding effect.

suggested by Katzenellenbogen showed that this A-B mechanism could be associated with the pick from 10.2 to 11.0 Å and the A-C mechanism with the pick from 7.7 to 9.7 Å. A third category of chemicals was found to have distances exceeding 11 Å (a pick from 11.3 to 14 Å). The last group of chemicals are classified into a new mechanism — A–D, which is assumed to correspond to an A–B–C mechanism of Katzenellenbogen. These distance ranges and anticipated interaction mechanisms are also assigned to chemicals from the lower activity bin — 0.1 < RBA < 10%. COREPA models are derived for chemicals belonging to each of the mechanisms and activity bins. The model for AB mechanism, for most active RBA ranges, is listed in Figure 10.5. As seen, the model is composed by a filtering block imposing parametric constraints for different mechanisms and a COREPA probabilistic model (see Figure 10.6) discriminating active from nonactive chemicals within same binding mechanism.

The approach for deriving the models for the AB binding mechanism and RBA > 10% and 0.1 < RBA < 10% are explained next. After the global prescreen for eliminating acyclic chemicals (under the assumption that cyclicity is an obligatory structural requirement for estrogenicity), 642 chemicals are submitted to prefiltering requirement for AB-mechanism: distance between hydroxyl-O (from C{aromatic}-OH) and O bound to C-cyclic to be in the range of 10.2–11.0 Å (see Figure 10.4). 72 chemicals meet this requirement, among them 17 active (RBA > 10%) and 55 inactive (RBA < 10%). Subsequently, the COREPA model discriminated successfully (concordance 89%) 16 active chemicals (sensitivity 94%) at the cost of seven false positives (specificity 87%) by making use of the functional parameter Q_Distance and log Kow. To build a model for a lower-activity bin, all chemicals from the higher-activity bins were eliminated from the training set. Thus, to derive the model for 0.1 < RBA < 10%, within the same AB mechanism, 66 chemicals with observed RBA > 10% were eliminated from the initial training set with 642 cyclic chemicals (see Figure 10.5). The remaining chemicals are submitted to investigate the distance range between nucleophilic sites. As can be seen,

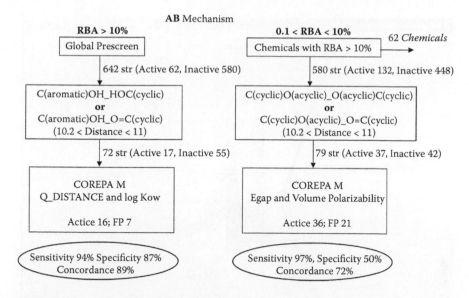

FIGURE 10.5 The common reactivity pattern (COREPA) models for RBA > 10% and 0.1 < RBA < 10% associated with the AB mechanism.

the prefiltering conditions for AB mechanism are practically the same (O_O distance from 10.2 to 11.0 Å) as for the RBA > 10% range. The molecular descriptors selected by the COREPA system to distinguish chemicals from each of the specified mechanisms and two most active bins are listed in Table 10.1A and Table 10.1B, respectively. The analysis of parameters in the COREPA model for the highest activity range (RBA > 10%) shows that the AB mechanism is probably associated with a contribution of stereoelectronic and global hydrophobic interactions; the functional parameter Q_Distance, and log Kow, are used as COREPA discriminating parameters.

Similarly, models were developed for the two most active bins (RBA > 10% and 0.1 < RBA < 10%) for AC and AD mechanisms. The analysis of the models for AC mechanisms showed that the same type of parameters (electronic and hydrophobic) hold for each of the specified mechanisms, which is an indication that the same interaction mechanisms act across the potency bins. Thus, bump parameters are always statistically selected to discriminate active chemicals for the AC mechanism. An interesting "bump" parameter with respect to the mechanistic interpretation of the results is the distance R_bump at which the hydrophobic "Bump" is situated from the most remote nucleophilic site A (see Figure 10.2). Three parameter ranges were defined for this parameter. The first is around 5.4 to 8.4 Å; the second, from 5.6 to 7.5 Å; and the third between 10.1 and 12 Å. The first two ranges correspond to the 7α and 11β position of the estradiol pharmacophore described in the literature [26]. Thus, the distance between the OH group in the third position and C7 in estradiol is 6.1 Å, whereas the distance between OH and C11 is 6.5 Å. The estimated volumes of the bump parameters were compared with those documented in the literature [26]. The estimated van der Waals volumes of the bumps listed in the third column of Table 10.2 show that they are comparable in magnitudes with those documented given the flexibility of the receptor site.

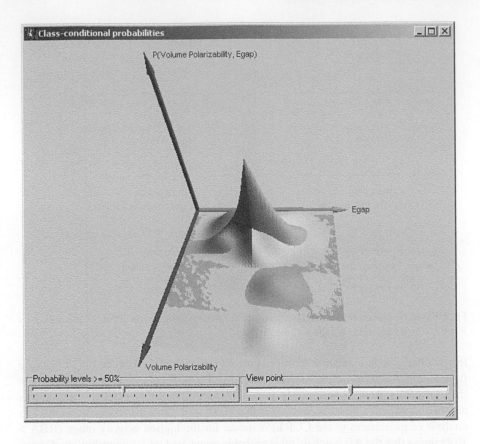

FIGURE 10.6 The common reactivity pattern (COREPA) probabilistic distributions discriminating active (0.1 < RBA < 10%) from inactive (0 < RBA < 0.1%) chemicals belonging to the AC mechanism (meeting the respective filtering condition). Energy gap (Egap) and volume polarizability (VolP) appeared to be discriminating parameters in the COREPA models.

The specific ER binding mechanisms — AB, AC, and AD (requiring two nucleophilic atoms as reactive sites) — were not identified at the lowest activity range. However, an important finding was that the range of 7.7 to 14.0 Å between two nucleophilic sites (O-atoms, bound to cyclic C and at least one of them to be hydroxyl oxygen), encompassing the distance screened associated with all three binding mechanisms (as illustrated in Figure 10.4) appeared to be a necessary requirement for eliciting the ER binding effect. This distance range discriminated successfully 67 active chemicals (43%) from the lowest-activity bin at the cost of 17 false positives (6%). The potency of this effect across activity bins and interaction mechanisms is modulated by specific steric, electronic, and hydrophobic factors combined with the distance screens. The fourth mechanism — A-only type, including mostly phenolic chemicals — has been specified for 0.1 < RBA < 10% and 0 < RBA < 0.1% bins [27] (Figure 10.7).

The rest of the chemicals eliciting activity in the lowest binding range were grouped into classes: halogen-containing chemicals, phthalates, and flavones; esters;

TABLE 10.1
Parameters of the Common Reactivity Pattern (COREPA) Models Discriminating Chemicals from Each Mechanism and Two Activity Bins: (A) RBA > 10%; and (B) 0.1 < RBA < 10% (Number of Chemicals Belonging to Each Mechanism Is Listed along with the Sensitivity, Specificity, and Concordance of Each Model)

Mechanisms	Distance Requirements	Number of Chemicals	Parameters in the COREPA Models	Sensitivity (%)	Specificity (%)	Concordance (%)
AB	10.2–11.0 Å	72	Q_DISTANCE logKow Local	94	87	89
AC	7.7–9.7 Å	159	Hydrophobicity (Bump), logKow	83	86	86
AD	11.3–14.0 Å and not having distances <10 Å	92	Diameff, E_{HOMO}, logKow	100	86	88
AB	10.2–11.0 Å	79	Egap, volume polarizability Local	97	50	72
AC	7.6–9.7 and not distances >11 Å	100	Hydrophobicity (Bump), E_{LUMO}	78	68	72
AD	11.6–14.0 Å	101	E_{LUMO}, PPSA3	77	77	77

and ketones. The modeling was performed for each of these groups. Four classes of chemicals analyzed for activity at the lowest-activity bin are listed in Figure 10.8.

More data are needed to better distinguish structural reasoning for the effects at the lowest-activity bin. Given that the majority of industrial chemicals elicit effects at the lowest-activity bin, it appears to be very important to refine the tests to identify extremely weak effects. Thus, according to the Chemicals Evaluation and Research

TABLE 10.2
Parametric Ranges for the "Bump" Parameter — BABumProp Associated with the AC Interaction Mechanism of Most Active ER Ligands (RBA > 10% as Defined by the COREPA Models)

Bump_H [Å]	R_Bump [Å]	Bump_vdW_Volume [Å³]	Log Kow [a.u.]
7.53–10.1	5.35–8.4	53.7–95.1	4.86–6.73
8.96–9.73	10.1–12.0	29.3–48.1	4.74–6.38
11.7–13.9	5.6–7.5	96.9–129.0	4.98–7.19

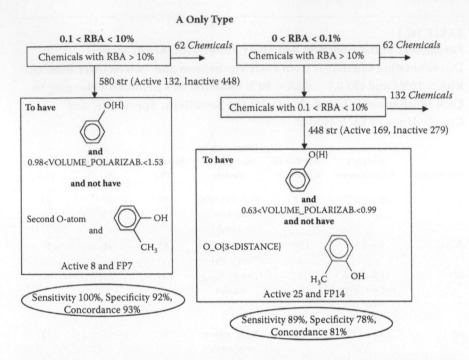

FIGURE 10.7 The models for the A-only-type interaction mechanism: (a) 0.1 < RBA < 10%, and (b) 0 < RBA < 0.1%. Sensitivity, specificity, and concordance of the model are listed.

Institute (CERI, Japan) protocol, chemicals are tested at a concentration up to 10^{-4}, which is set by medicinal chemists in their attempt to identify potent ER ligands. This threshold is justified by lack of interest in weak ER binders as potential drugs. However, for regulatory purposes, this concentration threshold could be considered as insufficient given that most industrial chemicals elicit effects that, although extremely low, are significant for risk assessment given their production volume and other fate

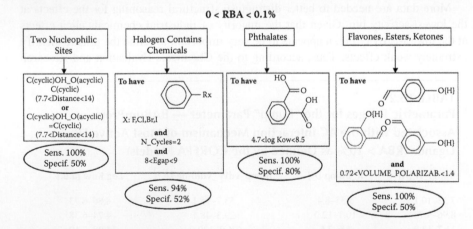

FIGURE 10.8 The four classes of chemicals analyzed for activity in the range of 0 < RBA < 0.1.

properties of industrial chemicals. Thus, testing performed at MED-Duluth on trout ER [44] showed that the number of negatives from the METI list elicits weak effects if chemicals are studied up to their solubility threshold. Hence, the present model could be considered as limited at the lowest-activity bin in terms of possibly underestimating the ER binding affinity of some low active chemicals.

The models for different mechanisms and activity ranges were combined in a screening battery, as shown in Figure 10.9.

The total performance of the model for training set chemicals is 0.86 in terms of Pearson's contingency coefficient and concordance 73%; the highest concordance was found for nonactive chemicals (81%) and most active chemicals (86%), whereas for the 0.1 < RBA < 10% and 0 < RBA < 0.1% ranges, the concordance is relatively low — 0.61 and 0.63, respectively.

The battery of ER binding models was used for screening 387 chemicals tested for hER binding affinity [28]. The results are listed in Table 10.3.

The screening results showed a concordance of 0.69 between the predicted and experimentally observed hER data, accounting for the distribution of chemicals

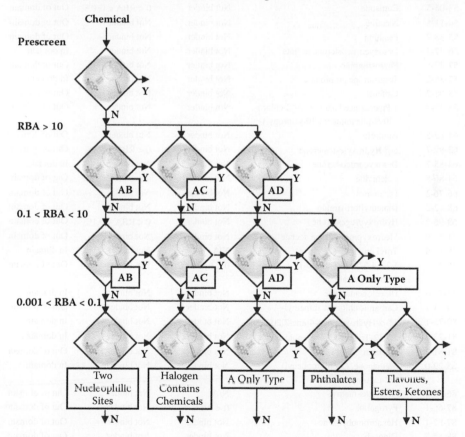

FIGURE 10.9 Screening battery for ER binding activity consisting of models for different mechanisms and activity bins.

TABLE 10.3
Screening Results for ER Binding Affinity of 387 hER Inventory

CAS	Chemical Name	Observed	Predicted	Domain
50-02-2	Dexamethasone	Not binder	0 < RBA < 0.1%	Out of domain
50-03-3	Hydrocortisone acetate	Not binder	0 < RBA < 0.1%	Out of domain
50-04-4	Cortisone acetate	Not binder	0 < RBA < 0.1%	Out of domain
50-06-6	Phenobarbital	Not binder	Not binder	Out of domain
50-23-7	Hydrocortisone	Not binder	0 < RBA < 0.1%	Out of domain
50-55-5	Reserpine	Not binder	0.1 < RBA < 10%	Out of domain
51-24-1	Tiratricol	0 < RBA < 0.1%	0 < RBA < 0.1%	Out of domain
51-26-3	3,3′,5-Triiodothyropropionic acid	0 < RBA < 0.1%	0 < RBA < 0.1%	Out of domain
51-49-0	Dextrothyroxine	Not binder	0 < RBA < 0.1%	Out of domain
51-55-8	Atropine	Not binder	Not binder	Out of domain
52-01-7	Spironolactone	Not binder	Not binder	Out of domain
52-86-8	Haloperidol	Not binder	Not binder	Out of domain
53-06-5	Cortisone	Not binder	0 < RBA < 0.1%	Out of domain
54-11-5	Nicotine	Not binder	Not binder	Out of domain
55-38-9	Fenthion	Not binder	Not binder	Out of domain
56-47-3	Desoxycorticosterone acetate	Not binder	Not binder	In domain
57-47-6	Physostigmine	Not binder	Not binder	Out of domain
57-85-2	Testosterone propionate	Not binder	Not binder	In domain
58-08-2	Caffeine	Not binder	Not binder	Out of domain
58-39-9	1-Piperazineethanol, 4-(3-(2-chloro-10 H-phenothiazin-10-yl)propyl)-	Not binder	Not binder	Out of domain
61-82-5	Amitrole	Not binder	Not binder	Out of domain
62-99-7	6-β-Hydroxytestosterone	Not binder	0 < RBA < 0.1%	Out of domain
64-85-7	Desoxycorticosterone	Not binder	Not binder	In domain
64-86-8	Colchicine	Not binder	0.1 < RBA < 10%	Out of domain
66-76-2	Dicumarol	Not binder	Not binder	Out of domain
68-12-2	Dimethylformamide	Not binder	Not binder	Out of domain
68-96-2	Hydroxyprogesterone	Not binder	0 < RBA < 0.1%	Out of domain
71-58-9	Medroxyprogesterone acetate	Not binder	Not binder	Out of domain
76-83-5	Trityl chloride	N/A	Not binder	In domain
77-47-4	1,2,3,4,5,5-Hexachloro-1,3-cyclopentadiene	0 < RBA < 0.1%	Not binder	Out of domain
77-73-6	Dicyclopentadiene	Not binder	Not binder	In domain
79-44-7	Carbamic chloride, dimethyl-	Not binder	Not binder	Out of domain
80-75-1	11-α-Hydroxypregn-4-ene-3,20-dione	Not binder	Not binder	In domain
81-23-2	3,7,12-Triketo-5β-cholanoic acid	Not binder	Not binder	In domain
81-81-2	Warfarin	Not binder	Not binder	Out of domain
85-29-0	Methanone, (2-chlorophenyl)(4-chlorophenyl)-	Not binder	Not binder	In domain
86-50-0	Azinphos-methyl	N/A	Not binder	Out of domain
87-66-1	Pyrogallol	0 < RBA < 0.1%	Not binder	Out of domain
87-82-1	Hexabromobenzene	Not binder	Not binder	Out of domain
88-85-7	Dinoseb	Not binder	Not binder	Out of domain
90-33-5	Hymecromone	Not binder	Not binder	In domain
90-90-4	4-Bromobenzophenone	Not binder	Not binder	In domain

TABLE 10.3 (CONTINUED)
Screening Results for ER Binding Affinity of 387 hER Inventory

CAS	Chemical Name	Observed	Predicted	Domain
92-66-0	1,1-Biphenyl, 4-bromo-	Not binder	0 < RBA < 0.1%	In domain
92-91-1	Ethanone, 1-(1,1'-biphenyl)-4-yl-	Not binder	Not binder	In domain
93-65-2	Mecoprop	Not binder	Not binder	Out of domain
93-72-1	Silvex	Not binder	Not binder	Out of domain
93-76-5	2,4,5-T	Not binder	Not binder	Out of domain
94-12-2	Risocaine; p-Aminobenzoic acid, propyl ester	Not binder	Not binder	In domain
94-25-7	Butyl 4-aminobenzoate	0 < RBA < 0.1%	Not binder	In domain
94-59-7	5-(2-propenyl)-1,3-benzodioxole	Not binder	Not binder	Out of domain
94-74-6	Acetic acid, (4-chloro-2-methylphenoxy)-	Not binder	Not binder	Out of domain
94-75-7	2,4-D	Not binder	Not binder	Out of domain
94-82-6	2,4-DB	Not binder	Not binder	Out of domain
94-96-2	Ethohexadiol	Not binder	Not binder	In domain
95-33-0	2-Benzothiazolesulfenamide, N-cyclohexyl-	0 < RBA < 0.1%	Not binder	Out of domain
95-53-4	2-Toluidine	Not binder	Not binder	Out of domain
95-55-6	2-Aminophenol	Not binder	Not binder	Out of domain
95-76-1	3,4-Dichloroaniline	Not binder	Not binder	Out of domain
95-80-7	2,4-Diaminotoluene	Not binder	Not binder	Out of domain
96-12-8	1,2-Dibromo-3-chloropropane	Not binder	Not binder	Out of domain
96-29-7	2-Butanone, oxime	Not binder	Not binder	Out of domain
96-45-7	Ethylenethiourea	Not binder	Not binder	Out of domain
97-02-9	2,4-Dinitroaniline	Not binder	Not binder	Out of domain
97-23-4	Dichlorophen	N/A	0 < RBA < 0.1%	Out of domain
97-74-5	Bis(dimethylthiocarbamyl)sulfide	0 < RBA < 0.1%	Not binder	Out of domain
97-77-8	Bis(diethylthiocarbamoyl)disulfide	0 < RBA < 0.1%	Not binder	Out of domain
99-77-4	Benzoic acid, 4-nitro-, ethyl ester	Not binder	Not binder	Out of domain
100-00-5	4-Chloronitrobenzene	Not binder	Not binder	Out of domain
100-44-7	Benzyl chloride	Not binder	Not binder	In domain
100-63-0	Phenylhydrazine	Not binder	Not binder	Out of domain
101-61-1	4,4'-Methylenebis[N,N-dimethyl]aniline	Not binder	Not binder	Out of domain
101-77-9	4,4'-Diaminodiphenylmethane	Not binder	Not binder	In domain
101-80-4	4,4'-Oxydianiline	Not binder	Not binder	Out of domain
103-23-1	Hexanedioic acid, bis(2-ethylhexyl)ester	Not binder	Not binder	In domain
105-60-2	Hexahydro-2h-azepin-2-one	Not binder	Not binder	Out of domain
107-21-1	Ethylene glycol	Not binder	Not binder	Out of domain
108-39-4	3-Methylphenol	N/A	Not binder	In domain
108-45-2	3-Phenylenediamine	Not binder	Not binder	Out of domain
108-93-0	Cyclohexanol	Not binder	Not binder	In domain
114-26-1	Phenol, 2-(1-methylethoxy)-, methylcarbamate	Not binder	Not binder	Out of domain
115-29-7	Endosulfan	0 < RBA < 0.1%	Not binder	Out of domain
115-86-6	Phosphoric acid, triphenyl ester	N/A	Not binder	In domain
115-96-8	Tris(β-chloroethylphosphate)	Not binder	Not binder	In domain

(continued)

TABLE 10.3 (CONTINUED)
Screening Results for ER Binding Affinity of 387 hER Inventory

CAS	Chemical Name	Observed	Predicted	Domain
116-06-3	2-Methyl-2-(methylthio) propionaldehyde, o-(methylcarbamoyl) oxime	Not binder	Not binder	Out of domain
117-82-8	1,2-Benzenedicarboxylic acid, bis (2-methoxyethyl) ester	Not binder	Not binder	In domain
118-48-9	2H-3,1-Benzoxazine-2,4(1H)-dione	Not binder	Not binder	Out of domain
118-55-8	Phenyl salicylate	Not binder	0 < RBA < 0.1%	Out of domain
118-60-5	Benzoic acid, 2-hydroxy-, 2-ethylhexyl ester	Not binder	0 < RBA < 0.1%	Out of domain
119-36-8	2-Hydroxybenzoic acid, methyl ester	Not binder	Not binder	Out of domain
120-36-5	Dichlorprop	Not binder	Not binder	Out of domain
120-48-9	Butyl 4-nitrobenzoate	Not binder	Not binder	Out of domain
120-78-5	Altax	0 < RBA < 0.1%	Not binder	Out of domain
120-80-9	Pyrocatechol	Not binder	Not binder	Out of domain
121-75-5	Butanedioic acid, [(dimethoxyphosphinothioyl)thio]-, diethylester	Not binder	Not binder	Out of domain
122-14-5	Fenitrothion	Not binder	Not binder	Out of domain
123-30-8	4-Aminophenol	0 < RBA < 0.1%	Not binder	Out of domain
126-73-8	Tributyl phosphate	Not binder	Not binder	In domain
127-18-4	Tetrachloroethene	Not binder	Not binder	In domain
127-31-1	Hydrocortisone-9α-fluoro	Not binder	0 < RBA < 0.1%	Out of domain
131-53-3	Methanone, (2-hydroxy-4-methoxyphenyl)(2-hydroxyphenyl)-	Not binder	0 < RBA < 0.1%	Out of domain
131-54-4	Methanone, bis(2-hydroxy-4-methoxyphenyl)-	N/A	0 < RBA < 0.1%	Out of domain
133-06-2	1H-Isoindole-1,3(2H)-dione, 3a,4,7,7 a-tetrahydro-2-[(trichloromethyl)thio]-	0 < RBA < 0.1%	Not binder	Out of domain
133-07-3	1H-Isoindole-1,3(2H)-dione, 2-[(trichloromethyl)thio]-	0 < RBA < 0.1%	Not binder	Out of domain
134-62-3	N,N-diethyl-3-methylbenzamide	Not binder	Not binder	Out of domain
134-83-8	Benzene, 1-chloro-4-(chlorophenylmethyl)-	N/A	Not binder	In domain
134-85-0	4-Chlorobenzophenone	Not binder	Not binder	In domain
137-26-8	Tetramethylthioperoxydicarbonicdiamide	0 < RBA < 0.1%	Not binder	Out of domain
140-56-7	Diazoben	Not binder	Not binder	Out of domain
145-14-2	Dihydroprogesterone	Not binder	Not binder	In domain
149-30-4	Captax	0 < RBA < 0.1%	Not binder	Out of domain
150-68-5	Urea, N'-(4-chlorophenyl)- N, N-dimethyl-	Not binder	Not binder	Out of domain
152-58-9	Cortodoxone	Not binder	0 < RBA < 0.1%	Out of domain
218-01-9	Chrysene	Not binder	Not binder	Out of domain
238-84-6	Benzo[a]fluorene	Not binder	Not binder	Out of domain
298-02-2	Phorate	Not binder	Not binder	Out of domain
299-84-3	Fenchlorphos	0 < RBA < 0.1%	Not binder	Out of domain

TABLE 10.3 (CONTINUED)
Screening Results for ER Binding Affinity of 387 hER Inventory

CAS	Chemical Name	Observed	Predicted	Domain
299-86-5	Crufomate	Not binder	Not binder	Out of domain
302-23-8	Hydroxyprogesterone acetate	Not binder	Not binder	Out of domain
302-79-4	Tretinoin	N/A	Not binder	Out of domain
303-26-4	n-(4-Chlorobenzhydryl)piperazine	Not binder	Not binder	Out of domain
303-45-7	Gossypol	Not binder	0 < RBA < 0.1%	Out of domain
319-85-7	β-Hexachlorocyclohexane	N/A	Not binder	In domain
324-74-3	4-Fluorobiphenyl	Not binder	0 < RBA < 0.1%	Out of domain
330-54-1	n'-(3,4-Dichlorophenyl)-n,n-dimethylurea	Not binder	Not binder	Out of domain
330-55-2	Linuron	Not binder	Not binder	Out of domain
333-41-5	Diazinon	N/A	Not binder	Out of domain
342-25-6	2,4'-Difluorobenzophenone	Not binder	0 < RBA < 0.1%	Out of domain
345-83-5	Methanone, (4-fluorophenyl)phenyl-	Not binder	Not binder	In domain
349-58-6	Phenol, 3,5-bis(trifluoromethyl)-	Not binder	Not binder	Out of domain
370-14-9	Pholedrine	Not binder	0 < RBA < 0.1%	Out of domain
382-44-5	Androst-4-ene-3,17-dione, 11β-hydroxy-(8CI)	Not binder	Not binder	In domain
382-45-6	Androst-4-ene-3,11,17-trione	Not binder	Not binder	In domain
427-51-0	Cyproterone acetate	N/A	0 < RBA < 0.1%	Out of domain
434-03-7	Ethisterone	N/A	0 < RBA < 0.1%	In domain
434-07-1	Oxymetholone	0 < RBA < 0.1%	0 < RBA < 0.1%	Out of domain
434-22-0	Nandrolone	0 < RBA < 0.1%	0 < RBA < 0.1%	In domain
434-90-2	Decafluorobiphenyl	Not binder	0 < RBA < 0.1%	Out of domain
437-64-9	4',5 -Dihydroxy-7-methoxyflavone	0 < RBA < 0.1%	0 < RBA < 0.1%	Out of domain
451-46-7	Ethyl 4-fluorobenzoate	Not binder	Not binder	In domain
455-19-6	Benzaldehyde, 4-(trifluoromethyl)-	Not binder	Not binder	Out of domain
455-24-3	Benzoic acid, 4-(trifluoromethyl)-	Not binder	Not binder	Out of domain
457-68-1	Benzene, 1,1'-methylenebis[4-fluoro-	Not binder	Not binder	In domain
466-37-5	Ethanone, tetraphenyl-	0 < RBA < 0.1%	Not binder	Out of domain
467-62-9	4,4',4"-Triaminotrityl alcohol	0 < RBA < 0.1%	Not binder	In domain
470-90-6	Clofenvinfos	N/A	Not binder	In domain
475-25-2	Hematein	0 < RBA < 0.1%	0 < RBA < 0.1%	Out of domain
480-40-0	4H-1-Benzopyran-4-one, 5,7-dihydroxy-2-phenyl-	Not binder	0 < RBA < 0.1%	In domain
480-44-4	4H-1-Benzopyran-4-one, 5,7-dihydroxy-2-(4-methoxyphenyl)-	Not binder	0 < RBA < 0.1%	In domain
481-30-1	Epitestosterone	Not binder	0 < RBA < 0.1%	In domain
485-72-3	Formononetin	0 < RBA < 0.1%	0 < RBA < 0.1%	In domain
487-26-3	2,3-Dihydroflavone	Not binder	Not binder	Out of domain
491-54-3	4'-Methoxy 3,5,7-trihydroxyflavone	Not binder	0 < RBA < 0.1%	In domain
491-67-8	Baicalein	0 < RBA < 0.1%	0 < RBA < 0.1%	Out of domain
491-78-1	5-Hydroxyflavone	Not binder	0 < RBA < 0.1%	Out of domain
510-64-5	19-Hydroxy-4-androstene-3,17-dione	Not binder	Not binder	In domain
514-36-3	Fludrocortisone 21-acetate	Not binder	0 < RBA < 0.1%	Out of domain

(continued)

TABLE 10.3 (CONTINUED)
Screening Results for ER Binding Affinity of 387 hER Inventory

CAS	Chemical Name	Observed	Predicted	Domain
516-15-4	11-Ketoprogesterone	Not binder	Not binder	In domain
517-28-2	Benz[b]indeno[1,2-d]pyran-3,4,6a,9,10(6H)-pentol, 7,11 b-dihydro-, cis-(+)-	0 < RBA < 0.1%	0.1 < RBA < 10%	Out of domain
519-73-3	Triphenylmethane	0 < RBA < 0.1%	Not binder	In domain
520-85-4	Medroxyprogesterone	Not binder	0 < RBA < 0.1%	Out of Domain
525-82-6	4H-1-Benzopyran-4-one, 2-phenyl-	Not binder	Not binder	Out of Domain
527-54-8	Phenol, 3,4,5-trimethyl-	Not binder	Not binder	Out of domain
528-29-0	o-Dinitrobenzene	Not binder	Not binder	Out of domain
528-48-3	3,3′,4′,7-Tetrahydroxyflavone; Fisetin; Flavone, 3,3′,4′,7-tetrahydroxy-	0 < RBA < 0.1%	0 < RBA < 0.1%	Out of domain
529-44-2	3,3′,4,4′,5′,7-Hexahydro-2-phenyl-4h-chromen-4-one	0 < RBA < 0.1%	0 < RBA < 0.1%	Out of domain
530-44-9	4-(Dimethylamino)Benzophenone	Not binder	Not binder	In domain
537-98-4	Trans-Ferulic acid	Not binder	Not binder	Out of domain
548-83-4	3,5,7-Trihydroxy-2-phenyl-4-benzopyrone	N/A	0 < RBA < 0.1%	In domain
564-35-2	Testosterone, 11-keto	N/A	0 < RBA < 0.1%	In domain
566-48-3	Androst-4-ene-3,17-dione, 4-hydroxy-	Not binder	0 < RBA < 0.1%	Out of domain
566-65-4	3,20-Allopregnanedione	Not binder	Not binder	In domain
571-22-2	Androstan-3-one, 17-hydroxy-, (5-β,17-β)- (9CI)	Not binder	0 < RBA < 0.1%	In domain
575-03-1	7-Hydroxy-4-trifluoromethylcoumarin	0 < RBA < 0.1%	Not binder	Out of domain
577-85-5	4H-1-Benzopyran-4-one, 3-hydroxy-2-phenyl-	Not binder	Not binder	Out of domain
579-39-5	4,4′-Difluorobenzil	Not binder	0 < RBA < 0.1%	Out of domain
584-79-2	Bioallethrin	N/A	Not binder	Out of domain
591-27-5	3-Aminophenol	Not binder	Not binder	Out of domain
595-91-5	Triphenylacetic acid	Not binder	Not binder	Out of domain
596-01-0	3,3-Bis(4-hydroxy-1-naphthyl) phthalide	0 < RBA < 0.1%	RBA > 10%	Out of domain
596-27-0	Phenolphthalein, 3′,3″-dimethyl-	0 < RBA < 0.1%	0 < RBA < 0.1%	Out of domain
596-43-0	2-Bromo-1,1,1-triphenylethane	N/A	Not binder	Out of domain
601-57-0	Cholest-4-en-3-one	Not binder	Not binder	In domain
603-48-5	Leucogentian violet	N/A	Not binder	Out of domain
604-59-1	2-Phenylbenzo[h]chromen-4-one	Not binder	Not binder	Out of domain
608-71-9	Pentabromophenol	Not binder	Not binder	Out of domain
611-79-0	3,3′-Diaminobenzophenone	N/A	Not binder	Out of domain
611-94-9	4-Methoxybenzophenone	Not binder	Not binder	In domain
611-95-0	4-Benzoylbenzoic acid	Not binder	Not binder	In domain
611-98-3	Methanone, bis(4-aminophenyl)-(9CI)	Not binder	Not binder	In domain
612-96-4	2-Phenylquinoline	Not binder	Not binder	Out of domain
630-56-8	Hydroxyprogesterone caproate	Not binder	Not binder	Out of domain
631-64-1	Dibromoacetic acid	Not binder	Not binder	Out of domain

TABLE 10.3 (CONTINUED)
Screening Results for ER Binding Affinity of 387 hER Inventory

CAS	Chemical Name	Observed	Predicted	Domain
632-51-9	Benzene, 1,1,1,1-(1,2-ethenediylidene)tetrakis-	0.1 < RBA < 10%	Not binder	In domain
640-87-9	17-α,21-Dihydroxypregn-4-ene-3, 20-dione 21-acetate	Not binder	0 < RBA < 0.1%	Out of domain
641-38-3	6H-Dibenzo[b,d]pyran-6-one, 3,7,9-trihydroxy-1-methyl-	0 < RBA < 0.1%	0 < RBA < 0.1%	Out of domain
641-77-0	21-Deoxycortisol	Not binder	0 < RBA < 0.1%	Out of domain
700-13-0	1,4-Benzenediol, 2,3,5-trimethyl-	N/A	Not binder	Out of domain
728-87-0	p,p'-Dimethoxybenzhydryl alcohol	Not binder	0.1 < RBA < 10%	In domain
732-11-6	Phosmet	N/A	Not binder	Out of domain
779-51-1	1,2-Diphenylpropene	Not binder	Not binder	Out of domain
787-70-2	[1,1'-Biphenyl]-4,4'-dicarboxylic acid	Not binder	Not binder	Out of domain
797-64-8	(-)-Norogestrel	0 < RBA < 0.1%	0 < RBA < 0.1%	In domain
828-27-3	Phenol, 4-(trifluoromethoxy)-	0 < RBA < 0.1%	Not binder	Out of domain
833-81-8	Stilbene, α-methyl-, (E)-	Not binder	Not binder	Out of domain
834-12-8	Ametryne	Not binder	Not binder	Out of domain
846-46-8	5-α-Androstane-3,17-dione	Not binder	Not binder	In domain
855-96-9	3',5-Dihydroxy-4',6, 7-trimethoxyflavone	Not binder	0.1 < RBA < 10%	Out of domain
855-97-0	3',4',5,7-Tetramethoxyflavone	Not binder	0.1 < RBA < 10%	Out of domain
900-91-4	3,3,3-Triphenylpropionic acid	Not binder	Not binder	In domain
919-86-8	Demeton-S-methyl	Not binder	Not binder	Out of domain
927-67-3	Propyl-2-thiourea	Not binder	Not binder	Out of domain
1014-70-6	Simetryn	Not binder	Not binder	Out of domain
1016-78-0	3-Chlorobenzophenone	Not binder	Not binder	Out of domain
1022-46-4	4H-3,1-Benzoxazin-4-one, 2-phenyl-	Not binder	Not binder	Out of domain
1031-07-8	Endosulfan sulfate	N/A	Not binder	Out of domain
1034-10-2	O-(4-Hydroxyphenyl)-DL-tyrosine	Not binder	Not binder	In domain
1045-69-8	(17β)-Hydroxyandrost-4-en-3-one acetate	Not binder	Not binder	In domain
1095-77-8	4,4'-(Hexafluoroisopropylidene) ditoluene	0 < RBA < 0.1%	Not binder	Out of domain
1095-78-9	4,4'-(Hexafluoroisopropylidene) dianiline	0 < RBA < 0.1%	Not binder	Out of domain
1097-51-4	16α,17-Epoxypregn-4-ene-3,20-dione	Not binder	Not binder	Out of domain
1113-02-6	Omethoate	Not binder	Not binder	Out of domain
1120-71-4	1,2-Oxathiolane, 2,2-dioxide	Not binder	Not binder	Out of domain
1137-41-3	4-Aminobenzophenone	N/A	Not binder	In domain
1144-74-7	4-Nitrobenzophenone	Not binder	Not binder	Out of domain
1171-47-7	4,4'-(Hexafluoroisopropylidene) bis(benzoic acid)	0 < RDA < 0.1%	Not binder	Out of domain
1180-25-2	Testosterone glucuronate	Not binder	0 < RBA < 0.1%	Out of domain
1214-39-7	N-(Phenylmethyl)-1H-purin-6-amine	Not binder	Not binder	Out of domain
1424-00-6	Mesterolone	0 < RBA < 0.1%	0 < RBA < 0.1%	In domain

(continued)

TABLE 10.3 (CONTINUED)
Screening Results for ER Binding Affinity of 387 hER Inventory

CAS	Chemical Name	Observed	Predicted	Domain
1450-63-1	1,1,4,4-Tetraphenylbuta-1,3-diene	0 < RBA < 0.1%	Not binder	Out of domain
1482-70-8	Androstane-3,11,17-trione, (5α)-	Not binder	Not binder	In domain
1524-88-5	Fludroxycortide	Not binder	Not binder	Out of domain
1553-56-6	5β-Cholanic acid-3-one	Not binder	Not binder	In domain
1563-66-2	2,3-Dihydro-2,2-dimethyl-7-benzofuranol, methylcarbamate	Not binder	Not binder	Out of domain
1582-09-8	Trifluralin	N/A	Not binder	Out of domain
1596-67-4	O-(4-Hydroxyphenyl)-L-tyrosine	Not binder	Not binder	In domain
1607-57-4	Bromotriphenylethylene	0.1 < RBA < 10%	Not binder	Out of domain
1662-06-2	4-Pregnene-17α,20β-diol-3-one	Not binder	0 < RBA < 0.1%	In domain
1675-54-3	2,2-Bis(4-glycidyloxyphenyl)propane	Not binder	Not binder	Out of domain
1689-83-4	Loxynil	Not binder	Not binder	Out of domain
1689-84-5	Bromoxynil	Not binder	Not binder	Out of domain
1702-17-6	3,6-Dichloropicolinic acid	Not binder	Not binder	Out of domain
1746-81-2	Monolinuron	Not binder	Not binder	Out of domain
1801-42-9	2,3-Diphenyl-1H-inden-1-one	0 < RBA < 0.1%	Not binder	Out of domain
1816-85-9	11-Hydroxytestosterone	Not binder	0 < RBA < 0.1%	In domain
1844-00-4	4-[1-(4-Hydroxyphenyl)-2-methylpropyl]Phenol	0.1 < RBA < 10%	0.1 < RBA < 10%	In domain
1889-71-0	Benzyl 4-chlorophenyl ketone	Not binder	0 < RBA < 0.1%	In domain
1929-82-4	N-serve	Not binder	Not binder	Out of domain
2005-08-5	4-Chlorophenyl benzoate	Not binder	0 < RBA < 0.1%	In domain
2033-89-8	3,4-Dimethoxyphenol	Not binder	Not binder	Out of domain
2051-90-3	Dichloro(diphenyl)methane	Not binder	Not binder	In domain
2104-96-3	Bromophos	0 < RBA < 0.1%	Not binder	Out of domain
2203-97-6	Hydrocortisone 21-(hydrogen succinate)	Not binder	0 < RBA < 0.1%	Out of domain
2212-67-1	Molinate	Not binder	Not binder	In domain
2268-98-6	Pregn-4-ene-3,20-dione, 11α-hydroxy-, acetate (8CI)	N/A	Not binder	In domain
2425-06-1	Captafol	0 < RBA < 0.1%	Not binder	Out of domain
2439-99-8	Glycine, N,N-bis(phosphonomethyl)-	Not binder	Not binder	Out of domain
2540-82-1	Formothion	Not binder	Not binder	Out of domain
2668-66-8	Medrysone	Not binder	0 < RBA < 0.1%	Out of domain
2772-45-4	Phenol, 2,4-bis(1-methyl-1-phenylethyl)-	0 < RBA < 0.1%	Not binder	In domain
2835-78-1	3-Aminobenzophenone	N/A	Not binder	Out of domain
2921-88-2	Phosphorothioic acid, O,O-diethyl O-(3,5,6-trichloro-2-pyridinyl) ester	0 < RBA < 0.1%	Not binder	Out of domain
3016-97-5	1,4-Dibenzoylbenzene	Not binder	Not binder	In domain
3093-35-4	Halcinonide	Not binder	0 < RBA < 0.1%	Out of domain
3218-36-8	[1,1'-Biphenyl]-4-carboxaldehyde	Not binder	Not binder	In domain
3457-48-5	Di-p-tolylethanedione	Not binder	Not binder	Out of domain
3604-87-3	(2β,3β,5β,22r)-2,3,14,22,25-Pentahydroxycholest-7-en-6-one	Not binder	0.1 < RBA < 10%	Out of domain
3739-38-6	3-Phenoxybenzoic acid	Not binder	Not binder	Out of domain
4143-64-0	3',4'-Dihydroxyflavone	0 < RBA < 0.1%	0 < RBA < 0.1%	Out of domain

TABLE 10.3 (CONTINUED)
Screening Results for ER Binding Affinity of 387 hER Inventory

CAS	Chemical Name	Observed	Predicted	Domain
4319-56-6	Desoxycortone 21-d-glucoside	Not binder	0 < RBA < 0.1%	Out of domain
4471-22-1	N-(Triphenylmethyl)aniline	N/A	Not binder	Out of domain
4759-48-2	Isotretinoin	N/A	Not binder	Out of domain
5300-03-8	Retinoic acid, 9-cis	Not binder	Not binder	Out of domain
5392-40-5	Citral	Not binder	Not binder	Out of domain
5447-02-9	3,4-Bis(benzyloxy)benzaldehyde	0 < RBA < 0.1%	Not binder	Out of domain
5817-39-0	3 ,3',5'-Triiodo-L-thyronine	Not binder	0 < RBA < 0.1%	Out of domain
6051-87-2	3-Phenyl-1h-naphtho[2,1-b] pyran-1-one	Not binder	Not binder	Out of domain
6164-98-3	Chlordimeform	Not binder	Not binder	Out of domain
6174-86-3	3-Chloro-7-hydroxy-4-methyl-2- benzopyrone	N/A	Not binder	Out of domain
6297-11-6	9H-Fluoren-9-one, 2,7-dichloro-	Not binder	Not binder	Out of domain
6554-98-9	p-Styrylphenol	0 < RBA < 0.1%	0 < RBA < 0.1%	In domain
6665-83-4	6-Hydroxy-2-phenyl-4-benzopyrone	0 < RBA < 0.1%	0 < RBA < 0.1%	Out of domain
6678-14-4	11β,17,21-Trihydroxypregn-4-ene- 3,20-dione 21-octanoate	Not binder	0 < RBA < 0.1%	Out of domain
6893-02-3	Liothyronine	Not binder	0 < RBA < 0.1%	Out of domain
6898-97-1	Phenol, 4,4'-(1,2-diethyl-1, 2-ethenediyl)bis-	RBA > 10%	RBA > 10%	In domain
6948-88-5	4-Hydroxy-α-(4-hydroxynaphthyl)- α-phenylnaphthalene-1-methanol	0.1 < RBA < 10%	RBA > 10%	Out of domain
7287-19-6	Prometryn	Not binder	Not binder	Out of domain
7786-34-7	Mevinphos	Not binder	Not binder	Out of domain
8001-35-2	Toxaphene	0 < RBA < 0.1%	Not binder	In domain
10192-62-8	4,4'-Isopropylidenediphenyl diacetate	N/A	0.1 < RBA < 10%	Out of domain
10215-74-4	21-Hydroxypregn-4-ene-3,20-dione 21-(hydrogen succinate)	Not binder	Not binder	In domain
10311-84-9	Dialifos	0 < RBA < 0.1%	Not binder	Out of domain
10453-86-8	Resmethrin	N/A	Not binder	Out of domain
10605-21-7	1H-Benzimidazol-2-yl carbamic acid, methyl ester	Not binder	Not binder	Out of domain
13036-02-7	Dimethyl 5-hydroxyisophthalate	Not binder	0 < RBA < 0.1%	Out of domain
13311-84-7	Flutamide	Not binder	Not binder	Out of domain
13593-03-8	Quinalphos	N/A	Not binder	Out of domain
13608-87-2	2',3',4'-Trichloroacetophenone	Not binder	Not binder	Out of domain
13609-67-1	11β,17,21-Trihydroxypregn-4-ene- 3,20-dione 17-butyrate	Not binder	0 < RBA < 0.1%	Out of domain
13680-35-8	Benzenamine, 4,4'-methylenebis 2,6-diethyl-	N/A	Not binder	Out of domain
14548-48-2	4-(4'-Chlorobenzoyl)pyridine	Not binder	Not binder	Out of domain
14816-18-3	3,5-Dioxa-6-aza-4-phosphaoct-6-ene-8- nitrile, 4-ethoxy-7-phenyl-, 4-sulfide	N/A	Not binder	Out of domain

(continued)

TABLE 10.3 (CONTINUED)
Screening Results for ER Binding Affinity of 387 hER Inventory

CAS	Chemical Name	Observed	Predicted	Domain
15262-86-9	17β-Hydroxyandrost-4-ene-3-one 4-methylvalerate	Not binder	Not binder	In domain
15972-60-8	Alachlor	Not binder	Not binder	Out of domain
16069-36-6	Icosahydrodibenzo[b,k][1,4,7,10,13,16] hexaoxacyclooctadecin	Not binder	Not binder	In domain
16245-79-7	4-Octylaniline	Not binder	Not binder	In domain
16355-28-5	4-Pregnen-6-beta,17,21-triol-3,11,20-trione	Not binder	0.1 < RBA < 10%	Out of domain
16752-77-5	Methomyl	Not binder	Not binder	Out of domain
17078-27-2	1,2-Bis(p-Dimethylaminophenyl)-1, 2-ethanedione	Not binder	Not binder	Out of domain
17138-28-2	Ethyl 4-hydroxyphenylacetate	Not binder	Not binder	In domain
17230-88-5	Danazol	0 < RBA < 0.1%	Not binder	Out of domain
17804-35-2	Benomyl	Not binder	Not binder	Out of domain
18144-43-9	Benzoic acid, 4-amino-, 1-methylethyl ester	Not binder	Not binder	In domain
19044-88-3	3,5-Dinitro-n(sup4), n(sup4)-dipropyl-sulfanilamide	N/A	Not binder	Out of domain
19315-93-6	2, 6-Quinolinediol	Not binder	Not binder	Out of domain
19471-12-6	3,3'-Methylenedianiline	Not binder	Not binder	Out of domain
19811-05-3	2,4-Dichlorobenzophenone	N/A	Not binder	In domain
21087-64-9	Metribuzin	Not binder	Not binder	Out of domain
21725-46-2	Cyanazine	Not binder	Not binder	Out of domain
22395-24-0	4H-1-Benzopyran-4-one, 2-(3,4-dimethoxyphenyl)-7-methoxy-	Not binder	0.1 < RBA < 10%	Out of domain
22494-42-4	[1,1-Biphenyl]-3-carboxylic acid, 2,4-difluoro-4-hydroxy-	Not binder	0 < RBA < 0.1%	Out of domain
23564-05-8	Carbamic acid, [1,2-phenylenebis-(iminocarbonothioyl)]bis-, dimethyl ester	Not binder	Not binder	Out of domain
23950-58-5	3,5-Dichloro-n-(1,1-dimethylprop-2-ynyl)benzamide	Not binder	Not binder	Out of domain
24602-86-6	Morpholine, 2,6-dimethyl-4-tridecyl	Not binder	Not binder	In domain
24758-49-4	4-Morpholinobenzophenone	Not binder	Not binder	Out of domain
25057-89-0	1H-2,1,3-Benzothiadiazin-4(3H)-one, 3-(1-methylethyl)-, 2,2-dioxide	Not binder	Not binder	Out of domain
26087-47-8	Iprobenfos	Not binder	Not binder	Out of domain
26644-46-2	N,N'-[(2-Methyl-1,4-piperazinediyl) bis(2,2,2-trichloroethylidene] bisformamide	Not binder	Not binder	Out of domain
26964-24-9	6-Methoxyflavone	Not binder	Not binder	Out of domain
27241-31-2	2-Pyrazolin-5-one, 3-amino-1-(2,4,6-trichlorophenyl)-	0 < RBA < 0.1%	0 < RBA < 0.1%	Out of domain
29082-74-4	Octachlorostyrene	Not binder	Not binder	Out of domain
29550-13-8	5,6-Dihydroxy-7-methoxyflavone	0 < RBA < 0.1%	0 < RBA < 0.1%	Out of domain
29752-43-0	Altenuene	N/A	0 < RBA < 0.1%	Out of domain
30560-19-1	Acephate	Not binder	Not binder	Out of domain

TABLE 10.3 (CONTINUED)
Screening Results for ER Binding Affinity of 387 hER Inventory

CAS	Chemical Name	Observed	Predicted	Domain
34256-82-1	2-Chloro-*n*-(ethoxymethyl)-*n*-(2-ethyl-6-methylphenyl)acetamide	Not binder	Not binder	Out of domain
34334-69-5	6,7-Dimethoxy-3′,4′,5-trihydroxyflavone	0 < RBA < 0.1%	0.1 < RBA < 10%	Out of domain
35323-91-2	(+)-Epicatechin	N/A	0.1 < RBA < 10%	Out of domain
35367-38-5	Benzamide, *N*-(4-chlorophenyl)amino carbonyl-2,6-difluoro-	Not binder	0 < RBA < 0.1%	Out of domain
35554-44-0	Enilconazole	Not binder	0 < RBA < 0.1%	Out of domain
38183-03-8	7,8-Dihydroxy-2-phenyl-4-benzopyrone	0 < RBA < 0.1%	0 < RBA < 0.1%	Out of domain
39300-45-3	Dinocap	N/A	Not binder	Out of domain
39634-42-9	*p*-[4-(Trifluoromethyl)phenoxy]phenol	0 < RBA < 0.1%	Not binder	Out of domain
40487-42-1	Pendimethalin	N/A	Not binder	Out of domain
40596-69-8	Methoprene	Not binder	Not binder	Out of domain
40615-36-9	Benzene, 1,1′-(chlorophenylmethylene) bis[4-methoxy-	Not binder	Not binder	Out of domain
42187-33-7	3,4-Dimethyl-3′-nitrobenzophenone	Not binder	Not binder	Out of domain
43121-43-3	1-(4-Chlorophenoxy)-3,3-dimethyl-1-(1,2,4-triazol-1-yl)butanone	Not binder	Not binder	Out of domain
49562-28-9	Fenofibrate	Not binder	0 < RBA < 0.1%	Out of domain
49757-42-8	4,4′,4″-Trimethoxytrityl chloride	0 < RBA < 0.1%	Not binder	Out of domain
51235-04-2	1,3,5-Triazine-2,4(1*H*,3*H*)-dione, 3-cyclohexyl-6-(dimethylamino)-1-methyl-	Not binder	Not binder	Out of domain
52918-63-5	Deltamethrin	N/A	Not binder	Out of domain
55219-65-3	Triadimenol	N/A	Not binder	Out of domain
56424-77-2	Ethyl 2-(ethoxycarbonylmethoxy) benzoate	Not binder	Not binder	Out of domain
57524-89-7	11*β*,17,21-Trihydroxypregn-4-ene-3,20-dione 17-valerate	Not binder	0 < RBA < 0.1%	Out of domain
60168-88-9	2,4′-Dichloro-*α*(pyrimidin-5-yl) benzhydryl alcohol	0 < RBA < 0.1%	Not binder	Out of domain
60207-90-1	1-[[2-(2,4-Dichlorophenyl)-4-propyl-1,3-dioxolan-2-yl]methyl]-1*h*-1,2,4-triazole	Not binder	Not binder	In domain
61630-32-8	4-Androsten-4-ol-3,17-dione acetate	Not binder	Not binder	Out of domain
64529-56-2	Ethiozin	Not binder	Not binder	Out of domain
64887-40-7	3-(2-Benzoxazolyl)umbelliferone	N/A	Not binder	Out of domain
65405-77-8	3-Hexenyl salicylate	Not binder	0 < RBA < 0.1%	Out of domain
66246-88-6	Penconazole	Not binder	Not binder	In domain
67845-93-6	Benzoic acid, 3,5-bis (1,1-dimethylethyl)-4-hydroxy-, hexadecyl ester	Not binder	Not binder	Out of domain
69806-50-4	Fluazifop-butyl	Not binder	Not binder	Out of domain
70458-96-7	3-Quinolinecarboxylic acid, 1,4-dihydro-1-ethyl-6-fluoro-4-oxo-7-(1-piperazinyl)	Not binder	Not binder	Out of domain

(continued)

TABLE 10.3 (CONTINUED)
Screening Results for ER Binding Affinity of 387 hER Inventory

CAS	Chemical Name	Observed	Predicted	Domain
73250-68-7	Acetamide, 2-(2-benzothiazolyloxy)-N-methyl-N-phenyl-	Not binder	Not binder	Out of domain
74115-24-5	3,6-Bis(2-chlorophenyl)-1,2,4,5-tetrazine	Not binder	Not binder	Out of domain
76578-14-8	Propanoic acid, 2-(4-((6-chloro-2-quinoxalinyl)oxy)phenoxy)-, ethyl es	Not binder	Not binder	Out of domain
76674-21-0	Flutriafol (PP450)	N/A	Not binder	Out of domain
78473-71-9	Enterolactone	0 < RBA < 0.1%	0.1 < RBA < 10%	Out of domain
79983-71-4	Hexaconazole	N/A	Not binder	In domain
82560-54-1	Benfuracarb	N/A	Not binder	Out of domain
82657-04-3	(2-Methyl{1,1'-biphenyl}-3-yl)methyl 3-(2-chloro-3,3,3-trifluoro-1-pro	N/A	Not binder	Out of domain
83055-99-6	Bensulfuron-methyl	Not binder	Not binder	Out of domain
84087-01-4	8-Quinolinecarboxylic acid, 3,7-dichloro-	Not binder	0 < RBA < 0.1%	Out of domain
84371-65-3	Mifepristone	0 < RBA < 0.1%	0 < RBA < 0.1%	Out of domain
85118-07-6	3,4-Difluorobenzophenone	Not binder	0 < RBA < 0.1%	Out of domain
88671-89-0	α-Butyl-α-(4-chlorophenyl)-1H-1,2,4-triazole-1-propanenitrile	Not binder	Not binder	In domain
94361-06-5	Cyproconazole	Not binder	Not binder	In domain
95333-64-5	1-(4-Hydroxybenzyl)imidazole-2-thiol	Not binder	0 < RBA < 0.1%	Out of domain
106246-33-7	Benzenamine, 4,4'-methylenebis[3-chloro-2,6-diethyl-	0 < RBA < 0.1%	0 < RBA < 0.1%	Out of domain
107534-96-3	1H-1,2,4-Triazole-1-ethanol, α-(2-(4-chlorophenyl)ethyl)-α-	Not binder	Not binder	In domain
114369-43-6	2-Cyano-2-phenyl-2-(β-p-chlorophenethyl)ethyl-1H-1,2,4-triazole	Not binder	Not binder	In domain
122931-48-0	N-((4,6-Dimethoxypyrimidin-2-yl)aminocarbonyl)-3-(ethylsulfonyl)-2-pyr	Not binder	Not binder	Out of domain
125-04-2	Hydrocortisone 21-(sodium succinate)	Not binder	0 < RBA < 0.1%	Out of domain
131-52-2	Phenol, pentachloro-, sodium salt	Not binder	Not binder	Out of domain
137-42-8	Sodium N-methyldithiocarbamate	N/A	Not binder	Out of domain
145-41-5	Sodium dehydrocholate	Not binder	Not binder	Out of domain
4418-26-2	2H-Pyran-2,4(3H)-dione, 3-acetyl-6-methyl-, ion(1-), sodium	Not binder	Not binder	Out of domain

across the potency bins. The concordance between predicted and observed hER was found to be 0.80 if one counts only chemicals falling in the model for the chemicals that are in the model applicability domain. As can be seen, a large portion of screened chemicals (75%) are out of model applicability domain.

The hER binding model was combined with a metabolism simulator to predict the metabolic activation of chemicals. The recently developed metabolic simulators for mammalian liver were used. To adjust the transformation probabilities and increase the reliability of transformations (aromatic ring hydroxylation and O-oxidative deal-kylation) predominantly involved in the activation of chemicals for eliciting ER binding affinity, 62 new documented maps were added to the training set with 332 mammalian liver metabolism maps. The expanded training set was used for rederiving the metabolism simulator. The resulting simulator has been combined with the hER binding battery and applied to 25 of these 62 chemicals for which data for ER binding activity were identified for parent structure and metabolites. The screening results are listed in Table 10.4, where the predicted ER binding affinities were compared with the observed RBA of these chemicals and their active metabolites.

As seen, 33 of all 41 active parent structures and metabolites (~80%) were correctly identified as active. In six cases, some of the documented active metabolites were not generated. The chemical with CAS# 109640-20-2, which is documented as active as parent structure, was predicted to be active after metabolic activation.

10.4   SUMMARY AND CONCLUSIONS

The current review of our modeling activity of ER binding affinity is an illustration of the category approach in QSAR. Here, defining the mechanistic domain of the model is a first step in its development. Once, in this mechanistic domain (category), the next step is to quantify the model for predicting the potency of the effect. In the current modeling, first, the COREPA analysis was used for defining the interaction mechanisms, and subsequently, 3D models were developed for each identified interaction type. Analysis of common reactivity patterns of most active ER binders, based on the distance between nucleophilic sites, resulted in identification of distinct interaction types, summarized as steroid-like A-B type, modulated by steric and electronic interactions; an A-C type, where the local hydrophobic interactions were found to be significant; and a mixed A-B-C (AD) type, modulated again by stereo-electronic parameters.

The combination of the distance ranges associated with each of the three interaction types was found to form a general requirement for eliciting a weak ER binding effect due to the presence of two nucleophilic sites. When combined with specific stereoelectronic and hydrophobic screens, this necessary requirement could yield much higher potency. The A-only type of binding was clearly identified for the phenolic type of chemicals only. The remaining three classes of chemicals are difficult to associate with any of the documented interaction mechanisms. The ultimate modeling platform for hER binding affinity was organized as a battery of such models. The performance of the models was illustrated by screening 387 chemicals tested for hER. A variety of molecular modeling approaches are presented and applied in the present chapter. They include the multivariate development of COREPA reactivity patterns, optimal analysis (saturation) of conformational space, and introduction of functional (multidimensional) parameters, such as local hydrophobicity parameters. The model was combined with a simulator of mammalian liver metabolism. The performance of the model accounting for metabolism activation of parent chemicals was evaluated.

TABLE 10.4
Observed and Predicted Data for 25 Chemicals and Some of Their Metabolites

CAS	Name	Smiles	RBA ER Observed	Predicted
80-05-7	2,2-Bis(4-hydroxyphenyl)propane	c1(C(C)(C)c2ccc(O)cc2)ccc(O)cc1	Active	Active
72-43-5	1,1,1-Trichloro-2,2'-bis(4-methoxyphenyl)ethane	C(Cl)(Cl)(Cl)C(c1ccc(OC)cc1)c1ccc(OC)cc1	Active	Unclassified
	metabolite 1	C(Cl)(Cl)(Cl)C(c1ccc(OC)cc1)c1ccc(O)cc1	Active	Active
	metabolite 2	C(Cl)(Cl)(Cl)C(c1ccc(O)cc1)c1ccc(O)cc1	Active	Active
613-37-6	4-Methoxybiphenyl	c1(-c2ccc(OC)cc2)ccccc1	Active	Active
	metabolite 1	c1(-c2ccc(O)cc2)ccccc1		
	3,5,7-Trihydroxyflavone	C1(O)C(=O)c2c(O)cc(O)cc2OC=1c1ccccc1	Active	Active
	metabolite 1	C1(O)C(=O)c2c(O)cc(O)cc2OC=1c1ccc(O)cc1	Active	Active
	metabolite 2	C1(O)C(=O)c2c(O)cc(O)cc2OC=1c1cc(O)c(O)cc1	Active	Active
548-83-4	5,7-Dihydroxyflavone	C1(=O)c2c(O)cc2OC2OC(c2ccccc2)=C1	Active	Active
	metabolite 1	C1(=O)c2c(O)cc(O)cc2OC(c2ccc(O)cc2)=C1	Active	Active
	metabolite 2	C1(O)C(=O)c2c(O)cc(O)cc2OC=1c1ccc(O)cc1	Active	Active
491-54-3	4'-Methoxy-3,5,7-trihydroxyflavone	C1(O)C(=O)c2c(O)cc(O)cc2OC=1c1ccc(OC)cc1	Active	Active
	metabolite 1	C1(O)C(=O)c2c(O)cc(O)cc2OC=1c1ccc(O)cc1	Active	Active
53-16-7	Estrone	C1(=O)C2(C)C(C3C(c4c(cc(O)cc4)CC3)CC2)CC1	Active	Active
	metabolite 1	C1(=O)C2(C)C(C3C(c4c(cc(O)cc4)CC3)CC2)CC1	Active	Active
50-28-2	Estradiol	c12c(C3C(C4C(C)(C(O)CC4)CC3)CC1)ccc(O)c2	Active	Not generated
	metabolite 1	**c12c(C3C(C4C(C)(C(O)CC4)CC3)CC1)ccc(O)c2**		

CAS / Compound	SMILES		
109640-20-2 [1-(<-(2-Dimethylaminoethoxy)phenyl]-1,2-diphenyl-1-butene	c1(C(c2ccc(OCCN(C)C)cc2)=C(c2ccccc2)CC)ccccc1	Active	Not active
metabolite 1	c1(C(c2ccc(OCCN(C)C)cc2)=C(c2ccccc2)CC)ccc(O)cc1	Active	Active
metabolite 2	c1(C(c2cc(O)ccc2)=C(c2ccccc2)CC)ccc(OCCN(C)C)cc1	Active	Not generated
1157-39-7 4',7-Dimethoxyisoflavone			
metabolite 1	c1(C2C(=O)c3c(cc(OC)cc3)OC=2)ccc(OC)cc1	Active	Active
491-80-5 5,7-Dihydroxy-4'-methoxyisoflavone			
metabolite 1	c1(C2C(=O)c3c(O)cc(O)cc3OC=2)ccc(OC)cc1	Active	Active
2196-14-7 7,4'-Dihydroxyflavanone			
metabolite 1	c1(C2C(=O)c3c(O)cc(O)cc3OC=2)ccc(O)cc1	Active	Active
56-53-1 Diethylstilbestrol			
metabolite 1	C1(=O)c2c(cc(O)cc2)OC(c2ccc(O)cc2)C1	Active	Active
94-41-7 Chalcone			
metabolite 1	c1(C(=O)C=Cc2ccccc2)ccccc1	Active	Active
metabolite 2	c1(C(=O)C=Cc2ccc(O)cc2)ccccc1	Active	Active
metabolite 3	c1(C(=O)C=Cc2ccccc2O)ccccc1	Active	Not generated
122-57-6 Benzylideneacetone			
metabolite 1	C(C)(=O)C=Cc1ccc(O)cc1	Active	Active
480-41-1 Naringenin			
metabolite 1	C1(=O)c2c(O)cc(O)cc2OC(c2ccc(O)cc2)=C1	Active	Not generated
485-72-3 Formononetin			
metabolite 1	c1(C2C(=O)c3c(cc(O)cc3)OC=2)ccc(OC)cc1	Active	Active
metabolite 2	c1(C2Cc3c(cc(O)cc3)OC2)ccc(O)cc1	Active	Not generated

(continued)

TABLE 10.4 (CONTINUED)
Observed and Predicted Data for 25 Chemicals and Some of Their Metabolites

CAS	Name	Smiles	RBA ER Observed	RBA ER Predicted
131-57-7	Oxybenzone	c1(C(=O)c2c(O)cc(OC)cc2)ccccc1		Active
	metabolite 1	c1(C(=O)c2c(O)cc(O)cc2)ccccc1	Active	Active
103-30-0	trans-1,2-Diphenylethylene	c1(C={t}Cc2ccccc2)ccccc1		
	metabolite 1	c1(C={t}Cc2ccc(O)cc2)ccc(O)cc1	Active	Active
92-52-4	Diphenyl	c1(-c2ccccc2)ccccc1		
	metabolite 1	c1(-c2ccc(O)cc2)ccccc1	Active	Active
	metabolite 2	c1(-c2ccc(O)cc2)ccc(O)cc1	Active	Active
	metabolite 3	**c1(-c2ccccc2)ccc(O)cc1**	**Active**	**Not generated**
100-41-4	Ethylbenzene	c1(CC)ccccc1		
	metabolite 1	c1(O)ccc(CC)cc1	Active	Active
446-72-0	Genistein	c1(C2C(=O)c3c(O)cc(O)cc3OC=2)ccc(O)cc1	Active	Active
520-18-3	Kempferol	C1(O)C(=O)c2c(O)cc(O)cc2OC=1c1ccc(O)cc1	Active	Active
50-27-1	Estriol	c12c(C3C(C4C(C)(C)(C)C(O)(C4)CC3)CC1)ccc(O)c2	Active	Active
117-39-5	Quercetin	C1(O)C(=O)c2c(O)cc(O)cc2OC=1c1cc(O)c(O)cc1	Active	Active

REFERENCES

[1] OECD, *Annexes to the Report on Principles for Establishing the Status of Development and Validation of (Quantitative) Structure-Activity Relationships [(Q)SARs]*, ENV/JM/TG(2004)27/ANN, 2004.

[2] N. Ing and B. O'Malley, *The steroid hormone receptor superfamily — Molecular mechanisms of action*, in *Molecular Endocrinology: Basic Concepts and Clinical Correlations*, B. Weintraub, ed., Raven Press, New York, 1995, pp. 195–215.

[3] OECD, *Final Report of the Sixth Meeting of the Task Force on Endocrine 530 Disrupters Testing and Assessment* (EDTA 6), ENV/JM/TG/EDTA/M, 2003.

[4] M. Mizutani, N. Tomioka, and A. Itai, *Rational automatic search method for stable docking models of protein and ligand,* J. Mol. Biol. 243 (1994), pp. 310–326.

[5] R.Cramer, D. Patterson, and J. Bunce, *Comparative molecular field analysis (CoMFA): Effect of shape on binding of steroids to carrier proteins,* J. Am. Chem. Soc. 110 (1988), pp. 5959–5967.

[6] L. Xing, W. Welsh, W. Tong, R. Perkins, and D. Sheehan, *Comparison of estrogen receptor alpha and beta subtypes based on comparative molecular field analysis (CoMFA),* SAR QSAR Environ. Res. 10 (1999), pp. 215–238.

[7] T. Gantchev, H. Ali, and J van Lier, *Quantitative-structure-activity relationships/comparative molecular field analysis (QSAR/CoMFA) for receptor-binding properties of halogenated estradiol derivates,* J. Med. Chem. 37 (1994), pp. 4164–4176.

[8] P. Schmieder, A. Aptula, E. Routledge, J. Sumpter, and O. Mekenyan, *Estrogenicity of alkylphenolic compounds: A 3-D structure-activity evaluation of geneactivation,* Environ. Toxicol. Chem. 19 (2000), pp.1727–1740.

[9] O. Mekenyan, N. Nikolova, and P. Schmieder, *Dynamic 3D QSAR techniques: Application in toxicology,* J. Mol. Struct. (THEOCHEM) 622 (2003), pp.147–165.

[10] P. Schmieder, O. Mekenyan, S. Bradbury, and G. Veith, *QSAR prioritization of chemical inventories for endocrine disruptor testing,* Pure Appl. Chem. 75 (2003), pp. 2389–2396.

[11] E.L. Eliel, *Chemistry in three dimensions,* in *Chemical Structures*, W. Warr, ed., Springer, Berlin, 1993, pp. 1–8.

[12] O. Mekenyan, J. Ivanov, G. Veith, and S. Bradbury, *Dynamic QSAR: A new search for active conformations and significant stereoelectronic indices,* Quant. Struct. Act. Relat. 13 (1994), pp. 302–307.

[13] O. Mekenyan, G. Veith, D. Call, and G. Ankley, *A QSAR evaluation of Ah receptors binding of halogenated aromatic xenobiotics,* Environ. Helth. Perspect. 104 (1996), pp. 1302–1310.

[14] O. Mekenyan, T. Schultz, G. Veith, and V. Kamenska, *"Dynamic" QSAR for semicarbazide-induced mortality in frog embryo,* J. Appl. Toxicol. 16 (1996), pp. 355–363.

[15] S. Bradbury, O. Mekenyan, and G. Ankley, *Quantitative Structure-Activity Relationships for polychlorinated hydroxybiphenyl estrogen receptor binding affinity: An assessment of conformer flexibility,* Environ. Chem. Toxicol. 15 (1996), pp. 1945–1954.

[16] S. Bradbury, O. Mekenyan, and G. Ankley, *The role of ligand flexibility in predicting biological activity: Structural-activity relationship for aryl hydrocarbon, estrogen and androgen receptor binding affinity,* Environ. Sci. Technol. 17 (1998), pp. 15–25.

[17] T. Wiese and S. Brooks, *Molecular modeling of steroidal estrogens: Novel conformations and their role in biological activity,* J. Steroid Biochem. Mol. Biol. 50 (1994), pp. 61–73.

[18] S. Dimitrov and O. Mekenyan, *Dynamic QSAR: Least squares fit with multiple predictors,* Chemom. Intell. Lab. Syst. 39 (1997), pp. 1–9.

[19] O. Mekenyan, J. Ivanov, S. Karabunarliev, S. Bradbury, G. Ankley, and W. Karcher, *A computationally based hazard identification algorithm that incorporates ligand flexibility*, Environ. Sci. Technol. 31 (1997), pp. 3702–3711.

[20] S. Bradbury, V. Kamenska, P. Schmieder, G. Ankley, and O. Mekenyan, *A computationally based identification algorithm for estrogen receptor ligands, Part I. Predicting hER binding affinity*, Toxicol. Sci. 58 (2000), pp. 253–269.

[21] O. Mekenyan, V. Kamenska, P. Schmieder, G. Ankley, and S. Bradbury, *Computationally based identification algorithm for estrogen receptor ligands, Part II. Evaluation of a hER binding affinity model*, Toxicol. Sci. 58 (2000), pp. 270–281.

[22] O. Mekenyan, D. Dimitrov, N. Nikolova, and S. Karabunarliev, *Conformational coverage by a genetic algorithm*, Chem. Inf. Comput. Sci. 39 (1999), pp. 997–1016.

[23] T. Pavlov, M. Todorov, G. Stoyanova, P. Schmieder, H. Aladjov, R. Serafimova, and O. Mekenyan, *Conformational coverage by a genetic algorithm: Saturation of conformational space*, J. Chem. Inf. Model. 47 (2007), pp. 851–863.

[24] M. Shelby, R. Newbold, D. Tully, K. Chae, and V. Davis, *Assessing environmental chemicals for estrogenicity using a combination of* in vitro *and* in vivo *assays*. Environ. Health Perspect. 104 (1996), pp. 1296–1300.

[25] J. Katzenellenbogen, R. Muthyala, and B. Katzenellenbogen, *Nature of the ligand-binding pocket of estrogen receptor α and β: The search for subtype-selective ligands and implications for the prediction of estrogenic activity*, Pure Appl. Chem. 75 (2003), pp. 2397–2403.

[26] G. Anstead, K. Carlson, and J. Katzenellenbogen, *The estradiol pharmacophore: Ligand structure-estrogen receptor binding affinity relationships and a model for the receptor binding site*, Steroids 62 (1997), pp. 268–303.

[27] R. Serafimova, M. Todorov, D. Nedelcheva, T. Pavlov, Y. Akahori, M. Nakai, and O. Mekenyan, *QSAR and mechanistic interpretation of estrogen receptor binding*, SAR QSAR Environ. Res. 18 (2007), pp. 389–421.

[28] METI, *Current status of testing methods development for endocrine disrupters*, Sixth Meeting of the Task Force on Endocrine Disrupters Testing and Assessment (EDTA), Tokyo, 2002. Available at www.meti.go.jp/english/report/data/gEndocappendix1e.pdf.

[29] J. Surfin, D. Dunn, and G. Marshall, *Steric mapping of the L-menthionine binding site of AYP-L-mentionine-S-adenosyl-transferase*, Molec. Pharmacol. 19 (1981), pp. 307–313.

[30] J. Katzenellenbogen, *Effectiveness of QSAR for prescreening of endocrine disruptor hazard*, SCOPE/IUPAC International Symposium on Endocrine Active Substances, Yokohama, Japan, 2002.

[31] SYBYL, Tripos Associates Inc., St, Louis, MO.

[32] R. Dammkoehler, S. Karasek, E. Shands, and G. Marshall, *Constrained search of conformational hyperspace,* J. Comput-Aided Mol. Des. 3 (1989), pp. 3–21.

[33] M. Lipton and W. Still, *The multiple minimum problem in molecular modeling. Tree searching internal coordinate conformational space*, J. Comput. Chem. 9 (1988), pp. 343–355.

[34] O. Mekenyan, T. Pavlov, V. Grancharov, P. Schmieder, and G. Veith, *2D-3D migration of large chemical inventories with conformational multiplication. Application of the genetic algorithm*, J. Chem. Inf. Model. 45 (2005), pp. 283–292.

[35] J. Steward, *MOPAC: A semiempirical molecular orbital program*, J. Comput.-Aided Mol. Des. 4 (1990), pp. 1–105.

[36] J. Steward, MOPAC 93. Fujitsu Limited, 9-3, Nakase 1-Chome, Mihama-ku. Chiba-City, Chiba 261, Japan and Stewart Computational Chemistry, 15210 Paddington Circle, Colorado Springs, CO.

[37] M. Connoly, *Analytical molecular surface calculation*, J. Appl. Cristallogr. 16 (1983), pp. 548–558.

[38] D. Stanton and P. Jurs, *Development and use of charged partial surface area structural descriptors in computer-assisted quantitative structure-property relationship studies*, Anal. Chem. 62 (1990), pp. 2323–2329.

[39] O. Mekenyan, N. Nikolova, P. Schmieder, and G. Veith, *COREPA-M: A multi-dimensional formulation of COREPA*, QSAR Comb. Sci. 23 (2004), pp. 5–18.

[40] S. Dimitrov, G. Dimitrova, T. Pavlov, N. Dimitrova, G. Patlewicz, J. Niemela, and O. Mekenyan, *A stepwise approach for defining the applicability domain of SAR and QSAR models*, J. Chem. Inf. Mod. 45 (2005), pp. 839–849.

[41] O. Mekenyan, S. Dimitrov, R. Serafimova, E. Thompson, S. Kotov, N. Dimitrova, and J. Walker, *Identification of the structural requirements for mutagenicity by incorporating molecular flexibility and metabolic activation of chemicals I: TA100 model*, Chem. Res.Toxicol. 17 (2004), pp. 753–766.

[42] O. Mekenyan, S. Dimitrov, T. Pavlov, and G. Veith, *A systematic approach to stimulating metabolism in computational toxicology. I. The TIMES heuristic modelling framework*, Curr. Pharm. Des. 10 (2004), pp. 1273–1293.

[43] R. Serafimova, M. Todorov, S. Kotov, E. Jacob, N. Aptula, and O. Mekenyan, *Identification of the structural requirements for mutagencitiy, by incorporating molecular flexibility and metabolic activation of chemicals II. General Ames mutagenicity model*, Chem. Res. Toxicol. 20 (2007), pp. 662–676.

[44] P. Schmieder, M. Tupper, and R. Kolanczyk, unpublished results, U.S. EPA, ORD, NHEERL, Mid-Continent Ecology Division, Duluth, MN.

11 Molecular Field Analysis Methods for Modeling Endocrine Disruptors

Jean-Pierre Doucet and Annick Panaye

CONTENTS

ABSTRACT

The wide dispersal into the environment of a huge number of chemicals potentially able to perturb the normal hormonal processes in human or wildlife focused interest on methods for *in silico* prediction of potential toxicity, to help in priority setting before costly and time-consuming *in vitro* and *in vivo* assays. In this field, CoMFA (comparative molecular field analysis) and CoMSIA (comparative molecular similarity analysis) methods proved to be efficient and highly reliable. We review some of their applications to quantitative evaluation of binding affinity of various classes of chemicals to some nuclear receptors (estrogen, androgen, progesterone, and aryl hydrocarbon receptors).

Compared with related approaches (receptor-based field analysis methods, similarity matrices analysis, classical two-dimensional [2D] or three-dimensional [3D] Quantitative Structure-Activity Relationship [QSAR], and spectroscopic QSAR) CoMFA and CoMSIA generally reach comparable or better performance. Furthermore, the display of steric and electrostatic contour plots gives some insight into the ligand–receptor interactions and may suggest what structural modifications to introduce on the molecular scaffold to modify binding affinity.

KEYWORDS

Androgens
Comparative molecular field analysis (CoMFA)
Comparative molecular similarity analysis (CoMSIA)
Endocrine disruptors
Estrogens
Nuclear receptors
Polyhalogenated aromatics
Progestagens
Quantitative Structure-Activity Relationship (QSAR)

11.1 INTRODUCTION

The large structural diversity of chemicals likely to act as potential endocrine disruptors (EDs) and, for many of them, their worldwide dispersal in the environment became for decades an international concern, leading several governments (U.S., Japan, etc.) to define protocols for hazard identification and risk assessment [1,2]. In this field, (Quantitative) Structure-Activity Relationships (SAR and QSAR) play a key role inasmuch as they largely avoid costly and time-consuming experimental measurements on a large scale and limit animal testing [1,3]. Among the various available modeling approaches, comparative molecular field analysis (CoMFA) developed by Cramer et al. [4] has been largely accepted as a good model to quantify ligand binding affinity. CoMFA has demonstrated its ability to provide accurate predictions and is now currently accepted as useful for hazard identification [5,6]. So, CoMFA (although computationally and person power intensive) has been integrated in a sequential "Four Phase" scheme of priority setting of endocrine disruptors [1]. On the other hand, CoMFA was also proposed as a rationale for synthesis of ligands of high and selective affinity, like selective androgen receptor modulators (SARMs) [7].

This review summarizes the use of CoMFA and other neighboring methods of molecular field analysis in ED screening. Attention will be focused on estrogen (ER), androgen (AR), and progesterone (PR) receptors. For other nuclear receptors (corticosteroid, retinoic, thyroid) less extensively studied, information may be found on recent reviews and articles [6,8–10].

Whatever the QSAR model may be, it is important to keep in mind some basic remarks:

- Any model cannot be better than the data employed to build it.
- Statistical methods may differ in their ability to establish the relationship.
- Some descriptors are better than others on a given chemical family.

Furthermore, for hazard assessment or risk management, in view of use in regulatory processes, careful examination of the prediction confidence and validation of the applicability area of the model (including domain extrapolation) are mandatory to assess the robustness of the model. In other words, caution is warranted in interpreting QSAR results for chemical classes that are not well represented in the training set [1,11]. The training set should cover the structural space, with a minimum of patterns, equally shared. However, it was indicated that some redundancy must exist to cope with possible single point errors [12]. As to validation of the results, leave-one-out (LOO) cross-validation has been frequently used, but it was noted that such a process corresponds to some kind of interpolation within the training set and has only limited ability to validate prediction for chemicals structurally different from those of the training set [1,13]. Using an external test set, ignored when building the model, is a more secure way to validate the approach.

Another important problem is the translation of *in vitro* results to *in vivo* conditions, due, for example, to problems of metabolization. So, Vinclozolin, which induces antiandrogenic effects *in vivo*, does not bind AR *in vitro*, its action being due to its metabolites M1 and M2 [14]. Conversely, in the search for SARMs, newly synthesized molecules, mimicking the steroid nucleus and looking promising for their binding affinity *in vitro* are ineffective *in vivo*, due to a too rapid metabolization [15]. Desolvation energy differences and H-bond properties may also complicate the comparison [16].

We will first briefly present the principle of the more largely widespread molecular field analysis methods CoMFA and CoMSIA. For more details, the reader may refer to specialized textbooks [12]. We will then present some of their applications for diverse nuclear receptors, and compare their results with other approaches (that will be briefly presented when encountered in the text). As a general remark, it was established that, for endocrine disruptors, *in vitro* affinity for receptor binding is a good predictor of *in vivo* activity. Binding affinity is generally defined, in competitive assays, as the negative log of the concentration (molar equivalent) necessary to displace from the receptor 50% of a bound reference compound, and it is expressed in relative values log RBA: [^{3}H] labeled estradiol, for example, was frequently used to characterize ER binders. Some molecules frequently cited in this review are represented in Figure 11.1.

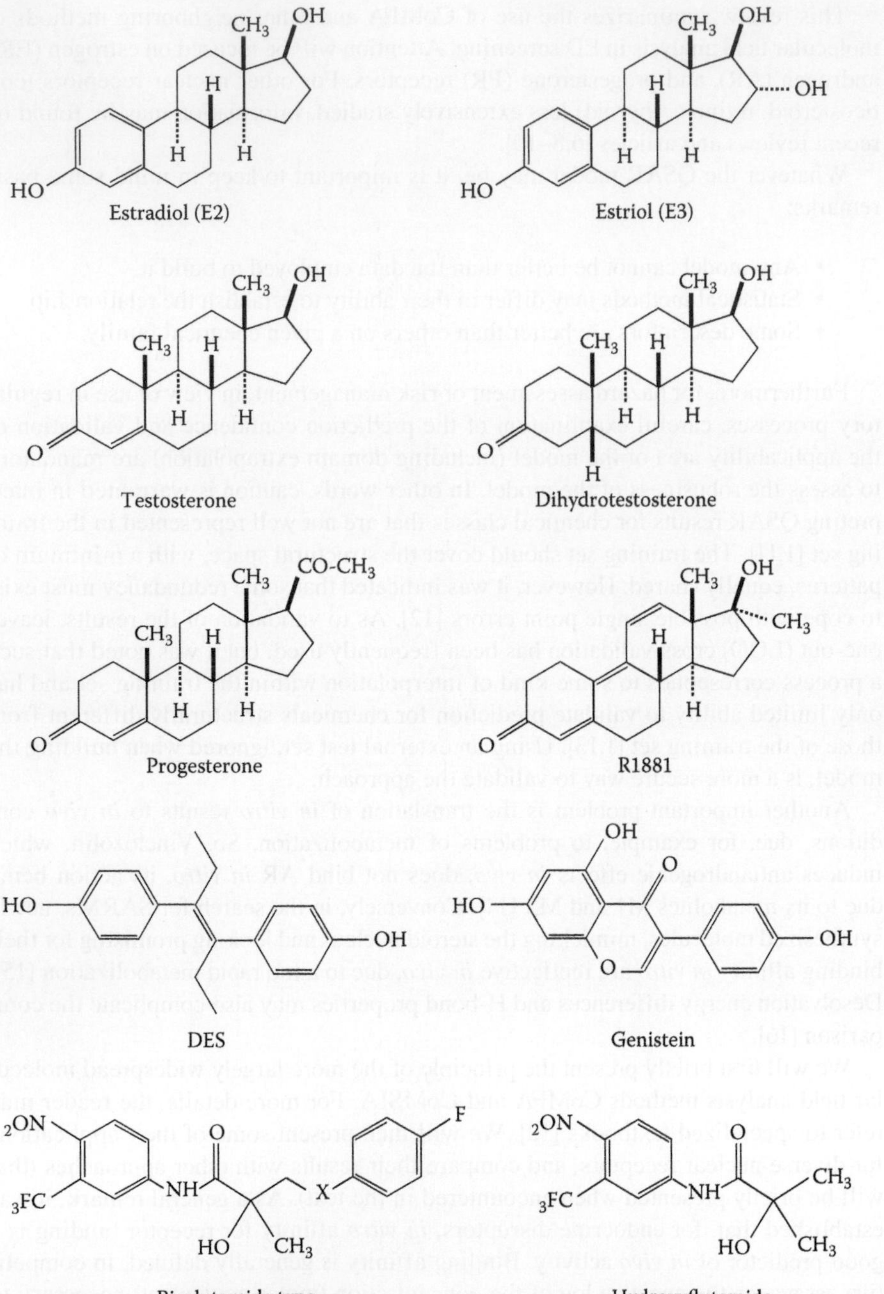

FIGURE 11.1 Some endocrine disruptors frequently cited in this review.

11.2 COMPARATIVE MOLECULAR FIELD ANALYSIS METHODS: CoMFA, CoMSIA

11.2.1 CoMFA: Basic Steps and Caveats

The basic premise, common to QSAR models, is that the biological response (in most cases) reflects the interactions between a receptor and the ligand that will bind to it. It is assumed that all molecules interact with the receptor at the same binding site in the same (or similar) manner. Variations in these interactions are related to the observed variations in the ligand affinity. CoMFA assumes that noncovalent forces dominate these interactions and can be modeled by the fields (electronic and steric) created by the ligand in its vicinity. So, differences in activity may be correlated with differences in these noncovalent fields. This contrasts with "classical" three-dimensional (3D) QSAR, where structural descriptors, encoding relevant features of the molecules under scrutiny, generally correspond to characteristics (local or global) directly attached to the molecular framework (such as heat of formation, molecular surface area or volume, atomic charges). CoMFA analyses, here reviewed, were carried out with the turnkey package included in SYBYL, and provided by TRIPOS® (St. Louis, Missouri). The successive steps in a CoMFA analysis are as follows:

- *Alignment*: Molecules are superimposed on a common template in a 3D lattice surrounding the molecules.
- *Field calculation*: Interaction energies (steric and electrostatic) are calculated locating a "probe atom" on the nodes of this 3D lattice.
- *Correlation with activity*: Interaction energies are correlated with values of the studied property, using partial least squares (PLS) method [17].
- *Contour plots*: From PLS coefficients, areas may be delineated, in the space surrounding all molecules, where an increase of steric bulk, respectively, an increase of positive potential, contributes to increase (or decrease) binding affinity.

However, to get significant results some precautions must be taken at each step. No doubt, alignment, sometimes time-consuming and requiring some chemical and biological knowledge, is the most crucial part of the treatment and often a real bottleneck [18]. It is generally carried out using as a template the most active molecule in the set, and choosing a rigid part of it, common to all molecules (if possible), as a "seed." For nonrigid molecules, the selection of the active conformation (that which binds the receptor) is, of course, a major problem. A systematic conformational search is therefore performed beforehand to define the minimum-energy conformation that will be used. The quality of the fit is determined by the root mean square error (RMSE) on atom locations (after superimposition). Of course, knowing the geometry of a template bounded to its receptor (from X-ray crystallography) is clearly a definite advantage.

But, using the crystal structure of the complex ligand binding domain (LBD) is not always without problems: close packing of residue side chains (as seen in the X-ray structure) may preclude binding in the receptor pocket or forbid H-bond formation [19]. Using energy-minimized conformations is not a panacea. Sometimes, the best aligned conformation is not that of minimal energy, as for bicalutamide.

Binding modes may not be the same for the different chemical families: in bisphenol A, the intuitive choice, adjusting the two hydroxyl group on those on the 3- and
17- group of E2 or DHT, must be discarded, a conformation with a hydrogen-bond
missing but with stronger hydrophobic interactions is preferred [20].

When the dataset encompasses different chemical families (for example, steroids
and more flexible molecules), it may be useful to choose some "secondary references": in each family, a lead is first aligned with the template and constitutes itself
a template for its own family.

In fact, CoMFA is based on the field values on the nodes of the lattice surrounding
the superimposed molecules. Rather than optimizing the location of atoms, the alignment may be refined in a "field-fit" minimization. In the energy minimization process
(usually by molecular mechanics), additional penalty terms are added to the force field
to reflect the degree of similarity of the steric and electrostatic fields (between template and fitted molecule). Internal coordinates of the fitted molecule are adjusted to
ensure optimal field and geometric overlap with the template. This causes an increase
of its potential energy and possibly some structural distortions. The fitted molecule is
therefore re–energy-minimized, without the "field fit" penalties, to relax to the nearest local minimum-energy structure. It was quoted that such a field-fit alignment may
avoid multiple possible solutions due to symmetry or presence of multiple rings [14].

For field calculation, the probe atom mimics an sp^3 hybridized carbon (radius 1.53
Å, charge +1.0). 6–12 Lennard Jones, and Coulombic potentials are respectively used.
Charges are generally derived from the Gasteiger-Marsili model (relying on partial electronegativity equalization), or calculated by semiempirical quantum mechanics programs
(such as AM1), with (often) a distance-dependent dielectric constant ($\varepsilon = r$). The lattice,
with a mesh of (usually) 2 Å, extends at least 4 Å in all directions beyond the common volume of the superimposed molecules (nodes internal to the common volume are ignored).
At short distances, these potentials may take very large values, and a truncation threshold
(generally ±30 kcal/mol) is fixed. Electrostatic interactions at "sterically forbidden" points
(high steric energy) are ignored, and a mean value for these nodes is retained.

For the correlation of field values with activity, interaction energies are gathered
in a spreadsheet where rows correspond to individual molecules and columns to individual nodes. Because there are many more values of potentials (for example, 300
nodes × 2 fields) that constitute the independent variables (x) for a dataset of (say)
50 activity values (dependent variable y), analysis is carried out using PLS method.
Basically, PLS extracts from the independent variables a reduced set of principal
components (linear combination of the original descriptor variables) to correlate the
observed activities. In fact, these latent variables are slightly skewed from the PC
for maximum intercorrelation [17]. In such analyses, columns (nodes) with small
variations of interaction energies may be discarded to reduce the computational task
("column filtering" makes the calculation faster by about 10 times) [14]. The number
of principal components to retain is determined from the cross-validated (usually in
LOO) determination coefficient q^2:

$$q^2 = 1 - \text{PRESS/SD}$$

$$\text{PRESS} = \sum (y_{obs} - y_{cal})^2$$

where SD is standard deviation, and PRESS is the sum of squared deviations between predicted and observed property values.

The model is then applied to the whole set to get the conventional determination coefficient r^2. Schematically, r^2 measures the model's goodness of fit and its internal consistency, whereas q^2 evaluates its robustness and its ability to interpolate within the training set. $r^2 > 0.9$ and $q^2 > 0.5$ are usually considered as significant [21]. A negative q^2 would mean that a "no model" (taking the mean of the observations) works better than the proposed model.

Results generally mention:

$$SPRESS = (PRESS/n-c-1)^{0.5}$$

where n is the number of compounds, and c is the number of components in the model.

SPRESS takes into account the complexity of the model. Otherwise, the (cross-validated) standard error of prediction and the (non-cross-validated) standard error of estimate are given:

$$SDEP_{(test)} = (PRESS/n)^{0.5}$$

$$SEE_{(training)} = (PRESS/n)^{0.5}$$

In fact, they more rely on RMSE than on standard error(s), and the terminology is not always unambiguous.

For prediction,

$$Pr-r^2 = 1-PRESS/SD$$

is calculated on the sum of squared differences between calculated values on objects in the test and the mean of the training set objects. Bootstrapping or y scrambling may be used as other criteria of model validity. At last, a variant of PLS, SAMPLS [12] accelerates the cross-validation process.

In addition to the usual prediction of activity, one of the interesting points of CoMFA is provided by the contour plots. They delineate areas in the space surrounding aligned molecules where an increase of steric bulk (brought about by larger training set molecules), respectively an increase of positive potential, contributes to increase (or decrease) activity. Schematically, the (very numerous) coefficients of the CoMFA treatment (giving the contribution of each lattice node to the PLS model) are used to define contour plots, corresponding to fixed values of the product "Standard Deviation x Coefficients." For example, for the steric field, a positive value on a node indicates that an increase of the field felt on this node would correspond to increased activity. An example of contour plots is given in Figure 11.2a and Figure 11.2b.

Contour plots may suggest what structural modifications introduce in those regions to modify activity (more bulky, more positively charged group and so on). Contour plots may be viewed as representing very roughly a complementary image of the binding domain of the receptor. However, the limited amount of information they bring involves extreme caution in such an interpretation. They may even be considered as a "self-fulfilling prophecy since totally dependent on alignment and composition of training set" [5]. Contour plots are directly related to the changes in structures that

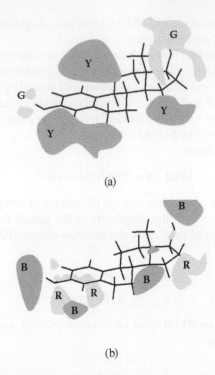

(a)

(b)

FIGURE 11.2 (a) Estradiol scaffold and comparative molecular field analysis (CoMFA) contour plots for estrogen receptor (ER) binding affinity. Light gray area represents regions (near the D-ring) where steric bulk should be increased for better affinity. In dark gray area (above and below the A-ring), bulk is detrimental. (b) Electrostatic contours: light gray indicates regions where more negative charge favors binding, whereas in dark gray regions (off A- and D-rings) positive charge is desired. Letters Y, G, B, R correspond to the conventional colors in CoMFA maps, which are yellow, green, blue, and red, respectively. (Adapted from J.D. McKinney, A. Richard, C. Waller, M.C. Newman, and F. Gerberick, *Toxicol. Sci.*, 56, 8, 2000. With permission.)

lead to changes in activity. Absence of a region in a contour does not mean this region is unimportant; it only indicates it is constant in the dataset [8]. Nevertheless, contour plots may give some indices as to the mechanisms intervening in binding and may also be used to inspect the necessary complementarity of bulk and charge for possible ligands, for example, to identify potentially active metabolites [14].

11.2.2 Improvements of CoMFA

To improve alignment, SEAL (Steric and Electrostatic ALignment) [22] proposes an automated method for a rigid body alignment. For each tested molecule, numerous randomly generated starting orientations are tried, with definition of a similarity index with the template, in order to maximize electrostatic and steric overlap.

Docking ligands to the receptor (derived from experiment or built by homology) gives an insight into their bound conformations. Then, these selected conformations are aligned with better confidence [7,23,24], Monte Carlo methods or molecular dynamics may also be used [19,25].

FLUFF (Flexible Ligand Unified Force Field) [26,27] is a semiautomatic, template-based, superimposition algorithm that aims to maximize the similarity of van der Waals and electrostatic fields. A superimposition force field evaluates the overlap between the template and the molecule under scrutiny. Negative superimposition ("no, like that...") may be considered. One original characteristic is the definition of "logical" molecules (one or several molecules, parts of real molecules, arbitrary set of atoms). Groups (such as C18-methyl, H) may be omitted to avoid barrier effects hindering superimposition. Flexibility can be introduced on the ligand or on both ligand and template. Weights on atoms may be added when alignment is not trivial (several possible solutions). In Korhonen et al. [26], FLUFF gave better superimpositions than SEAL, "presumably thanks to *a priori* information on weights of the template figures."

Due to problems posed by the alignment step in CoMFA, great interest focused on alternative "GRid-INDependent" approaches such as GRIND. GOLPE (Generating Optimal Linear PLS Estimations) avoids considering too many nonimportant grid points by carrying out comparative PLS analyses on subsets of the variables and selecting the most significant variables [28,29].

Region focusing solutions (CoMFA/q^2GRS) are also considered. It was shown [13] that the quality of the model may heavily depend on the overall orientation of the superimposed molecules in the lattice and can vary by as much as 0.5 units on q^2. To get rid of this problem, Cho and Tropsha [13] proposed a "q^2 Guided Region Selection (q^2GRS)". The basic idea was to eliminate regions where fields did not correlate well with changes in the activity. After a conventional CoMFA, the lattice embedding the aligned molecules was divided into 125 small regions where CoMFA was carried out with a smaller mesh (for example, 1 Å versus 2 Å for the usual treatment). Only regions where q^2 (from these analyses) was greater than a defined threshold were selected to create a master region file used to perform the final CoMFA treatment. See also Shi et al. [18] and Sadler et al. [30].

Another solution is the "All orientation search and all placement search" of Wang et al. [31]. With this, the molecular aggregate was rotated or translated systematically within the grid. A CoMFA was performed for each orientation or placement, and the best (highest q^2) was retained.

Introducing additional columns of descriptors (HOMO, LUMO, log P, dipole moment) to the usual field values may, in some cases, improve the results. In Shi et al. [18], adding log P was inefficient, whereas a "phenol indicator" increased q^2 up to 0.71 in place of 0.66 with traditional CoMFA. See Gantchev et al. [32].

Various changes in the parameterization of CoMFA have also been attempted, including extending the grid, imposing ring carbon constraints in field fit, and changing charges [32].

For a better analysis of H-bonding properties, changing probe to O and H atoms has been proposed |32|. Alternatively, beyond steric and electrostatic fields, other contributions may be considered. For example, Bohl et al. [7] added indicator and H-bonded fields representing H-bonding acceptor and donor components. See also Kroemer and Hecht [33] and Bohacek and McMartin [34]. Hydrophobic interactions may be taken into account with HINT [35] in which hydropathic interactions are evaluated *via* an empirical potential:

$$A_j = \sum a_i \, S_i \, R_{ij}$$

where a_i, S_i are hydrophobic atom constant and solvent-accessible surface area for atom i; R_{ij} is the Gaussian attenuation factor, function of the distance between atom i and the probe j.

Polarizability effects could also be introduced with a P/r^3 field (P atomic polarizability) as in the SOMFA method (Self-Organizing Molecular Field Analysis) [36].

11.2.3 COMPARATIVE MOLECULAR SIMILARITY ANALYSIS: CoMSIA

The interest for introducing different fields in order to have a more precise analysis of the different influences intervening is one of the raisons d'être of the Comparative Molecular Similarity Analysis, CoMSIA [37,38]. For example, entropic contributions (important in the immobilization at the binding site) may be insufficiently covered by van der Waals and Coulomb potentials. On another hand, the CoMFA potentials are very steep near the van der Waals surface. As a consequence, contour maps often present singularities, making them less clear. Furthermore, simultaneously using steric and electrostatic fields requires introducing an arbitrary scaling and different threshold values for these two components, leading to some information loss. To overcome these problems, Klebe et al. [37] proposed a new approach: the Comparative Molecular Similarity Analysis (CoMSIA).

After alignment and embedding of the molecules in the 3D lattice (as in CoMFA), the similarities between the atoms of the studied molecule and a probe are evaluated for five properties related to steric, electrosatic, hydrophobic, and H-bond donor or acceptor fields. For property k, the similarity index at grid point q for molecule j is calculated according to

$$A_{F,k}^q(j) = -\sum (W_{\text{probe},k} \cdot W_{ik}) \exp(-\alpha/r_{iq}^2) \qquad (11.1)$$

(summation over all atoms i of molecule j). r_{iq} is the distance between the probe and atom i of the tested molecule, and W_{ik} is the actual value of property k at atom i (volume for the steric part, AM1 charges, experiment-derived rules for H-bonds, and atom-based parameters for hydrophobicity). For the probe, radius r is 1 Å, and other properties are set to arbitrary value +1 (charge, hydrophobicity, H-bond-donating, H-bond-accepting). A Gaussian-type attenuation factor generally set at $\alpha = 0.3$ gives smoother variations, avoiding singularities on the contour plots and allowing for calculation inside the molecular surface. Data analysis is then carried out with SAMPLS.

According to the authors [37], using more fields than CoMFA does not necessarily increase prediction accuracy (fields may be intercorrelated). But it allows for a sharper analysis displaying space regions where the contributions of the different fields are important for the biological activity and indicating particular characteristics that may be useful for the design of new, improved ligands. Particularly, similarity indices may be calculated on nodes within the molecular volume (which is not possible with CoMFA). In CoMFA, contour plots indicate regions of space where molecules would favorably (or not) interact with a possible environment. In contrast, CoMSIA field contributions indicate nodes where a particular property may be given a higher weight to enhance an activity, making contour maps easier to interpret. Difference maps may indicate discriminating regions monitoring selectivity. It was also shown, on examples, that CoMSIA was less sensitive than CoMFA to changes in the orientation of the superimposed molecules, variations of q^2 on translations and rotations of the superimposed molecules remaining inferior to 0.10 in log RBA.

11.3   BINDING THE ESTROGEN RECEPTORS

Among the diverse models proposed for endocrine disruption processes, binding affinity to the estrogen receptors (ERs) is the most documented. Particularly, the CoMFA approach was compared to various other 2D or 3D methods. In addition to the statistical quality of the results and the cost of the calculation, great attention was devoted to two facets: the need (or not) for a (geometry-optimized) 3D structure, and the need (or not) for alignment of the molecules constituting the dataset.

Among the first CoMFA models proposed in the field, Gantchev et al. [32] considered 44 halogenated derivatives of estradiol (E2), used for receptor imaging in cases of endocrine cancers. The relative binding affinity (RBA) was measured in a competitive binding assay with [3H] estradiol. Alignment was carried out on the steroid A, B, and C rings. Ten chemicals were chosen as test, but among them, binding affinity was only available for five. Various CoMFA parameters were systematically modified in order to obtain the highest q^2 (in a cross-validation over 25 groups). In the best CoMFA model (11 components), q^2 and r^2 were respectively equal to 0.895 and 0.988 (SEP 0.48; SEE 0.16 log units).

Affinities of polychlorinated hydroxybiphenyls (formerly used as commercial insulators and recognized as persistent environmental contaminants) were examined in [16], based on a competitive mouse uterine cytosol assay. Alignment was achieved on the A ring of estradiol and the parasubstituted ring, the field-fit option allowing to choose between several solutions (if that was the case). After elimination of three chemicals (ionized in the conditions of experiment), 11 molecules constituted the training set. Inclusion in the CoMFA model of the hydropathic HINT field was considered, but the best model was obtained with the steric field only, leading to $q^2 = 0.544$; $r^2 = 0.974$ with two components (SEP 1.12; SEE 0.27).

To overcome the lack of structural diversity, in some previous studies, Waller et al. [5] examined a population of 55 molecules corresponding to 8 structurally diverse subsets (phenols, DDTs, DESs, PCBs, phthalates, phytoestrogens, steroids, pesticides), on 6.5 log units of binding affinity (measured on mouse uterine cytosol assays). Rigid body alignment to the template estradiol by SEAL avoids subjective hypotheses. Adding to the usual steric and electrostatic fields, a hydrophobic contribution (HINT) led to a better model, for both internal prediction and robustness, than usual CoMFA or HINT alone ($q^2 = 0.590$; $r^2 = 0.881$). It may be noted that phenolic rings were generally superimposed on ring A of estradiol. But for the 2,4,6 trichloro-4'-biphenylol, the phenol ring was oriented over the estradiol D ring and its

17-OH group. This position provided better hydrophobic interactions in the central part of the molecules (over B and C rings). Interestingly, external predictive ability was examined excluding a whole structural family and training the model on the other ones. Errors on these "unbiased" predictions were typically on the order of 1.0 to 2.0 log units depending on the structural extrapolation from the training set (whereas in the fitted mode, error was about 1 log unit). LOO predictions were not significantly better. This may indicate that the training set already encompassed a sufficient structural diversity to be able to treat new families.

The binding of nonsteroidal ligands to ER was also studied by Sadler et al. [30] on 30 chemicals (DES metabolites, Indenestrol analogs), with binding affinity measured on mouse cytosol uterus. Various alignments were tested based on the oxygen atoms in 3- and 17- positions of estradiol and either the centroids of rings A and D or carbons of ring A (the best model), with field-fit or not. With region focusing, the best model led to $q^2 = 0.796$ with three principal components, compared to 0.720 (2 PC) with traditional CoMFA. As in other studies, electrostatic contribution predominated (55%) in agreement with an area of favored negative charges near the 3- and 17-OH groups of E2.

From these studies [5,30], convergent indications could be gained from the contour plots that schematically indicated that affinity is favored by two electronegative centers separated by a rigid hydrophobic scaffold.

Steric bulk is tolerated near positions 2 or 3 of ring A and mainly near the 17-substituent on ring D, but is detrimental out of plane of ring A or near atoms 15 and 16 on D ring, as indicated in Figure 11.2a. Preference for a positive potential near positions 3 and 17 reflects the orientation of the hydroxyl groups (frequently present in the dataset). But preferences for negative charge also exist near atoms 3 and 17, as shown in Figure 11.2b. Hydrophobic bulk is beneficial in the vicinity of the A ring but not near the surface close to 1-, 4-, and 16- positions. A hydrophobic pocket is suggested in the 11-β region of estradiol. The proposed mode of binding was validated by comparison of relatively rigid DESs or Indenestrol analogs: pairs of S (more active)/R enantiomers or EE (more active)/ZZ isomers, highlighting the importance of beneficial hydrophobic interactions in the 11-β region of the template and steric bulk intolerance below the D ring, as already noted in [5]. Some of these compounds are represented in Figure 11.3.

11.3.1 Comparison with "Classical QSAR"

Tong et al. [39] compared CoMFA to CODESSA (COmprehensive DEscriptors for Structural and Statistical Analysis) models [40,41] on a set of structurally diverse chemicals (53 compounds) using data from von Angerer et al. [42] on 2-phenylindoles, plus 3 steroids, 2 triphenylethylenes, and Hexestrol. Relative binding affinities were calculated from a calf uterine ER competitive binding assay. The test set was constituted of 16 steroidal estrogenic compounds [43]. Their RBA values were obtained on human ER, but for four compounds they were also measured on calf ER. The quite comparable RBA values obtained in the two assays suggested that accurate predictions may also be achieved "across species."

CODESSA is a "classical QSAR" package, calculating hundreds of descriptors (constitutional, topological, geometrical, electrostatic, and quanto-chemical) from the (optimized) 3D molecular geometry. A stepwise process then selects the relevant

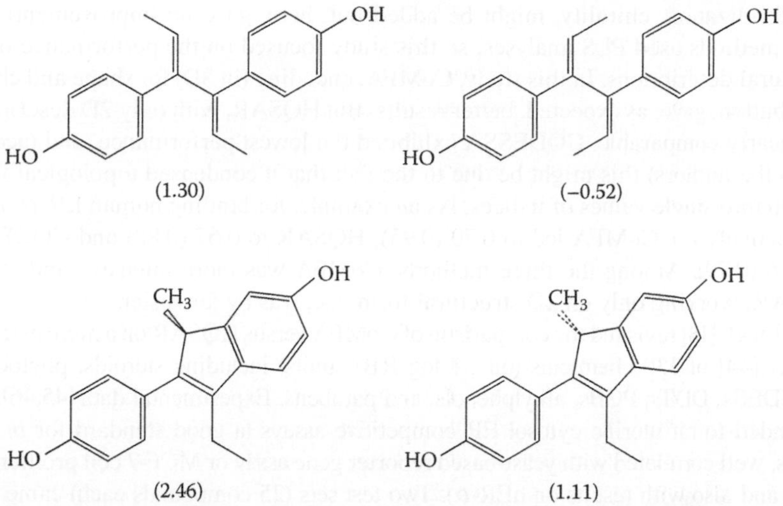

(1.30)   (−0.52)

(2.46)   (1.11)

FIGURE 11.3 Variations of a binding affinity to an estrogen receptor (ER) ligand-binding domain (LBD) for pairs of isomers. Log RBA (relative binding affinity) values are indicated in parentheses (reference Estradiol, log RBA = 2). (Adapted from B.R. Sadler, S.J. Cho, K.S. Ishaq, K. Chae, and K.S. Korach, *J. Med. Chem.*, 41, 2261, 1998. With permission.)

descriptors to be included in a PLS or MLR model. In this study, CoMFA ($q^2 = 0.61$, $r^2 = 0.97$) outperformed CODESSA ($q^2 = 0.54$, $r^2 = 0.68$) that seemed unable to capture some structural factors (for example, the position of the OH group on the indole phenyl ring). However, for prediction, results were more comparable. The authors also stressed that CODESSA (although requiring a 3D structure) was less prone to arbitrariness (alignment choice) and was easier to implement.

CoMFA contour plots indicated that increasing the negative charge near the 3′ position of 2-phenylindoles would favor a decrease in RBA, in agreement with the trend observed for an OH substitution on this position. Binding appeared to be sensitive to the length of substituents at the indole N1-position: a short alkyl chain (ethyl group) increased affinity (as for the B- ring of steroids), but a longer chain (butyl group) was detrimental.

In another publication [44], the CoMFA method was compared to CODESSA and Hologram QSAR (HQSAR) TRIPOS® (St Louis, Missouri) on three datasets, from competitive binding assays with labeled endogenous E2. The first two sets encompassed 31 compounds (among them 19 steroids) binding human ER-α and rat ER-β, respectively. The third encompassed 47 compounds (mainly congeners of 2-phenylindole from [42]) studied with calf ER.

HQSAR systematically generates all possible fragments of varied lengths (here four to seven atoms). These fragments are counted in bins of a fixed length array (hashing process) to form a molecular hologram and selected on the basis of their performance. The process is analogous to the generation of molecular fingerprints. Bin occupancies are the structural descriptors encoding 2D molecular information. Here only element- and bond-type information was considered (additional 3D information on hybridization, chirality, might be added but, here, gave no improvement). The three methods used PLS analyses, so this study focused on the performance of the structural descriptions. In this study, CoMFA, encoding (in 3D) for shape and charge distribution, gave, as expected, better results. But HQSAR, with only 2D descriptors, was nearly comparable. CODESSA exhibited the lowest performance, and (according to the authors) this might be due to the fact that it condensed topological information into single values of indices. As an example, for binding human ER-α, for q^2 (respectively r^2), CoMFA led to 0.70 (0.95), HQSAR to 0.67 (0.88) and CODESSA to 0.46 (0.79). Among the three methods, CoMFA was more intensive and tricky; HQSAR, working only on 2D structural formulae, was by far easier.

Shi et al. [18] revisited the comparison of CoMFA versus HQSAR on a more extended dataset [44] of 130 chemicals (on 7.1 log RBA units) including steroids, phytoestrogens, DESs, DDTs, PCBs, alkylphenols, and parabens. Experimental data [45,46], corresponded to rat uterine cytosol ER competitive assays (a good standard for *in vitro* assays, well correlated with yeast-based reporter gene assay or MCF-7 cell proliferation assay and also with results on hER-α). Two test sets (25 compounds each) came from the data of Kuiper et al. [47] and Waller et al. [5]. Alignments were based on the crystal structure of four ligands, representative of the chemical family. Standard CoMFA gave $q^2 = 0.665$, $r^2 = 0.908$, whereas with HQSAR q^2 was only 0.585 (and r^2 0.756). Electrostatic field played the major role (57%) consistent with results on smaller sets [39,44]. Neither "region focusing" [13] nor the "all orientation, all placement search" of Wang et al. [31], or the introduction of log P, gave significant improvement. On the other hand, adding a "phenol indicator" raised q^2 and r^2 to 0.707 and 0.903 (with no significant change on SEE), an improvement consistent with the well-known importance of H-bonding from C-3. Robustness of the model was established on extensive runs (100) of leave-N-out cross-validation. With the two external test sets, $q^2 pred.$ reached 0.71 and 0.62 (with phenolic indicator) in place of 0.15 and 0.22 for HQSAR.

However, a parallel study [48] to identify significant descriptors associated with ER binding affinity was carried out on 131 compounds (same origin as the compounds studied by Shi et al. [18]). Starting from 151 descriptors (quantum mechanical, graph theoretical indicators, log P, and so forth), a rigorous selection by stepwise regression had a predictive power $q^2 = 0.627$ comparable to the CoMFA result 0.623 (in a leave-25%-out cross-validation) [18].

Comparison of CoMFA with CoMSIA and HQSAR was also investigated by Coleman et al. [20] on 25 derivatives of bisphenol A, a weak ER agonist intervening at a large scale in polycarbonate and epoxy resin production. E2 and DES were added to the dataset, as reference compounds, leading to an affinity scale of 4 log units. Four structures were chosen as the test set. For CoMFA, alignment was carried out in two steps: first the most active bisphenol was fitted to DES bound to hER; then all structures were fitted to it. The HINT field [35] was introduced. Three bioassays were used: competitive ER binding, gene induction, and cell proliferation. Consistent results were obtained, suggesting that ER binding data alone constituted a good indicator. For this population, the three methods gave nearly similar results, with a slight advantage to HQSAR ($q^2 = 0.542$) versus CoMFA (0.514) or CoMSIA (0.513) at the price of a more complex model (6 PC versus 2). Unlike the preceding examples, electrostatic contribution was not the most important (36% in CoMFA).

A graphical display of CoMFA contours was consistent with the fact that ER could accommodate ligands with phenolic rings (preferably unencumbered) bound to a hydrophobic core in the central portion of the binding pocket. Interestingly, it was shown that alignment based on E2 as template produced bad models, placing bisphenol in an extended position, in an effort to match the OH groups with those of E2. DES, utilizing more features of the ER, was a better template, suggesting one phenol ring as in DES, the other and the central substituents lying in the same region as the ethyl groups of DES to take advantage of hydrophobic interactions.

Another comparison [49] concerned the performance in ER binding affinity prediction of CoMFA, HQSAR, and FRED/SKEYS. The FRED/SKEYS approach MDL® (Information Systems, San Leandro, California) uses 166 substructural keys to encode the presence of singular atoms, rings, or more complex patterns, and then select by FRED (Fast Random Elimination of Descriptors) an evolutionary algorithm. The dataset (58 compounds), nearly identical to that of Waller et al. [5], but three compounds, encompassed eight chemical families. The CoMFA model included hydropathic interactions (HINT) and used SEAL for alignment on estradiol. For test, as in [5], each chemical family was in turn excluded. Performances varied somewhat according to the chemical family tested. But, as a whole, according to the authors, CoMFA and HQSAR displayed similar cross-validated results (q^2 about 0.580), FRED/SKEYS being better ($q^2 = 0.700$). The better result of CoMFA in a non-cross-validated treatment ($r^2 = 0.893$ versus 0.805 and 0.783) was attributed to the larger number of variables the model contained. An interesting point was that HQSAR might identify beneficial or detrimental substructures. However, working on dissimilar structures (with substructures not represented in the training set) might be difficult.

The k nearest neighbor QSAR method (kNN-QSAR), initially proposed by Zheng and Trophsa [50], applied on the same 58 estrogen receptor ligands, gave $q^2 = 0.77$ for a ten descriptor model. In this method, the activity of a compound is evaluated as a weighted average of the activity of its k most similar neighbors. Similarity is calculated by the Euclidian distance between the descriptor vectors. In the paper of Asikainen et al. [51], the kNN approach was coupled with a consensus process. First, the huge number of descriptors generated by Dragon (about 1,200) [52] was reduced by simulated annealing (random subsets of descriptors were progressively selected according to a Metropolis Monte Carlo scheme). From 50 models of 250 descriptors each (this number being not really decisive), a consensus result was proposed. The method was applied to five datasets (some of them previously studied): binding to calf [39], rat [18], mouse [49], and human α and β. As to internal predictions, this method (presented as simple and robust) gave results better than the other approaches, and results were confirmed by leave-some-out cross-validation. It was noted that the discriminating power of the descriptors was more important than their modeling power, so that a rather large pool of variables was recommended.

11.3.2 COMPARISON WITH "SPECTROSCOPIC QSAR"

Other approaches in field analysis rely on the "Spectroscopic QSAR methods" such as EVA (EigenVAlue) [53], EEVA (Electronic EigenVAlue) [54], and so on. The basic idea is that spectroscopic quantities reflecting intrinsic physicochemical properties of molecules (implicitly related to their 3D structure) may be considered as molecular descriptors. These descriptors are "physical observables" and not some artificial calculated descriptions of the molecular structure, and no alignment is necessary when using them in QSAR [55]. These are constituted of the eigenvalues of the corresponding Hamiltonian (normal modes frequencies, in the infrared [IR], orbital energies in ultraviolet [UV]-visible spectra). The approach was then extended to nuclear magnetic resonance (NMR) shifts in CoSA (Comparative Spectra Analysis) [55]. The common process consists in determining the (usually calculated) spectroscopic quantities (MO energies, vibrational frequencies, NMR shifts), converting them to a bounded scale, introducing a Gaussian shape factor over each "peak," and adding the "intensity" at regular intervals to build up the "spectrum." A semiempirical quantum method (after geometry optimization) was claimed to be sufficient for evaluating MO energies and vibrational frequencies, and evaluation of NMR shifts requires high-level *ab initio* calculations.

A comparative study was carried out by Asikainen et al. [56] on 36 estrogens previously examined [30,57,58] and the quality of the models carefully checked by external validation on a large number of randomized test sets. According to the authors, CoSA (^{13}C) had good predictive ability ($q^2 = 0.69$), whereas EEVA ($q^2 = 0.42$) was a borderline case, and EVA or CoSA (^{1}H) only gave semiquantitative information. But CoSA remained inferior to standard CoMFA ($q^2 = 0.80$) [30].

The SOMFA method of Robinson et al. [59] was also used on the same dataset [56]. SOMFA is a grid-based, alignment-dependent method. Molecules (aligned on E2) are embedded in a lattice of nodes (as in CoMFA). To each grid node are given values of a shape indicator (1 inside the van der Waals volume, 0 otherwise) and of electrostatic potential. But the important point is that, at every node, these values, for a given molecule, are multiplied by the mean centered activity for that molecule (so as to give less interest to molecules close to the mean activity). A QSAR relating a property (shape, polarizability) to activity is derived by MLR.

For the 36 compounds of Asikainen et al. [56], the best SOMFA model, obtained using only the molecular shape led to $q^2 = 0.76$ (SPRESS = 0.63). The (somewhat modest) results of SOMFA in this study, as in others, have been attributed to the built-in regression tool, SOR, that was shown to be equivalent to SIMPLS and NIPALS with one principal component (whereas more than one component is obviously necessary for mapping complex datasets) [60]. Replacing SOR by external tools as MCSOR (MultiComponent Self-Organizing Regression) or SIMPLS improved the performance that came comparable to that of CoMFA [36]. For the Sadler dataset, SOMFA ($q^2 = 0.698$) became better than receptor interaction energies ($q^2 = 0.570$) and comparable to basic CoMFA ($q^2 = 0.720$), but still lower than "sophisticated" CoMFA with region focusing ($q^2 = 0.796$) or GRID with receptor alignment and region focusing ($q^2 = 0.921$) [57,58]. But the authors stressed that some of these studies lacked extended validation.

In the CoSCoSA (Comparative Structural Connectivity Spectra Analysis) approach [61], structural descriptors are constituted by the ^{13}C-^{13}C COSY spots that may be considered as one-bond C-C fragments augmented by the chemical shift values (indirect information on their connectivity, hybridization, and environment). The model, applied to 130 compounds [18], led to $q^2 = 0.78$, $r^2 = 0.827$, comparable to CoMFA ($q^2 = 0.707$, $r^2 = 0.903$). Good results were also obtained with two external test sets. However, for about 50 compounds of the training set (weak binders), with no hit in any of the selected bins in the COSY map, activity was predicted constant, whereas experimental values spanned a range of about 4 log units.

Although appealing, these spectroscopic QSAR present some limitations: chirality (not considered) or symmetry (reducing the number of signals) may cause problems. From a practical point of view, recording experimental spectra suffers from several constraints and requires availability of the product. In view of general application, these methods are therefore largely dependent on reliable spectra simulation systems, covering a wide structural range. This is particularly true for CoSCoSA, due to the low natural abundance of the ^{13}C nucleus and its intrinsic low sensitivity, although ^{13}C-^{13}C COSY spectra can be partially reconstructed from more sensitive ^{1}H-^{1}H COSY and HETCOR ^{13}C-^{1}H sequences [62].

In Asikainen et al. [63], for 30 estradiol derivatives (from Napolitano et al. [64]), the performances of EVA and EEVA were compared to a classical Hansch-type relationship using molecular refraction (MR), hydrophobic parameter π, and an indicator (for the presence of a 16-α OH group) yielding $r^2 = 0.821$, but without any cross-validation [65], and to the MEDV-4 (molecular electronegativity distance vector) approach of Sun et al. [66]. The principle of MEDV-4 is to split nonhydrogen atoms into four categories according to their connectivity. Ten descriptors are calculated corresponding to the different pairs i, j (with $0 < i, j \leq 4$):

$$M_{ij} = \sum (q_i \, q_j)/d_{ij}^2$$

where q_i, q_j are the relative electronegativities, and d_{ij} is the (topological) interatomic distance.

Then an MLR is performed. For the quoted example, an optimized model (with only six terms) gave $q^2 = 0.747$ ($r^2 = 0.852$), but no external validation was performed.

11.3.3   RELATED LIGAND-BASED MODELS AND RECEPTOR-BASED MODELS

In connection with the superposition tool FLUFF, Korhonen et al. [27] proposed a new QSAR approach, BALL (boundless adaptive localized ligand), based on a local coordinate system. One advantage is that the ligands and the template can choose the best common conformation. To establish a QSAR model, BALL computes the similarity between the template (given as a logical molecule) and the molecules on both van der Waals and electrostatic volumes. A sparse localized grid tied to the template is created (with vertices placed at the atoms of the template, which allows for adaptation to conformational changes). Ligands are described as soft functions, Gaussian primitives, and electrostatic potentials (so that a "molecule does not end brutally but slowly fades away"). This coarse grid has been compared [27] to a sort of region focusing as in Cho and Tropsha [13]. Then usual statistical methods may be used. The scarcity of the grid points (and the "fuzziness" of molecules) does not allow for drawing contour plots. However, as the grid is tied to the template, the importance of each site of the template can be estimated.

In addition to a test on the "classical" steroid benchmark [27], the FLUFF-BALL software was tested on 245 xenoestrogen molecules from EDKB (Endocrine Disruptor

Knowledge Base: http://edkb.fda.gov) with five different ER (calf, human-α, human-β, mouse, rat) [26]. This dataset, amounting a total of 374 RBA values [51], recovered six diverse chemical subfamilies (biphenyls, phenols, other phenyl compounds, steroids, indolines, others). 17-β Estradiol was chosen as the template molecule (possibly with a preferential weight on its A-ring). The performance of different pairs associating a superimposition tool (FLUFF or SEAL) and a QSAR model (CoMFA or BALL) was investigated on the five datasets. In all cases but one, FLUFF-BALL worked the best, the last one being SEAL-CoMFA. "y-Scrambling" and numerous partitions (training/external validation) sets confirmed the robustness of the approach. As a typical example, with calf ER (53 compounds), the best q^2 obtained was 0.824 (FLUFF-BALL) compared to SEAL-CoMFA (0.117). FLUFF-CoMFA and SEAL-BALL gave intermediate values (0.530 and 0.223, respectively). However, these results contrast with the conclusions of Tong et al. [39] that gave for CoMFA, $q^2 = 0.60$. Similar remarks may be made for the other populations investigated.

CoMFA was also compared to the GRIND approach [28] on Raloxifene analogs (from [67]). With the GRID program [68], interaction fields are calculated with diverse probes (water, amide-nitrogen, carbonyl-oxygen) and positions where (for a certain ligand) interactions with a potential receptor are favorable ("virtual receptor site") are identified. The trick is that, rather than raw interaction values, correlograms, calculated from them, may be used as structural GRid INdependent Descriptors (GRIND) [28]. They implicitly convey 3D information, and alignment is no longer necessary. As an example, for 11 compounds (on an activity range of 2.7 log units), GRIND ($q^2 = 0.920$) outperformed CoMFA (0.584) and CoMSIA (0.835). Comparable results were obtained on an enlarged set (39 chemicals) [69].

Parallel to CoMFA or COMSIA methods, directly built from the ligands, in the receptor-based approach, docking programs allow for accurate calculation of the position and orientation of a ligand in a receptor binding domain. But it is still difficult to evaluate the free energy of binding with crude potential functions. The approach proposed by Sippl [57,58] takes advantage of a docking process with the receptor (either experimentally resolved or obtained by homology) to guide alignment before performing a 3D QSAR (here with GRID/GOLPE [29]). The study concerned 30 ER agonists from Sadler et al. [30]. The crystal structure ER-α (bound to E2 or DES) [70] was chosen as a model of the receptor LBD. Docking was performed with atom potentials calculated on a grid (as in [22]) originating from the receptor's supposed fix, but the ligand might change its conformation. Then ligands (in their so-determined geometry and orientation) were aligned and a QSAR was built, calculating ligand interaction energies with a water probe. The model led to $q^2 = 0.921$ and $r^2 = 0.992$ with 4 PC. Validation with a leave-20%-out process and three external test sets gave a similar performance. Another model of classical ligand-based alignment with flexible superposition on a template (E2), with no reference to a receptor structure, was slightly inferior: $q^2 = 0.851$, $r^2 = 0.971$.

However, using directly the interaction energies obtained with a crude potential function to correlate binding affinities led to bad results ($q^2 = 0.570$, $r^2 = 0.617$). This was not unexpected, presumably because of the neglect of solvation energy or entropic terms: their correct evaluation required heavy calculations (such as molecular dynamics, free energy perturbation, or thermodynamic integration) [25,71].

In the approach by Akahori et al. [72], the binding affinity was predicted from the energy changes upon binding. Interaction energies were calculated on a grid allowing flexibility for the ligand and reorientation of the neighboring side chains of the receptor. The originality of the model is that desolvation effects and loss of degrees of freedom of the ligand are explicitly taken into account. A two-step model was built, first to distinguish binders from nonbinders and then to predict binding affinity. For a dataset gathering alkylphenols, phthalates, diphenylethanes, and benzophenones, 87% good prediction (in LOO) was achieved to distinguish binders and nonbinders by LDA; and, in an MLR correlation, q^2 reached 0.75 (on a range of 5.7 log units). However, no good model was obtained for phthalates and benzophenones (presumably because of their low RBA values and their small range of variations).

The QUASAR approach of Vedani et al. was also applied to the 116 chemicals of Blair et al. [46] in 4D and 5D levels (see below). A receptor-mediated alignment protocol was used, based on the crystal structure of bound DES. r^2 reached 0.908 in training (93 compounds) and 0.907 in test (23 chemicals) [73]. For the 106 compounds of the ToxDataBase [10,74], q^2 reached 0.895 and r^2 0.892 at the 5D level. Other models were proposed separately for α and β receptors.

11.4 BINDING THE ANDROGEN RECEPTOR

Binding the androgen receptor deserved less numerous papers than ER binding.

In a pioneering paper, Loughney and Schwender [75] examined by CoMFA the binding affinity for both the androgen receptor (AR) and the progesterone receptor (PR) of 41 steroids. For AR, on a range of 2.5 log units (data from Delettré et al. [76]), r^2 reached 0.919, with nine principal components. But q^2 was only 0.545 (using RBA values), not very far from the limit admitted for a significant model. Electrostatic contribution was clearly predominant (74%).

A few years later, Waller et al. [14] considered 28 natural and synthetic AR binders studied in competitive assays with [^{3}H] R1881. In order to examine the potential utility of CoMFA for structurally diverse chemical families, the set was composed of 9 steroids, 2 synthetic steroids, and 10 diverse compounds, including DES, DDT, and kepone. The A-ring of DHT was chosen as the template for field-fit alignment. A two field model (electrostatic plus steric), although slightly less internally consistent than a model with the electrostatic field only, was preferred for its superior external predictive ability. On 21 chemicals in training, q^2 amounted 0.792 (SEP = 1.01) and $r^2 = 0.989$ ($s = 0.24$) with 4 PC. For the external test set (7 compounds), the average absolute error was 0.58 log unit (max error of 1.7 log units).

Binding the androgen receptor has been revisited by Hong et al. [77] on an extended set of 146 chemicals (from the data of Fang et al. [78] covering a range of 6 log RBA units, and determined in competitive assays with [^{3}H] R1881 on a recombinant rat AR ligand binding domain protein). Alignment was performed using as template R1881 (one of the highest affinity ligands, with a rigid structure, and a bound conformation with AR LBD known from X-ray crystallography [79]). The CoMFA model had $r^2 = 0.902$ ($q^2 = 0.571$) with a standard error of estimate (SEE) of 0.39. The contribution of the electrostatic field (48%) is similar to that noted for binding to the estrogen receptor [18,39,44]. On a test set of eight chemicals (from Waller et al. [14]),

CoMFA predictions (after conversion to the scale of Waller, via a linear relationship) led to an average absolute error of 0.63 log RBA unit.

CoMFA contour plots indicated that binding was favored by negative charges around the 3-keto and the 17β-OH groups, suggesting H bonding with AR or electrostatic interaction between negative charges of the ligand and positive charges of residues in the receptor. Consistent with these observations, the crystal structure of the complex R1881-hAR LBD [79] suggested two H-bonds of the 17β-OH with Asn705 and Thr 877 and, at the other end, an H-bond of the 3-keto group with Arg752 and Gln 711.

Regions where more bulky groups were predicted as favoring binding extend near C1, above C17, and on a smaller area under C3. They corresponded to empty regions of the LBD, where an increase in van der Waals interactions would increase binding affinity.

Nearly the same dataset (excluding 28 compounds retained for the test set) was considered by Zhao et al. [80] in various QSAR treatments. With five descriptors, among the hundreds generated by CODESSA, the best result was obtained with a support vector machine (SVM) yielding $q^2 = 0.76$.

CoMFA was proposed as a rationale for the synthesis of high-affinity AR ligands, in the search for SARMs [7]. Nonsteroidal SARMs may offer many advantages over conventional steroid therapies and their drawbacks (lack of selectivity due to crossreactivity with other steroid receptors, poor oral bioavailability, and side effects). Considered were 122 ligands, mainly nonsteroidal compounds, such as bicalutamide and hydroxyflutamide analogs, on a range of 5.4 log RBA units. For alignment, a model of hAR LBD was built by homology with the human progesterone receptor [19]. Six representative templates (one per each chemical family investigated) were first docked into the receptor, and their AR bound-conformation constituted secondary references to fit their analogs, using common atoms as alignment points. This docking methodology, comparable to that of Marhefka et al. [19], is more specific to the binding of nonsteroidal ligands but significantly differs from Waller's approach, based on the A-ring of DHT. It was quoted that crystal structure may give inaccurate models due to crystal packing effects [7]. Furthermore, two-ring ligands, such as bicalutamide, adopted a conformation very different from the MOPAC optimization. The CoMFA treatment, including indicator- and H-bonded fields, representing the acceptor and donor components [33,34] gave $q^2 = 0.593$, $r^2 = 0.974$ (with RMSE of 0.74 and 0.26, respectively) excluding corticosterone. The electrostatic component was slightly predominant (61%). For the test set (10 compounds), the determination coefficient r^2 was 0.953 (RMSE = 0.34).

CoMFA contours were consistent with contact sites for important residues identified in mutational analysis and crystal structure. A common region of favorable negative charge corresponds to the para position of the A-ring of bicalutamide and hydroxyflutamide and the 3-keto group of DHT (although according to Waller, the A-ring of steroids lies in a different region). This observation was supported by the location of residues Arg752 and Gln711 that act as H-donors (in crystal) to the 3-keto group of R1881 [79] and to the 4-NO2 of hydroxyflutamide in the homology model of Marhefka [19], and was confirmed by the loss of activity resulting from a mutation of Arg752 [79]. On the other hand, hydrophobic interactions (possibly with Val746 and Met742) favor binding by overlap with the 3-CF3 group of bicalutamide, the A-ring of steroids (carbons 5, 6, 7) or the methyl group of tricyclic quinolinones. The chiral OH group of bicalutamide analogs interacts with Asn705 in a region that also overlaps the 17-OH of steroids. This interaction is essential because enantiomers devoid of such interaction exhibit a weaker affinity.

A region disfavoring bulky substituents appears near the linkage group of bicalutamide derivatives, similar to that near C-17 of steroids. This is an important region discriminating AR binders from estrogens or progestagens bearing bulky C17-substituents. An additional subpocket of the LBD bordering the B-ring is also highlighted thanks to the use of bicalutamide. A difference was noted between the homology model and the crystal structure, where Met780 would forbid docking, due to a closer packing in that place than that indicated in the homology model [19,79].

In a similar concern for SARMs, Söderholm et al. [23] carried out a CoMSIA treatment on 70 widely diverse AR binding compounds with only three steroids: derivatives of flutamide, nilutamide, and bicalutamide, corresponding to diverse scaffolds.

Docking the ligands in the crystal structure of a mutated AR LB protein complexed with a steroidal agonist (9α-fluorocortisol) [81] led, for nonsteroidal compounds, to an unrealistic "sandwich" structure. To maintain an essential H-bonding pattern (with Arg752, at the 3-keto group and Asn705 and Thr877 at the steroidal OH [82]), the authors resorted to a mutated LBP structure giving room enough for bulky nonsteroidal ligands, biased docking using DHT as a guide, and manually adjusted some compounds. Their CoMSIA model (excluding six compounds, due to problems in modeling or the presence of singular substituents) gave, with only hydrophobic (51%) and H-bond acceptor fields, $q^2 = 0.656$, $r^2 = 0.911$ with SDEP = 0.58 and SEE = 0.29 for 55 chemicals. For an external set of 9 compounds, the predictive r^2 was 0.800 with SEE = 0.37. Robustness of the model was also checked on several leave-some-out cross-validations and examination of newly synthesized compounds with similar performance.

11.4.1 LIGAND-INDUCED FIT

In the docking techniques of Bohl et al. [7] or Söderholm et al. [23], the receptor was kept rigid. However, experimental evidence [83] showed that AR (like PR) could accommodate larger compounds than endogenous ligands, suggesting some ligand-induced fit avoiding steric clashes. In usual 3D QSAR models, in the absence of a "true biological" receptor, the receptor is featured by means of a grid or a van der Waals surface built from all the aligned ligands, with properties mapped on it. This surface may be viewed as a mirror of the binding site, but it only constitutes something like an "averaged surrogate" of the true receptor [73,83,84]. In fact, this mean envelope adjusts its geometry to each ligand (ligand-induced receptor fit). This results in changes in the fields it generates. For residues bearing a conformationally flexible H-bond donor- or acceptor-group a flip-flop change may even occur.

After RAPTOR [73], the QUASAR [84] approach provides a multiple representation of the ligand topology, conformation, and orientation. (Adding additional information on the 3D shape is known as "4D QSAR.") The model also allows for characterization of the ligand-induced receptor fit: a problem relying on the "fifth dimension" in QSAR.

Flexible docking (with Monte Carlo search and energy minimization) is first used for selecting the favored binding modes, and a molecular dynamics simulation is carried out to generate an ensemble of configurations for each binding mode retained (at least for a representative of each chemical family). Then each ligand–receptor complex is refined. Alignment is carried out and local induced-fit is simulated by mapping the "mean" envelope on a transiently generated envelope that closely accommodates the individual ligand. Points on the receptor surface are populated with atomic properties. The RMS deviation between the envelopes gives an estimate of the energy cost. Different processes may be considered where adaptation for each ligand is governed on the basis, for example, of steric, electrostatic, H-bonding fields. Estimation of the relative free energy of binding includes terms corresponding to ligand–receptor interactions, changes (upon binding) in internal energy of the ligand, ligand desolvation, variation of ligand entropy, plus the adaptation of the receptor envelope. Various simulation protocols may be used to generate an ensemble of models (up to several hundreds) that represent the various configurational states of the true receptor. Individual models are then averaged with Boltzmann criterion to give the binding affinity.

This strategy reduces the bias associated with the choice of the bioactive conformation, of ligand alignment and induced fit. However, it supposes some guess-work about the mechanism of adaptation. The 5D QSARs introduce the possibility to consider, at a time, different adaptation protocols [10,74]. This new development of QUASAR allows for simultaneous consideration of a family of models, corresponding to different protocols with possible dynamic interchange between them. A genetic algorithm controls the evolution of the system until convergence to a single induced-fit model. QUASAR was applied to evaluate affinity to diverse endocrine disruptors LBD [10,85]. For each receptor, details are given in the corresponding paragraphs of this review.

As to AR binding, a dataset of 119 chemicals (measured in a competitive binding assay with recombinant rat protein [78] and covering a range of 5.5 in log RBA) was split into a training (88) and a test (31) set [83]. After alignment, guided by the DHT-AR LBD complex [84], ten receptor models were built, giving an averaged r^2 of 0.858 in training and a predictive r^2 of 0.792 (26 test compounds). Similar results were obtained in the 5D approach of Vedani et al. [10] (ToxDataBase).

11.4.2 AR BINDING SITE

Although not directly involved in the derivation of CoMFA models, additional information about the human AR binding site was gained with the homology model of TES bound to hAR [19], built on the basis of crystal structure of human PR LBD and compared to the crystal structure hAR-R1881 [79]. Refinement by a molecular dynamics simulation (with explicit solvent) also afforded some insights into the dynamic behavior of the protein–ligand complex.

Key features are a stable H-bond of the 17-OH with Asn705, Thr877, and a more dynamic one (presumably direct rather than water mediated) between the 3-keto group of TES and Arg752 and mainly Gln711, although this is later not seen in the crystal structure of hAR-R1881. Binding to Asn705 is important in binding hydroxyflutamide derivatives (by H bonding with their α-hydroxy group). Enantioselectivity of chiral derivatives is also interpreted by the lack of H-bonds with Thr877 and steric interactions. Several mutation studies are in agreement with the model. It was also suggested that for nonsteroidal ligands, it is necessary to have an aromatic ring, not necessarily electron deficient, but bearing an H- bond acceptor group in position 4 (to mimic the 3- keto group of TES), the presence of a hydrophobic group in position 3 being also favorable to binding.

11.4.3 AGONISTS VERSUS ANTAGONISTS

Tamura et al. [2] have stressed that several recent publications on AR binding were based on rat AR competitive binding assays. Although it is useful to predict whether a chemical binds or not to the AR, these assays could not distinguish between agonists and antagonists, an ambiguity that made these models difficult to adapt for rational drug design. From *in vitro* reporter gene assays, the authors carried out CoMFA analyses on 49 agonists/antagonists. The optimized conformation of testosterone in AR LBD [19] was chosen for alignment, with flutamide and bisphenol A as secondary references. For the whole set, $q^2 = 0.446$, $r^2 = 0.944$ with SEP = 0.644 and SEE = 0.21 (5 PC).

Contour plots revealed characters consistent with some common patterns of steric and electronic features involved in AR binding affinity [7,23,77,83]: ligands have strong H-bonding or electrostatic interaction ability in a position comparable to that of the 3-keto group of DHT, hydrophobic substituents (CF3, CH3) in position 3 increasing activity. Agonists must have an H-bond acceptor or donor in a position corresponding to C-17 of DHT, which implies a distance of about 10 Å to insure these two interactions (in agreement with a previous conclusion of the authors [86]). More interesting, sterically unfavorable regions appear near Asn705, Arg752, Thr877 in AR LBD for pure antagonists. Ligands with a bulky group interact with these residues, resulting in a transcriptionally inactive form of the AR–ligand complex.

It may be noted that residues involved in hydrophobic interactions differ between this model and that of Bohl et al. [7]. The difference comes from the fact that Bohl et al. used for docking a hAR-LBD based on homology to the human progesterone receptor, whereas for Tamura et al. it was the natural crystal structure hAR-LBD.

Some ER agonists (DES, bisphenol A) act as pure AR antagonists. This may be a consequence of a smaller size of the AR LBD compared to ER LBD. For example, DES is well accommodated in the ER-LBD, whereas its two ethyl groups, protruding out the AR LBD in a sterically forbidden region (suggested also in CoMFA) near Asn705 and Thr877, cause AR antagonist activity.

11.5 BINDING THE PROGESTERONE RECEPTOR

QSAR on progestagens (compounds mimicking progesterone) are less numerous, despite the interest these chemicals arouse for the development of contraceptives or drugs against certain cancers or gynecological disorders. In a pioneering work, Loughney and Schwender [75] investigated the binding affinity of 48 steroids to AR and PR from CoMFA (see above). As to progesterone receptor, the best model gave a $q^2 = 0.725$, with 7 components for the relative binding affinity, but, with only 3 PC,

q^2 was still 0.601, a fairly good result. Introduction of supplementary descriptors (dipole moment, log P or molecular refractivity) did not improve the model. However, using log RBA (on 2.7 log units), q^2 was only 0.48. A comparison between the androgen and progesterone receptors was also carried out from the CoMFA contour plots (see Section 11.6).

Bursi et al. [55] compared CoMFA to the CoSA approach on a set of 45 steroidal progestagens, on a range of 2.8 log units in RBA (data from [87]). Experimental or simulated spectra were considered (depending on the technique). According to the authors, spectral descriptors led to better performance than CoMFA, either with rigid alignment or field-fit option, both for training and test sets. In training, the best CoMFA led to $q^2 = 0.550$, $r^2 = 0.871$, whereas EVA (simulated IR spectra) yielded $q^2 = 0.638$, $r^2 = 0.986$, and similar results were obtained from the mass spectra. However predictions were mediocre for ^{13}C (simulated) or IR (experimental). Interestingly, combining spectral descriptors with others (spectral or molecular fields) generally provided better models than using individual descriptors: (CoMFA + ^{1}Hexp), (IRsim. + ^{1}Hexp), or (^{1}Hexp + ^{13}Csim) gave nearly similar results (q^2 about 0.62 or better), s about 0.46, r^2, SEE better than 0.980, 0.10 and for the test, RMSE about 0.49.

Another population of 33 steroids (on a range of 2.8 log units for RBA values) was investigated by Chen et al. [24], taking advantage of the known crystal structure of a complex PR-ligand (PDB code 1A28).

The CoMFA and CoMSIA models were built using the "docking-guided conformation selection" (see Section 11.1). Another model, using the "all-orientation all-placement searching" [31] slightly improved the results for CoMFA yielding to $q^2 = 0.773$ ($r^2 = 0.997$). An oxygen probe was also used with similar conclusions (and a slightly enhanced electrostatic contribution). For the test set (7 compounds), $r^2 = 0.995$ (CoMFA) and 0.951 (CoMSIA) with SEE = 0.06 and 0.16. It can also be noted that in the CoMSIA model, the H-bond fields intervene about 40%.

Docking scheme and CoMFA or CoMSIA contour plots are consistent with the results of Tanenbaum et al. [88] on the structure of progesterone complex of hPR LBD. Contour plots indicate favored electrostatic interactions near the 3-keto group, in agreement with the strong H bonds of this group with Gln725 and Arg766. Near position-11, a deep cavity of the receptor (near Glu723) can accommodate medium-chain substituents via hydrophobic interactions. Around position 17, a wide, shallow cavity may accept small substituents, whereas more bulky groups are detrimental.

A set of 56 steroids known to bind PR *in vitro* on a 2.8 log units in RBA was the subject of a comparative study involving varied approaches [89–91]. The 52 structural descriptors encoded 43 characteristics of substituents (or individual atoms): charge, volume, surface, plus nine whole-molecule properties: dipole moment, HOMO or LUMO energy, heat of formation, and so forth (from AM1 calculations). In a preliminary study, CoMFA outperformed a 10-10-1 back-propagation neural network (BNN) (that was clearly overfitted) [89]. In subsequent publications, diverse methods for descriptor selection were compared, from the same dataset. The best results were obtained with a genetic function approximation (GFA) regression and an artificial neural network (ANN) used for descriptor selection and model building, respectively. The r^2 values for the training ($n = 43$) and testing ($n = 11$) sets were equal to 0.64 and 0.49, respectively. Using the ten most frequently used descriptors in a 10/5/1 ANN led to slightly improved results (respectively, 0.880 and 0.570) [90]. Comparison was pursued [91] on the same dataset, but with only eight descriptors to maintain a 6/1 ratio between the number of patterns and descriptors. We examined subjective selection from known models, forward stepping regression, GFA regression, where a genetic algorithm selected the descriptors' input in a nonlinear regression (with, *inter alia*, splines) and two models where data were mapped from a neural network, and descriptors selected either by simulating annealing, in generalized simulated annealing (GSA) or by a genetic algorithm (GA), in genetic neural network (GNN). It was then established [91] that an 8-2-1 BNN combined to GSA or GA gave the best results, respectively, $q^2 = 0.626$ and 0.635 and $r^2 = 0.722$, 0.760. The more limited performance of GFA may result from its difficulty to treat nonlinearity. However, GFA ($q^2 = 0.626$, $r^2 = 0.722$) or stepwise regression ($q^2 = 0.535$, $r^2 = 0.721$) may be useful in preliminary studies because it is easier to implement. On the other hand, GAs often give several models of comparable quality, making the interpretation more difficult. It may be noticed that, here, 5 descriptors were common in the selection by GSA and GFA, and that half of the descriptors retained concerned the "key" positions 11, 13, 17 of the steroid scaffold. Niculescu and Kaiser [92] compared these results to the performance of a probabilistic neural network (PNN) with the same descriptors. On learning, PNN ($r^2 = 0.863$) was comparable to GSA ($r^2 = 0.860$) and slightly lower than GNN ($r^2 = 0.880$) but a little better in test (0.769 versus 0.488 and 0.610 on 10 compounds).

11.6 BINDING THE ARYL HYDROCARBON RECEPTOR

Halogenated aromatic hydrocarbons represent an important group of largely widespread and persistent contaminants (particularly owing to their liphophilic character). Modeling their toxic effects generally concerns their affinity to an intracellular cytosolic protein (the aryl hydrocarbon receptor [AhR]) [93]. It is generally expressed as pIC_{50} (the negative logarithm of the molecular concentration necessary to displace 50% of radiolabeled 2,3,7,8-tetrachlorodibenzo-*p*-dioxin (TCDD) from the Ah receptor. Data are also available on induction of aryl hydrocarbon hydroxylase (AHH) and 7-ethoxyresorufin-*O*-deethylase (EROD), but they seem more difficult to rationalize because they are related to the much more complicated biological response of enzyme induction [93,94]. Another important remark is that the structure of the receptor is still unknown and homology models are difficult to build (poor sequence identity), at the level of difference of the cases just discussed.

Ambiguity due to symmetry is reduced thanks to International Union of Pure and Applied Chemistry (IUPAC) nomenclature convention. But the possibility of multiple binding modes must also be considered (see below).

A largely investigated dataset [94] concerns the binding affinity to cytosolic AhR for 78 halogenated aromatic hydrocarbons (25 polyhalogenated dibenzo-*p*-dioxins PCDDs, 39 dibenzofurans PCDFs, and 14 chlorobiphenyls PCBs) on a range of 6.3 log units. After alignment on TCDD, CoMFA treatments, carried out for every family and for the whole set, yielded similar performance: for the whole set, $q^2 = 0.724$, with 6 PC and $r^2 = 0.878$, SEE = 0.53 [94]. It was also noted that fitting fields would force biphenyls to an unrealistic planar geometry. In a subsequent study [95], biphenyls were removed from the training set and used as a test set to evaluate the predictive ability of the model.

The same dataset was revisited [96], using as descriptors autocorrelation vectors of the hydrophobicity potential for points randomly distributed (with a preset density) on the van der Waals surface. With 12 autocorrelation coefficients, q^2 reached 0.83 ($r^2 = 0.89$). The method needs no alignment, but the drawback is that the original information cannot be retrieved from the autocorrelation vectors.

On the same set, a PLS treatment was carried out with WHIM (weighted holistic invariant molecular) descriptors [97] that encode size, shape, and electrostatic distribution. Basically, six molecular properties were calculated on points scattered on the molecular surface: unitary (referring to atomic positions), molecular electrostatic potential (positive or negative), and with binary values, H-bond-acceptor, respectively, -donor capacity, and hydrophobicity. For each property, three principal component axes were defined. Molecular descriptors were the values (on these axes) of the variance, variance proportion, skewness, and kurtosis, that respectively correspond to size, shape, symmetry, and distribution. This amounted to $6 \times 3 \times 4 = 72$ descriptors per molecule. WHIM descriptors (invariant in geometrical transformations) do not require alignment. Unfortunately, no contour plots are available. Interpretation is difficult because information is global and not local as in field analysis approaches. The best selected model retained 4 properties (48 descriptors), including H-bonding capacity and electrostatic potential, and led to $q^2 = 0.732$, RMSE $= 0.76$; $r^2 = 0.794$; RMSE $= 0.66$, with two components. Here, q^2 was calculated not on LOO but on a less optimistic process: the mean of 100 leave-20%-out.

Affinity to AhR receptor was also approached by So and Karplus [98] with similarity matrices, on the same set, after elimination of redundancies and incorrect values (73 polyhalogenated aromatics). After alignment (as in [94]), every molecule was characterized by its similarity to the others in the set, both at size and electrostatic level. For two molecules A, B the electrostatic similarity index was calculated by the Hodgkin formula:

$$\text{Hab} = 2 \sum p_A \, p_B \Big/ \left(\sum p_A^2 + \sum p_B^2 \right)$$

where p is the electrostatic potential calculated at regular grid points. Similarly, the shape index was:

$$S_{AB} = U_{AB} / (T_A T_B)^{0.5}$$

counting grid points inside the volume of A and B ($T_A T_B$) and in the union (A U B), U_{AB}.

A genetic neural network (GNN) fed with these descriptors gave $q^2 = 0.72$ with the electrostatic index, and 0.85 with the shape index ($r^2 = 0.89$ with 6 descriptors), consistent with the limited importance of electrostatic effects in that series. Considering both indices at a time gave no improvement. These results favorably compete with those of Waller et al. [94] $q^2 = 0.72$ and Wagener et al. [96] 0.83. The authors suggested that evaluating similarity eliminates the (very critical) phase of alignment. Because contour plots (as in CoMFA) are not available, considering subgrids as in [13] might indicate the most important regions.

The spectroscopic QSAR technique EEVA, a modification of the EVA approach [53], using MO energies in place of vibrational frequencies, was applied to the same initial set (but 5 benzofurans) with $q^2 = 0.818$ (8 PC), SPress = 0.69, $r^2 = 0.912$, SEE = 0.48 [99]. The model was then used to predict all possibly existent PCDDs and PCDFs (210 compounds in all). The authors suggested that alignment-free EEVA is well adapted to reflect electronic substituent effects, but there is a loss in efficiency for nonplanar structures where these effects are partially hindered (as in PCBs).

This basic set was extended to naphthalenes and indol [3, 2-bicarbazoles] amounting to 95 chemicals (6.3 log units) [93]. As previously [94], alignment was carried out on the lateral positions of TCDD (and atoms 3, 4, 3′, 4′ for PCBs). For all compounds, $q^2 = 0.574$ ($s = 0.98$); $r^2 = 0.831$ ($s = 0.62$); with 41% contribution for electrostatic field. Discarding 9 outliers raised q^2 to 0.631 and r^2 to 0.90. CoMSIA (with all fields) gave $q^2 = 0.711$, $r^2 = 0.873$, and $s = 0.58$, but similar results were obtained with only a hydrophobic field combined with either steric or electrostatic fields or both.

In the study of Lo Piparo et al. [100], the affinity of 93 aromatics (from the 95 of Waller and McKinney [95]) to the AhR receptor was approached using CoMFA, Volsurf [101], and HQSAR. Volsurf calculates energetically favorable interaction sites to produce a 3D grid map, thereafter compressed in a few 2D descriptors. With 5 probes (water, hydrophobic, carbonyl oxygen, carboxy oxygen, amphipathic), 118 descriptors were generated characterizing molecular weight, volume, surface, size of the hydrophilic and hydrophobic regions, H-bond properties, local interaction energy minima, and integy moment vectors (vectors pointing from the center of mass to the centers of hydrophilic and hydrophobic regions). log P was added as a crude measure of desolvation energy and bioavailability (in HQSAR). A binary indicator for torsionally constrained compounds (intervening about 5%) was introduced in the CoMFA and HQSAR treatments. CoMFa and HQSAR gave similar results ($q^2 = 0.62$), whereas Volsurf was slightly inferior. But a hybrid model, combining Volsurf and HQSAR, gave the best q^2 (0.70), although r^2 was not improved. The test set encompassed 9 compounds (on the upper half of the affinity range). The r^2 (CoMFA) value was only 0.42 in place of 0.69 to 0.72 for the other methods. In this example, the authors stressed the good results obtained with Volsurf, which is alignment free, and HQSAR, which works on 2D descriptors only, two methods less intensive than CoMFA.

CoMFA contour plots published in the literature [93,94,100], although seemingly slightly different, present some convergent observations: Steric bulk is beneficial on lateral positions, 2, 3, (7), 8, but detrimental on medial positions (on C9 and between carbons 4 and 6), or (in CoMSIA maps) on medial positions of the second ring.

As to electrostatic field, a negative charge is favored near positions 2, 3 of TCDD and 3, 4 in biphenyls (increased activity with halogen atoms on these positions). Positive charge is favorable near C4 and the neighboring oxygen (naphthalene and biphenyl, without O atom in medial, show some activity) [93].

Activity is favored by lateral hydrophobic substitution on the first ring (that has greater impact), whereas hydrophilic groups may be accommodated in the medial region and above the second ring. H bonding interactions from this ring would also increase activity.

However, it was suggested that the favorable presence of lateral halogens might reflect polarizability effects rather than steric influences [94]. Consistent with CoMFA, Volsurf indicated increased activity with the presence of high hydrophilic regions, and with the delocalization of the hydrophobicity in few areas of the molecular surface [100].

More limited because different series were investigated. A reduced set of 29 compounds (7 PCDDs, 10 PCDFs, 12 PCBs) was examined by Koyano et al. [102]. Several original points in his CoMFA model may be noted: Geometry optimization was carried out at a high level (HF/6-31*), and CHelpG atomic charges were fitted to the (exact) electrostatic potential. Furthermore, alignment used as template 1,2,3,7,8-PeCDD rather than the symmetrical 2,3,7,8-TCDD, and largely twisted PCBs were extended from one side of PCDD to the other. With three components, q^2 equaled 0.955 on a training set of 18 chemicals. This high value was attributed to the alignment rule and sophisticated quantum calculations. However, the dataset was relatively limited. Prediction (for new compounds) was fairly good for low affinity chemicals, whereas the very reactive ones were largely underestimated.

Wang et al. [103] examined, by CoMFA and CoMSIA, the binding affinity of 18 polybrominated diphenyl ethers (PBDEs) from Chen et al. [104]. Such compounds are widely distributed as they are used as flame retardants. Molecules were aligned (with flexible fitting) on the template 2,2′,3,4,4′-bromo-diphenyl ether (the most active compound). On an affinity range of 2.9 log units, CoMFA led to $q^2 = 0.580$ ($r^2 = 0.995$) with 6 PCs, and electrostatic field contributing to 69%. With CoMSIA (steric + electrostatic + hydrophobic fields), $q^2 = 0.680$ ($r^2 = 0.982$). These models were used to predict the RBA values of other 46 PBDEs, on account of their possible role in environmental pollution, but, for lack of experimental values, external predictivity could not be checked. For this population, contour plots are presented in Figure 11.4 and the color insert.

Although fairly similar, maps are easier to interpret for CoMSIA than CoMFA. Bulky substituents are beneficial near positions 2,5,5′,6, but they are detrimental near positions 3, 5, 4′. As to the electrostatic field, activity would increase with negatively charged groups above ring B, whereas positively charged groups are beneficial in 3,5,2′,3′,4′. Similarly, hydrophilic substituents at 3,4,5,4′ increase activity, as do hydrophobic groups in 2 or near ring B. See also Henry et al. [105] for 11 flavones binding AhR.

The quantum topological molecular similarity method (QTMS) [106] was also used. QTMS relies on the theory of atoms in molecules (AIM) in which an atom system is partitioned into atomic constituents on the basis of electron density (ρ). In this approach, a bond is described by five components, evaluated at bond critical points ($\nabla\rho = 0$): electron density, density Laplacian $\nabla^2\rho$, ellipticity (null for a cylindrically symmetrical bond), kinetic energy density, plus the equilibrium bond length. These descriptors were used in a PLS treatment for correlating the pIC_{50} values of 13 PCDDs. At the sophisticated HF/3-21G(d)//HF/3-21G(d) level, bond critical points descriptors led to $q^2 = 0.57$ ($r^2 = 0.74$). But surprisingly, using only bond lengths (calculated at AM1 level), q^2 reached 0.89 ($r^2 = 0.84$).

Localizing "active centers" also evidenced the importance of halogen substitution on lateral positions.

At the 4D and 5D levels (see above), the QUASAR approach was applied to 121 chemicals [10,74,85,107]. Due to (partial) symmetry, for each compound, 4 orientations were considered that cannot be trivially deduced from 2D or 3D formulae; this multiple representation reduces the bias. In training (91 compounds), $q^2 = 0.857$ (4D) or 0.838 (5D), whereas for the test set (31 compounds), $r^2 = 0.795$ versus 0.832 in 5D. More recent results in the ToxDataBase indicated that for 140 compounds

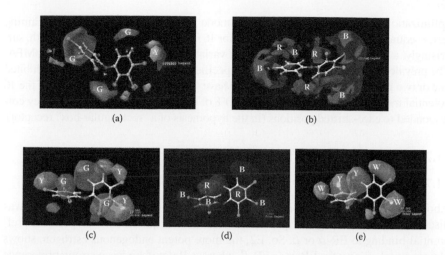

(a) (b)

(c) (d) (e)

FIGURE 11.4 Comparative molecular field analysis (CoMFA) and comparative molecular similarity analysis (CoMSIA) contour maps. 2,2′,3,4,4′-bromodiphenylether is shown only as reference structure. (a) CoMFA steric contour map. Green contours (G on black and white figure) indicate regions where a relatively bulky substitution would increase the binding affinity, whereas yellow contours (Y) indicate areas where a bulkier substituent would decrease the binding affinity. (b) CoMFA electrostatic contour map. Red contours (R) indicate regions where a negative-charged substitution will increase affinity, whereas blue contours (B) show areas where a negative-charged substitution would decrease affinity. (c) CoMSIA steric contour map and (d) CoMSIA electrostatic contour map: same graphic convention, as for CoMFA. Greater values of the binding affinity are correlated with more bulk near green or less bulk near yellow, and similarly, more positive charge near blue and more negative charge near red. (e) CoMSIA hydrophobic contour map. Yellow (Y) contours indicate region where hydrophobic group will increase affinity, whereas white contours (W) show areas where a hydrophilic group favors affinity. (Reproduced with permission, from Y. Wang, H. Liu, C. Zhao, H. Liu, Z. Cai, and J. Jiang, *Environ. Sci. Technol.*, 35, 4961, 2005.) **(See color insert following page 244.)**

$q^2 = 0.824$ (in training on 105 molecules and a four-group cross-validation for which $r^2 = 0.769$) [10].

Lukacova and Balaz [108] considered the possibility of multimode binding of PCDFs in the receptor site. Such a situation may occur when ligands can bind in single but different modes or when a ligand binds in an unknown unique mode among several that are *a priori* plausible. The approach is similar to that of competitive complexation reactions from a single substrate, where the global association constant is the sum of the individual constants. This competition makes the relationship between the binding energy and the probe interaction energies nonlinear. But a linearized form (from a Taylor expansion) may be iteratively solved until self consistency is achieved.

For 34 polychlorodibenzofurans (PCDFs), various binding modes (2, 4, or 16) were investigated, corresponding from an alignment of medial atoms of PCDF (atoms 2,4,6,9) and TCDD to flip- (left/right or up/down) and shift-motions toward Cl7, Cl8 and H1, H9 (for better overlap on the edges). Then GOLPE selected the variables before PLS optimization. Assuming a unique binding mode for the 22 compounds of the training test, r^2 equaled 0.963 ($q^2 = 0.786$); whereas for 16 modes, $q^2 = 0.961$, $r^2 = 0.999$ with, surprisingly, a significantly limited number of variables as compared to "usual" CoMFA. The prevalence of the binding modes was specified for each compound. Most exhibited one or two modes, but 1,2,3,7-PCDF would have 4 significant binding modes. Of the 16 potential modes, three represented about 60% of the whole. And surprisingly, they corresponded to edge-shifted positions (in the hypothesis of a "rectangular-box" receptor).

11.7 COMPARISON BETWEEN RECEPTORS

11.7.1 DIFFERENTIATION OF SUBTYPES ER-α AND ER-β

About 10 years after the cloning of ER-α, another subtype (receptor ER-β) was discovered [109], raising the question of structural requirements for possible preferential binding to ER-α or β. So, E2, the more potent endogenous estrogen, shows almost equal affinity for ER-α or ER-β, whereas Raloxifene has a strong preference for ER-α and Genistein (an isoflavone abundant in soy products) for ER-β. Affinity to both human ER-α and rat ER-β subtypes was investigated by Tong et al. [110] for 31 chemicals (19 steroids, synthetic estrogens, phytoestrogens, and two environmental estrogens) with the alignment rules of Tong et al. [39]. The two series displayed similar results, with $q^2 = 0.70$ (human ER-α) and 0.60 (rat ER-β). Contour plots showed a very close similarity as to (for example) the influence of position 3 and 17-α of steroids. However, a minute difference appeared near the 17α-position.

The bulk-favored region was more extended for ER-α, indicating (in agreement with experiment) that steric bulk enhanced affinity more for ER-α than for ER-β.

Recently, Zhu et al. [111] examined the binding affinity in a competitive assay with [^{3}H] E2 on recombinant human ER-α and ER-β proteins. The dataset encompassed 74 natural or synthetic estrogens with more than 50 analogs of E2 and estrone E, and including mainly metabolites endogenously formed from the diverse rings. The template was E2 bound to ER-α [70]. CoMFA treatments on two series of 47 compounds (46 being common for the two sets) led to similar results (with $q^2 = 0.531$ and 0.634).

Contour maps for ER-α and ER-β (Figure 11.5 and the color insert) show common features: bulky groups near C16, C17 tend to enhance affinity, as do a negative charge near C2 region of the A-ring. But sizable differences appear: introduction of a polar group (OH for example) near C17 and C16 modifies the binding preferences, possibly related to the change of residue Met421 (in ER-α) to hydrophobic Ile373 (in ER-β). Also, increase of steric bulk near C2 favors binding to ER-β. It is interesting to note that E2, although the more potent endogenous estrogen (with almost equal affinity for ER-α or ER-β), is not the major constituent in the organism due to its easy conversion to E1 (Estrone) favoring ER-α activation or to E3 (16α-hydroxyestradiol or Estriol), predominant during pregnancy, favoring ER-β activation. Similar studies may guide the design of selective ER-β ligands [112].

11.7.2 COMPARISON PR VERSUS AR AND PR VERSUS ER-α

Comparison between AR and PR was initially approached by Ojasoo et al. [76] from homology building. In the two structures, empty space was found around 3-keto,

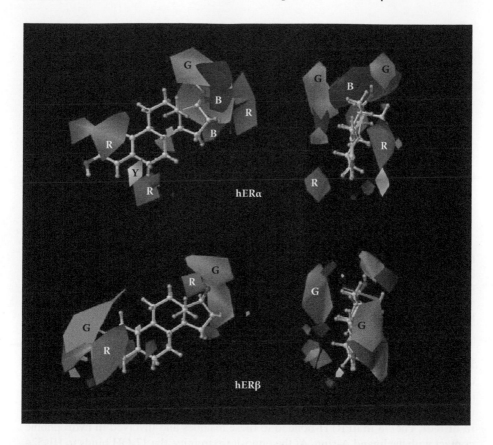

FIGURE 11.5 Comparative molecular field analysis (CoMFA) contour plots for human ERα and ERβ. E$_2$ is shown only as reference structure. Color code is the same as in Figure 11.4. (Reproduced with permission from T. Zhu, G.Z. Han, J.Y. Shim, Y. Wen, and X.R. Jiang, *Endocrinology*, 147, 4132, 2006. Copyright 2006, The Endocrine Society.) **(See color insert.)**

7-α, 11-β, and, for PR, C21. In the study of Loughney and Schwender [75], the better results obtained with PR compared to AR suggested possible differences in receptor selectivity, which can be approached from the CoMFA contour plots. For both receptors, steric contours indicate a volume detrimental to activity near the A-ring. Its larger extension for PR may indicate a greater sensitivity to steric bulk. Differences are more evident in the region of C17-α, favorable for progestagens but detrimental for androgens, in agreement with the fact that AR ligands bear only small groups on C17 (OH in β, H or Me in α). Both receptors have electrostatic contours corresponding to a favorable interaction with a negative charge on the D-ring region. But an additional contour for AR suggests a charge located closer to the C17 position for greater activity.

The structure of hPR and hER-α was examined by Tanenbaum et al. [88]. Generally speaking, the extremities of the ligands are involved in H-bonds and, for their central part, in hydrophobic contacts (Figure 11.6).

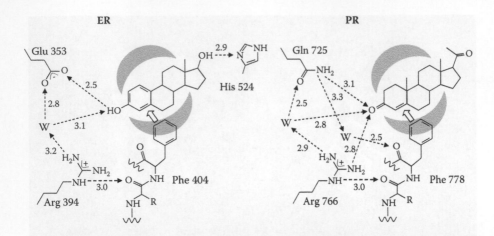

FIGURE 11.6 Hydrogen bonding network for human ER-α LBD (estrogen receptor α ligand-binding domain) and human PR LBD (progesterone receptor ligand-binding domain). The gray areas symbolize the hydrophobic contacts. For these, only Phe residue is shown, on account of its anchoring role. (Adapted from D.M. Tanenbaum, Y. Wang, S.P. Williams, and P.B. Sigler, *Proc. Natl. Acad. Sci. USA*, 95, 5998, 1998. With permission.)

For the A-ring, the estrogen receptor has a glutamate (Glu353) that accepts an H-bond from the 3-OH group, whereas in hPR LBD a glutamine (Gln725) in the corresponding position gives an H-bond to the 3-keto group of progesterone. An arginine (Arg394 in hER-α, Arg766 in hPR) maintains this organization via water-mediated H-bonds. A supplementary cohesion is brought by a phenylalanine (Phe404 in hER-α, Phe778 in hPR) H-bonded to these residues and involved in hydrophobic contact with the ligand A-ring. At the opposite extremity, the 17-OH binds to His524 in ER, but things are less clear for PR [88]. In the central part, residues implicated in hydrophobic interactions with the ligand are highly conserved.

11.8 CONCLUDING REMARKS

Twenty years after the introduction of CoMFA in QSAR, molecular field analysis methods proved to be reliable and efficient tools for ED modeling. Continuous developments have largely improved performance. Introducing new fields gives better insight into the mechanisms intervening in receptor binding. Comparison with crystal structures of ligand-ligand LBD complexes now available, or with homology-built receptors, and docking processes (where ligand flexibility may be considered), allow for more secure alignments, the most critical step of the CoMFA treatment. In the neighboring GRIND method, alignment is even avoided by using auto- or cross-correlation to generate "grid-independent" descriptors. Receptor-based methods constitute an alternative complementary approach. With the more sophisticated QUASAR package, the adaptation of the receptor to individual ligands is taken into account.

Fragmental methods (HQSAR, FRED/SKEYS, for example) working from the 2D structural formulae give a first level of analysis with less intensive calculations.

Finally, as stated by Tong et al. [1], QSAR models may be considered as "living models" relying on the concept of "active learning": a recursive process where the current model suggests new molecules to be tested. The results, incorporated into the dataset, help to refine the model. The updated model in turn suggests new experiments, and the cycle goes on.

The end-user has now at his or her disposal a full range of methods of increasing complexity to delineate the potential action of chemicals on environment and human health. In connection with experimental effort, these models would allow, with limited resources, increased performance in the definition of hazard identification and risk assessment.

REFERENCES

[1] W. Tong, H. Fang, H. Hong, Q. Xie, R. Perkins, J. Anson, and D.M. Sheehan, *Regulatory application of SAR/QSAR for priority setting of endocrine disruptors: A perspective,* Pure Appl. Chem. 75 (2003), pp. 2375–2388.

[2] H. Tamura, Y. Ishimoto, T. Fujikawa, H. Aoyama, H. Yoshikawa, and M. Akamatsu, *Structural basis for androgen receptor agonists and antagonists: Interaction of SPEED 98-listed chemicals and related compounds with the androgen receptor based on an in vitro reporter gene assay and 3D QSAR,* Bioorg. Med. Chem. 14 (2006), pp. 7160–7174.

[3] J.D. McKinney, A. Richard, C. Waller, M.C. Newman, and F. Gerberick, *The practice of structure activity relationships (SAR) in toxicology,* Toxicol. Sci. 56 (2000), pp. 8–17.

[4] R.D. Cramer III, D.E. Patterson, and J.D. Bunce, *Comparative molecular field analysis (CoMFA). 1. Effect of shape on binding of steroids to carrier proteins,* J. Am. Chem. Soc. 110 (1988), pp. 5959–5967.

[5] C.L. Waller, T.I. Oprea, K. Chae, H.K. Park, K.S. Korach, S.C. Laws, T.E. Wiese, W.R. Kelce, and L.E. Gray Jr., *Ligand-based identification of environmental estrogens,* Chem. Res. Toxicol. 9 (1996), pp. 1240–1248.

[6] H. Fang, W. Tong, W.J. Welsh, and D.M. Sheehan, *QSAR models in receptor-mediated effects: The nuclear receptor superfamily,* J. Mol. Struct. Theochem. 622 (2003), pp. 113–125.

[7] C.E. Bohl, C. Chang, M.L. Mohler, J. Chen, D.D. Miller, P.W. Swaan, and J.T. Dalton, *A ligand-based approach to identify quantitative structure-activity relationships for the androgen receptor,* J. Med. Chem. 47 (2004), pp. 3765–3776.

[8] R. Bursi and M.B. Groen, *Application of (quantitative) structure-activity relationships to progestagens: From serendipity to structure-based design,* Eur. J. Med. Chem. 35 (2000), pp. 787–796.

[9] J. Devillers, N. Marchand-Geneste, A. Carpy, and J.M. Porcher, *SAR and QSAR modeling of endocrine disruptors,* SAR QSAR Environ. Res. 17 (2006), pp. 393–412.

[10] A. Vedani, A.V. Descloux, M. Spreafico, and B. Ernst, *Predicting the toxic potential of drugs and chemicals in silico: A model for peroxisome proliferator-activated receptor γ (PPARγ),* Toxicol. Let. 173 (2007), pp. 17–23.

[11] W. Tong, Q. Xie, H. Hong, L. Shi, H. Fang, and R. Perkins, *Assessment of prediction confidence and domain extrapolation of two structure-activity relationship models for predicting estrogen receptor binding activity,* Environ. Health Perspect. 112 (2004), pp. 1249–1254.

[12] H. Kubinyi, *Comparative molecular field analysis,* in *Handbook of Chemoinformatics,* J. Gasteiger, ed., Wiley-VCH, Weiheim, 2003, pp. 1555–1575.

[13] S.J. Cho and A. Tropsha, *Cross validated R² guided region selection for comparative molecular field analysis: A simple method to achieve consistent results*, J. Med. Chem 38 (1995), pp. 1060–1066.

[14] C.L. Waller, B.W. Juma, L.E. Gray, and W.R. Kelce, *Three-dimensional quantitative structure-activity relationships for androgen receptor ligands*, Toxicol. Appl. Pharmacol. 137 (1996), pp. 219–227.

[15] P.M. Burden, T.H. Ai, H.Q. Lin, M. Akinci, M. Costandi, T.M. Hambley, and G.A.R. Johnston, *Chiral derivatives of 2-cyclohexylideneperhydro-4,7-methanoindenes, a novel class of nonsteroidal androgen receptor ligand: Synthesis, X-ray analysis and biological activity*, J. Med. Chem. 43 (2000), pp. 4629–4635.

[16] C.L. Waller, D.L. Minor, and J.D. McKinney, *Using the three-dimensional quantitative structure-activity relationships to examine estrogen-receptor binding affinities of polychlorinated hydroxybiphenyls*, Environ. Health Perspect. 103 (1995), pp. 702–707.

[17] S. Wold, C. Albano, W.J. Dunn III, U. Edlund, K. Esbensen, P. Geladi, S. Helberg, E. Johansson, W. Lindberg, and M. Sjostrom, *Multivariate data analysis in chemistry*, in *Chemometrics: Mathematics and Statistics in Chemistry*, B. Kowalski, ed., Reidel, Dordrecht, The Netherlands, 1984.

[18] L.M. Shi, H. Fang, W. Tong, J. Wu, R. Perkins, R.M. Blair, W.S. Branham, S.L. Dial, C.L. Moland, and D.M. Sheehan, *QSAR models using a large diverse set of estrogens*, J. Chem. Inf. Comput. Sci. 41 (2001), pp. 186–195.

[19] C.A. Marhefka, B.M. Moore II, T.C. Bishop, L. Kirkovsky, A. Mukherjee, J.T. Dalton, and D.D. Miller, *Homology modeling using multiple molecular dynamics simulations and docking studies of the human androgen receptor ligand binding domain bound to testosterone and nonsteroidal ligands*, J. Med. Chem. 44 (2001), pp. 1729–1740.

[20] K.P. Coleman, W.A. Toscano Jr., and T.E. Wiese, *QSAR models of the in vitro estrogen activity of bisphenol A analogs*, QSAR Comb. Sci. 22 (2003), pp. 78–88.

[21] R.D. Cramer III, J.D. Bunce, and D.E. Patterson, *Cross-validation, bootstrapping, and partial least squares compared with multiple regression in conventional QSAR studies*, Quant. Struct.-Act. Relat. 7 (1988), pp. 18–25.

[22] S.K. Kearsley and G.M. Smith, *An alternative method for the alignment of molecular structures: Maximizing electrostatic and steric overlap*, Tetrahedron Comput. Methodol. 3 (1990), pp. 613–633.

[23] A.A. Söderholm, P.T. Lehtovuori, and T.H. Nyrönen, *Three dimensional structure-activity relationships of nonsteroidal ligands in complex with androgen receptor ligand binding domain*, J. Med. Chem. 48 (2005), pp. 917–925.

[24] H.F. Chen, Q. Li, X.J. Yao, B.T. Fan, S.G. Yuan, A. Panaye, and J.P. Doucet, *3D-QSAR and docking study of the binding mode of steroids to progesterone receptor in active site*, QSAR Comb. Sci. 22 (2003), pp. 604–613.

[25] N. Marchand-Geneste, M. Cazaunau, A.J.M. Carpy, M. Laguerre, J.M. Porcher, and J. Devillers, *Homology model of the rainbow trout estrogen receptor (rtERα) and docking of endocrine disrupting chemicals (EDCs)*, SAR QSAR Environ. Res. 17 (2006), pp. 93–106.

[26] S.P. Korhonen, K. Tuppurainen, R. Laatikainen, and M. Peräkylä, *Comparing the performance of FLUFF-BALL to SEAL CoMFA with a large diverse estrogen data set: From relevant superpositions to solid predictions*, J. Chem. Inf. Model. 45 (2005), pp. 1874–1883.

[27] S.P. Korhonen, K. Tuppurainen, R. Laatikainen, and M. Peräkylä, *FLUFF-BALL A template based grid independent superposition and QSAR technique: Validation using a benchmark steroid dataset*, J. Chem. Inf. Comput. Sci. 43 (2003), pp. 1780–1793.

[28] M. Pastor, G. Cruciani, I. McLay, S. Pickett, and S. Clementi, *GRid INdependent Descriptors (GRIND): A novel class of alignment-independent three-dimensional molecular descriptors*, J. Med. Chem. 43 (2000), pp. 3233–3243.

[29] G. Cruciani and K.A. Watson, *Comparative molecular field analysis using GRID force-field and GOLPE variable selection methods in a study of inhibitors of glycogen phosphorylase b*. J. Med. Chem. 37 (1994), pp. 2589–2601.

[30] B.R. Sadler, S.J. Cho, K.S. Ishaq, K. Chae, and K.S. Korach, *Three dimensional quantitative structure-activity relationship study of nonsteroidal estrogen receptor ligands using the comparative molecular field analysis/cross-validated r²-guided region selection approach,* J. Med. Chem. 41 (1998), pp. 2261–2267.

[31] R.X. Wang, Y. Gao, L. Liu, and L.H. Lai, *All-orientation search and all-placement search in Comparative Molecular Field Analysis*, J. Mol. Model. 4 (1998), pp. 276–283.

[32] T.G. Gantchev, H. Ali, and J.E. van Lier, *Quantitative structure-activity relationships/ comparative molecular field analysis (QSAR/CoMFA) for receptor-binding properties of halogenated estradiol derivatives,* J. Med. Chem. 37 (1994), pp. 4164–4176.

[33] R.T. Kroemer and P. Hecht, *Replacement of the steric 6-12 potential derived interaction energies by atom-based indicator variables in CoMFA leads to models of higher consistency,* J. Comput.-Aided Mol. Des. 9 (1995), pp. 205–212.

[34] R.S. Bohacek and C. McMartin, *Definition and display of steric hydrophobic and hydrogen-bonding properties of ligand binding sites in proteins using Lee and Richards accessible surface: Validation of a high-resolution graphical tool for drug design,* J. Med. Chem. 35 (1992), pp. 1671–1684.

[35] G.E. Kellogg, S.F. Semus, and D.J. Abraham, *HINT: A new method of empirical hydrophobic field calculation for CoMFA,* J. Comput.-Aided Mol. Des. 5 (1991), pp. 545–552.

[36] S.P. Korhonen, K. Tuppurainen, A. Asikainen, R. Laatikainen, and M. Peräkylä, *SOMFA on a large diverse xenoestrogen data set: The effect of superposition algorithms and external regression tools*, QSAR Comb. Sci. 26 (2007), pp. 809–819.

[37] G. Klebe, U. Abraham, and T. Mietzner, *Molecular similarity indices in a comparative analysis (CoMSIA) of drug molecules to correlate and predict their biological activity,* J. Med. Chem. 37 (1994), pp. 4130–4146.

[38] M. Böhm, J. Stürzebecher, and G. Klebe, *Three-dimensional quantitative structure-activity relationship analyses using comparative molecular field analysis and comparative molecular similarity indices analysis to elucidate selectivity differences of inhibitors binding to trypsin, thrombin and factor Xα,* J. Med. Chem. 42 (1999), pp. 458–477.

[39] W. Tong, R. Perkins, R. Strelitz, E.R. Collantes, S. Keenan, W.J. Welsh, W.S. Branham, and D.M. Sheehan, *Quantitative structure-activity relationships (QSARs) for estrogen binding to the estrogen receptor: Prediction across species,* Environ. Health Perspect. 105 (1997), pp. 1116–1124.

[40] A.R. Katrizky, V.S. Lobanov, and M. Karelson, *CODESSA, Reference manual*, University of Florida, Gainesville, 1995.

[41] A.R. Katrizky, V.S. Lobanov, and M. Karelson, *CODESSA, Training manual*, University of Florida, Gainesville, 1995.

[42] E. von Angerer, J. Prekajac, and J. Strohmeier, *2-Phenylindoles. Relationship between structure, estrogen receptor affinity, and mammary tumor inhibiting activity in the rat*, J. Med. Chem. 27 (1984), pp. 1439–1447.

[43] J.A. VanderKuur, T. Wiese, and S.C. Brooks, *Influence of estrogen structure on nuclear binding and progesterone reporter induction by the receptor complex*, Biochem. 32 (1993), pp. 7002–7008.

[44] W. Tong, D.R. Lowis, R. Perkins, Y. Chen, W.J. Welsh, D.W. Goddette, T.W. Heritage, and D.M. Sheehan, *Evaluation of quantitative structure-activity relationship methods for large-scale prediction of chemical binding to the estrogen receptor,* J. Chem. Inf. Comput. Sci. 38 (1998), pp. 669–677.

[45] W.S. Branham, S.L. Dial, C.L. Moland, B.S. Hass, R.M. Blair, H. Fang, L. Shi, W. Tong, R.G. Perkins, and D.M. Sheehan, *Phytoestrogens and mycoestrogens bind to the rat uterine estrogen receptor,* J. Nutr. 132 (2002), pp. 658–664.

[46] R. Blair, H. Fang, W.S. Branham, B. Hass, S.L. Dial, C.L. Moland, W. Tong, L. Shi, R. Perkins, and D.M. Sheehan, *Estrogen receptor relative binding affinities of 188 natural and xenochemicals: Structural diversity of ligands,* Toxicol. Sci. 54 (2000), pp. 138–153.

[47] G.G. Kuiper, J.G. Lemmen, B. Carlsson, J.C. Corton, S.H. Safe, P.T. van der Saag, B. van der Burg, and J.A. Gustafsson, *Interaction of estrogenic chemicals and phytoestrogens with estrogen receptor β,* Endocrinology 139 (1998), pp. 4252–4263.

[48] T. Ghafourian and M.T.D. Cronin, *The impact of variable selection on the modelling of oestrogenicity,* SAR QSAR Environ. Res. 16 (2005), pp. 171–190.

[49] C.L. Waller, *A comparative QSAR study using CoMFA, HQSAR and FRED/SKEYS paradigms for estrogen receptor binding affinities of structurally diverse compounds,* J. Chem. Inf. Comput. Sci. 44 (2004), pp. 758–765.

[50] W. Zheng and A. Tropsha, *Novel variable selection quantitative structure-property relationship approach based on the k-nearest neighbor principle,* J. Chem. Inf. Comput. Sci. 40 (2000), pp. 185–194.

[51] A.H. Asikainen, J. Ruuskanen, and K.A.Tuppurainen, *Consensus kNN QSAR: A versatile method for predicting the estrogenic activity of organic compounds in silico. A comparative study with five estrogen receptors and a large diverse set of ligands,* Environ. Sci. Technol. 38 (2004), pp. 6724–6729.

[52] R. Todeschini and V. Consoni, *Handbook of Molecular Descriptors, Methods and Principles in Medicinal Chemistry,* Vol. 11, Wiley-VCH, Weinheim, 2000.

[53] D.B. Turner, P. Willett, A.M. Ferguson, and T.W. Heritage, *Evaluation of a novel infrared range vibration-based descriptor (EVA) for QSAR studies. I. General application,* J. Comput.-Aided Mol. Des. 11 (1997), pp. 402–422.

[54] K. Tuppurainen, *EEVA (electronic eigenvalues): A new QSAR/QSPR descriptor for electronic substituent effects based on molecular orbital energies,* SAR QSAR Environ. Res. 10 (1999), pp. 39–46.

[55] R. Bursi, T. Dao, T. van Wijk, M. de Goyeer, E. Kellenbach, and P. Verwer, *Comparative spectra analysis (CoSA): Spectra as three dimensional molecular descriptors for the prediction of biological activities,* J. Chem. Inf. Comput. Sci. 39 (1999), pp. 861–867.

[56] A. Asikainen, J. Ruuskanen, and K. Tuppurainen, *Spectroscopic QSAR methods and self-organizing molecular field analysis for relating molecular structure and estrogenic activity,* J. Chem. Inf. Comput. Sci. 43 (2003), pp. 1974–1981.

[57] W. Sippl, *Receptor-based 3D QSAR analysis of estrogen receptor ligands — Merging the accuracy of receptor based alignments with the computational efficiency of ligand-based methods,* J. Comput.-Aided Mol. Des. 14 (2000), pp. 559–572.

[58] W. Sippl, *Binding affinity prediction of novel estrogen receptor ligands using receptor-based 3D QSAR,* Bioorg. Med. Chem. 10 (2002), pp. 3741–3755.

[59] D.D. Robinson, P.J. Winn, P.D. Lyne, and W.G. Richards, *Self organizing molecular field analysis: A tool for structure-activity studies,* J. Med. Chem. 42 (1999), pp. 573–583.

[60] S.P. Korhonen, K. Tuppurainen, R. Laatikainen, and M. Peräkylä, *Improving the performance of SOMFA by use of standard multivariate methods,* SAR QSAR Environ. Res. 16 (2005), pp. 567–579.

[61] R.D. Beger, K.J. Holm, D.A. Buzatu, and J.G. Wilkes, *Using simulated 2D ^{13}C NMR nearest neighbor connectivity spectral data patterns to model a diverse set of estrogens*, Internet Electron. J. Mol. Des. 2 (2003), pp. 435–453.

[62] C. Epouhe, B.T. Fan, S.G. Yuan, A. Panaye, and J.P. Doucet, *Contribution to structural elucidation: Behaviours of substructures partially defined from 2D-NMR*, Chin. J. Chem. 21 (2003), pp. 1268–1274.

[63] A.H. Asikainen, J. Ruuskanen, and K.A. Tuppurainen, *Alternative QSAR models for selected estradiol and cytochrome P450 ligands: Comparison between classical, spectroscopic, CoMFA and GRID/GOLPE methods*, SAR QSAR Environ. Res. 16 (2005), pp. 555–565.

[64] E. Napolitano, R. Fiaschi, K.E. Carlson, and J.A. Katzenellenbogen, *11β-substituted estradiol derivatives, potential high-affinity carbon-11 labeled probes for the estrogen receptor: A structure-affinity relationship study*, J. Med. Chem. 38 (1995), pp. 429–434.

[65] H. Gao, J.A. Katzenellenbogen, R. Garg, and R. Hansch, *Comparative QSAR analysis of estrogen receptor ligands*, Chem. Rev. 99 (1999), pp. 723–744.

[66] L. Sun, Y. Zhou, L. Genrong, and S.Z. Li, *Molecular electronegativity-distance vector (MEDV-4): A two-dimensional QSAR method for the estimation and prediction of biological activities of estradiol derivatives*, J. Mol. Struct. Theochem. 679 (2004), pp. 107–113.

[67] T.A. Grese, L.D. Pennington, J.P. Sluka, M.D. Adrian, H.W. Cole, T.R. Fuson, D.E. Magee, D.L. Phillips, E.R. Rowley, P.K. Setler, L.L. Short, M.Venugopalan, N.N. Yang, M. Sato, A.L. Glasebrook, and H.U. Bryant, *Synthesis and pharmacology of conformationally restricted raloxifenes analogues: Highly potent selective estrogen receptor modulators*, J. Med. Chem. 41 (1998), pp. 1272–1283.

[68] P.J. Goodford, *A computational procedure for determining energetically favorable binding sites on biologically important macromolecules*, J. Med. Chem. 28 (1985), pp. 849–857.

[69] I.R.A. Menezes, A. Leitão, and C.A. Montanari, *Three-dimensional models of non-steroidal ligands: A comparative molecular field analysis*, Steroids 71 (2006), pp. 417–428.

[70] A.M. Brzozowski, A.C. Pike, Z. Dauter, R.E. Hubbard, T. Bonn, O. Engström, L. Ohman, G.L. Greene, J.A. Gustafsson, and M. Carlquist, *Molecular basis of agonism and antagonism in the oestrogen receptor*, Nature 389 (1997), pp. 753–758.

[71] B.C. Oostenbrink, J.W. Pitera, M.M.H. van Lipzig, J.H.N. Meerman, and W.F. van Gunsteren, *Simulations of the estrogen receptor ligand binding domain: Affinity of natural ligands and xenoestrogens*, J. Med. Chem. 43 (2000), pp. 4594–4605.

[72] Y. Akahori, M. Nakai, Y. Yakabe, M. Takatsuki, M. Mizutani, M. Matsuo, and Y. Shimohigashi, *Two-step models to predict binding affinity of chemicals to human estrogen receptor α by three-dimensional quantitative structure-activity relationships (3D-QSARs) using receptor–ligand docking simulation*, SAR QSAR Environ. Res. 16 (2005), pp. 323–337.

[73] M.A. Lill, A. Vedani, and M. Dobler, *Raptor: Combining dual-shell representation, induced-fit simulation, and hydrophobicity scoring in receptor modelling: Application towards the simulation of structurally diverse ligand sets*, J. Med. Chem. 47 (2004), pp. 6174–6186.

[74] A. Vedani and M. Dobler, *Multidimensional QSAR: Moving from three- to five dimensional concepts*, Quant. Struct.-Act. Relat. 21 (2002), pp. 382–390.

[75] D.A. Loughney and C.F. Schwender, *A comparison of progestin and androgen receptor binding using the CoMFA technique*, J. Comput.-Aided Mol. Des. 6 (1992), pp. 569–581.

[76] J. Delettré, J.P. Mornon, G. Lepicard, T. Ojasoo, and J.P. Raynaud, *Steroid flexibility and receptor specificity*, J. Steroid Biochem. 13 (1980), pp. 45–59.

[77] H. Hong, H. Fang, Q. Xie, R. Perkins, D.M. Sheehan, and W. Tong, *Comparative molecular analysis (CoMFA) model using a large diverse set of natural chemicals for binding to the androgen receptor*, SAR QSAR Environ. Res. 14 (2003), pp. 373–388.

[78] H. Fang, W. Tong, W.S. Branham, C.L. Moland, S.L. Dial, H.X. Hong, Q. Xie, R. Perkins, W. Owens, and D.M. Sheehan, *Study of 202 natural, synthetic, and environmental chemicals for binding to the androgen receptor,* Chem. Res. Toxicol. 16 (2003), pp. 1338–1358.

[79] P.M. Matias, P. Donner, R. Coelho, M. Thomaz, C. Peixoto, S. Macedo, N. Otto, S. Joschko, P. Scholtz, A. Wegg, S. Bäsler, M. Schäfer, U. Egner, and M.A. Carrondo, *Structural evidence for ligand specificity in the binding domain of the human androgen receptor. Implications for pathogenic gene mutation*, J. Biol. Chem. 275 (2000), pp. 26164–26171.

[80] C.Y. Zhao, R.S. Zhang, H.X. Zhang, C.X. Xue, H.X. Liu, M.C. Liu, Z.D. Hu, and B.T. Fan, *QSAR study of natural, synthetic and environmental endocrine disrupting compounds for binding to the androgen receptor*, SAR QSAR Environ. Res. 16 (2005), pp. 349–367.

[81] P.M. Matias, M.A. Carrondo, R. Coelho, M. Thomaz, X.Y. Zhao, A. Wegg, K. Crusius, U. Egner, and P. Donner, *Structural basis for the glucocorticoid response in a mutant human androgen receptor (AR^{ccr}) derived from an androgen-independent prostate cancer,* J. Med. Chem. 45 (2002), pp. 1439–1446.

[82] N. Poujol, J.M. Wurtz, B. Tahiri, S. Lumbroso, J.C. Nicolas, D. Moras, and C. Sultan, *Specific recognition of androgens by their nuclear receptor. A structure-function study,* J. Biol. Chem. 275 (2000), pp. 24022–24031.

[83] M.A. Lill, F. Winiger, A. Vedani, and B. Ernst, *Impact of induced fit on ligand binding to the androgen receptor: A multidimensional QSAR study to predict endocrine-disrupting effects of environmental chemicals,* J. Med. Chem. 48 (2005), pp. 5666–5674.

[84] A. Vedani, H. Briem, M. Dobler, K. Dollinger, and D.R. McMasters, *Multiple-conformation and protonation-state representation in 4D-QSAR: The neurokinin-1 receptor system*, J. Med. Chem. 43 (2000), pp. 4416–4427.

[85] M.A. Lill, M. Dobler, and A. Vedani, *In silico prediction of receptor-mediated environmental toxic phenomena: Application to endocrine disruption*, SAR QSAR Environ. Res. 16 (2005), pp. 149–169.

[86] H. Tamura, H. Yoshikawa, K.W. Gaido, S.M. Ross, R.K. DeLisle, W.J. Welsh, and A.M. Richard, *Interaction of organophosphate pesticides and related compounds with the androgen receptor,* Environ. Health Perspect. 111 (2003), pp. 545–552.

[87] E.W. Berking, F. van Meel, E.W. Turpijn, and J. van der Vies, *Binding of progestagens to receptor proteins in MCF-7 cells*, J. Steroid Biochem. 19 (1983), pp. 1563–1570.

[88] D.M. Tanenbaum, Y. Wang, S.P. Williams, and P.B. Sigler, *Crystallographic comparison of the estrogen and progesterone receptor's ligand binding domains*, Proc. Natl. Acad. Sci. 95 (1998), pp. 5998–6003.

[89] S.P. van Helden and H. Hamersma, *3D QSAR of the receptor binding of steroids. A comparison of multiple regression neural networks and comparative molecular field analysis,* in *Proceedings of the 10th Symposium on Structure-Activity Relationships, QSAR and Molecular Modelling Concepts, Computational Tools and Biological Applications*, F. Sanz, J. Giralkdo, and D F. Manaut, eds., Prous Science Publishers, Barcelona, 1995, pp. 481–483.

[90] S.P. van Helden, H. Hamersma, and V.J. van Geerestein, *Prediction of the progesterone receptor binding of steroids using a combination of genetic algorithms and neural networks,* in *Genetic Algorithms in Molecular Modeling*, J. Devillers, ed., Academic Press, London, 1996, pp. 159–192.

[91]   S.S. So, S.P. van Helden, V.J. van Geerestein, and M. Karplus, *Quantitative structure-activity relationship studies of progesterone receptor binding steroids*, J. Chem. Inf. Comput. Sci. 40 (2000), pp. 762–772.

[92]   S.P. Niculescu and K.L.E. Kaiser, *Modeling the relative binding affinity of steroids to the progesterone receptor with probabilistic neural networks*, Quant. Struct.-Act. Relat. 20 (2001), pp. 223–226.

[93]   A. Ashek, C. Lee, H. Park, and S.J. Cho, *3D QSAR studies of dioxins and dioxin-like compounds using CoMFA and CoMSIA*, Chemosphere 65 (2006), pp. 521–529.

[94]   C.L. Waller and J.D. McKinney, *Comparative molecular field analysis of polyhalogenated dibenzo-p-dioxins, dibenzofurans and biphenyls*, J. Med. Chem. 35 (1992), pp. 3660–3666.

[95]   C.L. Waller and J.D. McKinney, *Three-dimensional quantitative structure-activity relationships of dioxins and dioxin-like compounds: Model validation and Ah receptor characterization*, Chem. Res. Toxicol. 8 (1995), pp. 847–858.

[96]   M. Wagener, J. Sadowski, and J. Gasteiger, *Autocorrelation of molecular surface properties for modelling corticosteroid binding globulin and cytosolic Ah receptor activity by neural networks*, J. Am. Chem. Soc. 117 (1995), pp. 7769–7775.

[97]   G. Bravi and J.H. Wikel, *Application of MS-WHIM descriptors. 1. Introduction of new molecular surface properties and 2. Prediction of binding affinity data*, Quant. Struct.-Act. Relat. 19 (2000), pp. 29–37.

[98]   S.S. So and M. Karplus, *Three-dimensional quantitative structure-activity relationships from molecular similarity matrices and genetic neural networks. 2 Applications*, J. Med. Chem. 40 (1997), pp. 4360–4371.

[99]   K. Tuppurainen and J. Ruuskanen, *Electronic eigenvalue (EEVA): A new QSAR/QSPR descriptor for electronic substituent effect based on molecular orbital energies. A QSAR approach to the Ah receptor binding affinity of polychlorinated biphenyls (PCBs), dibenzo-p-dioxins (PCDDs) and dibenzofurans(PCDFs)*, Chemosphere 41 (2000), pp. 843–848.

[100] E. Lo Piparo, K. Koehler, A. Chana, and E. Benfenati, *Virtual screening for aryl hydrocarbon receptor binding prediction*, J. Med. Chem. 49 (2006), pp. 5702–5709.

[101] G. Cruciani, P. Crivori, P.-A. Carrupt, and B. Testa, *Molecular fields in quantitative structure-permeation relationships: The VolSurf approach*, J. Mol. Struct. THEOCHEM 503 (2000), pp. 17–30.

[102] K. Koyano, T. Nakano, and T. Kaminuma, *Electrostatic potentials and CoMFA analysis of toxicity of dioxins*, Chem-Bio. Informatics J. 1 (2001), pp. 60–72.

[103] Y. Wang, H. Liu, C. Zhao, H. Liu, Z. Cai, and G. Jiang, *Quantitative structure-activity relationship models for prediction of the toxicity of polybrominated, diphenyl ether congeners*, Environ. Sci. Technol. 39 (2005), pp. 4961–4966.

[104] G. Chen, A.D. Konstantinov, B.G. Chittim, E.M. Joyce, N.C. Bols, and N.J. Bunce, *Synthesis of polybrominated diphenyl ethers and their capacity to induce CYP1A by the Ah receptor mediated pathway*, Environ. Sci. Technol. 35 (2001), pp. 3749–3756.

[105] E.C. Henry, A.S. Kende, G. Rucci, M.J. Totleben, J.J. Willey, S.D. Dertinger, R.S. Pollenz, J.P. Jones, and T.A. Gasiewicz, *Flavone antagonists bind competitively with 2,3,7,8-tetrachlorodibenzo-p-dioxin (TCDD) to the aryl hydrocarbon receptor but inhibit nuclear uptake and transformation*, Mol. Pharmacol. 55 (1999), pp 716–725.

[106] P.L.A. Popelier, U.A. Chaudry, and P.J. Smith, *Quantum topological molecular similarity. Part 5. Further development with an application to the toxicity of polychlorinated dibenzo-para-dioxins (PCDDs)*, J. Chem. Soc. Perkin. Trans. 2 (2002), pp. 1231–1237.

[107] A.Vedani and M. Dobler, *5D-QSAR: The key for simulating induced fit?*, J. Med. Chem. 45 (2002), pp. 2139–2149.

[108] V. Lukacova and S. Balaz, *Multimode ligand binding in receptor site modeling: Implementation in CoMFA*, J. Chem. Inf. Comput. Sci. 43 (2003), pp. 2093–2105.

[109] S. Mosselman, J. Polman, and R. Dijkema, *ERβ: Identification and characterization of a novel human estrogen receptor*, FEBS Lett. 392 (1996), pp. 49–53.

[110] W. Tong, R. Perkins, L. Xing, W.J. Welsh, and D.M. Sheehan, *QSAR models for binding of estrogenic compounds to estrogen receptor α and β*, Endocrinology 138 (1997), pp. 4022–4025.

[111] B.T. Zhu, G.Z. Han, J.Y. Shim, Y. Wen, and X.-R. Jiang, *Quantitative structure-activity relationship of various endogenous estrogen metabolites for human estrogen receptor α and β subtypes: Insights into the structural determinants favoring a differential subtype binding*, Endocrinology 147 (2006), pp. 4132–4150.

[112] E.S. Manas, R.J. Unwalla, Z.B. Xu, M.S. Malamas, C.P. Miller, H.A. Harris, C. Hsiao, T. Akopian, W.-T. Hum, K. Malakian, S. Wolfrom, A. Bapat, R.A. Bhat, M.L. Stahl, W.S. Somers, and J.C. Alvarez, *Structure-based design of estrogen receptor-β selective ligands*, J. Amer. Chem. Soc. 126 (2004), pp. 15106–15119.

12 Structure-Activity Modeling of a Diverse Set of Androgen Receptor Ligands

*James Devillers, Jean-Pierre Doucet,
Annick Panaye, Nathalie Marchand-Geneste,
and Jean-Marc Porcher*

CONTENTS

ABSTRACT

Numerous chemicals released into the environment can interfere with normal, hormonally regulated biological processes to adversely affect development and reproductive functions. SARs and QSARs are powerful screening tools to detect potential endocrine disruptors and to prioritize them for more intensive and costly evaluations based on *in vitro* and *in vivo* assays.

In this context, androgen-receptor binding data (active/inactive) for a large set of about 200 structurally diverse chemicals, described by CODESSA descriptors encoding topological and physicochemical properties, were used for deriving structure-activity models. Classification and Regression Tree (CART) analysis, different types of artificial neural networks (linear artificial neural network, three-layer perceptron, radial basis function neural network, probabilistic neural network, learning vector quantization) and a support vector machine with a Gaussian kernel function were tested as statistical tools. The comparison exercise was performed on the basis of the same learning and testing sets as well as from the same set of selected descriptors. The simulation performances of the models designed from the probabilistic neural network or support vector machine were better than those computed from the other statistical tools.

KEYWORDS

Androgen receptor
Binding assays
Endocrine disruptors
Linear and nonlinear methods
Quantitative Structure-Activity Relationship (QSAR)

12.1   INTRODUCTION

In recent years, various agricultural, industrial, and household chemicals have been shown to directly or indirectly interfere with the endocrine system of wildlife species and humans [1]. Because these chemicals, called endocrine disruptors, represent a potential threat to the male and female reproductive functions, there has been a rising scientific and regulatory interest in their identification [2]. In the past, most attention has been paid to the estrogenic potential of chemicals [3–8]. Conversely, androgenic and antiandrogenic activities of xenobiotics have only recently been a focus of interest. These chemicals interact with the androgen receptor (AR) belonging to the nuclear receptor family and may lead to important disorders in the reproductive system of males [1]. Androgenic and antiandrogenic activities have been described in rivers contaminated by pulp and paper mill effluents and other human activities [9–13], in diesel exhaust particles [14,15], and in components of sunscreens [16]. More specifically, various structurally diverse chemicals such as chlordecone, o,p'-DDT, p,p'-DDT, p,p'-DDD, p,p'-DDE, linuron, fenthion, methoxychlor, metabolites of vinclozolin, prochloraz, HPTE (p-hydroxyphenyl-trichloroethylene), nonylphenol, octylphenol, and zearalenone are known to interact with the AR [17–21].

To address the concerns about the ability of xenobiotics to disrupt endocrine functions and to recommend potential screening strategies, regulatory agencies and international organizations have proposed testing strategies to screen for large numbers of chemicals. Thus, different assays have been developed for evaluating the affinity of chemicals to the AR (see, for example, [22–24]). Unfortunately, due to time and cost constraints, they cannot be used for estimating the endocrine disruption potential of the huge number of chemicals that can potentially contaminate ecosystems. *In silico* estimations based on Quantitative Structure-Activity Relationship (QSAR) modeling allow us to rise above this problem. However, if the number of models designed for estimating the estrogenic activity of chemicals is rather high, the number of structure-activity models in relation with the AR is limited [25]. Thus, for example, a comparative molecular field analysis (CoMFA) was used by Waller et al. [26] to examine AR-binding affinities of a series of 28 structurally diverse natural and synthetic compounds. The model was derived from a learning set of 21 molecules, and the remaining compounds were used as the testing set. These authors showed that the combined (steric and electrostatic) two-field model was less internally consistent than the electrostatic-field-only model ($q^2 = 0.792$, 4 PCs, SEP = 1.012 versus $q^2 = 0.828$, 5 PCs, SEP = 0.952). However, the two-field model yielded the greatest external predictive ability and, hence, was selected. The relative binding affinity of 48 steroids to AR and progesterone receptor (PR) was also investigated by Loughney and Schwender [27] from CoMFA. The androgen model was of poor quality, only reaching a cv-r^2 of 0.545 after nine components [27]. Another CoMFA model was derived by Hong et al. [28] but from a larger learning set of 146 chemicals tested according to the same conditions with an AR competitive binding assay [29]. Eight PCs were selected ($r^2 = 0.902$, SE = 0.389). The relative contributions of the steric and electrostatic fields were 0.522 and 0.478, respectively. The predictive power of the model was estimated from an external testing set of eight chemicals.

Recently, Zhao et al. [30] compared the performances of regression analysis, radial basis function network (RBFN) [31], and support vector machine (SVM) [32] for modeling the database of 146 molecules used by Hong et al. [28] for CoMFA. The database was split into a learning set and an external testing set of 118 and 28 chemicals, respectively. Chemicals were described from the CODESSA software [33]. The best simulation results were obtained with the nonlinear statistical tools, especially the SVM. However, it is noteworthy that their RBFN model was overfitted because it yielded poor performances with the external testing set. Even though, in addition to this kind of methodological problem, Hong et al. [28] and Zhao et al. [30] used only a part of the valuable results produced by Fang et al. [29] in their modeling process, an attempt was made to derive a new AR-binding structure-activity model presenting a larger domain of application. To reach this goal, seven different statistical methods commonly used in (Q)SAR were experienced.

12.2 MATERIALS AND METHODS

12.2.1 AR Binding Assay Data

The AR binding affinity of 202 natural and man-made chemicals was determined by Fang et al. [29] from a rapid and inexpensive recombinant AR competitive binding test. Briefly, the binding activity of a chemical was determined by competing with [^{3}H]-R1881 (methyltrienolone) for AR and was expressed as relative binding affinity (RBA), which was calculated by dividing the IC50 of R1881 by the IC50 of the studied chemical and by multiplying by 100 (RBA = 100 for R1881). Forty-eight chemicals that failed to compete [^{3}H]-R1881 in binding were classified as nonbinders (NBs), and eight compounds that showed binding but without reaching 50% inhibition at a maximum concentration were designated as slight binders (SBs). Because the goal of Hong et al. [28] and Zhao et al. [30] was to derive models in the form log RBA = f(descriptors), the NBs and SBs were excluded from their modeling process.

This methodological choice led to voluntary reduction of the domain of application of the models by excluding chemicals presenting interesting structures and functional groups. Consequently, structure-activity models were derived from the whole dataset. In fact, 4-aminosalicylic acid sodium salt (NB), 1,3-diphenyltetramethyldisiloxane (log RBA = −3.13), 1,3-dibenzyltetramethyldisiloxane (NB), and amaranth (NB) were excluded from the modeling exercise because the software used for calculating molecular descriptors was not able to run with them. A log RBA value of −2.30 was selected as cutoff value. This yielded the constitution of a dataset of 100 active (1) and 98 inactive (0) compounds (Table 12.1). The dataset was further split into different training and testing sets (178/20) accounting for the different chemical families tested by Fang et al. [29] as well as the range of activities.

12.2.2 MOLECULAR DESCRIPTORS

The 198 chemical structures were geometry-optimized with the AM1 semiempirical method in HyperChem (Hypercube, Gainesville, Florida) and exported to MOPAC. The optimized structures were then introduced into CODESSA software [33] to compute 132 topological and physicochemical descriptors. In addition, the 1-octanol/water partition coefficients (log P) calculated by Fang et al. [29] were collected and used as hydrophobic parameters.

A standardized PCA (zero mean and unit variance) was used for feature selection. This allowed us to select the 15 following descriptors: log P, Kier shape index of second order (T12), Kier flexibility index (T13), average information content of zero order (T14), Balaban index (T19), YZ shadow (T28), molecular volume (T32), minimum partial charge for a hydrogen atom (U04), maximum partial charge (Q_{max}) (U07), polarity parameter/square distance (U10), relative positive charge (RPCG) (U28), HOMO energy (V02), LUMO energy (V03), HOMO–LUMO energy gap (V05), and minimum nucleophilic reactivity index for an oxygen atom (V39).

It is noteworthy that when stereochemistry was not specified, the lowest energy stereoisomer was chosen (for example, double bond in *trans* position).

12.2.3 STATISTICAL TOOLS

12.2.3.1 Classification and Regression Tree (CART) Analysis

CART analysis is also known as binary recursive partitioning. The term "binary" implies that each parent node can only be split into exactly two child nodes in the decision tree. The term "recursive" refers to the fact that the process is repeated by treating each child as a parent when further splitting is possible. Last, the term "partitioning" indicates that the whole dataset under study is partitioned into clusters of different sizes. In practice, the tree consists of a root node containing all objects to classify. This root node is split into two child nodes on the basis of a threshold value of the most discriminating variable. At least one of the child nodes becomes a parent node that is split by a threshold value of another variable selected by a stepwise procedure. Splitting criteria are determined by statistical analysis of each value of each variable. A node for

TABLE 12.1

Observed and Calculated Androgenic Binding Activity of Chemicals

Number	Chemical Name	log RBA	AC[a]	C1	C2	C3	C4	C5	C6	C7	C8	C9
1	5α-Androstan	−3.32	0	0	1	0	0	0	0	0	1	0
2	Androsterone*	−2.12	1	1	1	1	1	1	1	1	1	1
3	5,6-Didehydroisoandrosterone	−1.98	1	1	1	1	1	1	1	1	1	1
4	5α-Androstane-3,11, 17-trione	−1.64	1	1	0	1	0	1	1	1	1	1
5	Epitestosterone	−1.00	1	1	0	1	1	1	1	1	1	1
6	3α-Androstanediol	−0.81	1	1	1	1	1	1	1	1	1	1
7	T propionate	−0.79	1	1	1	1	1	1	1	1	1	1
8	5α-Androstan-3β-ol	−0.74	1	1	1	1	1	1	1	1	1	1
9	Androstenediol	−0.66	1	1	1	1	1	1	1	1	1	1
10	4-Androstenedione	−0.62	1	1	1	1	1	1	1	1	1	1
11	4-Androstenediol	−0.31	1	1	1	1	1	1	1	1	1	1
12	Etiocholan-17β-ol-3-one*	−0.10	1	1	1	1	1	1	1	1	1	1
13	DHT benzoate	0.07	1	1	1	1	1	1	1	1	1	1
14	3β-Androstanediol	0.36	1	1	1	1	1	1	1	1	1	1
15	11-Keto-testosterone	0.54	1	1	1	1	1	1	1	1	1	1
16	Methyltestosterone	1.28	1	1	1	1	1	1	1	1	1	1
17	Testosterone (T)	1.28	1	1	1	1	1	1	1	1	1	1
18	5α-Androstan-17β-ol	1.45	1	1	1	1	1	1	1	1	1	1
19	Methyltrienolone (R1881)	2.00	1	1	1	1	1	1	1	1	1	1
20	Trenbolone	2.05	1	1	0	0	1	0	1	1	1	1
21	DHT	2.14	1	1	1	1	1	1	1	1	1	1
22	Mibolerone*	2.27	1	1	1	1	1	1	1	1	1	1
23	Estriol (E₃)	−3.15	0	0	1	0	0	1	1	1	1	1
24	17α-Estradiol	−2.40	0	1	1	1	1	1	1	1	1	1
25	3-Methylestriol	−2.25	1	1	1	1	1	1	1	1	1	1
26	17-Deoxyestradiol	−2.13	1	1	1	1	1	1	1	1	1	1
27	16β-OH-16α-Me-3-Me-estradiol	−2.08	1	1	1	1	1	1	1	1	1	1
28	2-OH-estradiol	−1.44	1	1	1	0	1	1	1	1	1	1
29	Ethynylestradiol (EE)	−1.42	1	1	1	1	1	1	1	1	1	1
30	4-OH estradiol	−0.91	1	1	1	0	1	1	1	1	1	1
31	Estradiol (E₂)	−0.12	1	1	1	1	1	1	1	1	1	1
32	3-Deoxyestradiol*	0.54	1	1	1	1	1	1	1	1	1	1
33	ICI 182,780	NB[b]	0	0	1	0	1	0	0	0	0	0
34	Moxestrol	SB	0	0	1	1	1	1	1	1	1	1
35	Estrone (E₁)	SB	0	0	1	1	1	1	1	1	1	1
36	ICI 164,384	SB	0	0	1	0	1	0	0	0	0	0
37	Cortisol	−2.77	0	0	0	0	0	0	0	0	1	0
38	Dexamethasone	−2.42	0	0	0	0	0	0	0	0	1	0
39	Corticosterone	−1.87	1	1	0	1	0	1	0	0	1	0
40	Norethynodrel	−0.70	1	1	1	1	1	1	1	1	1	1
41	Progesterone	−0.70	1	1	1	1	1	1	1	1	1	1
42	Promegestone*	−0.64	1	0	0	1	1	1	1	1	1	1

(continued)

TABLE 12.1 (CONTINUED)
Observed and Calculated Androgenic Binding Activity of Chemicals

Number	Chemical Name	log RBA	AC[a]	C1	C2	C3	C4	C5	C6	C7	C8	C9
43	6α-Me-17α-OH-progesterone	−0.41	1	1	0	1	1	1	1	1	1	1
44	Spironolactone	−0.35	1	1	1	1	1	1	1	1	1	1
45	Cyproterone acetate	−0.32	1	1	1	1	1	1	1	1	1	1
46	Norethindrone	0.41	1	1	1	1	1	1	1	1	1	1
47	6α-Me-17α-OH-progesterone acetate	0.94	1	1	1	1	1	1	1	1	1	1
48	Norgestrel	1.22	1	1	1	1	1	1	1	1	1	1
49	Aldosterone	NB	0	0	0	0	0	0	0	0	1	0
50	Prednisolone	NB	0	0	0	0	0	0	0	0	1	0
51	Pregnenolone	NB	0	1	1	1	1	1	1	1	1	1
52	Cholesterol*	NB	0	1	1	0	1	0	0	0	1	0
53	Sitosterol	NB	0	1	1	0	1	0	0	0	1	0
54	Triamcinolone acetonide	SB	0	0	1	0	1	0	0	0	1	0
55	4,4′-Dihydroxystilbene	−2.44	0	0	0	0	0	0	0	0	0	0
56	trans-4-Hydroxystilbene	−2.13	1	0	1	1	0	0	1	1	0	1
57	3,3′-Dihydroxyhexestrol	−2.08	1	1	0	1	1	1	1	1	1	1
58	3,4-Diphenyltetrahydrofuran	−1.98	1	1	1	1	1	1	1	1	1	1
59	Dimethylstilbestrol	−1.66	1	1	0	1	1	1	1	1	0	1
60	DES	−1.66	1	1	1	1	1	1	1	1	1	1
61	Hexestrol monomethyl ether	−1.63	1	1	1	1	1	1	1	1	1	1
62	Clomiphene*	−1.64	1	1	1	1	1	1	1	1	1	1
63	Nafoxidine	−1.63	1	1	1	1	1	1	1	1	1	1
64	Tamoxifen	−1.59	1	1	1	1	1	1	1	1	1	1
65	4-Hydroxy-tamoxifen	−1.49	1	1	1	1	0	1	1	1	0	1
66	6-Hydroxyflavone	−2.77	0	0	0	1	0	0	0	0	1	0
67	4′-Hydroxyflavanone	−2.48	0	0	0	0	0	0	0	1	0	0
68	Genistein	−2.44	0	0	0	0	0	0	0	1	0	0
69	Flavone	−2.40	0	0	0	0	1	0	0	0	0	0
70	Equol	−2.39	0	0	1	0	0	0	1	1	0	1
71	Chalcone	−2.32	0	0	0	0	0	0	0	0	0	0
72	4′-Hydroxychalcone*	−2.27	1	0	0	1	0	1	0	1	1	0
73	Flavanone	−2.25	1	0	1	0	1	0	1	1	0	0
74	4-Hydroxychalcone	−2.19	1	0	1	1	0	0	0	1	1	1
75	Zearalanone	−2.14	1	1	0	1	1	1	1	1	1	1
76	β-Zearalenol	−2.09	1	1	0	1	1	1	1	1	1	1
77	6-Hydroxyflavanone	−1.78	1	0	0	0	0	0	0	1	0	0
78	β-Zearalanol	−1.72	1	1	1	1	1	1	1	1	1	1
79	Zearalenol	−1.64	1	1	0	1	1	1	1	1	1	1
80	Coumestrol	NB	0	0	0	0	0	0	0	0	0	0
81	7-Hydroxyflavone	NB	0	0	0	0	0	0	0	1	1	0
82	Naringin*	SB	0	0	0	0	0	0	0	0	0	0
83	3-Chlorophenol	−3.17	0	0	0	0	0	0	0	0	0	0
84	Propylparaben	−3.00	0	0	0	0	0	0	0	0	0	0
85	4-Benzyloxyphenol	−2.89	0	0	0	0	0	0	0	0	0	0
86	Isoeugenol	−2.81	0	0	0	0	0	0	0	0	0	0

TABLE 12.1 (CONTINUED)
Observed and Calculated Androgenic Binding Activity of Chemicals

Number	Chemical Name	log RBA	AC[a]	C1	C2	C3	C4	C5	C6	C7	C8	C9
87	4-*tert*-Butylphenol	−2.67	0	0	0	0	0	0	0	0	0	0
88	4-Chloro-2-methyl phenol	−2.59	0	0	0	0	0	0	0	0	0	0
89	2-*sec*-Butylphenol	−2.52	0	0	0	0	0	0	0	0	0	0
90	4-*sec*-Butylphenol	−2.44	0	0	1	0	0	0	0	0	0	0
91	4-*tert*-Amylphenol	−2.39	0	0	0	0	0	0	0	0	0	0
92	4-Dodecylphenol*	−1.81	1	0	1	1	1	1	1	1	1	1
93	4-*n*-Octylphenol	−1.80	1	1	1	1	1	1	1	1	1	1
94	Igepal CO-210	−1.78	1	0	0	1	0	1	1	1	1	1
95	4-Heptyloxyphenol	−1.69	1	1	0	1	0	0	1	1	0	1
96	Nonylphenol	−1.57	1	1	1	1	1	1	1	1	1	1
97	Vanillin	NB	0	0	0	0	0	0	0	0	0	0
98	Phenol	NB	0	0	0	0	0	0	0	0	0	0
99	Methyl paraben	NB	0	0	0	0	0	0	0	0	0	0
100	2-Chlorophenol	NB	0	0	0	0	0	0	0	0	0	0
101	4-Ethylphenol	SB	0	0	0	0	0	0	0	0	0	0
102	4′-Chloroacetoacetanilide*	−3.46	0	0	0	0	0	0	0	0	0	0
103	Procymidone	−2.61	0	0	1	1	0	0	0	0	0	0
104	Metolachlor	−2.61	0	1	0	0	0	0	0	0	1	0
105	Vinclozolin	−2.50	0	0	1	1	1	1	0	0	0	0
106	Flutamide	−2.42	0	0	0	0	0	0	0	0	0	0
107	Linuron	−2.25	1	0	0	1	0	0	1	0	0	1
108	Propanil (DCPA)	−2.22	1	0	0	0	0	0	1	1	0	1
109	Fenpiclonil	−1.61	1	0	0	0	0	0	1	1	0	1
110	*p*-Lactophenetide	NB	0	0	0	0	0	0	0	0	0	0
111	4-Hydroxybenzophenone	−2.78	0	0	0	0	0	0	0	0	0	0
112	4,4′-Dihydroxybenzophenone*	−2.67	0	0	0	0	0	0	0	0	0	0
113	Benzophenone	−2.63	0	0	0	0	0	0	0	0	0	0
114	2,4-Dihydroxybenzophenone	−2.53	0	0	0	0	0	0	0	1	0	0
115	Bisphenol A	−2.39	0	0	0	0	1	1	1	1	0	1
116	*p*-Cumyl phenol	−2.11	1	1	1	1	1	1	1	1	1	1
117	Bisphenol B	−2.09	1	1	0	1	1	1	1	1	0	1
118	*o,p′*-DDE	−1.81	1	1	0	1	1	1	1	1	1	1
119	*p,p′*-DDT	−1.76	1	1	1	1	1	1	1	1	1	1
120	*p,p′*-DDD	−1.70	1	1	1	1	1	1	1	1	1	1
121	*p,p′*-DDE	−1.70	1	1	1	1	1	1	1	1	0	1
122	*o,p′*-DDT*	−1.69	1	1	1	1	1	1	1	1	1	1
123	*o,p′*-DDD	−1.52	1	1	1	1	1	1	1	1	1	1
124	*p,p′*-Methoxychlor olefin	−2.20	1	1	1	1	1	1	1	1	1	1
125	*p,p′*-Methoxychlor	−1.94	1	1	1	1	1	1	1	1	1	1
126	Monohydroxymethoxychlor olefin	−1.84	1	1	0	1	1	1	1	1	1	1
127	HPTE	−1.47	1	1	1	1	1	1	1	1	1	1
128	Dihydroxymethoxychlor olefin	−1.31	1	1	0	1	1	1	1	1	0	1

(continued)

TABLE 12.1 (CONTINUED)
Observed and Calculated Androgenic Binding Activity of Chemicals

Number	Chemical Name	log RBA	AC[a]	C1	C2	C3	C4	C5	C6	C7	C8	C9
129	3,3′,5,5′-Tetrachloro-4, 4′-biphenyldiol	−2.10	1	1	1	1	1	1	1	1	0	1
130	2,2′,4,4′-Tetraclorobiphenyl	−1.74	1	1	0	1	1	1	1	1	0	1
131	2,3,4,5-Tetrachloro-4′-biphenylol	−1.73	1	1	1	1	1	1	1	1	0	1
132	2,4′-Dichlorobiphenyl*	−1.72	1	0	1	0	0	0	0	0	0	0
133	4-Hydroxybiphenyl	−1.43	1	0	0	1	0	0	1	1	0	1
134	4,4′-Dichlorobiphenyl	NB	0	0	0	0	0	0	0	0	0	0
135	2,4,5-T	−3.18	0	0	1	1	0	0	0	0	0	0
136	Lindane (γ-HCH)	−2.12	1	1	1	1	1	1	1	1	0	1
137	Aldrin	−2.02	1	1	1	1	1	1	1	1	1	1
138	Endosulfan	−1.87	1	1	1	1	1	1	1	1	1	1
139	Heptachlor	−1.64	1	1	1	1	1	1	1	1	0	1
140	Kepone	−1.58	1	1	1	1	1	1	1	1	1	1
141	Chlordane	−1.51	1	1	1	1	1	1	1	1	1	1
142	2,4-D (2,4-dichlorophenoxy-acetic acid)*	NB	0	0	0	0	0	0	0	0	0	0
143	Hexachlorobenzene	NB	0	0	0	0	0	0	0	0	0	0
144	Mirex	NB	0	1	1	0	0	0	0	0	1	0
145	Diisononylphthalate	−3.56	0	0	1	0	0	0	0	0	0	0
146	Diethylphthalate	−3.44	0	0	0	0	0	0	0	0	0	0
147	Bis(n-octyl)phthalate	−3.28	0	0	1	0	0	0	0	0	0	0
148	di-i-Butylphthalate (DIBP)	−2.22	1	1	0	1	1	1	1	1	1	1
149	Butylbenzylphthalate	−2.07	1	1	1	1	1	1	1	1	1	1
150	di-n-Butylphthalate (DBuP)	−1.95	1	1	0	1	1	1	1	1	1	1
151	Bis(2-ethylhexyl)phthalate	SB	0	0	0	0	0	0	0	0	0	0
152	Triphenylethylene*	−1.98	1	0	1	0	0	0	0	0	1	0
153	sec-Butylbenzene	NB	0	0	0	0	0	0	0	0	0	0
154	n-Butylbenzene	NB	0	0	0	0	0	0	0	0	0	0
155	1,6-Dimehtylnaphthalene	NB	0	0	0	0	0	0	0	0	0	0
156	1,3-Butadiene,trans,trans-1, 4-diphenyl	NB	0	0	0	0	0	0	0	0	0	0
157	Chrysene	NB	0	0	0	0	0	0	0	0	0	0
158	1,1,2-Triphenylpropane	NB	0	0	1	0	1	1	0	0	1	0
159	Diisobutyl adipate	−2.84	0	0	0	0	0	0	0	0	0	0
160	Dibutyl adipate	−2.73	0	0	0	0	0	0	0	0	0	0
161	Spermidine	NB	0	0	0	0	0	0	0	0	0	0
162	Suberic acid*	NB	0	0	0	0	0	0	0	0	0	0
163	2-Ethyl-1,3-hexanediol	NB	0	0	0	0	0	0	0	0	0	0
164	1,2-Octanediol	NB	0	0	0	0	0	0	0	0	0	0
165	1,8-Octanediol	NB	0	0	0	0	0	0	0	0	0	0
166	1-Octen-3-ol	NB	0	0	0	0	0	0	0	0	0	0
167	Palmitic acid	NB	0	0	0	0	0	0	0	0	0	0
168	Di-2-Ethylhexyl adipate	NB	0	0	0	0	0	0	0	0	0	0
169	4-Amino butylbenzoate	−2.85	0	0	0	0	0	0	0	0	0	0

TABLE 12.1 (CONTINUED)
Observed and Calculated Androgenic Binding Activity of Chemicals

Number	Chemical Name	log RBA	AC[a]	C1	C2	C3	C4	C5	C6	C7	C8	C9
170	4-Heptyloxybenzoic acid	−2.74	0	1	0	0	0	0	0	0	0	0
171	Salicylamide	NB	0	0	0	0	0	0	0	0	0	0
172	Cinnamic acid*	NB	0	0	0	0	0	0	0	0	0	0
173	Methyl salicylate	NB	0	0	0	0	0	0	0	0	0	0
174	1-Methoxy-4-[1-propenyl] benzene	−3.19	0	0	0	0	0	0	0	0	0	0
175	Carbaryl	−3.12	0	0	0	0	0	0	0	0	0	0
176	Nordihydroguaiaretic acid	−2.28	1	1	0	1	1	1	1	1	1	1
177	4-(3,5-Diphenylcyclohexyl) phenol	−2.27	1	1	1	1	1	1	1	1	1	1
178	2,6-Dihydroxyanthraquinone	NB	0	0	0	0	0	0	0	0	0	0
179	2-Naphthol	NB	0	0	0	0	0	0	0	0	0	0
180	2-Benzyl-isoindole-1,3-dione	−3.12	0	0	1	0	1	0	0	1	0	0
181	2-(4-OH-benzyl)isoindole-1,3-dione	−2.76	0	0	1	1	1	0	0	1	1	0
182	2-(4-Nitro-benzyl)isoindole-1,3-dione*	−2.46	0	1	1	1	1	0	0	0	0	0
183	Methylparathion	−2.26	1	0	1	1	1	1	1	1	0	1
184	Ethylparathion	−2.05	1	1	1	1	1	1	1	1	1	1
185	Triphenyl phosphate	−1.69	1	1	1	1	1	1	1	1	1	1
186	Triphenylsilanol	−2.05	1	1	1	1	1	1	1	1	1	1
187	Simazine	NB	0	0	0	0	0	0	0	0	0	0
188	Atrazine	NB	0	0	0	0	0	0	0	0	0	0
189	Prometon	SB	0	0	0	0	0	0	0	0	0	0
190	Aurin	−1.7	1	0	0	1	1	1	1	1	1	1
191	Phenol red	NB	0	0	1	0	0	0	0	1	1	0
192	Phenolphthalin*	NB	0	1	1	1	0	1	1	1	1	1
193	Folic acid	NB	0	0	0	0	0	0	0	0	1	0
194	Caffeine	NB	0	0	0	0	0	0	0	0	0	0
195	Melatonin	NB	0	0	0	0	0	0	0	0	0	0
196	4,4′-Methylenebis(N, N-dimethylaniline)	NB	0	0	1	0	1	1	0	0	0	0
197	Doisynoestrol	NB	0	1	1	1	1	1	1	1	1	1
198	4,4′-Sulfonyldiphenol	−3.09	0	0	0	0	0	0	0	0	0	0

Note: * Denotes chemicals belonging to the testing set.

[a] Actual classification (AC), calculated classification with Classification and Regression Tree (CART) (C1), linear neural network (LNN) model (C2), three-layer perceptron (TLP) models (C3 and C4), radial basis function network (RBFN) model (C5), probabilistic neural network (PNN) models (C6 and C7), learning vector quantization (LVQ1) model (C8), and support vector machine (SVM) model (C9).

[b] NB, nonbinder; SB, slight binder.

which no further split is necessary becomes a terminal node [34,35]. CART analysis was performed from SCAN (Minitab, State College, Pennsylvania).

12.2.3.2 Artificial Neural Networks (ANNs)

ANNs are now commonly used in Structure-Activity Relationship (SAR) and Quantitative Structure-Activity Relationship (QSAR) to find complex nonlinear relationships between sets of descriptors and biological data. Among all the available paradigms [36,37], five different types of ANNs, were selected mainly because they present various levels of complexity. Their characteristics are briefly described in the subparagraphs below. All the neural network models were derived from Statistica (StatSoft, Paris) after a pretreatment of the data consisting in a classical min/max transformation.

12.2.3.2.1 Linear Neural Network (LNN)

According to the Ockham's Razor principle, also called the law of parsimony, a simple model should always be chosen in preference to a complex model if the performances of the latter do not significantly outperform those of the former. The simplest ANN is a linear neural network (LNN) without a hidden layer and an output with dot product synaptic function and identity activation function. It is equivalent to a classical discriminant analysis. The LNN was trained from the standard pseudo-inverse (SVD) linear optimization algorithm [38]. Briefly, this algorithm uses the singular value decomposition technique to calculate the pseudo-inverse of the matrix needed to set the weights in the linear output layer, so as to find the least mean squared solution. In this study, the LNN was mainly used as a benchmark against which to compare the performances of other more complex ANNs.

12.2.3.2.2 Three-Layer Perceptron (TLP)

The TLP is perhaps the most popular ANN in use not only in QSAR (for example, [36,39–46]) but also in a huge number of other disciplines [47]. This ANN, like a LNN, presents one input layer with a number of neurons corresponding to the number of selected molecular descriptors and one output layer of one neuron in numerous problems including the case of a binary classification (active versus inactive). In addition, it includes a hidden layer with an adjustable number of neurons which is determined from a trial and error procedure but which is basically linked to the number of neurons on the input layer as well as the training set size. Too many connections often yield overfitting; hence, to avoid problems, it is necessary to limit the number of connections within the network. The neurons of each layer are connected in the forward direction (that is, input to output) and are activated by means of activation functions. Each connection is associated to a weight. The weights are adjusted during the learning process aiming at minimizing an error computed from the target and calculated outputs. Numerous learning algorithms are available, and among them, the back-propagation alone or in combination with the conjugated gradient descent or Levenberg-Marquardt algorithms [45] were tested. The two parameters necessary to accurately tune the TLP are the learning rate (η) and the momentum (α). A large η value corresponds to a rapid learning but might also result in oscillations.

If η is set too low, the convergence is difficult and the risk of falling into and remaining in local minima is high. The goal of α is to prevent oscillations [45].

12.2.3.2.3   Radial Basis Function Network (RBFN)

RBFNs are three-layer feedforward ANNs also trained using a supervised training algorithm. They have shown their interest in QSAR (for example, [48–50]). RBFNs are typically configured with a single hidden layer of neurons whose activation function is selected from a class of functions called radial basis functions, usually a Gaussian or some other kernel function. Each hidden unit acts as a locally tuned processor that computes a score for the match between the input vector and its connection weights or centers. The weights connecting the basis units to the outputs are used to take linear combinations of the hidden units to produce the final classification or output. More precisely, in an RBFN, the weights into the hidden layer basis units are usually set before the second layer of weights is adjusted. As the input moves away from the connection weights, the activation value falls off. This is why the term "center" is allocated to the first-layer weights [31]. These center weights can be computed by different statistical techniques. In the present study, the K-means algorithm was used. They are then used to set the areas of sensitivity for the RBF hidden units, which then remain fixed. Once the hidden layer weights are set, a second phase of training allows the adjustment of the output weights [31]. An RBFN trains much faster than a classical back-propagation three-layer perceptron and does not suffer from local minima.

12.2.3.2.4   Probabilistic Neural Network (PNN)

A PNN is a supervised ANN that features a feedforward architecture. In environmental QSAR, the PNNs have been widely popularized by Kaiser [51–55]. Typically, a PNN is particularly suited for classification problems. It includes an input, a radial, and an output layer. The radial units are copied directly from the training data, one per case. Each models a Gaussian function centered at the training case. There is one output unit per class. Each is connected to all the radial units belonging to its class, with zero connections from all other radial units. Consequently, the output units simply add up the responses of the units belonging to their own class. The outputs are each proportional to the kernel-based estimates of the probability density functions of the various classes. By normalizing these to sum to 1.0, estimates of class probability are produced. The technique is extremely fast, because training a PNN actually consists mostly of copying training cases into the network. Another great advantage of a PNN is the fact that the output is probabilistic, yielding its interpretation easily [56]. With enough training data, a PNN is guaranteed to converge to a Bayesian classifier. Last, a PNN algorithm allows data to be added or deleted from the training set without lengthy retraining, while this is not the case with classical ANNs such as a back-propagation TLP [57].

12.2.3.2.5   Learning Vector Quantization (LVQ)

LVQ, which is only marginally used in QSAR [58,59], can be seen as a supervised version of the Kohonen self-organizing map (KSOM) [60]. Briefly, KSOM performs a mapping from an n-dimensional input vector onto a two-dimensional (2D) array of nodes displayed in a rectangular or hexagonal lattice. This mapping preserves the topology of the input data. This means that input vectors that are similar to each other are mapped to neighboring regions of the 2D output lattice. Each node in the output lattice is associated to an n-dimensional reference vector of weights. KSOM works by comparing the relationship, in terms of Euclidean distance, between each input vector and each reference vector in an iterative process during which the reference vectors are adjusted. The closest reference vector to a given input vector (i.e., the winning reference vector) is updated to match at best the input vector. LVQ uses the same internal architecture as KSOM but the learning algorithm is supervised. During the learning phase, the input data are tagged with their correct class and each output neuron represents a known category. There are several LVQ algorithms characterized by different learning rules. In this study, LVQ1, rather similar to the KSOM training algorithm, was used [60].

12.2.3.3 Support Vector Machine (SVM)

SVMs are learning methods originally designed to perform classifications, but they are now increasingly used for real function approximation (regression estimation) tasks [32,61,62]. SVMs are commonly selected as a statistical tool for designing (Q)SAR models because they compete favorably with a lot of linear and nonlinear methods [48–50,63–69]. SVM classification is rooted in the kernel trick, introduced by Aizeman and coworkers [70] which is a method allowing the linear separation of objects after their mapping into a higher-dimensional space. In the case of a SVM, a maximum margin hyperplane is constructed that separates the two clouds of points and that is at equal distance from the two. A cost parameter (C), which is necessary to tune the system, allows some flexibility in separating the categories by creating a soft margin that permits some data points to push their way through the margin of the separating hyperplane without affecting the final result [71].

In this work, a Gaussian kernel function, commonly used for classification problems, was employed. The spread of the Gaussian kernel was monitored by means of the γ parameter.

SVM models were derived from in-house MATLAB software (The MathWorks, Natick, Massachusetts) and from the e1071 R-package [72] after standardization of the data (zero mean and unit variance).

12.3   RESULTS AND DISCUSSION

A CART analysis performed from the 15 selected molecular descriptors yielded a rather simple tree constructed with only seven of them (Figure 12.1). Among these descriptors, three are topological indices: the molecular volume (T32), which is at the origin of the tree; the Kier flexibility index (T13); and the average information content of zero order (T14). It is also interesting to note that the 1-octanol/water partition coefficient (log P) appears twice within the tree. The tree displayed in Figure 12.1 represents the best compromise between the number of descriptors and the modeling performances obtained with the learning and testing sets. Thus, with this tree, 18

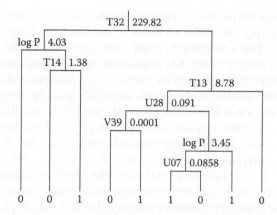

FIGURE 12.1 Classification tree obtained with the Classification and Regression Tree (CART) model (C1).

chemicals belonging to the learning set of 178 chemicals are badly predicted (C1 in Table 12.1). This represents 89.9% of good classifications. The results are worse for the testing set of 20 chemicals because eight of them are incorrectly predicted by the decision tree (Table 12.1) yielding only 60% of good predictions.

The best linear neural network (LNN) model included the whole set of descriptors as inputs. The percentages of correct predictions are 71.9 and 75 for the learning set and testing set, respectively. They correspond to 50 and 5 chemicals incorrectly predicted for the learning and testing sets, respectively (C2 in Table 12.1). It is noteworthy that the five chemicals of the testing set that are badly predicted by the LNN model are also incorrectly classified by the decision tree (Table 12.1). Although the predictions obtained on the external testing set are acceptable and significantly better than those produced by the decision tree, it is the converse regarding the learning set. The results obtained with the LNN clearly reveal that the linear methods are not suited to model the complex relationships existing between the AR-binding activity of the chemicals and the set of selected topological and physicochemical descriptors.

Consequently, a TLP was first used as a statistical engine to find a nonlinear relationship between the classes of activity of the chemicals and the molecular descriptors. A 15/10/1 (input/hidden/output layer) TLP model provided interesting results after only 135 cycles (100 cycles with the back-propagation algorithm and 35 with the conjugate gradient descent algorithm). The learning rate (η) and momentum (α) were equal to 0.01 and 0.3, respectively. The synaptic functions for the three layers were all linear, and the activation functions were linear, hyperbolic, and logistic from the input layer to the output layer.

Inspection of Table 12.1 shows that 17 chemicals belonging to the learning set are badly predicted by this TLP model. This represents 90.4% of good classifications. Regarding the testing set, four chemicals are incorrectly predicted yielding 80% of good predictions (C3 in Table 12.1). With a so limited number of epochs, undoubtedly the model is not overtrained. Conversely, because the network includes

160 connections and the number of sample examples in the learning set is 178, some problems of overfitting cannot be excluded. Consequently, a huge number of runs were performed to find a model with fewer connections but, in the meantime, that showed acceptable performances. It is noteworthy that a pruning algorithm was used to find the best compromise between the number of connections within the ANN and the simulation performances of the model, especially with the external testing set. Thus, the best model was a 8/10/1 TLP obtained after 501 cycles (100 cycles with the back-propagation algorithm and 401 with the conjugate gradient descent algorithm). The input neurons were the following molecular descriptors: log P, T13 (Kier flexibility index), T14 (average information content of zero order), T19 (Balaban index), T28 (YZ shadow), U04 (minimum partial charge for a hydrogen atom), U10 (polarity parameter/square distance), and V39 (minimum nucleophilic reactivity index for an oxygen atom). The η and α parameters were equal to 0.01 and 0.3, respectively. The synaptic functions for the three layers were also all linear and the activation functions were linear, hyperbolic, and logistic from the input layer to the output layer. Inspection of Table 12.1 (C4) shows that with this new TLP model, 28 chemicals belonging to the learning set are badly predicted. This represents 84.3% of correct classifications. Regarding the testing set, five chemicals are incorrectly predicted (Table 12.1) yielding 75% of good predictions.

It is worth noting that the two above TLP models were selected as the most predictive among about 6,500 models designed by changing the architecture and parameters of the ANNs. Obviously, each change was tested on a batch of ten runs. Consequently, even if the results provided by the TLP are better than those obtained with CART and the LNN, we claim that they remain a little bit disappointing due to the time and effort spent to obtain them.

Conversely, the simulation results obtained with an RBFN are more interesting (Table 12.1). The best model selected by using a pruning algorithm was a 5/16/1 ANN. The five input neurons were log P, T13 (Kier flexibility index), T14 (average information content of zero order), T19 (Balaban index), and T28 (YZ shadow). These descriptors were also chosen as inputs in the previous 8/10/1 TLP model. The synaptic functions for the three layers were also all linear and the activation functions were linear, negative of the exponential, and linear from the input layer to the output layer. With this model, 20 learning set chemicals are badly predicted (C5 in Table 12.1) representing 88.8% of good classifications. The results are not as good for the external testing set because three chemicals over a total of 20 are badly predicted by the RBFN model yielding 85% of good predictions. In addition to the good results obtained with the RBFN, it is worth mentioning that this ANN is easier to tune than a TLP. This advantage is even more true for a PNN.

A first PNN model was obtained from the whole set of descriptors. The synaptic functions for the three layers were all linear and the activation functions were linear, negative of the exponential, and unitsum from the input layer to the output layer. The unitsum function normalizes the outputs to sum to 1.0. It is used in a PNN to allow the outputs to be interpreted as probabilities. With this model, 93.8% and 80% of good classifications are recorded for the learning and testing sets, respectively. This only represents 11 chemicals badly classified in the learning set and four in the testing set (C6 in Table 12.1). Although this PNN model outperforms the RBFN model regarding the learning set, it is not the case for the testing set. Consequently, attempts were made to increase the simulation performances of the model for this set. A PNN model including seven descriptors as input neurons allowed us to reach this goal. The seven descriptors were the following: log P, T14 (average information content of zero order), T19 (Balaban index), T28 (YZ shadow), U10 (polarity parameter/square distance), V02 (HOMO energy), and V39 (minimum nucleophilic reactivity index for an oxygen atom). The synaptic and activation functions were the same as in the previous PNN model. With this new configuration, three chemicals are badly predicted with the testing set. This corresponds to 85% of good predictions (C7 in Table 12.1). Conversely, among the learning set of 178 chemicals, 17 are badly classified yielding 90.4% of good predictions (Table 12.1).

Among the different ANNs tested in this study, the learning vector quantization (LVQ1) network was undoubtedly the most difficult to parameterize. A lot of LVQ1 models with different architectures and characteristics were tested. The most predictive model included the seven following descriptors as inputs: T12 (Kier shape index of second order), T13 (Kier flexibility index), T19 (Balaban index), T28 (YZ shadow), T32 (molecular volume), U10 (polarity parameter/square distance), and V05 (HOMO–LUMO energy gap). It was obtained after 500 epochs with a learning rate equal to 0.001. The synaptic functions were linear and radial and the activation functions were linear and square root. It is noteworthy that the square root is used to transform the square distance activation in the LVQ network to the actual distance as an output. Although the model fails only in the prediction of three testing set chemicals (85% of correct predictions), 40 chemicals belonging to the learning set are badly predicted yielding only 77.5% of good predictions (C8 in Table 12.1).

The last statistical tool submitted to the comparison exercise was an SVM. Again, a huge number of configurations were tested to obtain a model presenting optimized performances on both sets. This was achieved with a model including the whole set of descriptors as inputs and a Gaussian kernel ($C = 3$ and $\gamma = 0.128$). Thus, inspection of Table 12.1 (C9) reveals that with this SVM model, 93.8% and 80% of good predictions are obtained for the learning and testing sets, respectively. These percentages correspond to 11 and 4 chemicals incorrectly predicted for the learning and testing sets of 178 and 20 chemicals, respectively.

A summary of the percentages of correct classifications obtained with the different models for the learning and testing sets is displayed in Table 12.2. The percentages of false positives (FPs) and false negatives (FNs) are also given. An FP suggests that the chemical is active, but this is actually not the case. For an FN, it is the converse. Even if in the design process of models the numbers of FNs and FPs both have to be minimized as far as possible, in the frame of the hazard and risk assessment of chemicals, the FNs are more dangerous than the FPs. Consequently, models favoring FN predictions have to be eliminated in priority. Thus, inspection of Table 12.2 shows that the PNN (C7) model presents the highest percentage of good predictions for the external testing set and the lowest percentage of FNs. Conversely, the percentage of FPs is rather high. Consequently, this protective model can be selected for the detection of chemicals acting on the AR especially in environmental decision-making processes rooted in the precautionary principle. The other PNN

TABLE 12.2
Simulation Performances of the Selected Models

Model[a]	LS(%)[b]	TS(%)	FP(%)	FN(%)
CART (C1)	89.9	60	5.1	8.1
LNN (C2)	71.9	75	13.6	14.1
TLP (C3)	90.4	80	6.1	4.5
TLP (C4)	84.3	75	9.1	7.6
RBFN (C5)	88.8	85	5.6	6.1
PNN (C6)	93.8	80	4.5	3.0
PNN (C7)	90.4	85	8.1	2.0
LVQ1 (C8)	77.5	85	11.6	10.1
SVM (C9)	93.8	80	4.5	3.0

[a] See text for the significance of the acronyms characterizing the different models.
[b] Percentages of good predictions for the learning set (LS) and the testing set (TS). FP,
false positive; FN, false negative.

model (C6) and the SVM (C9) model, which provide a good compromise between the
percentages of correct predictions on both sets and low percentages of FNs and FPs
(Table 12.2), can also be selected. At the opposite, the LNN (C2) and the LVQ1 (C8)
models have to be eliminated due to their poor performances. This is particularly
true for the former.

Even if Table 12.2 reveals important variations in the performances of the
models, about 58% of the chemicals (that is, 114/198) are correctly predicted by all
the selected models (Table 12.1). At the opposite, only 17α-estradiol (chemical number 24), pregnenolone (chemical number 51), and doisynoestrol (chemical number
197) are always badly predicted by the models (Table 12.1).

It is important to note that different learning and testing sets of 178/20 chemicals
were modeled with the seven selected statistical methods, and the results presented
here correspond to those providing the worst results on the external testing set. This
was the unique way to perform a fair comparison of the linear and nonlinear methods and to discuss the androgenic binding activity of the chemicals.

Last, it is worth mentioning that the 1-octanol/water partition coefficient (log P),
the Kier flexibility index (T13), the average information content of zero order (T14),
and the Balaban index (T19) are very often found as molecular descriptors in the
models. The importance of log P to AR binding was stressed by Fang et al. [28],
especially for phytoestrogens and phenols. Its presence as input in all the models
except in the LVQ1 model confirms this finding. Regarding the topological indices,
we claim that their selection in the models has to be related to their ability to discriminate chemicals having very different structures. Indeed, topological indices are
particularly suited to encode size and branching information within the molecules
[73]. Among them, the Balaban index (T19) was undoubtedly the most interesting
descriptor to discriminate the active and nonactive compounds.

12.4   CONCLUSIONS

The models selected in this study show a domain of application superior to those of the 3D and 2D QSAR models designed by Hong et al. [28] and Zhao et al. [30] which totally or partially excluded important chemical families such as aromatic acids, triazines, or triphenylmethanes.

On the basis of the learning and testing sets presented in Table 12.1 but also from the results obtained by selecting other sets of 178/20 chemicals, the modeling performances of the seven studied statistical methods are as follows: PNN ~ SVM > RBFN > TLP > CART > LVQ1 > LNN. In the models, the 1-octanol/water partition coefficient (log P) and topological indices are very often selected as molecular descriptors. These descriptors are easily interpretable and can be computed for any kind of organic molecules.

More generally, the results obtained in this study are rather good, the models generally presenting high percentages of good predictions. However, the modeling target was not too complex. Indeed, the design of a model allowing the discrimination of chemicals between two categories is not too difficult especially when nonlinear learning techniques are used and when the learning set is large and correctly designed. Consequently, the next step will be to envision the design of models allowing us to classify the chemicals into three and more categories determined from their AR binding activity.

ACKNOWLEDGMENTS

This study was partially granted by the French Ministry of Ecology and Sustainable Development (PNETOX-N°24-B/2004-N°CV 04000175). A part of this work was presented at the European SETAC meeting (Porto, Portugal, 20-24/05/07). The whole study was presented at the CMTPI 2007 meeting (Moscow, Russia, 01-05/09/07).

REFERENCES

[1]   J. Lintelmann, A. Katayama, N. Kurihara, L. Shore, and A. Wenzel, *Endocrine disruptors in the environment*, Pure Appl. Chem. 75 (2003), pp. 631–681.

[2]   L.E. Gray, J. Ostby, V. Wilson, C. Lambright, K. Bobseine, P. Hartig, A. Hotchkiss, C. Wolf, J. Furr, M. Price, L. Parks, R.L. Cooper, T.E. Stoker, S.C. Laws, S.J. Degitz, K.M. Jensen, M.D. Kahl, J.J. Korte, E.A. Makynen, J.E. Tietge, and G.T. Ankley, *Xenoendocrine disrupters-tiered screening and testing. Filling key data gaps*, Toxicology 181/182 (2002), pp. 371–382.

[3]   W.S. Baldwin, D.L. Milam, and G.A. LeBlanc, *Physiological and biochemical perturbations in* Daphnia magna *following exposure to the model environmental estrogen diethylstilbestrol*, Environ. Toxicol. Chem. 14 (1995), pp. 945–952.

[4]   E.J. Routledge and J.P. Sumpter, *Estrogenic activity of surfactants and some of their degradation products assessed using a recombinant yeast screen*, Environ. Toxicol. Chem. 15 (1996), pp. 241–248.

[5]   J.E. Harries, D.A. Sheahan, S. Jobling, P. Matthiessen, P. Neall, J.P. Sumpter, T. Tylor, and N. Zaman, *Estrogenic activity in five United Kingdom rivers detected by measurement of vitellogenesis in caged male trout*, Environ. Toxicol. Chem. 16 (1997), pp. 534–542.

[6] J. Koistinen, M. Soimasuo, K. Tukia, A. Oikari, A. Blankenship, and J.P. Giesy, *Induction of EROD activity in HEPA-1 mouse hepatoma cells and estrogenicity in MCF-7 human breast cancer cells by extracts of pulp mill effluents, sludge, and sediment exposed to effluents*, Environ. Toxicol. Chem. 17 (1998), pp. 1499–1507.

[7] S.H. Safe and K. Gaido, *Phytoestrogens and anthropogenic estrogenic compounds*, Environ. Toxicol. Chem. 17 (1998), pp. 119–126.

[8] Y. Allen, A.P. Scott, P. Matthiessen, S. Haworth, J.E. Thain, and S. Feist, *Survey of estrogenic activity in United Kingdom estuarine and coastal waters and its effects on gonadal development of the flounder* Platichthys flesus, Environ. Toxicol. Chem. 18 (1999), pp. 1791–1800.

[9] A.K. Hotchkiss, J. Furr, E.A. Makynen, G.T. Ankley, and L.E. Gray, *In utero exposure to the environmental androgen trenbolone masculinizes female Sprague-Dawley rats*, Toxicol. Lett. 174 (2007), pp. 31–41.

[10] L.G. Parks, C.S. Lambright, E.F. Orlando, L.J. Guillette, G.T. Ankley, and L.E. Gray, *Masculinization of female mosquitofish in Kraft mill effluent-contaminated Fenholloway River water is associated with androgen receptor agonist activity*, Toxicol. Sci. 62 (2001), pp. 257–267.

[11] R.J. Ellis, M.R. van den Heuvel, E. Bandelj, M.A. Smith, L.H. McCarthy, T.R. Stuthridge, and D.R. Dietrich, In vivo *and* in vitro *assessment of the androgenic potential of a pulp and paper mill effluent*, Environ. Toxicol. Chem. 22 (2003), pp. 1448–1456.

[12] E.F. Orlando, D.E. Bass, L.M. Caltabiano, W.P. Davis, L.E. Gray, and L.J. Guillette, *Altered development and reproduction in mosquitofish exposed to pulp and paper mill effluent in the Fenholloway River, Florida, USA*, Aquat. Toxicol. 84 (2007), pp. 399–405.

[13] R. Urbatzka, A. van Cauwenberge, S. Maggioni, L. Vigano, A. Mandich, E. Benfenati, I. Lutz, and W. Kloas, *Androgenic and antiandrogenic activities in water and sediment samples from the River Lambro, Italy, detected by yeast androgen screen and chemical analyses*, Chemosphere 67 (2007), pp. 1080–1087.

[14] R. Kizu, K. Okamura, A. Toriba, A. Mizokami, K.L. Burnstein, C.M. Klinge, and K. Hayakawa, *Antiandrogenic activities of diesel exhaust particle extracts in PC3/AR human prostate carcinoma cells*, Toxicol. Sci. 76 (2003), pp. 299–309.

[15] C. Li, S. Taneda, A.K. Suzuki, C. Furuta, G. Watanabe, and K. Taya, *Estrogenic and anti-androgenic activities of 4-nitrophenol in diesel exhaust particles*, Toxicol. Appl. Pharmacol. 217 (2006), pp. 1–6.

[16] T. Suzuki, S. Kitamura, R. Khota, K. Sugihara, N. Fujimoto, and S. Ohta, *Estrogenic and antiandrogenic activities of 17 benzophenone derivatives used as UV stabilizers and sunscreens*, Toxicol. Appl. Pharmacol. 203 (2005), pp. 9–17.

[17] W.R. Kelce and L.E. Gray, *Endocrine disruptors: Effects on sex steroid hormone receptors and sex development*, in *Drug Toxicity in Embryonic Development II*, R.J. Kavlock and G.P. Daston, eds., Springer, Berlin, 1997, pp. 435–474.

[18] S.C. Maness, D.P. McDonnell, and K.W. Gaido, *Inhibition of androgen receptor-dependent transcriptional activity by DDT isomers and methoxychlor in HepG2 human hepatoma cells*, Toxicol. Appl. Pharmacol. 151 (1998), pp. 135–142.

[19] C. Lambright, J. Ostby, K. Bobseine, V. Wilson, A.K. Hotchkiss, P.C. Mann, and L.E. Gray, *Cellular and molecular mechanisms of action of linuron: An antiandrogenic herbicide that produces reproductive malformations in male rats*, Toxicol. Sci. 56 (2000), pp. 389–399.

[20] A.M. Vinggaard, C. Nellemann, M. Dalgaard, E.B. Jorgensen, and H.R. Andersen, *Antiandrogenic effects* in vitro *and* in vivo *of the fungicide prochloraz*, Toxicol. Sci. 69 (2002), pp. 344–353.

[21] S. Kitamura, T. Suzuki, S. Ohta, and N. Fujimoto, *Antiandrogenic activity and metabolism of the organophosphorus pesticide fenthion and related compounds*, Environ. Health Perspect. 111 (2003), pp. 503–508.

[22] P. Roy, H. Salminen, P. Koskimies, J. Simola, A. Smeds, P. Saukko, and I.T. Huhtaniemi, *Screening of some anti-androgenic endocrine disruptors using a recombinant cell-based in vitro bioassay*, J. Steroid Biochem. Molec. Biol. 88 (2004), pp. 157–166.

[23] A. Freyberger and H.J. Ahr, *Development and standardization of a simple binding assay for the detection of compounds with affinity for the androgen receptor*, Toxicology 195 (2004), pp. 113–126.

[24] N. Araki, K. Ohno, M. Nakai, M. Takeyoshi, and M. Iida, *Screening for androgen receptor activities in 253 industrial chemicals by in vitro reporter gene assays using AR-EcoScreen™ cells*, Toxicol. *In Vitro* 19 (2005), pp. 831–842.

[25] J. Devillers, N. Marchand-Geneste, A. Carpy, and J.M. Porcher, *SAR and QSAR modeling of endocrine disruptors*, SAR QSAR Environ. Res. 17 (2006), pp. 393–412.

[26] C.L. Waller, B.W. Juma, L.E. Gray, and W.R. Kelce, *Three-dimensional quantitative structure-activity relationships for androgen receptor ligands*, Toxicol. Appl. Pharmacol. 137 (1996), pp. 219–227.

[27] D.A. Loughney and C.F. Schwender, *A comparison of progestin and androgen receptor binding using the CoMFA technique*, J. Comput.-Aided Mol. Des. 6 (1992), pp. 569–581.

[28] H. Hong, H. Fang, Q. Xie, R. Perkins, D.M. Sheehan, and W. Tong, *Comparative molecular field analysis (CoMFA) model using a large diverse set of natural, synthetic and environmental chemicals for binding to the androgen receptor*, SAR QSAR Environ. Res. 14 (2003), pp. 373–388.

[29] H. Fang, W. Tong, W.S. Branham, C.L. Moland, S.L. Dial, H. Hong, Q. Xie, R. Perkins, W. Owens, and D.M. Sheehan, *Study of 202 natural, synthetic, and environmental chemicals for binding to the androgen receptor*, Chem. Res. Toxicol. 16 (2003), pp. 1338–1358.

[30] C.Y. Zhao, R.S. Zhang, H.X. Zhang, C.X. Xue, H.X. Liu, M.C. Liu, Z.D. Hu, and B.T. Fan. *QSAR study of natural, synthetic and environmental endocrine disrupting compounds for binding to the androgen receptor*, SAR QSAR Environ. Res. 16 (2005), pp. 349–367.

[31] J.P. Bigus, *Data Mining with Neural Networks*, McGraw-Hill, New York, 1996.

[32] N. Cristianini and J. Shawe-Taylor, *An Introduction to Support Vector Machines and Other Kernel-Based Learning Methods*, Cambridge University Press, London, 2000.

[33] A.R. Katritzky, R. Petrukhin, S. Perumal, M. Karelson, I. Prakash, and N. Desai, *A QSPR study of sweetness potency using the CODESSA program*, Croat. Chem. Acta 75 (2002), pp. 475–502.

[34] L. Breiman, J. Friedman, R. Olshen, and C. Stone, *Classification and Regression Trees*, Chapman & Hall, London, 1984.

[35] R.J. Lewis, *An introduction to classification and regression tree (CART) analysis*, Annual Meeting of the Society for Emergency Medicine, San Francisco, California, 2000.

[36] J. Devillers, *Neural Networks in QSAR and Drug Design*, Academic Press, London, 1996.

[37] R. Leardi, *Nature-Inspired Methods in Chemometrics: Genetic Algorithms and Artificial Neural Networks*, Elsevier, Amsterdam, 2003.

[38] G. Golub and W. Kahan, *Calculating the singular values and pseudo-inverse of a matrix*, J. SIAM Numer. Anal. Ser. B 3 (1965), pp. 205–224.

[39] J. Devillers, *A general QSAR model for predicting the acute toxicity of pesticides to Lepomis macrochirus*, SAR QSAR Environ. Res. 11 (2001), pp. 397–417.

[40] J. Devillers, M.H. Pham-Delègue, A. Decourtye, H. Budzinski, S. Cluzeau, and G. Maurin, *Structure-toxicity modeling of pesticides to honey bees*, SAR QSAR Environ. Res. 13 (2002), pp. 641–648.

[41] J. Devillers, M.H. Pham-Delègue, A. Decourtye, H. Budzinski, S. Cluzeau, and G. Maurin, *Modeling the acute toxicity of pesticides to* Apis mellifera, Bull. Insect. 56 (2003) pp. 103–109.

[42] J. Devillers, *Prediction of toxicity of organophosphorus insecticides against the midge,* Chironomus riparius, *via a QSAR neural network model integrating environmental variables*, Toxicol. Methods 10 (2000), pp. 69–79.

[43] J. Devillers, *QSAR model for predicting the acute toxicity of pesticides to gammarids*, in *Nature-Inspired Methods in Chemometrics: Genetic Algorithms and Artificial Neural Networks*, R. Leardi, ed., Elsevier, Amsterdam, 2003, pp. 323–339.

[44] J. Devillers, *Linear versus nonlinear QSAR modeling of the toxicity of phenol derivatives to* Tetrahymena pyriformis, SAR QSAR Environ. Res. 15 (2004), pp. 237–249.

[45] J. Devillers, *Strengths and weaknesses of the backpropagation neural network in QSAR and QSPR studies*, in *Neural Networks in QSAR and Drug Design*, J. Devillers, ed., Academic Press, London, 1996, pp. 1–46.

[46] K. Elkhou, A. I. Afifi, M. Kabbaj, D. Villemin, and D. Cherqaoui, *QSAR analysis of estrogen receptor ligands using neural networks*, ACH Models in Chemistry 137 (2000), pp. 633–642.

[47] R.C. Eberhart and R.W. Dobbins, *Neural Network PC Tools. A Practical Guide*, Academic Press, San Diego, California, 1990.

[48] X.J. Yao, A. Panaye, J.P. Doucet, R.S. Zhang, H.F. Chen, M.C. Liu, Z.D. Hu, and B.T. Fan, *Comparative study of QSAR/QSPR correlations using support vector machines, radial basis function neural networks, and multiple linear regression*, J. Chem. Inf. Comput. Sci. 44 (2004), pp. 1257–1266.

[49] X.J. Yao, A. Panaye, J.P. Doucet, H.F. Chen, R.S. Zhang, B.T. Fan, M.C. Liu, and Z.D. Hu, *Comparative classification study of toxicity mechanisms using support vector machines and radial basis function neural networks*, Anal. Chim. Acta 535 (2005), pp. 259–273.

[50] A. Panaye, B.T. Fan, J.P. Doucet, X.J. Yao, R.S. Zhang, M.C. Liu, and Z.D. Hu, *Quantitative structure-toxicity relationships (QSTRs): A comparative study of various nonlinear methods. General regression neural network, radial basis function neural network and support vector machine in predicting toxicity of nitro- and cyano- aromatics to* Tetrahymena pyriformis, SAR QSAR Environ. Res. 17 (2006), pp. 75–91.

[51] K.L.E. Kaiser and S.P. Niculescu, *On the PNN modeling of estrogen receptor binding data for carboxylic acid esters and organochlorine compounds*, Water Qual. Res. J. Canada 36 (2001), pp. 619–630.

[52] K.L.E. Kaiser and S.P. Niculescu, *Using probabilistic neural networks to model the toxicity of chemicals to the fathead minnow (*Pimephales promelas*): A study based on 865 compounds*, Chemosphere 38 (1999), pp. 3237–3245.

[53] S.P. Niculescu, K.L.E. Kaiser, and T.W. Schultz, *Modeling the toxicity of chemicals to* Tetrahymena pyriformis *using molecular fragment descriptors and probabilistic neural networks*, Arch. Environ. Contam. Toxicol. 39 (2000), pp. 289–298.

[54] K.L.E. Kaiser, S.P. Niculescu, and T.W. Schultz, *Probabilistic neural network modeling for the toxicity of chemicals to* Tetrahymena pyriformis *with molecular fragment descriptors*, SAR QSAR Environ. Res. 13 (2002), pp. 57–67.

[55] K.L.E. Kaiser and S.P. Niculescu, *Modeling acute toxicity of chemicals to* Daphnia magna*: A probabilistic neural network approach*, Environ. Toxicol. Chem. 20 (2001), pp. 420–431.

[56] P.D. Wasserman, *Advanced Methods in Neural Computing*, van Nostrand Reinhold, New York, 1993.

[57] J. Devillers, *QSAR modeling of large heterogeneous sets of molecules*, SAR QSAR Environ. Res. 12 (2001), pp. 515–528.

[58] N. Baurin, J.C. Mozziconacci, E. Arnoult, P. Chavatte, C. Marot, and L. Morin-Allory, *2D QSAR consensus prediction for high-throughput virtual screening. An application to COX-2 inhibition modeling and screening of the NCI database*, J. Chem. Inf. Comput. Sci. 44 (2004), pp. 276–285.

[59] A. Asikainen, M. Kolehmainen, J. Ruuskanen, and K. Tuppurainen, *Structure-based classification of active and inactive estrogenic compounds by decision tree, LVQ and kNN methods*, Chemosphere 62 (2006) pp. 659–673.

[60] T. Kohonen, *Self-Organizing Map*, Springer, Berlin, 1995.

[61] V.N. Vapnik, *The Nature of Statistical Learning Theory,* Springer, Berlin, 1995.

[62] C. Cortes and V. Vapnik, *Support vector networks*, Machine Learning 20 (1995), pp. 273–297.

[63] E. Byvatov, U. Fechner, J. Sadowski, and G. Schneider, *Comparison of support vector machine and artificial neural network systems for drug/nondrug classification*, J. Chem. Inf. Comput. Sci. 43 (2003), pp. 1882–1889.

[64] V.V. Zernov, K.V. Balakin, A.A. Ivaschenko, N.P. Savchuk, and I.V. Pletnev, *Drug discovery using support vector machines. The case studies of drug-likeness, agrochemical-likeness, and enzyme inhibition predictions*, J. Chem. Inf. Comput. Sci. 43 (2003), pp. 2048–2056.

[65] P. Mahé, L. Ralaivola, V. Stoven, and J.P. Vert, *The pharmacophore kernel for virtual screening with support vector machines*, J. Chem. Inf. Model. 46 (2006), pp. 2003–2014.

[66] Q. Li, A. Bender, J. Pei, and L. Lai, *A large descriptor set and a probabilistic kernel-based classifier significantly improve druglikeness classification*, J. Chem. Inf. Model. 47 (2007), pp. 1776–1786.

[67] J. Huang, G. Ma, I. Muhammad, and Y. Cheng, *Identifying P-glycoprotein substrates using a support vector machine optimized by a particle swarm*, J. Chem. Inf. Mod. 47 (2007), pp. 1638–1647.

[68] S. Yuan, M. Xiao, G. Zheng, M. Tian, and X. Lu, *Quantitative structure-property relationship studies on electrochemical degradation of substituted phenols using a support vector machine*, SAR QSAR Environ. Res. 17 (2006), pp. 473–481.

[69] J.P. Doucet, F. Barbault, H.R. Xia, A. Panaye, and B.T. Fan, *Nonlinear SVM approaches to QSPR/QSAR studies and Drug Design,* Cur. Comput. Aided Drug Des. 3 (2007), pp. 263–289.

[70] M. Aizerman, E. Braverman, and L. Rozonoer, *Theoretical foundations of the potential function method in pattern recognition learning*, Autom. Remote Control 25 (1964), pp. 821–837.

[71] W.S. Noble, *What is a support vector machine?* Nature Biotechnol. 24 (2006), pp. 1565–1567.

[72] E. Dimitriadou, K. Hornik, F. Leisch, D. Meyer, and A. Weingessel, *e1071: Misc Functions of the Department of Statistics*, TU Wien. R package version 1.5–16.

[73] J. Devillers and A.T. Balaban, *Topological Indices and Related Descriptors in QSAR and QSPR*, Gordon and Breach Science Publishers, The Netherlands, 1999.

13 SAR and QSAR Analyses of Substituted Dibenzoylhydrazines for Their Mode of Action as Ecdysone Agonists

Toshio Fujita and Yoshiaki Nakagawa

CONTENTS

ABSTRACT

A series of our Structure-Activity Relationship (SAR) and Quantitative Structure-Activity Relationship (QSAR) studies of synthetic molting hormone agonists, *N,N'*-dibenzoyl-*N-t*-butylhydrazines (DBHs) exhibiting insecticidal/larvicidal activity are reviewed in this chapter. We prepared a number of analogs where various substituents are introduced into the two benzene rings of DBH and measured their activity using various biological systems. Larvicidal activity was against larvae of the rice stem borer *Chilo suppressalis* and the molting hormone activity was in terms of the stimulation of *N*-acetylglucosamine incorporation in a cultured integument system of the same insect species. Binding affinity to the ecdysone receptor was assayed with intact Sf-9 cell lines in which the absorption, distribution, metabolism, and excretion (ADME) processes are negligible as well as using receptor proteins obtained by *in vitro* translation of the responsible cDNA cloned from cell-free preparation of integumentary tissue of *C. suppressalis*. Variations in the biological activity indices were either correlated between two types of activity or correlated using physicochemical molecular and substituent parameters in terms of the classical QSAR. Three-dimensional QSAR (comparative molecular field analysis [CoMFA]) for the activity of the receptor response using cultured cells of *Bombyx mori* has also been explored. Comparisons among correlations and with recently revealed X-ray cocrystallographic findings clearly indicate the physicochemical meaning of parameters significant in the correlation equations to help understand the molecular mechanism of the molting hormonal action especially for *lepidopteran* insect larvae.

KEYWORDS

Ecdysone agonists
Ecdysone receptor
Insecticides
Larvicides
N,N'-dibenzoyl-*N-t*-butylhydrazines
Quantitative Structure-Activity Relationship (QSAR)
Receptor binding

13.1 INTRODUCTION

Among *N,N'*-dibenzoyl-*N-t*-butylhydrazines (DBHs) are molting hormone agonists exhibiting insecticidal (or larvicidal) activity associated with premature abnormal molting that is ultimately lethal [1]. The first member of this series of agonists is RH-5849, which was discovered by scientists at Rohm and Haas in the late 1980s [2–4]. DBHs share the molecular target, the molting hormone receptor (EcR), with endogenous/endocrine-active ecdysteroids working in arthropods and nonarthropod invertebrates. The EcR is a member of superfamily of nuclear receptors, and one of ligand-dependent transcription factors. Because of simplicity of the structure and uniqueness of the action of DBHs, a number of analogs mostly with various combinations of substituents on two benzene rings have been explored for practical use. Four compounds, tebufenozide, methoxyfenozide, chromafenozide, and halofenozide (Figure 13.1), are currently used for pest insect control depending upon their selective activity spectra [5].

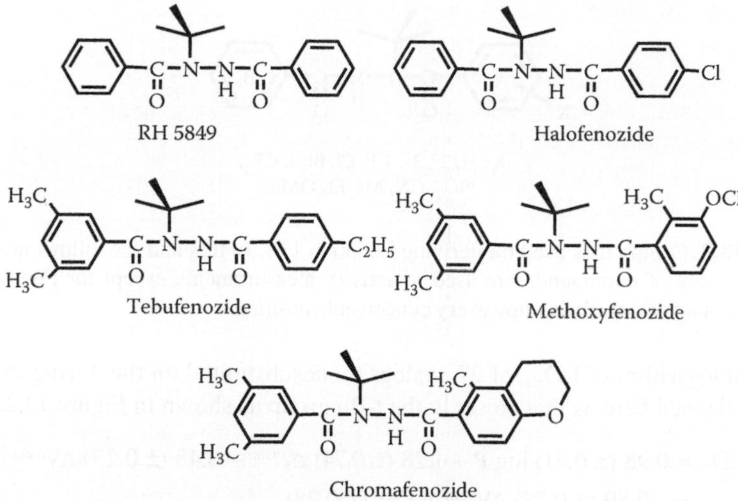

FIGURE 13.1 Structure of dibenzoylhydrazine larvicides.

We have been studying this series of compounds since 1991 [5]. A number of substituted DBHs and related compounds have been synthesized and their biological activities have been measured at such biological levels as whole body (larvicidal), tissue, cell, and receptor protein mostly prepared from rice stem borers, *Chilo suppressalis*. We have analyzed potency variations observed in these activities using physicochemical molecular and substituent parameters quantitatively in terms of the classical QSAR, and relationships between sets of activity indices at various levels (activity-activity relationships [AARs]). We also analyzed classical and three-dimensional (3D) Quantitative Structure-Activity Relationships (QSARs) of these compounds for their EcR activation effect measured using cloned cells from silkworm, *Bombyx mori*. We virtually constructed the 3D structure of *B. mori* EcR protein using a homology modeling procedure. With outcomes of these studies, the mechanism of action of DBH series of compounds has been elucidated with a considerable depth of understanding at the molecular level. The purpose of this chapter is to review our studies proposing intimate relationships between larvicidal activity and receptor binding for rice stem borers and physicochemical consistency of QSARs with homology-modeled receptor structure for silkworms. The original articles should be consulted for detailed experimental procedures.

13.2 BIOLOGICAL ACTIVITIES OF DBHS TO RICE STEM BORERS, *CHILO SUPPRESSALIS*

13.2.1 CLASSICAL QSAR ANALYSIS OF INSECTICIDAL ACTIVITY

The insecticidal activity to rice stem borer larvae was measured by topical application to the dorsal part in terms of LD_{50} (in mmol/insect) using 20 heads in a set for a number of substituted DBHs. Equation 13.1 was derived for the pLD_{50} value, the

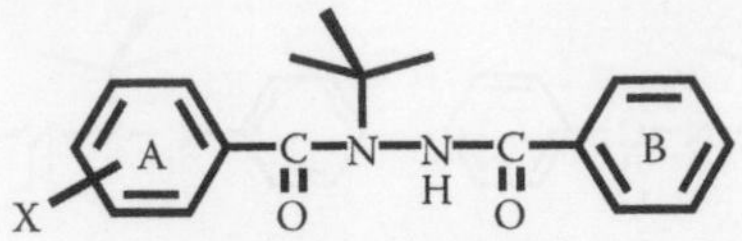

X : H, 2-, 3-, 4-F, Cl, Br, I, CF$_3$,
NO$_2$,CN, Me, Et, OMe

FIGURE 13.2 Compounds used for deriving Equation 13.1. In this and the following figures for the structure of compounds are used in activity measurements except for Figure 13.8, substituents not necessarily occupy every conceivable position.

reciprocal logarithm of LD$_{50}$, of 27 analogs monosubstituted on the A-ring [6]. The A-ring is defined here as that closer to the t-Bu group as shown in Figure 13.2.

$$pLD_{50} = 0.98 \, (\pm 0.31) \log P + 1.28 \, (\pm 0.74) \, \sigma_I^{ortho} - 0.48 \, (\pm 0.27) \, \Delta V^{meta}$$
$$- 0.89 \, (\pm 0.29) \, \Delta V^{para} + 3.61 \, (\pm 0.28)$$
$$n = 27, s = 0.300, r = 0.899 \tag{13.1}$$

In this and the following equations, n is the number of compounds, s is the standard deviation, r is the correlation coefficient, and the figures in parentheses are the 95% confidence intervals of the regression coefficient and intercept. Log P is the hydrophobicity parameter of compounds in terms of 1-octanol/water partition coefficient P. σ_I^{ortho} is for the inductive/field electronic effect of ortho substituents, and V represents the van der Waals' volume estimated after Bondi. The ΔV value used here is the difference from the reference V value of hydrogen and scaled by 0.1 to make the scale comparable to that of other parameters. Superscripts associated with ΔV parameters indicate the substituent positions.

Equation 13.1 was selected as that of the best quality considering both the physicochemical meaning and the statistical significance. Preliminary examinations for substituent effects at each position separately indicated a common participation of the hydrophobic effect of a comparable magnitude. The log P term in Equation 13.1 is compatible with those observed empirically in a number of QSAR examples. The coefficient close to unity means that the (sub)molecular moiety with structural variations undergoes hydrophobic interactions so as to be engulfed almost completely in hydrophobic milieu. Electronic effect specific to ortho substituents is well understood by considering to work in a proximity to the "side-chain" carbonyl group inductively (through space but not through bonds). The negative ΔV terms indicate unfavorable steric effects of meta and para substituents in terms of their volume. It should be recognized, however, that certain colinearities exist among various sets of steric parameters, such as ΔV, STERIMOL, and extended E_s values.

A number of analogs multisubstituted at the A-ring were also tested. The effect of multiple substitutions was analyzed by examining the difference between observed value and that calculated by Equation 13.1 assuming the additivity of substituent effects. In 2,4-, 3,4-, and 3,5-disubstituted compounds, the additivity model was approximately approved, and in 2,3-, 2,5-, and 2,6-di- and 2,3,5- and 2,3,6-trisubstituted derivatives, effects specific to substitution patterns to lower the activity were shown to be operative,

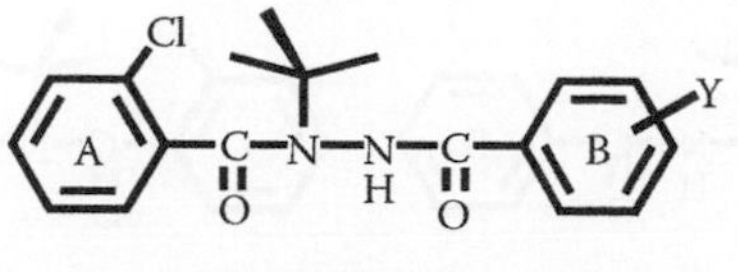

Y : H, 2-, 3-, 4-F, Cl, Br, I, CF$_3$, NO$_2$,
CN, Me, Et, Pr, Bu, OMe

FIGURE 13.3 Compounds used for deriving Equation 13.2.

but their physicochemical meaning was not exactly understood. In Equation 13.1, some ortho substituted compounds are not included because of their very low (inaccurate) activity. This was attributable to the size of substituents bulkier than a threshold.

In the next set of compounds, the A-ring substituent is fixed as the 2-Cl, because the 2-chloro compound is one of the most active compounds included in Equation 13.1. For compounds shown in Figure 13.3 in which the B-ring substituent is variously changed singularly, Equation 13.2 was formulated [7].

$$pLD_{50} = 0.72 \ (\pm 0.21) \log P - 0.88 \ (\pm 0.22) \Delta L^{ortho} - 0.98 \ (\pm 0.24) \Delta V^{meta}$$
$$- 0.59 \ (\pm 0.19) \Delta L^{para} + 4.92 \ (\pm 0.26)$$
$$n = 30, \ s = 0.254, \ r = 0.912 \tag{13.2}$$

In Equation 13.2, L is the STERIMOL length parameter, which represents the length of substituents (in Å) along the axis connecting the α atom of substituents with the rest of the molecule, and ΔL is the difference from that of the hydrogen atom. The hydrophobic effect (the log P term) nonspecific to positions and unfavorable steric effects of position dependence are similar to those for the A-ring substituents. The effect of multiple substitutions was also examined. Effects of such substitution patterns as 2,5-, 2,6-, 3,4-, and 3,5-disubstitutions were shown to be nearly additive. For 2,3- and 2,4- disubstituted compounds, the activity was suggested to be independent of the steric effect of 3- and 4-substituents.

13.2.2 CLASSICAL QSAR ANALYSIS OF THE *N*-ACETYLGLUCOSAMINE INCORPORATION INTO THE INTEGUMENT

Excised fragments of a certain size were prepared from the integument tissue of diapause larvae of the rice stem borer. Under certain culture conditions, the stimulative effect of DBHs on the incorporation of ^{14}C-labeled *N*-acetyl-glucosamine into fragments was estimated as a tissue-level activity and represented as the pEC_{50} value. EC_{50} (in M) is the concentration of DBHs required to exhibit 50% the maximum incorporation. DBHs of the types included in both Equation 13.1 and Equation 13.2 were used together here (Figure 13.4) to give Equation 13.3 [8].

$$pEC_{50} = 1.02 \ (\pm 0.30) \log P + 1.40 \ (\pm 0.99) \ \sigma_I^{ortho} (X) - 0.52 \ (\pm 0.35) \Delta V^{meta} (X)$$
$$- 1.08 \ (\pm 0.36) \Delta V^{para} (X) - 0.90 \ (\pm 0.33) \Delta V^{ortho} (Y)$$
$$- 1.17 \ (\pm 0.35) \Delta V^{meta} (Y) - 0.77 \ (\pm 0.33) \Delta V^{para} (Y) + 4.06 \ (\pm 0.85)$$
$$n = 37, \ s = 0.339, \ r = 0.900 \tag{13.3}$$

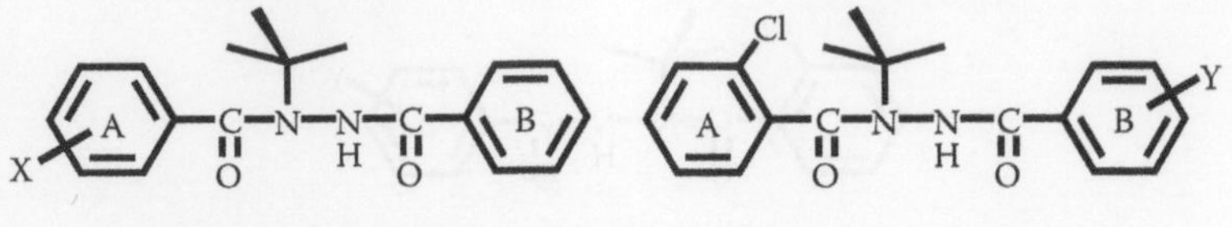

X : H, 2-, 3-, 4-F, Cl, I, CF$_3$,   Y : 2-, 3-, 4-F, Cl, Br, I, CF$_3$,
NO$_2$, CN, Me, OMe   NO$_2$, CN, Me, Et, OMe

FIGURE 13.4 Compounds used for *N*-acetylglucosamine incorporation.

In Equation 13.3, X and Y indicate the substituent locations, A- and B-ring, respectively. Each component of this equation corresponds very well to the corresponding terms in Equation 13.1 and Equation 13.2. The log P term of Equation 13.3 looks similar to that of Equation 13.1 for the effect of X substituents better than that of Equation 13.2, but they overlap within the 95% confidence intervals. In Equation 13.2 for the larvicidal effect of B-ring (Y) substituents, the steric effect of ortho and para substituents is represented by ΔL, the substituent length parameter, whereas in Equation 13.3, they are expressed by ΔV, the volume parameter. As mentioned above, there is some colinearity between two sets of steric parameters. Depending upon the substituents selected in each correlation, the best parameter set could be switched over from one to another. The relative importance of the unfavorable steric effect as the sequence of meta > ortho > para is common between two equations (after the scale adjustment). The larvicidal activity induced by premature molting and the stimulation of *N*-acetylglucosamine incorporation into a new cuticular system are closely linked. Substituents of two benzene rings seem to function so that the hydrophobic effect is almost nonspecific to positions, the unfavorable effect of steric "bulkiness" is position dependent, and the electronic effect is specific to ortho substituents of the A-ring.

13.2.3 THE AAR AND CLASSICAL QSAR OF THE CELLULAR-LEVEL ACTIVITY

We examined the cellular-level activity of molting hormone agonists using the Sf9 cells. The cell line was originally isolated in 1970s from pupal ovarian tissue of the fall armyworm, *Spodoptera frugiperda*, belonging to the same order, *Lepidoptera*, as the rice stem borer. The cells have often been utilized with baculovirus expression vectors for producing foreign recombinant proteins of large size [9].

We measured the IC$_{50}$ concentration (in M) to show the 50% inhibition of the binding (and incorporation) of ^{3}H-labeled ponasterone A (in)to intact Sf9 cells as suspension under certain culture conditions. Ponasterone A is a potent steroidal agonist of plant origin. The pIC$_{50}$ (intact) value is used as an index of the binding affinity. In Figure 13.5, the pEC$_{50}$ values for the tissue-level effect on *N*-acetyl-glucosamine incorporation are plotted against the more fundamental cellular pIC$_{50}$ for 34 compounds (Figure 13.6). They include natural steroidal agonists (n = 7), DBHs (n = 19), and benzoylalkanoylhydrazines (n = 8) [10]. Compounds in the last group are also larvicidal as found originally in Rohm and Haas [1,4]. They have alkyl groups

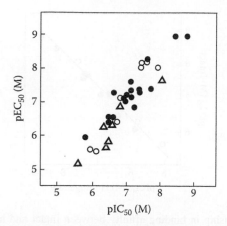

FIGURE 13.5 Relationship between N-acetylglucosamine incorporation and cellular binding affinity (closed circle: N,N'-dibenzoyl-N-t-butylhydrazines; open circle: benzoylalkanoylhydrazines; open triangle: steroidal agonists).

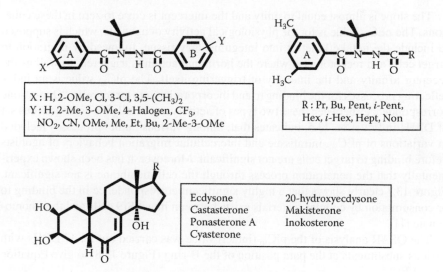

X : H, 2-OMe, Cl, 3-Cl, 3,5-(CH$_3$)$_2$
Y : H, 2-Me, 3-OMe, 4-Halogen, CF$_3$,
NO$_2$, CN, OMe, Me, Et, Bu, 2-Me-3-OMe

R : Pr, Bu, Pent, i-Pent,
Hex, i-Hex, Hept, Non

Ecdysone	20-hydroxyecdysone
Castasterone	Makisterone
Ponasterone A	Inokosterone
Cyasterone	

FIGURE 13.6 Structure of agonists used for drawing Figure 13.5.

in place of the B-ring of DBHs. Some scatters are observed in Figure 13.5 as represented by Equation 13.4 with a somewhat large s value.

$$pEC_{50} = 1.20 \ (\pm 0.19) \ pIC_{50} \ (\text{intact}) - 1.34 \ (\pm 1.31)$$
$$n = 34, \ s = 0.38, \ r = 0.92 \tag{13.4}$$

For 19 DBHs, the correlation was improved to give Equation 13.5:

$$pEC_{50} = 1.00 \ (\pm 0.20) \ pIC_{50} \ (\text{intact}) + 0.17 \ (\pm 1.45)$$
$$n = 19, \ s = 0.29, \ r = 0.93 \tag{13.5}$$

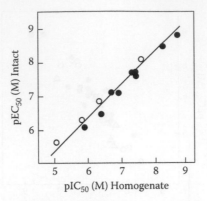

FIGURE 13.7 Relationship in binding affinity between intact and homogenized Sf9 cells (closed circle: N,N'-dibenzoyl-N-t-butylhydrazines, open: steroidal agonists). (Reproduced from C. Minakuchi, Y. Nakagawa, and H. Miyagawa, *J. Pestic. Sci.* 28, 55, 2003. With the permission of the Pesticide Science Society of Japan.)

The slope is almost equal to unity and the intercept is close to zero in these equations. The pEC_{50} value is for the physiological activity of agonists, which is supposed to include the uptake kinetics into integumentary tissue, followed by migration to target cells and receptor sites where the hormonal action is triggered, as well as the receptor affinity and the intensity of triggering itself. The pIC_{50} value is an index reflecting processes for the binding to and incorporation into target cells. "One-to-one correspondence" between these two types of activity, especially for homogeneous set of DBHs in Equation 13.5, indicates that, among various "unit" processes included in variations of pEC_{50}, intratissue and intercellular migration behaviors of agonists before binding to target cells are not significant. Moreover, it has been shown experimentally that the penetration process through the cell membrane is not significant. Figure 13.7 clearly shows that a highly significant correspondence in the binding to or consumption by cellular materials exists between intact Sf9 cells and their homogenate [11].

The QSAR analysis of the pIC_{50} (intact) value was carried out for 17 DBHs with various substituents at the para position of the B-ring (Figure 13.8) to give Equation 13.6 [12]:

$$pIC_{50} \text{ (intact)} = 0.61 \ (\pm 0.24) \log P - 0.82 \ (\pm 0.53) \ \sigma$$
$$- 0.37 \ (\pm 0.44) \ \Delta B_1 + 5.50 \ (\pm 0.72)$$
$$n = 17, \ s = 0.24, \ r = 0.96 \tag{13.6}$$

The B_1 is one of the STERIMOL parameters for the minimum width of substituents from the axis connecting the α atom with the rest of the molecule. ΔB_1 means the difference from that of the reference hydrogen atom. In Equation 13.2 and Equation 13.3 for the rice stem borer activities, such steric parameters as ΔL and ΔV are required to give the best correlations instead of ΔB_1 for para substituents of the B-ring. Although the ΔB_1 term is justified only at the 91% level of significance, the

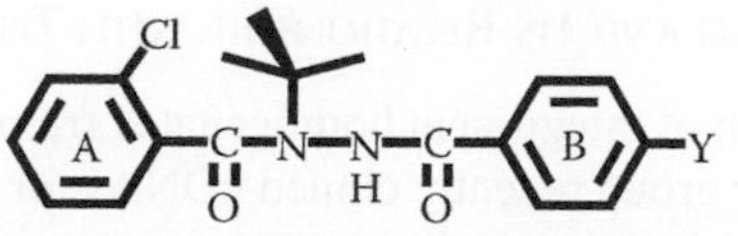

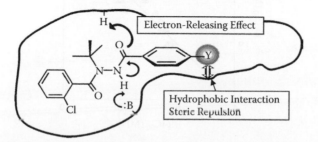

Y : H, 4-F, Cl, Br, I, CF$_3$, NO$_2$, CN, Me, Et,
Pr, i-Pr, Bu, t-Bu, OMe, SO$_2$Me, Ac

FIGURE 13.8 Compounds used for formulating Equation 13.6.

steric inhibitory nature of substituents does not vary among Equation 13.2, Equation 13.3, and Equation 13.6.

The log P term in Equation 13.6 corresponds with that in Equation 13.2 and Equation 13.3, but the σ term does not. The most likely reason is the participation of oxidative metabolism under whole body and tissue-level test conditions. We previously demonstrated, in QSAR studies of a series of chitin-synthesis-inhibiting benzoylphenylureas [13], that the coefficient of the electronic parameter significantly shifts toward the negative direction when the metabolic inhibitors are applied together with test compounds for the larvicidal as well as integumental assay using the rice stem borer larvae. We consider that the electron-donating property is ultimately favorable even in the whole body and tissue-level activities of DBHs. But, this effect is hidden in Equation 13.2 and Equation 13.3, because the electron-donating property also works to enhance the oxidative metabolism of the ring system, reducing the activity canceling the σ term. Moreover, in Equation 13.6, the contribution of $\rho\sigma$ product values of the size of −0.6 to +0.3 to variations of pIC$_{50}$ of a range of 2.8 is rather narrow. This range of variations could be able to distinguish itself with significance in compact sets of accurate data. In larger sets of compounds, such as Equation 13.2 and Equation 13.3, it may easily sink into noise of the data giving no significant σ term.

Because the tissue-level activity is also related to the binding affinity to the EcR as shown below, the situation can be regarded as what is happening at the receptor site. The physicochemical factors illustrated in Equation 13.6 are depicted in Figure 13.9.

FIGURE 13.9 Interaction of N,N'-dibenzoyl-N-t-butylhydrazines (DBHs) with the "receptor" cavity surface as expected from Equation 13.6. (Reproduced from T. Ogura, Y. Nakagawa, C. Minakuchi, and H. Miyagawa, *J. Pestic. Sci.* 30, 1, 2005. With the permission of the Pesticide Science Society of Japan.)

13.2.4 Receptor Binding and Its Relationship with Tissue-Level Activity

Using a cell-free preparation of integument homogenates from rice stem borer larvae in the wandering stage, our group recently cloned cDNAs for the EcR and the ultra-spiracle (USP), and expressed corresponding proteins using an *in vitro* transcription/translation system. USP is the invertebrate homolog of the mammalian retinoid X receptor, and EcR forms with USP a heterodimer that triggers the molting hormonal activity upon binding agonists [14].

The hormone-specific binding affinity to the heterodimer was estimated in terms of the pIC_{50} value, the IC_{50} being the 50% inhibitory concentration (in M) of agonists against the binding of ^{3}H-labeled ponasterone A [14]. In Figure 13.10, the pIC_{50} values of seven DBHs and five steroidal agonists among those shown in Figure 13.6 are compared with their pEC_{50} values for the tissue-level activity promoting the incorporation of N-acetylglucosamine. Between these two types of biological activity, there is a very good linear relationship with the one-to-one correspondence. Both types of activity are biological preparations from rice stem borer larvae. The significance of this relationship outweighs that of the relationship shown in Figure 13.5 where indices are for the activity to biological preparations of different origin. Variations in intercellular distribution behaviors within integumentary tissue as well as intracellular migration processes in the target cellular phase are not critical in defining variations in tissue-level activity. Variations in tissue-level activity are governed almost exclusively by variations in the binding affinity to the EcR protein. Because the tissue-level activity is closely related to the larvicidal activity as shown by Equations 13.1, 13.2, and 13.3, it is reasonable to consider that structure-activity relationship of DBHs as larvicides is governed almost conclusively by the structure-activity patterns of the binding to EcR at least as far as the rice stem borer is concerned.

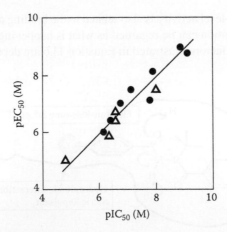

FIGURE 13.10 Relationship between N-acetylglucosamine incorporation and receptor affinity (closed circle: N,N'-dibenzoyl-N-t-butylhydrazines, open triangle: steroidal agonists). (Reproduced from C. Minakuchi, Y. Nakagawa, M. Kamimura, and H. Miyagawa, *Eur. J. Biochem.* 270, 4095, 2003. With the permission of Blackwell Publishing Ltd.)

It should be noted that steroidal agonists are not larvicidal under our experimental conditions. This is probably due to the fact that a mechanism such as an active transport associated with appropriate membrane transporters or, at least, facilitated diffusion of the steroidal hormone does not work in the route through the integumental tissue from the topically applied dorsal surface. The existence of such transport mechanisms of steroids to the nuclear receptor sites has been postulated recently [15]. In addition, if they are incorporated into cells and tissues, they are metabolized oxidatively and by other mechanisms without exerting lethal action under our test conditions, the duration of which is 7 days. On the target cellular level, the response to steroidal hormones has been recognized to occur very rapidly [15].

13.2.5 VALIDATION OF THE (Q)SAR SIGNIFICANCE WITH X-RAY CRYSTALLOGRAPHY OF THE LIGAND-EcR COMPLEXES OF THE TOBACCO BUDWORM

Recently, Billas and coworkers [16] achieved X-ray crystallographic analyses of the ligand-binding domain (LBD) of heterodimeric EcR prepared from a lepidopteran insect species, the tobacco budworm *Heliothis virescens*. They disclosed the crystallographic structures of LBD complexed with ponasterone A and BYI 06830, a member of DBHs (Figure 13.11). Although their studies are with the budworm EcR, the information could well be extrapolated to LBD in complex with ligands of the rice stem borer belonging to the same order, *Lepidoptera*.

According to them [17], the LBD of the budworm complexed with the DBH has a V-shaped cavity with certain flexibility. Ligand molecules fit into the cavity with steric constraints to various extents depending upon the orientation of submolecular moieties (substituent positions in DBHs). There are potential rooms and barriers for increasing in the steric bulk of substituents. The unfavorable situation is reflected by negative steric parameter terms depending upon the substituent positions in Equations 13.1, 13.2, 13.3, and 13.6. The ligand molecule forms an extensive network of hydrophobic interaction with surrounding LBD. The A-ring moiety is sandwiched between two adjacent methionine residues (Met 380, 381) and also neighbored by Val 384 and Ile 339. The B-ring is surrounded by such hydrophobic amino acid residues as Met 413, 507, Val 416, Leu 420, 500, 511, and Trp 526. The situation is consistent with the size of coefficient of the log P term in Equations 13.1, 13.2, 13.3, and 13.6.

The crystallographic structure showed that there are functional polar residues within LBD capable of H-bonding to CO and NH groups of the ligand other than nonpolar residues. The position of Thr 343 is such that its OH group forms an H-bond with the carbonyl group next to the A-ring. The carbonyl next to the B-ring is likely to interact with the

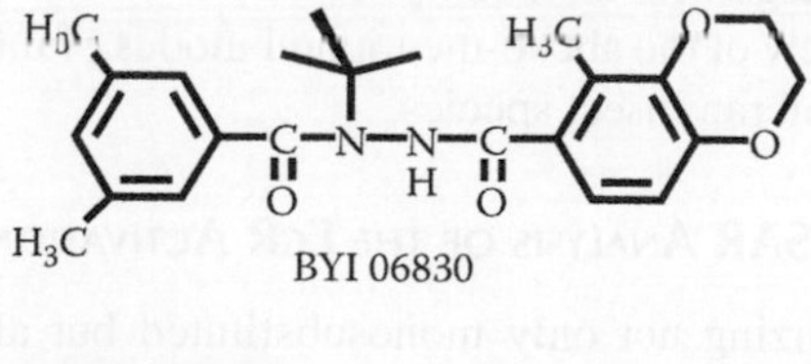

FIGURE 13.11 Structure of BYI 06830.

amide NH_2 hydrogen of Asn 504 and the NH hydrogen with phenolic OH oxygen of Tyr 408. In Equations 13.1 and 13.3, σ_I^{ortho} effect is supposed to increase the electron density of the "A-ring" carbonyl oxygen through space inductively to stabilize the H-bond from Thr 343. Perhaps because of the bulky t-Bu group, the A-ring would rotate so that the ortho substituent takes a conformation toward "parallel" to the carbonyl group. In this situation, the A-ring carbonyl group would be affected easily by a through-space inductive effect of the ortho substituent simulated by the σ_I^{ortho} term in Equation 13.1 and Equation 13.3. In Equation 13.6, the size of the coefficient of σ is close to -1.0. Probably, an electron-donating interaction of the "B-ring" carbonyl oxygen with the amide hydrogen occurring at the third "bond position" from the benzene ring is the driving force in the "concerted" H-bonding interactions as shown in Figure 13.9.

The most important aspect among findings of Billas et al. is that the EcR-LBD complexed with steroidal and nonsteroidal agonists exhibits different and only partially overlapping ligand-binding cavities. Whereas the cavity has a rather bulky V-shape for DBHs, it possesses a lengthy and thin L-shaped form for ponasterone A. As a matter of fact, the heterodimer of the wild-type EcR and USP was crystallized with ponasterone A, but that of a mutant EcR and USP with an increased solubility was crystallized with the nonsteroidal BYI 06830. The X-ray crystallographic data of these EcR complexes have been registered in the Protein Data Bank under the entry code of 1R1K and 1R20, respectively. Mutated residues (four amino acids) are located at the surface of the LBD, and the mutation does not affect the overall structure of EcR so as to modify its function [16]. The superimposition of BYI 06830 (Figure 13.11) and ponasterone A as found in cavities suggests that, whereas the t-butyl group together with the A-ring moiety of nonsteroidal agonists overlap the hydroxylated side chain at the C17 position of steroidal agonists, the B-ring part does not overlie any of the steroidal substructures. This superimposition denies previous predictions made by modeling studies considering a single binding niche.

13.3 QSAR EXAMINATIONS FOR THE ACTIVATION OF SILKWORM ECR AND HOMOLOGY MODELING OF ITS 3D ARCHITECTURE

We published a couple of papers for 3D-QSAR studies about tissue- and cellular-level activities of molting hormone agonists, including both DBHs and steroidal compounds using the CoMFA (comparative molecular field analysis) procedure. In these studies, we assumed a single common cavity in which two types of agonists are oriented so as to overlap as far as possible [18,19]. Although these studies would be of no value at the moment, our recent CoMFA study [20] dealing with only DBHs and closely related analogs with an assumption of a common "binding cavity" could be of significance in view of the above-mentioned modes of interaction of DBH-type agonists for the lepidopteran insect species.

13.3.1 CLASSICAL QSAR ANALYSIS OF THE ECR ACTIVATION

We have been synthesizing not only monosubstituted but also a number of multisubstituted DBHs and aliphatic acyl analogs (see Sections 13.2.1 and 13.2.3) as well as compounds in which the N-t-butyl group and the double-amide bridging moiety are modified variously. Swevers and coworkers measured the EcR activation effect of our total set of compounds [20]. They used cloned cells, Bm5, originally isolated from the ovary tissue of silkworms, *Bombyx mori* [21], into the genome of which an ecdysone-inducible green-fluorescent-protein (GFP) reporter-cassette had been incorporated [22]. The Bm5 cells, like Sf9 cells, have been used as baculovirus expression vectors for production of foreign proteins [23]. With the Bm5 assay system, the activation of EcR by agonists is amplified resulting in 500- to 2,000-fold expression of the GFP reporter gene. The EC_{50} concentration (in M) of agonists required to exhibit 50% the maximum GFP-fluorescent emission was measured using a high-throughput procedure. The pEC_{50} (Bm5) value represents a cellular level activity. We expected, however, that the cell membrane penetration process is insignificant in governing the variations in the EcR activation potency in a way similar to the case of Sf9 cells (also from ovarian tissue of lepidopteran species, see Section 13.2.3) as indicated in Equation 13.4 and Equation 13.5. The pEC_{50} (Bm5) value could be used here as being a parameter at the receptor level to activate the silkworm EcR.

We examined the classical QSAR for variations in the pEC_{50} (Bm5) value with physicochemical substituent parameters preliminarily. To make the situation simpler, we selected sets of DBH analogs monosubstituted on either of the aromatic moieties. After correcting parameter sets used in the previous publication [20], we reanalyzed to recognize that variations in the pEC_{50} (Bm5) value of A-ring analogs could possibly be governed not only by position-specific steric, hydrophobic, and electronic effects but also by position-specific H-bonding formation of substituents. Because of too many parameter terms for the number of compounds, it is not advisable to analyze directly the entire set of compounds together using every possible parameter. We tried to set aside compounds with such substituents capable of H-bonding as OMe and NO_2 at each position to give Equation 13.7. The effect of H-bond formation of OCH_3 and NO_2 substituents of the A-ring seems to enhance the activity at the ortho position but to reduce it at meta and para positions (the correlation not formulated). Similar to Equation 13.1 for the stem borer larvicidal activity, the Bm5 activity of ortho substituted compounds with Ph, OCH_2Ph, and Os-Bu is so low that they are also not included in the analysis.

$$pEC_{50}(Bm5) = 1.18\,(\pm 1.02)\,\sigma_I^{ortho} - 0.42\,(\pm 0.26)\,\Delta V^{ortho}$$
$$+ 3.11\,(\pm 1.34)\,\pi_{o,m,p} - 2.91\,(\pm 1.01)\,\pi^2_{o,m,p} + 6.26\,(\pm 0.36)$$
$$n = 21,\ s = 0.325,\ r = 0.886 \tag{13.7}$$

Equation 13.8 is for the B-ring analogs with the 2-Cl fixed on the A-ring. Some OCH_3 and NO_2 compounds are also not included because of their outlying behavior perhaps due to specific H-bond formation.

$$pEC_{50}(Bm5) = 0.72\,(\pm 0.24)\,\pi_{m,p} - 0.43\,(\pm 0.23)\,\Delta V^{ortho}$$
$$- 0.87\,(\pm 0.25)\,\Delta V^{meta} + 6.94\,(\pm 0.23)$$
$$n = 30,\ s = 0.370,\ r = 0.889 \tag{13.8}$$

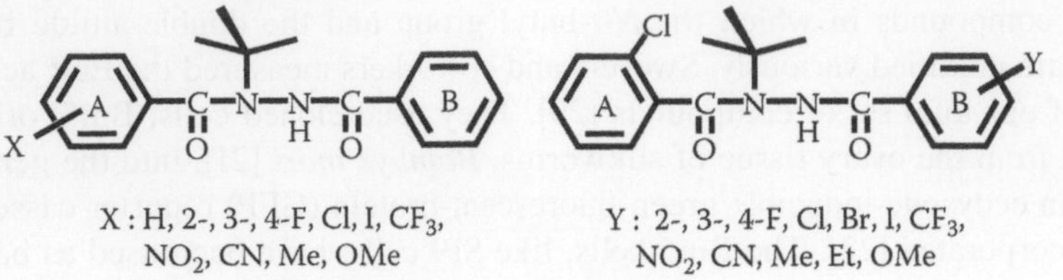

X : H, 2-, 3-, 4-F, Cl, I, CF$_3$,   Y : 2-, 3-, 4-F, Cl, Br, I, CF$_3$,
NO$_2$, CN, Me, OMe   NO$_2$, CN, Me, Et, OMe

FIGURE 13.12 *N,N'*-Dibenzoyl-*N-t*-butylhydrazine compounds used for formulating Equation 13.7 and Equation 13.8.

The positive σ_I term in Equation 13.7 for the inductive effect of ortho substituents is matched well with those in Equation 13.1 and Equation 13.3. The π constant is a hydrophobicity parameter for substituents defined as the difference of log P between substituted and nonsubstituted analogs. Although Equation 13.7 indicates an optimum hydrophobicity for A-ring substituents (π_{opt} = 0.6), Equation 13.8 shows that the higher hydrophobicity of meta and para substituents in B-ring contributes to the activity enhancement. No common hydrophobic effect is operative between A- and B-ring substituents. The steric effect is position specific so that the bulk of ortho substituents on the A-ring and ortho and meta substituents on the B-ring are unfavorable to the activity. Equation 13.7 and Equation 13.8 are to some extent similar to but different in detail from Equation 13.1 and Equation 13.2, respectively, because of the species difference. The structures of compounds included in Equation 13.7 and Equation 13.8 are shown in Figure 13.12.

13.3.2 3D-QSAR (CoMFA) ANALYSIS OF THE RECEPTOR ACTIVATION

Despite a reasonable quality, Equation 13.7 and Equation 13.8 are not satisfactory with a good number of analogs not included. Using 3D QSAR analysis, the entire sets of diacylhydrazines and analogs can be analyzed together, although empirical and tangible physicochemical information about substituent effects in classical QSARs is mostly lost. We used the Sybyl software for the analysis [24]. The structure of all compounds was generated by modifying the X-ray crystallographic structure of RH-5849 (Figure 13.1). After the geometry of compounds was fully optimized, the C-C-N-N-C-C bridge in the common skeleton, R$_A$-CO-NR$_C$-NH-(CO)-R$_B$, was superposed (R$_A$, R$_B$: A- and B-ring and their respective surrogates, R$_C$: *t*-Bu and its alternatives) to make the RMSs (root mean squares) of distances of six atomic positions from corresponding atomic positions of RH-5849 as small as possible. The electrostatic and steric potential energy fields in the space surrounding molecules at defined lattice points were calculated according to Coulomb and Lennard-Jones potential functions under default conditions. For 158 compounds (Figure 13.13), Equation 13.9 was obtained by partial least squares (PLS) regression procedure. The "*m*" is the optimum number of "principal components" in the PLS regression analysis, *press* is the leave-one-out "cross-validated" standard deviation, and *q* is the "cross-validated" correlation coefficient:

$$pEC_{50} \text{ (Bm5)} = \text{[CoMFA field descriptor terms]} + 4.041$$

$$n = 158, s = 0.554, r = 0.859, [m = 4, press = 0.78, q^2 = 0.447] \qquad (13.9)$$

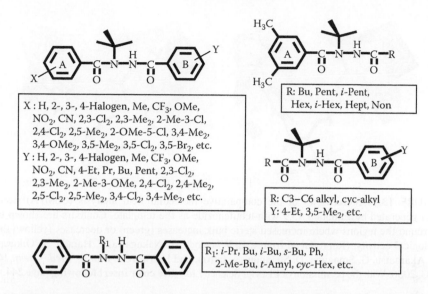

X : H, 2-, 3-, 4-Halogen, Me, CF$_3$, OMe,
NO$_2$, CN, 2,3-Cl$_2$, 2,3-Me$_2$, 2-Me-3-Cl,
2,4-Cl$_2$, 2,5-Me$_2$, 2-OMe-5-Cl, 3,4-Me$_2$,
3,4-OMe$_2$, 3,5-Me$_2$, 3,5-Cl$_2$, 3,5-Br$_2$, etc.
Y : H, 2-, 3-, 4-Halogen, Me, CF$_3$, OMe,
NO$_2$, CN, 4-Et, Pr, Bu, Pent, 2,3-Cl$_2$,
2,3-Me$_2$, 2-Me-3-OMe, 2,4-Cl$_2$, 2,4-Me$_2$,
2,5-Cl$_2$, 2,5-Me$_2$, 3,4-Cl$_2$, 3,4-Me$_2$, etc.

R: Bu, Pent, i-Pent,
Hex, i-Hex, Hept, Non

R: C3–C6 alkyl, cyc-alkyl
Y: 4-Et, 3,5-Me$_2$, etc.

R$_1$: i-Pr, Bu, i-Bu, s-Bu, Ph,
2-Me-Bu, t-Amyl, cyc-Hex, etc.

FIGURE 13.13 N,N'-Dibenzoyl-N-i-butylhydrazines and related analogs used for three-dimensional Quantitative Structure-Activity Relationship (3D QSAR) for the silkworm cellular EcR activation.

Describing approximately 74% (r^2) in variations of the biological activity ranging over about ten thousand times, Equation 13.9 is acceptable for the effect of diacylhydrazines and analogs. In the CoMFA analysis, "correlation equations" such as Equation 13.9 give practically no physicochemical information about structure-activity relationship. Instead, the contour "surfaces" for steric and electrostatic fields are visualized according to (latent) regression coefficients and standard deviations of molecular field parameter terms at lattice points. Each surface is connecting and covering contour points for the value of defaulted lattice indices surrounding submolecular moieties [25]. In other words, the surfaces are drawn where the "slope" for variations is "steep" enough so that significant differences in the value of lattice field indices for steric and electrostatic properties are recognizable.

Figure 13.14 represents contour surfaces for the steric energy parameter with tebufenozide placed inside as the reference. The green surfaces denote contours that enclose volumes within which increases in bulk favor higher receptor activation. Yellow surfaces cover regions within which increases in the bulk disfavor the receptor activation conversely. A sterically favorable (green) region is widely covering the 4-position of the B-ring, whereas a sterically unfavorable (yellow) surface is surrounding broadly the 4-position of the A-ring. In addition, sterically unfavorable fields are observed around the 3- and 5-positions of the B-ring.

The electrostatic contour diagrams are shown in Figure 13.15. The blue contours represent surfaces of volumes within which increases in positive charge are favorable to the receptor activation. The red surfaces indicate that increases in negative charge are favorable so that they prefer to cover electronegative substructural moieties. Figure 13.15 shows positive (blue) fields around the 2- and 3-positions of the A-ring,

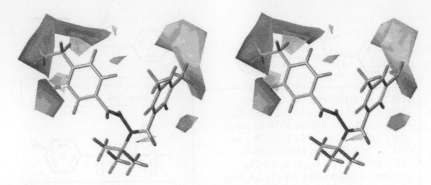

FIGURE 13.14 Stereoviews of the comparative molecular field analysis (CoMFA) steric field generated by Equation 13.9 with tebufenozide as the template. Contours are shown to surround the regions where increased steric bulk increases (green) or decreases (yellow) the biological activity. (Reproduced from C.E. Wheelock, Y. Nakagawa, T. Harada, N. Oikawa, M. Akamatsu, G. Smagghe, D. Stefanou, K. Iatrou, and L. Swevers, *Bioorg. Med. Chem.* 14, 1143, 2006. With the permission of Elsevier Science Ltd.) **(See color insert following page 244.)**

covering the 2-, 3-, and 4-positions of the B-ring, and over the NH hydrogen of the bridge. The negative (red) fields are over the C=O oxygen atoms of the bridge and around the 2-, 5-, and 6-positions of the B-ring.

These CoMFA findings do not contradict those from the classical QSAR Equation 13.7 and Equation 13.8, although the number and variety of compounds included in the CoMFA analysis are much higher, and the classical QSAR information may mostly be buried into that according to contour diagrams. The blue contours covering

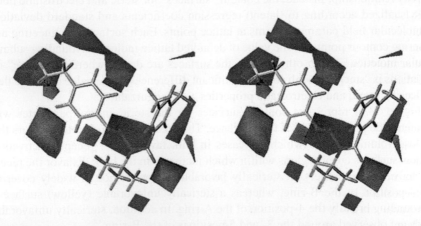

FIGURE 13.15 Stereoviews of the comparative molecular field analysis (CoMFA) electrostatic field generated by Equation 13.9 with tebufenozide as the template. Contours are shown to surround the regions where a positive (blue) or negative (red) electrostatic potential increases the biological activity. (Reproduced from C.E. Wheelock, Y. Nakagawa, T. Harada, N. Oikawa, M. Akamatsu, G. Smagghe, D. Stefanou, K. Iatrou, and L. Swevers, *Bioorg. Med. Chem.* 14, 1143, 2006. With the permission of Elsevier Science Ltd.) **(See color insert.)**

A- and B-rings in Figure 13.15 perhaps signify that the fringe of the aromatic substituents is electropositive. This would reflect an ultimate advantage of electron donating substituents for the activity, being reminiscent of the discussion in Section 13.2.3. The inductive effect of the ortho substituent of the A-ring could be represented by a small red region found close to the corresponding position. The absence of negative ΔV^{para} term in Equation 13.8 is supposed to be in accord with the bulk-permissible green surface covering the para position of the B-ring in Figure 13.14.

No field is visualized around the position corresponding to the t-butyl group of tebufenozide in Figure 13.14 and Figure 13.15. The variations in physicochemical properties in this region are not significant. That is, structural varieties of the N-substituent are limited ($n = 10$) and no significant activity was observed for most of the compounds except for tebufenozide and the N-t-amyl analog. Because the hydrophobic effect is not considered explicitly in the CoMFA procedure, and also because some component of hydrophobicity is thought to overlap inherently with the bulk of substituents, the substituent hydrophobic effect could considerably be included inside the bulk effects viewed in Figure 13.14. The quadratic π terms in Equation 13.7 suggesting the existence of optimum hydrophobicity could, in part, be reproduced by the sterically limiting yellow barriers around substituent positions of A-ring in Figure 13.14.

13.3.3 HOMOLOGY MODELING OF THE SILKWORM EcR

We performed homology modeling studies for the LBD of *B. mori* EcR to understand further physicochemical outcomes of CoMFA analysis [20]. We first searched in databases for template proteins, of which amino acid sequence in EcR-LBD is known and crystal structure is established, using an alignment procedure with the amino acid sequence of EcR-LBD of *B. mori* [26,27]. We identified two proteins, both being the EcR-LBD protein of *H. virescens* as expected. Their crystal structures have been registered under the entry code, 1R1K and 1R20, as described in Section 13.2.5. The protein more homologous to that from *B. mori* (in terms of amino acid sequence) was 1R1K. The sequence similarity between the two types of *H. virescens* EcR is 98%. Their EcR-LBD crystallographic structures are, however, only partially overlapped because the structure of ligands with which the crystalline complex is to be built is different. Thus, 1R20 was considered to be better for the modeling of the *B. mori* EcR-LBD, with which the sequence similarity is 89% and DBHs are supposed to be complexed.

Second, the 3D structure of the *B. mori* EcR was constructed on the basis of the structure of 1R20 and optimized using the simulated annealing method. Coordinates of $C\alpha$-atoms and conformation of main and side chains were refined automatically with use of a full automatic homology modeling system, PDFAMS [28], developed by Umeyama and his group [29]. The original ligand of 1R20, BYI 06830, was accommodated to optimize the protein structure. The coordinates of BYI 06830 were fixed during optimization. The cavity surface at the LBD site of the homology-modeled *B. mori* EcR was represented by the MOLCAD module of the SYBYL software [24] as shown in Figure 13.16a.

Finally, the cavity surface of the *B. mori* EcR-LBD was superimposed over the CoMFA steric boundary (Figure 13.16b). The reference ligand, tebufenozide, is shown

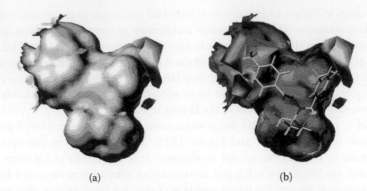

(a)   (b)

FIGURE 13.16 (a) Surface of the EcR ligand binding cavity of *Bombyx mori* for *N,N'*-dibenzoyl-*N-t*-butylhydrazine-type ecdysone agonists. (b) Superimposition of comparative molecular field analysis (CoMFA) steric boundary over the *B. mori* EcR cavity, with tebufenozide as the template; a part of the cavity surface was cut down to see the inside. (Reproduced from C.E. Wheelock, Y. Nakagawa, T. Harada, N. Oikawa, M. Akamatsu, G. Smagghe, D. Stefanou, K. Iatrou, and L. Swevers, Bioorg. Med. Chem. 14, 1143, 2006. With the permission of Elsevier Science Ltd.) **(See color Insert.)**

to fit just inside the LBD. There is potentially room for substituents with increased steric bulk for promoting the activity around the position corresponding with the 4-position of the B-ring (upper left-hand side of Figure 13.16b). The ligand orientation in this cavity as shown in Figure 13.16b enables us to examine hydrophobic and H-bonding interactions between ligand and component amino acids in LBD. The region of the *t*-butyl group in tebufenozide is surrounded by a number of hydrophobic amino acids including M409, M503, L507, and L514. The side-chain H-bonding groups (NH_2 and OH) of amino acids T339, Y404, and N500 are located in positions to form H-bonds with the counterparts on the double amide bridge within a distance of 2.0 to 2.5 Å. These results are consistent with precedent examples observed by X-ray crystallography for other lepidopteran ligand-EcR binding investigations [17,30].

It is important to notice that the limits on steric bulk indicated by yellow CoMFA contours coincide well with the boundaries of the modeled LBD. The CoMFA boundary was derived from 3D QSAR examinations of the Bm5 activity, whereas the cavity surface was from a computational procedure with the EcR-LBD amino acid sequence of *B. mori* and the geometry of the template EcR-LBD of *Heliothis virescens*. The information derived from different origins and procedures correspond very well here about EcR-LBD. The above issues could also be in accord with the assumption for the Bm5 cell-level activity to be simulated as the activity at the receptor level.

13.4 CONCLUSIONS

We conducted structure-activity and molecular modeling studies to show important achievements. With preparations from rice stem borers, *Chilo suppressalis*, we showed that, for combined sets of steroidal agonists and DBHs, variations in tissue-level activity are governed almost solely by variations in the binding affinity to the EcR. For DBHs, variations in larvicidal activity are governed almost solely by variations in the binding affinity to the EcR. It should be recognized that the difference in the orientation at the LBD between two types of agonists could be combined with the difference in their behaviors before triggering the hormonal action. Despite the difference in the mode of orientation on the EcR activation site, the hormonal response should almost exclusively be dependent on the binding affinity as far as the rice stem borers are concerned. The situation is likely to be not much changed in other lepidopteran insect species, such as *S. frugiperda, H. virescens,* and *B. mori.* Our group has shown quite recently, however, that, in the Colorado potato beetle, *Leptinotarsa decemlineata,* the story is not so simple, as no close relationship is observed between larvicidal activity and binding affinity with the EcR protein of the beetles [31]. The analysis of differences in AAR and QSAR between two insect orders including EcR functions is believed to be another important research target for the molecular mechanism of ecdysone agonists.

It was gratifying to see that the "size and shape" of *B. mori* EcR-LBD that derived by homology modeling corresponded very well with that deduced from biological QSAR. With use of the computational procedure as well as bioinformatics methodology, present studies are believed to support, albeit indirectly and virtually, the proposition of Billas and coworkers [17] that nuclear receptors and their ligands can mutually adapt in their 3D architectures and fit to each other.

ACKNOWLEDGMENTS

The authors express their sincere thanks to Drs. Craig Wheelock, Nobuhiro Oikawa, Chieka Minakuchi, and Takehiko Ogura, and Messrs. Kazunari Hattori and Toshiyuki Harada for their devoted collaboration. We are also grateful to Professors Hisashi Miyagawa and Guy Smagghe for their helpful comments and discussions.

A part of this chapter was taken from T. Fujita and Y. Nakagawa, *QSAR and mode of action studies of insecticidal ecdysone agonists,* SAR QSAR Environ. Res. 22 (2006), pp. 34–56.

REFERENCES

[1] K.D. Wing, *RH 5849, a nonsteroidal ecdysone agonist: Effects on a* Drosophila *cell line,* Science 241 (1988), pp. 467–469.

[2] A.C.-T. Hsu, *1,2-Diacyl-1-alkylhydrazines — A new class of insect growth regulators,* in *Synthesis and Chemistry of Agrochemicals II,* D.R. Baker, J. Fenyes, and W.K. Moberg, eds., American Chemical Society, Washington, DC, 1991, pp. 478–490.

[3] K.D. Wing, R.A. Slawecki, and G.R. Carlson, *RH 5849, a nonsteroidal ecdysone agonist: Effects on larval Lepidoptera,* Science 241 (1988), pp. 470–472.

[4] A.C.-T. Hsu, T.T. Fujimoto, and T.S. Dhadialla, *Structure-activity study and conformational analysis of RH-5992, the first commercialized nonsteroidal ecdysone agonist,* in *Phytochemicals for Pest Control,* P.A. Hedin, R.M. Hollingworth, E.P. Masler, and J. Miyamoto, eds., American Chemical Society, Washington, DC, 1997, pp. 206–219.

[5] Y. Nakagawa, *Nonsteroidal ecdysone agonists*, Vitamins and Hormones 73 (2005), pp. 131–173.

[6] N. Oikawa, Y. Nakagawa, K. Nishimura, T. Ueno, and T. Fujita, *Quantitative structure-activity analysis of larvicidal 1-(substituted benzoyl)-2-benzoyl-1-t-butylhydrazines against* Chilo suppressalis, Pestic. Sci. 41 (1994), pp. 139–148.

[7] N. Oikawa, Y. Nakagawa, K. Nishimura, T. Ueno, and T. Fujita, *Quantitative structure-activity studies of insect growth regulators 10: Substituent effects on larvicidal activity of 1-t-butyl-1-(2-chlorobenzoyl)-2(substituted benzoyl)-hydrazines against* Chilo suppressalis *and design synthesis of potent derivatives,* Pestic. Biochem. Physiol. 48 (1994), pp. 135–144.

[8] Y. Nakagawa, Y. Soya, K. Nakai, N. Oikawa, K. Nishimura, T. Ueno, and T. Fujita, *Quantitative structure-activity studies of insect growth regulators 11: Stimulation and inhibition of N-acetylglucosamine incorporation in a cultured integument system by substituted N-t-butyl-N,N'-dibenzoylhydrazines,* Pestic. Sci. 43 (1995), pp. 339–345.

[9] G.E. Smith, G. Ju, B.L. Ericson, J. Moschera, H.-W. Lahm, R. Chizzionite, and M.D. Summers, *Modification and secretion of human interleukin 2 produced in insect cells by a baculovirus expression vector*, Proc. Natl. Acad. Sci. USA 82 (1985), pp. 8404–8408.

[10] Y. Nakagawa, K. Takahashi, H. Kishikawa, T. Ogura, C. Minakuchi, and H. Miyagawa, *Classical and three-dimensional QSAR for the inhibition of [^{3}H]ponasterone A binding by diacylhydrazine-type ecdysone agonists to insect Sf9 cells*, Bioorg. Med. Chem. 13 (2005), pp. 1333–1340.

[11] C. Minakuchi, Y. Nakagawa, and H. Miyagawa, *Validity analysis of a receptor binding assay for ecdysone agonists using cultured intact insect cells*, J. Pestic. Sci. 28 (2003), pp. 55–57.

[12] T. Ogura, Y. Nakagawa, C. Minakuchi, and H. Miyagawa, *QSAR for binding affinity of substituted dibenzoylhydrazines to intact Sf9 cells*, J. Pestic. Sci. 30 (2005), pp. 1–6.

[13] Y. Nakagawa, T. Akagi, H. Iwamura, and T. Fujita, *Quantitative structure-activity studies of benzoylphenylurea larvicides 6: Comparison of substituent effects among activities against different insect species*, Pestic. Biochem. Physiol. 33 (1989), pp. 144–157.

[14] C. Minakuchi, Y. Nakagawa, M. Kamimura, and H. Miyagawa, *Binding affinity of nonsteroidal ecdysine agonists against the ecdysone receptor complex determines the strength of their molting hormonal activity*, Eur. J. Biochem. 270 (2003), pp. 4095–4104.

[15] E. Marcinkowska and A. Wiedlocha, *Steroid signal transduction activated at the cell membrane: From plants to animals*, Acta Biochim. Polon. 49 (2002), pp. 735–745.

[16] I.M.L. Billas, T. Iwema, J.M. Garnier, A. Mitschler, N. Rochel, and D. Moras, *Structural adaptability in the ligand-binding pocket of the ecdysone hormone receptor*, Nature 426 (2003), p. 91.

[17] I.M.L. Billas and D. Moras, *Ligand-binding pocket of the ecdysone receptor*, Vitamins and Hormones 73 (2005), pp. 101–129.

[18] Y. Nakagawa, B. Shimizu, N. Oikawa, M. Akamatsu, K. Nishimura, N. Kurihara, T. Ueno, and T. Fujita, *Three-dimensional quantitative structure-activity analysis of steroidal and dibenzoylhydrazine-type ecdysone agonists*, in *Classical and Three-Dimensional QSAR in Agrochemistry*, C. Hansch and T. Fujita, eds., American Chemical Society, Washington, DC, 1995, pp. 288–301.

[19] Y. Nakagawa, K. Hattori, B. Shimizu, M. Akamatsu, H. Miyagawa, and T. Ueno, *Quantitative structure-activity studies of insect growth regulators 14: Three-dimensional quantitative structure-activity relationship of ecdysone agonists including dibenzoylhydrazine analogs*, Pestic. Sci. 53 (1998), pp. 269–277.

[20] C.E. Wheelock, Y. Nakagawa, T. Harada, N. Oikawa, M. Akamatsu, G. Smagghe, D. Stefanou, K. Iatrou, and L. Swevers, *High-throughput screening of ecdysone agonists using a reporter gene assay followed by 3-D QSAR analysis of the molting hormonal activity*, Bioorg. Med. Chem. 14 (2006), pp. 1143–1159.

[21] T.D.C. Grace, *Establishment of a line of cells from the silkworm*, Bombyx mori, Nature 216 (1967), p. 613.

[22] L. Swevers, L. Kravariti, S. Ciolfi, M. Xenou-Kokoletsi, N. Ragoussis, G. Smagghe, Y. Nakagawa, B. Mazomenos, and K. Iatrou, *A cell-based high-throughput screening system for detecting ecdysteroid agonists and antagonists in plant extracts and libraries of synthetic compounds*, FASEB J. 18 (2004), pp. 134–136.

[23] K. Iatrou and L. Swevers, *Transformed lepidopteran cells expressing a protein of the silkmoth fat body display enhanced susceptibility to baculovirus infection and produce high titers of budded virus in serum-free media*, J. Biotech. 120 (2005), pp. 237–250.

[24] Sybyl (ver. 6.91; Tripos Associates, Inc., St. Louis, Missouri).

[25] R.D. Cramer III, S.A. DePriest, D.A. Patterson, and P. Hecht, *The developing practice of comparative molecular field analysis*, in *3D QSAR in Drug Design — Theory, Methods, and Application*, H. Kubinyi, ed., Escom Science Publishers, Leiden, 1993, pp. 443–485.

[26] L. Swevers, J.R. Drevet, M.D. Lunke, and K. Iatrou, *The silkmoth homolog of the Drosophila ecdysone receptor (B1 isoform): Cloning and analysis of expression during follicular cell differentiation*, Insect Biochem. Mol. Biol. 25 (1995), pp. 857–866.

[27] M. Kamimura, S. Tomita, and H. Fujiwara, *Molecular cloning of an ecdysone receptor (B1 isoform) homologue from the silkworm*, Bombyx mori, *and its mRNA expression during wing disc development*, Comp. Biochem. Physiol. B. Biochem. Mol. Biol. 113 (1996), pp. 341–347.

[28] PDFAMS (ver. 2.0; In-Silico Sciences, Inc., Tokyo).

[29] K. Ogata and H. Umeyama, *An automatic homology modeling method consisting of database searches and simulated annealing*, J. Mol. Graph. Model. 18 (2000), pp. 258–272.

[30] J.A. Carmichael, M.C. Lawrence, L.D. Graham, P.A. Pilling, V.C. Epa, L. Noyce, G. Lovrecz, D.A. Winkler, A. Pawlak-Skrzecz, R.E. Eaton, G.N. Hannan, and R.J. Hill, *The X-ray structure of a hemipteran ecdysone receptor ligand-binding domain: Comparison with a lepidopteran ecdysone receptor ligand-binding domain and implications for insecticide design*, J. Biol. Chem. 280 (2005), pp. 22258–22269.

[31] T. Ogura, C. Minakuchi, Y. Nakagawa, G. Smagghe, and H. Miyagawa, *Molecular cloning, expression analysis and functional confirmation of ecdysone receptor and ultraspiracle from the colorado potato beetle*, Leptinotarsa decemlineata, FEBS J. 272 (2005), pp. 4114–4128.

14 e-Endocrine Disrupting Chemical Databases for Deriving SAR and QSAR Models

Nathalie Marchand-Geneste, James Devillers,
Jean-Christophe Doré, and Jean-Marc Porcher

CONTENTS

ABSTRACT

There is increasing evidence that numerous chemicals released into the environment by human activities have the potential to alter the normal functions of the endocrine system in wildlife. These xenobiotics are called endocrine disrupting chemicals (EDCs). The aim of this chapter was to catalog the different EDC database resources (biological data, chemical descriptor data, and so forth) available on the Internet for deriving structure-activity models. *In vitro* and *in vivo* experimental data from nuclear receptor binding assays, *in vitro* test methods for detecting EDCs, and comparison between biological assays have been critically analyzed.

KEYWORDS

Binding assays
Database
Endocrine disrupting chemicals (EDCs)
Endocrine disruptors
Quantitative Structure-Activity Relationship (QSAR)

14.1   INTRODUCTION

There is growing concern about the potential of structurally diverse chemicals to produce changes in the functioning of the endocrine systems of humans and animals. Endocrine disruption attracts a lot of public interest and is subject to worldwide discussions between experts, governmental organizations, academics, and industry. Known natural hormones as well as relatively unknown environmental pollutants seem to have the ability to potentially disrupt the endocrine systems of species living in ecosystems in such a way that harmful effects on their development and reproduction can occur. The issue of identifying chemicals that may elicit endocrine disruption has grown immeasurably in importance in the last decade. It was estimated that more than 87,000 chemicals might need to be screened for endocrine disruption potential [1]. Batteries of laboratory bioassays exist to test chemicals for their potential effects on the endocrine system [2]. However, due to time and cost limitations, they cannot be used to test all the chemicals that can be found in the environment. Structure-Activity Relationship (SAR) and Quantitative Structure-Activity Relationship (QSAR) models are now increasingly used to overcome these problems. This has made the need to screen the databases of existing chemicals a priority. The resources available in these databases provide a wealth of information on endocrine disruption and allow us to reduce dependency upon time-consuming and expensive animal experiments. There is a real effort to improve public access to chemical toxicity information resources by means of these databases.

The aim of this chapter is to focus on the different EDC databases available on the Internet that could be used for deriving structure-activity models. The goal is first to provide addresses allowing us to obtain qualitative information about potential EDC mode of action (that is, agonist effect or antagonist effect). Second, interesting addresses to retrieve quantitative biological data (for example, binding data) that could be used to elaborate QSAR models have also been compiled.

14.2   e-EDC DATABASES

14.2.1   SAR Information

Scorecard is a free Internet database maintained by a U.S. nonprofit organization Environmental Defense that collects resources for information about pollution problems and toxic chemicals [3]. It provides detailed information on more than 11,200 chemicals, including all the chemicals used in large amounts in the United States and all the chemicals regulated under major environmental laws. Scorecard makes

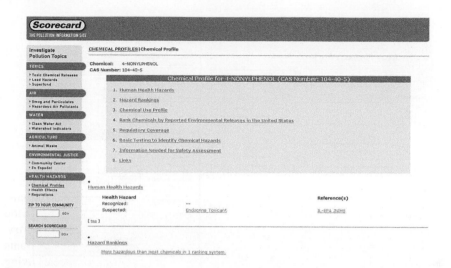

FIGURE 14.1 Chemical profile obtained with Scorecard.

hazard identification information accessible by means of an extensive review of scientific literature and toxicological databases. A chemical profile is accessible by typing the corresponding chemical name or its Chemical Abstracts Services registry number (CAS RN) in the home page. The results of the search yield a list gathering the following information: human health hazards, hazard ranking, chemical use profile, rank chemicals by reported environmental releases in the United States, regulatory coverage, basic testing to identify chemical hazards and information needed for safety assessment (Figure 14.1). The Scorecard "health effects" section provides links to the definitions of each of these health effects, a list of chemicals recognized or suspected of causing these health effects, and the references used to compile these lists. Chemicals are identified as recognized toxicants based on the hazard identification efforts of authoritative scientific and regulatory agencies. Among the twelve health effect categories defined by Scorecard, the suspected endocrine toxicity is accessible for a selected compound or for a list of compounds belonging to the Scorecard database. This list can be downloaded as a CSV file, and all the references used to compile this list of suspected endocrine toxicants are available. Useful links to U.S. Environmental Protection Agency (EPA) Web sites are consultable.

The Relational Database of Information on Potential Endocrine Disrupters (REDIPED) is an endocrine disrupter–related resource developed by the Institute for Environment and Health (IEH), University of Leicester (U.K.) [4]. The REDIPED CD-ROM is not free, but a free demonstration version is downloadable on the IEH Web site. This free version contains examples of the information available in REDIPED for two chemicals: bisphenol A and diadzein (Figure 14.2). The commercial version contains information on 79 potential EDCs including naturally occurring compounds, industrial chemicals, agrochemicals, and pharmaceuticals. The user is able to use a drop-down menu to select chemicals or to enter the name or CAS RN. Based on Microsoft Access, REDIPED has been designed to contain a wide range

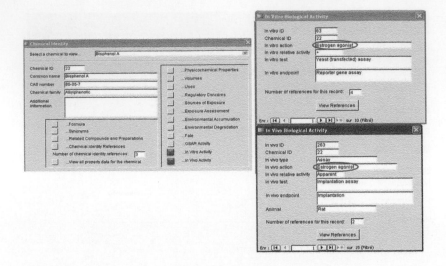

FIGURE 14.2 REDIPED program screenshot for bisphenol A.

of information on suspected endocrine disruptors, including data on physicochemical properties, production volumes and uses, environmental exposure, accumulation and fate, and relevant *in vivo* and *in vitro* biological effects (Figure 14.2). Standard or complex queries to extract specific information can be used. REDIPED collects over 2,000 searchable references. The main weakness of this resource is that only qualitative *in vivo* and *in vitro* biological effects (for example, agonist or antagonist) are available. Another downside of the resource is that like most databases, it is no longer updated.

A book entitled *Environmental Endocrine Disruptors: A Handbook of Property Data* by L.H. Keith provides information on chemical, physical, and toxicological properties of known and suspected environmental endocrine disruptors [5]. This resource is commercially available in print and electronic forms. This resource offers Boolean searching facilities and links to access analytical method data from EPA's pesticide methods and references. A review on this resource was published by Magos [6].

As many pesticides are suspected to be EDCs, the Pesticides Action Network (PAN) has proposed the PAN Pesticides Database that collects diverse information on about 6,400 pesticide active ingredients and their transformation products from many different sources of information [7]. The database provides human toxicity (chronic and acute), ecotoxicity, reproductive and developmental toxicity, endocrine disruption, and regulatory information. Most of the toxicity information comes directly from official sources, such as the EPA, World Health Organization (WHO), National Toxicology Program (NTP), National Institutes of Health (NIH), International Agency for Research on Cancer (IARC), the European Union (EU), and the State of California. The PAN suspected endocrine disruptor designation is based on the Illinois EPA list, the Danish EPA list, the EU prioritizing list, the Colborn list, the Keith list, and the Benbrook list [5,8–12]. The PAN Pesticides database ranks

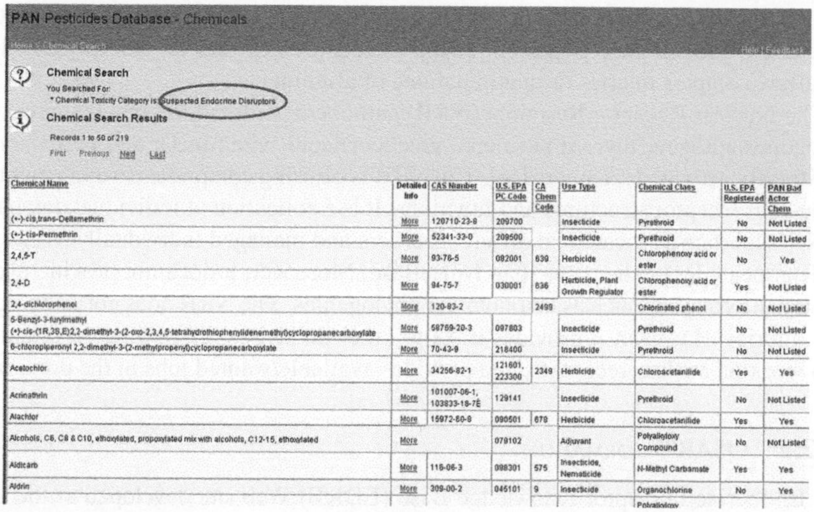

Chemical Name	Detailed Info	CAS Number	U.S. EPA PC Code	CA Chem Code	Use Type	Chemical Class	U.S. EPA Registered	PAN Bad Actor Chem
(+)-cis,trans-Deltamethrin	More	120710-23-8	209700		Insecticide	Pyrethroid	No	Not Listed
(+)-cis-Permethrin	More	52341-33-0	209500		Insecticide	Pyrethroid	No	Not Listed
2,4,5-T	More	93-76-5	082001	639	Herbicide	Chlorophenoxy acid or ester	No	Yes
2,4-D	More	94-75-7	030001	636	Herbicide, Plant Growth Regulator	Chlorophenoxy acid or ester	Yes	Not Listed
2,4-dichlorophenol	More	120-83-2		2499		Chlorinated phenol	No	Not Listed
5-Benzyl-3-furylmethyl (+)-cis-(1R,3S,E)2,2-dimethyl-3-(2-oxo-2,3,4,5-tetrahydrothiophenylidenemethyl)cyclopropanecarboxylate	More	58769-20-3	097803		Insecticide	Pyrethroid	No	Not Listed
6-chloropiperonyl 2,2-dimethyl-3-(2-methylpropenyl)cyclopropanecarboxylate	More	70-43-9	218400		Insecticide	Pyrethroid	No	Not Listed
Acetochlor	More	34256-82-1	121601, 223300	2349	Herbicide	Chloroacetanilide	Yes	Yes
Acrinathrin	More	101007-06-1, 103833-18-7È	129141		Insecticide	Pyrethroid	No	Not Listed
Alachlor	More	15972-60-8	090501	678	Herbicide	Chloroacetanilide	Yes	Yes
Alcohols, C6, C8 & C10, ethoxylated, propoxylated mix with alcohols, C12-15, ethoxylated	More		079102		Adjuvant	Polyalkyloxy Compound	No	Not Listed
Aldicarb	More	116-06-3	098301	575	Insecticide, Nematicide	N-Methyl Carbamate	Yes	Yes
Aldrin	More	309-00-2	045101	9	Insecticide	Organochlorine	No	Yes
						Polyalkyloxy		

FIGURE 14.3 PAN results showing chemicals obtained from the "Suspected Endocrine Disruptors" category.

a chemical as a suspected endocrine disruptor if this chemical is listed as being potentially endocrine disrupting by any of the previous sources (Figure 14.3). The accuracy of the data reported into the database has been ensured by a peer review process.

The Nuclear Receptor database (NuReBase) is a bioinformatic database of nuclear receptors developed by J. Duarte and hosted on the Pôle Bioinformatique Lyonnais (PBIL) server (France) [13]. The database contains protein and DNA sequences, as well as protein alignments, phylogenies, and taxonomy for all nuclear receptors. NuReBase provides search options for retrieving gene sequences or gene families, and for each group of homologous genes, a phylogenetic tree and a protein alignment are provided. NuReBase also contains EMBL sequences and comments for proteins and DNA, enriched with nuclear hormone receptor specific information. NuReBase is a valuable tool for researchers investigating the process of signaling.

NucleaRDB database stores sequences, mutations, and ligand-binding data for nuclear receptors [14]. Since April 2005, NucleaRDB contains 2,013 NR sequences, of which 123 have structural (3D) information of the ligand-binding domain. Such sequences are imported from the SWISS-PROT and TrEMBL databases [15]. The data are organized by means of a hierarchical list of known families based on a pharmacological classification of receptors. The database provides useful links to display structural information such as PDBreport database [16]. Multiple sequence alignments are performed with WHAT IF [17] and can be generated in several formats for families, subfamilies, groups, and so on. It is noteworthy that phylogenic trees are available to visualize the relationships between sequences in a family. This database offers quick access to high-quality data (sequences, mutations, and structures) about nuclear receptors. NucleaRDB is similar to NuReBase, including all nuclear receptors, with alignments and trees per family and subfamily. Unlike NuReBase, data in NucleaRDB are not reviewed, but rather are automatically generated by a system originally developed for G protein-coupled receptors. Moreover, NucleaRDB does not allow complex queries or manipulations of alignments.

The Nuclear Receptor Resource (NRR) gathers different individual resources for estrogen, androgen, thyroid hormone, glucocorticoid, vitamin D, and peroxisome-proliferator activated receptors [18]. The NRR is offering comprehensive information on nuclear receptor structures and functions. It is a collection of individual databases on NR superfamily located on different servers and managed individually. However, NRR appears less exhaustive than NuReBase. Moreover, it does not provide evolutionary information, such as taxonomy or phylogenies. The NRR does not allow complex queries. However, it provides links to other databases, gives a list of scientists who work on nuclear receptors, and provides available/wanted jobs in the domain.

14.2.2 QSAR INFORMATION

The Endocrine Disruptor Knowledge Base (EDKB) Web site developed at the U.S. Food and Drug Administration (FDA) National Center for Toxicological Research (NCTR) by W. Tong and coworkers consists of a biological activity database, with relevant literature citations and QSAR models to predict binding affinity of compounds to the estrogen and androgen nuclear receptors [19]. Data for more than 3,200 chemicals and some 2,000 relevant citations are available. Experimental data measuring estrogenic endpoints include *in vitro* assays for estrogen receptor competitive binding affinity, cell proliferation, reporter gene assays, and *in vivo* assays for uterotrophic activity. The database provides chemical structure search with Boolean operators, graphical and table displays, and data export functions. A major element of the EDKB program was the development of computer-based predictive models to predict affinity for the binding of compounds to the estrogen and androgen receptors. The estrogen receptor (ER) binding dataset containing 131 ER binders and 101 non-ER binders is available to download on the EDKB Web site, as is the androgen receptor (AR) binding dataset based on 146 AR binders and 56 non-AR binders. This database is one of the largest existing free databases providing quantitative endocrine endpoint values, similarity or substructure search through graphical interfaces, and useful links to toxicological databases.

The Distributed Structure-Searchable Toxicity (DSSTox) Database Network is a project of EPA's Computational Toxicology Program that aims to build a public database for improved structure-activity and predictive toxicology capabilities [20]. The DSSTox Web site provides a public forum for publishing downloadable, standardized chemical structure files associated with toxicity data. It is an expanded version of the NCTR ER database, which is contained within EDKB, where the chemical class fields and SAR information abstracted from Fang et al. [21] were added. The original NCTR ER database included the experimental ER binding results for developing QSAR models to predict ER binding affinities. *In vitro* rat uterine cytosol ER competitive binding assay results were collected for 232 structurally diverse chemicals covering most of the known chemical classes and spanning a wide range of biological activities [22,23]. It is noteworthy that a structure and similarity searching tool called "EPA DSSTox Structure-Browser" was recently developed from available

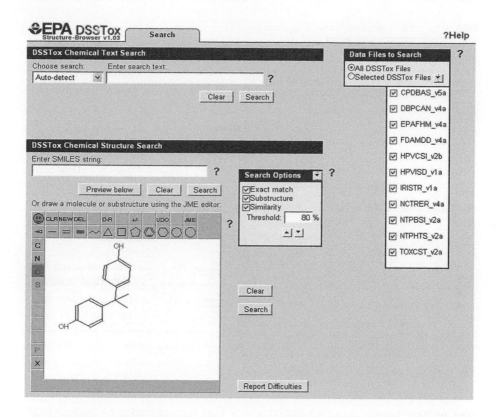

FIGURE 14.4 DSSTox search engine including a Java applet for drawing molecules and substructure searching. The search can also be done from the SMILES strings. The field "Chemical Text Search" offers various types of text searching (chemical name, CAS RN, InChI code, and formula).

structure viewing freeware and open source programming tools. It allows the user to screen through the chemical inventory of published DSSTox data files (Figure 14.4). In the DSSTox database, standard chemical and toxicity fields are used to summarize the general chemical information content of DSSTox Structure Data File (DSSTox SDF) for each compound. An alphabetically central listing of all fields contained in the most recent version of all DSSTox SDF files is consultable [24]. Among the interesting fields for EDCs, there are the measured ER relative binding affinity, activity category (active strong, active medium, active weak, slight binder, and inactive), and the possibility to visualize the two-dimensional (2D) and three-dimensional (3D) structures of the compounds (Figure 14.5). This database is open to any person or organization interested in publishing a database on the DSSTox Web site. Hence, to assist the draft documentation and data file design, the EPA Web site provides useful tools and templates.

The Endocrine Disruptor Priority-Setting Database (EDPSD v.2 Beta) was developed by the EPA on the recommendations of the Endocrine Disruptor Screening and Testing Advisory Committee (EDSTAC) [25]. It provides input to the priority-setting

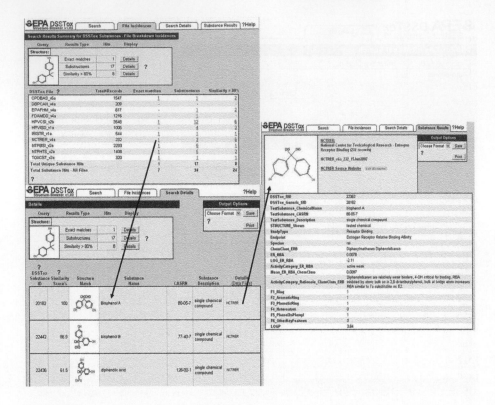

FIGURE 14.5 DSSTox results for bisphenol A. A table lists the number of matches (exact or partial) for the diverse databases available at the U.S. Environmental Protection Agency (EPA). Details on the NCTRER_v4a (National Center of Toxicology Research — Estrogen Receptor) database results are presented with a link to DSSTox Standard Chemical Structure Fields for each match.

step of the Endocrine Disruptor Screening Program using currently available data (for example, existing exposure and effects related information). This program relies on a tiered approach for predicting whether a substance may have an effect in humans that is analogous to an effect produced by naturally occurring estrogen, androgen, or thyroid hormones. EDPSD contains a variety of data sources related to chemical exposure potential, such as ambient and human biological monitoring data. It also provides data on chemicals that may be used in consumer, commercial, and food product chemicals and contains toxicological effect data related to the endocrine system or to endocrine mechanisms associated with specific chemicals. EDPSD includes information for over 140,000 chemicals. The database allows us to view the summarized information available for each data source, compartment, and category. Although the EDPSD database is well documented and tutorials are provided on the Web site, the program is not easy to use. Moreover, the drawback of this database is that the last update of the Web site was in 2002. More information on the EDPSD program can be found on the EPA Web site [26].

The ED-North database developed by the Research Group Environmental Toxicology, Laboratory of Environmental Toxicology and Aquatic Ecology, Ghent University (Belgium), deals with the data gathered during the SPSD I research project that aimed at establishing a synopsis of the increasing volume of available scientific literature on endocrine disruption [27,28]. Based on the available scientific literature, an electronic database of chemicals with (potential) endocrine disruptive activity was developed. This free database contains information on the hormone disrupting potential, including effects and physicochemical properties of these chemicals. The ED-North database is an MS Access relational database that contains 765 chemical compounds, 2,355 test results, and a large number of references. A search interface provides the possibility to investigate a chemical and/or a described effect. The entire list of compounds can be obtained by leaving the search fields blank. The results are collected into a table containing the CAS RN of the substance, the chemical formula, the number of tests described, and physicochemical properties (for example, boiling point, vapor pressure, solubility, log Kow, and so forth). The endocrine effect studies including the tested organism, the observed effect (agonist or antagonist), the relative binding affinity to 17β-estradiol, and all experimental details of the related study are also presented (Figure 14.6).

The U.S. National Cancer Institute (NCI) Developmental Therapeutics Program (DTP) collected and tested more than half a million natural and synthetic organic compounds since 1955 [29,30]. These data are entirely in the public domain and are commonly called the "Open NCI Database" or the "NCI Database" (NCI DB) [31]. The Enhanced NCI Database Browser is a Web-based graphical user interface developed to conduct rapid searches by numerous criteria among the 250,000

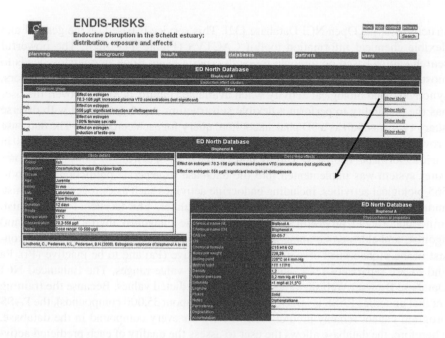

FIGURE 14.6 ENDIS-RISKS results obtained for bisphenol A.

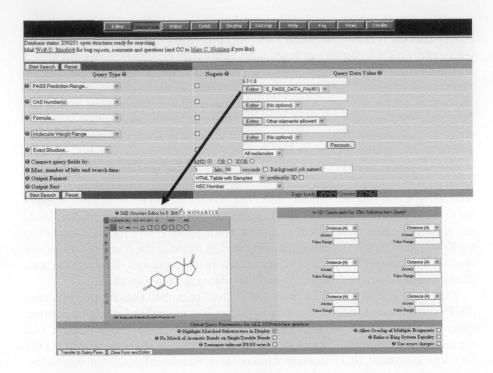

FIGURE 14.7 Enhanced NCI Database Browser home page.

structures of the Open NCI Database [32]. This service is based on the general and flexible chemical information toolkit CACTVS (Figure 14.7) [33]. The powerful search engine based on multiple research criteria using Boolean operators leads to mostly computed data, such as log Kow values, hydrogen bond donors and acceptors, synonyms, and biological activity when available (Figure 14.8). Moreover, the user has the possibility to perform a flexible substructure search combined with diverse query types, as well as a 3D pharmacophore query for each compound. The database provides 2D and 3D visualization options and the possibility to export results in various formats. The PASS (Prediction of Activity Spectrum of Substances) computer system was implemented into the database to generate predictions for up to 565 biological activities, including endocrine activities such as estrogen (ant)agonist, androgen (ant)agonist, estrogen receptor modulator, glucocorticoid (ant)agonist, mineralocorticoid, progestin agonist, progesterone antagonist, thyroid hormone (ant) agonist, and aldosterone antagonist [34]. The predictions calculated by PASS consist in the probabilities of a compound to be active (Pa) and to be inactive (Pi). Pa and Pi are separately searchable by probability value ranges. The Enhanced NCI Database Browser Web site offers 64,188,212 predicted values. Because the training set that underlies PASS is large but still limited (about 35,000 compounds), the PASS program cannot reliably predict each activity for every compound in the database. Therefore, the database allows the user to assess the quality of each predicted activity by having a look at the leave-one-out cross-validation results. Recently, the PASS

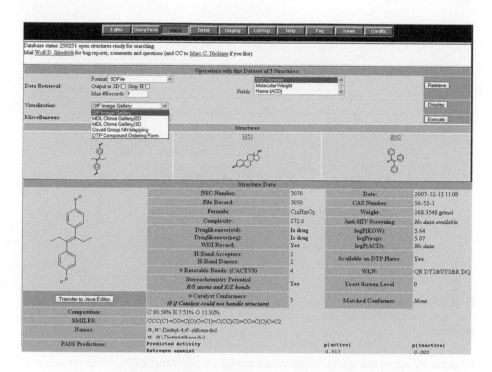

FIGURE 14.8 Enhanced NCI Database Browser result screenshot showing the diversity of the information provided by this database.

endocrine disruption profile of a large dataset of 11,416 molecules retrieved from the Enhanced NCI Database was analyzed from different multivariate methods and graphical tools, yielding the identification of specific and nonspecific structural features and allowing us to explain the endocrine activities of the chemicals [35].

The free database "e-experimental endocrine disruptor binding assays" (e³dba) was designed to improve and to centralize the flow of information on experimental endocrine disruptor *in vitro* binding assays on nuclear receptors [36]. All *in vitro* binding assay data were extracted from an NIEHS report [37]. The main objective of this report was to provide comprehensive summaries of the published and publicly available unpublished data on the scientific basis and performance of *in vitro* assays used to test substances for their ability to bind to the estrogen receptor (ER). ER binding data were collected for 638 substances tested in competitive binding studies. These substances have been classified into chemical classes, such as polychlorinated biphenyls, triphenylethylenes, organochlorines, PAHs, phenols, and bisphenols. They have been assigned to a product class, such as pharmaceuticals, pesticides, chemical intermediates, natural products, and plasticizers. The "Search" tool offers different kinds of queries, such as assay type, experiment name, compound name, CAS RN, chemical class, product class, and experimental measures. The option "Select All" in all fields allows us to obtain the four experimental binding values (IC$_{50}$, Ki, RBA, and log RBA) for all the compounds recorded in the e³dba database. It is also

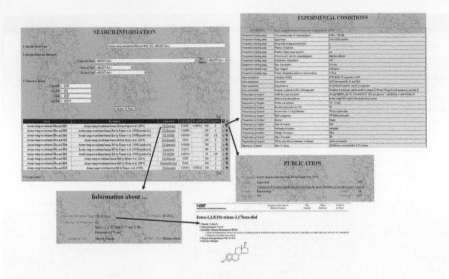

FIGURE 14.9 Organization of the e^3dba database from the search information interface to the display of the results and information on their origin and conditions in which they were obtained.

possible to refine the query using constraint(s) on the value (>, <, =) or to select a value range among the different fields. The results of the request are displayed as a table with seven columns representing the designation of the assay, the designation of the experiment, the chemical name of the endocrine disruptor, and, when available, the four experimental binding assay data (Figure 14.9).

On the Japanese Ministry of Economy, Trade and Industry (METI) Web site, a downloadable report allows the user to collect binding affinity data of 948 substances to the human ERα and data of human ERα agonistic activity of 177 chemicals [38]. For each compound, the chemical name, CAS registry number, relative binding affinity (RBA), and test year are given.

Competitive radiometric binding data for 645 chemicals can be obtained from the Web site of Pr. Mekenyan's group [39]. The downloadable file gathers the CAS registry number, the chemical name, the observed RBA value to the estrogen receptor (ER), and database affiliation of training set chemicals for their ER binding affinity model.

Hazardous Substances Data Bank (HSDB) is a free U.S. toxicological data file on the National Library of Medicine's (NLM) Toxicology Data Network (TOXNET®) focusing on the toxicology of potentially hazardous chemicals [40]. The search engine uses the chemical name, chemical name fragment, CAS RN, and/or subject terms. HSDB results consist of a list of information on human health effects, animal toxicity studies, industrial hygiene, emergency handling procedures, environmental fate/exposure, and physicochemical properties. HSDB is organized into individual chemical records and contains peer-reviewed toxicological data from books, government and technical reports, and scientific literature for over 5,000 chemicals. Search results can easily be viewed, printed, or downloaded.

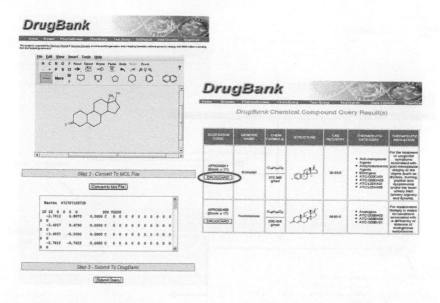

FIGURE 14.10 DrugBank structure query tool and substance query results with a link to a "DrugCard" that includes structural and biological information.

DrugBank is a free Canadian bioinformatics and cheminformatics resource, which was designed by Dr. Wishar and his colleagues at the University of Alberta (Canada) [41]. This resource pulls together a comprehensive amount of information on drugs (chemicals) and drug targets (proteins). It contains data on over 1,000 marketed and over 3,000 additional experimental chemical substances and over 14,000 protein sequences. The DrugBank home page displays links to multiple search options and the possibility of downloading the data, such as the structure files, substance information, protein sequence, and DNA sequence data. Information can be accessed in a number of ways ranging from simple text queries to more sophisticated structure queries using a ChemSketch applet or SMILES string input, and structure similarity search tool (Figure 14.10). The "Data Extractor" search engine allows the user to construct complex or constrained queries. The results are collected in a "DrugCard" entry containing 80 data fields with substance-related information, such as therapeutic category and mechanism of action. Information on the drug target including function and protein sequence information are also recorded. Many data fields are hyperlinked to other databases (KEGG, PubChem, ChEBI, PDB, Swiss-Prot and GenBank) and a variety of structure viewing applets makes the display of the 2D and 3D structures of compounds and proteins easier. DrugBank is a comprehensive Web database including quantitative physicochemical, pharmaceutical, and biological data for thousands of well-studied compounds and proteins.

KiBank is a free database of inhibition constant (K_i) values with 3D structures of target proteins and chemicals and many useful links [42]. The KiBank database contains 16,000 K_i values, 9,279 chemicals, and 101 target proteins covering 310 subtypes. It is updated on a daily basis [43]. It is noteworthy that 5,984 compounds and

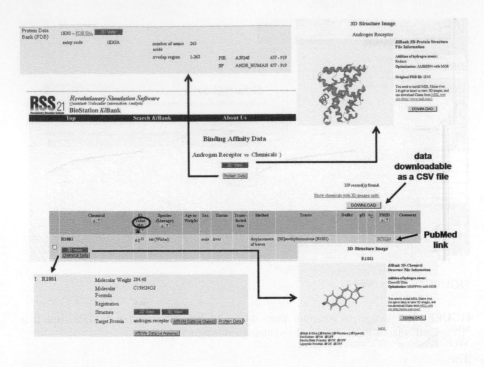

FIGURE 14.11 KiBank results and graphical interfaces allowing the manipulation of molecules.

54 protein 3D structures with hydrogen atoms were energy minimized and stored in MDL (MOL) and Protein Data Bank (PDB) format, respectively. Moreover, these 3D structures can be downloaded directly from the KiBank Web site. The search engine is very user friendly and complete. The search can be performed by selecting the receptor functions (for example, metabotropic, ionotropic membrane, nuclear) and protein target name, and/or by typing the compound name. The table gathering the results collects all the experimental details of the test (for example, K_i value, species, age or weight, gender, tissue, method, tracer, pH). Moreover, numerous links allow the user to display the 3D structure of the selected compound and/or protein and the chemical data of the selected substance and protein (Figure 14.11). These results can be downloaded as a CSV file. The weakness of this database is that it does not supply the three-dimensional structural information for the ligand-protein complexes.

The PDBbind database was developed from a collaboration between the groups of S. Wang at the University of Michigan and R. Wang at the Shanghai Institute of Organic Chemistry (China) [44,45]. It contains a collection of experimentally measured binding affinity data (K_d, K_i, and IC_{50}) exclusively for the ligand-protein complexes available in the PDB [46]. The 2007 release provides binding data for 3,214 ligand-protein complexes. The PDBbind Web site also supports interactive substructure/similarity search (Figure 14.12). For each ligand-protein complex, the database provides a "fact sheet" summarizing its four-letter PDB code, its binding affinity (K_d, K_i, or IC_{50}), the resolution of the structure, and the protein name (Figure 14.13). This

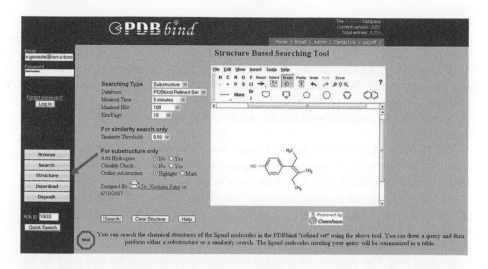

FIGURE 14.12 PDBbind interactive substructure/similarity search screenshot.

sheet provides a link to the corresponding page of this complex on the RCSB PDB Web site from which one can conveniently get information regarding the complex. The primary reference of this complex in which the experimentally measured binding affinity was cited is linked on this sheet to the corresponding Web page in the NCBI PubMed online literature database. A graphical interface allows the user to see the 3D display of the ligand-protein complex. However, theoretical models are not considered by PDBbind. A total of 1,300 ligand-protein complexes with high-quality binding data and structures are selected to form a refined set and to serve as a

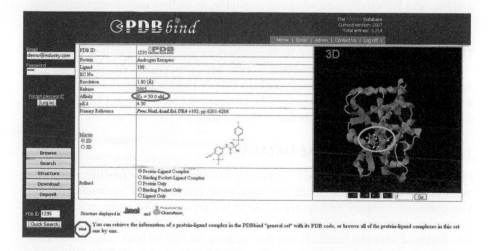

FIGURE 14.13 Screenshot of a result on PDBbind.

high-quality standard dataset for theoretical studies on ligand-protein binding. For each complex in this refined set, a sheet also provides 2D and 3D sketches of the chemical structure of the ligand molecule. This database emphasizes the link between the binding affinities and the structures of the ligand-protein complexes in PDB and provides convenient Web-based tools for data retrieval and analysis. The PDBbind database, including its downloadable contents, is free of charge, under a license agreement, for researchers working for academic, governmental, or other nonprofit institutions.

The PDSP K_i database is a public domain resource providing information on the ability of substances to interact with an expanding number of molecular targets [47]. The database includes 4,600 K_i values for a large number of substances interacting with numerous ion channels, nuclear receptors (for example, estrogen, androgen, thyroid, mineralocorticoid, glucocorticoid, and progesterone receptors) and enzymes. In some cases, a link to PubChem, which is a freely accessible database created by NIH in 2004, allows the user to get chemical information on the tested compound [48]. The database is regularly updated and data can be sent directly by the user to the database. However, the weakness of this database is the lack of 3D graphical interface displaying the compounds and/or the protein and the impossibility of exporting the data.

14.3 e-ECDYSTEROID DATABASE

Ecdybase is the third edition of *The Ecdysone Handbook* electronic version, originally created by R. Lafont and I.D. Wilson [49]. The aim of Ecdybase is to provide general data on all natural ecdysteroids through a search engine, based on the use of the compound name or partial name, the molecular weight, the formula, the species name where the compound can be found, and a browser. It allows the user to access a data file gathering biological, chemical, structural, spectroscopic (ultraviolet [UV], infrared [IR], mass spectra [MS], nuclear magnetic resonance [NMR]), chromatographic, and bioactivity data whenever available, as well as the relevant references. The number of compounds belonging to the ecdysteroid family is still growing (170 compounds in 1992, 312 compounds in 2000, and 407 compounds in August 2007). The Ecdybase is still considered by the authors as a developing source that can be continuously improved and extended. Forms are available for the electronic submission of additional data for new ecdysteroids or to provide more information on already known ones. This database is the only one devoted to insect EDCs. All the data collected in it could be useful to elaborate structure-activity models.

14.4 e-PHYTOCHEMICAL DATABASES

Phytoestrogens are plant-derived compounds that structurally or functionally mimic mammalian estrogens and hence may induce adverse effects. Conversely, they are also considered to play an important role in the prevention of some cancers, heart disease, menopausal symptoms, and osteoporosis [50]. Ethnobotany Database is a free database of phytochemicals and their actions, medicinal uses, and ethnobotany, compiled by Dr. J. Duke of the Green Pharmacy Garden and the U.S. Department of

Agriculture (USDA) [51,52]. J. Duke has chemically analyzed natural products and, based on their chemical constituents, listed the known effects of the ingredients, such as androgenic, estrogenic, mineralocorticoid, and progestational effects. The search engine allows us to list phytochemicals having one or multiple endocrine activities, the plants used for estrogenic activity, or the number of chemicals in plants having an endocrine activity. The database has several key strengths: a large number of plants (80,000 records of plant uses worldwide), complete data on most database entries, and chemical information on a variety of plants. It also has several critical weaknesses, most notably the lack of public input, and limited support for a variety of ethnobotanic data. Finally, the Ethnobotany Database is a closed system, which means our data cannot be added to a shared repository.

A similar free phytochemical database designed by USDA allows us to search for biological activities of phytochemicals, which are classified into several disease conditions so that their medicinal uses are quickly searched [53]. In this database, representative chemical groups and phytochemical classes are chosen and indexed for searching phytochemicals easily based on their chemical structures. This classification is continually updated for a better search.

14.5 e-PHARMACEUTICAL DATABASE

BIAM is a French pharmaceutical database that provides a large amount of information on drugs as well as the ingredients used in the formulation of pharmaceutical products [54]. The search engine allows the user to list the pharmaceuticals having an endocrine activity, such as androgen (ant)agonist, estrogen (ant)agonist, thyroid hormone (ant)agonist, aldosterone antagonist, mineralocorticoid agonist, progestin agonist, progesterone antagonist, glucocorticoid agonist, and growth hormone simulation. However, since 2001, the BIAM has not been updated.

14.6 CONCLUSIONS

The Internet search for "endocrine disruptor" on Google yields about 65,000 hits. From such a high number of responses, it could be expected that there exist enough resources to find data for QSAR modeling. Unfortunately, this is not the case, and a cross searching with additional keywords such as "binding data" or "database" induces a drastic drop in the number of available Internet resources on EDs. This is not surprising because a large number of Web sites have been mainly designed as information tools dedicated to the general public, the health professionals, the scientists, and so on.

The aim of this work was first to catalog the existing free databases available on the Internet and then to critically analyze them. In a last step, an attempt was made to see whether these resources could be useful for deriving (Q)SAR models.

Thus, our study clearly shows that most of the free databases come from universities and government organizations. Undoubtedly, until now, the United States produced the greatest effort for designing ED databases. When data are provided, they principally deal with binding data, especially for the estrogen and androgen receptors. Consequently, there is a need for more experimental data on other endocrine targets.

Most of the resources allow the user to access to the bibliographical references at the origin of the information displayed.

One of the major differences between the databases deals with 3D structural information and visualization availability. When different stereoisomers exist for one compound, it is of great importance to know the isomer responsible for the activity. A 3D representation of the molecules can avoid such structural ambiguity.

Last, the main drawback encountered in the databases is the update procedure, which is very often not specified.

ACKNOWLEDGMENT

This study was partially granted by the French Ministry of Ecology and Sustainable Development (PNETOX-N°24-B/2004-N°CV 04000175).

REFERENCES

[1] M. Patlak, *A testing deadline for endocrine disrupters*, Environ. Sci. Technol. 30 (1996), pp. 540–544.

[2] L.E. Gray, J. Ostby, V. Wilson, C. Lambright, K. Bobseine, P. Hartig, A. Hotchkiss, C. Wolf, J. Furr, M. Price, L. Parks, R.L. Cooper, T.E. Stoker, S.C. Laws, S.J. Degitz, K.M. Jensen, M.D. Kahl, J.J. Korte, E.A. Makynen, J.E. Tietge, and G.T. Ankley, *Xenoendocrine disrupters — Tiered screening and testing. Filling key data gaps*, Toxicology 181–182 (2002), pp. 371–382.

[3] Available at www.scorecard.org/chemical-profiles/ (accessed 18/07/07).

[4] Available at www.silsoe.cranfield.ac.uk/ieh/databases/rediped.html (accessed 18/07/07).

[5] L.H. Keith, *Environmental Endocrine Disruptors: A Handbook of Property Data*, Wiley Interscience, New York, 1997.

[6] L. Magos, *Environmental endocrine disruptors — A handbook of property data chemical brain injury*, J. Appl. Toxicol. 19 (1999), pp. 73–74.

[7] Available at www.pesticideinfo.org/Search_Chemicals.jsp#ChemSearch (accessed 18/07/07).

[8] "Endocrine Disrupting Chemicals" report, Illinois EPA, 1997.

[9] "Auxiliary Matters with Estrogenic Effects" report, Danish EPA, 2000.

[10] European Commission DG Environment, *Towards the establishment of a priority list of substances for further evaluation of their role in endocrine disruption*, Final report, BKH Engineers, Delft, the Netherlands, 2000. Available at http://ec.europa.eu/environment/docum/01262_en.htm (accessed 18/07/07).

[11] T. Colborn, F.S. Vom Saal, and A.M. Soto, *Developmental effects of endocrine-disrupting chemicals in wildlife and humans*, Environ. Health Perspect. 101 (1993), pp. 378–384.

[12] C.M. Benbrook, *Growing doubt: A primer on pesticides identified as endocrine disruptors and/or reproductive toxicants*, National Campaign for Pesticide Policy Reform, Washington, DC, 1996.

[13] J. Duarte, G. Perrière, V. Laudet, and M. Robinson-Rechavi, *NUREBASE: Database of nuclear hormone receptors*, Nucleic Acids Res. 30 (2002), pp. 364–368. Available at www.ens-lyon.fr/LBMC/laudet/nurebase/nurebase.html (accessed 27/07/07).

[14] F. Horn, G. Vriend, and F.E. Cohen, *Collecting and harvesting biological data: The GPCRDB & NucleaRDB databases*, Nucleic Acids Res. 29 (2001), pp. 346–349. Available at www.receptors.org/NR/ (accessed 20/10/07).

[15] B. Boeckmann, A. Bairoch, R. Apweiler, M.C. Blatter, A. Estreicher, E. Gasteiger, M.J. Martin, K. Michoud, C. O'Donovan, I. Phan, S. Pilbout, and M. Schneider, *The SWISS-PROT protein knowledgebase and its supplement TrEMBL in 2003*, Nucleic Acids Res. 31 (2003), pp. 365–370.

[16] R.W.W. Hooft, G. Vriend, C. Sander, and E.E. Abola, *Errors in protein structures*, Nature 381 (1996), p. 272.

[17] G. Vriend, *WHAT IF — A molecular modelling and drug design program*, J. Molec. Graphics 8 (1990), pp. 52–56.

[18] E. Martinez, D.D. Moore, E. Keller, D. Pearce, J.P. Vanden Heuvel, V. Robinson, B. Gottlieb, P. MacDonald, S. Simons Jr., E. Sanchez, and M. Danielsen, *The Nuclear Receptor Resource: A growing family*, Nucleic Acids Res. 26 (1998), pp. 239–241.

[19] W. Tong, R. Perkins, H. Fang, H. Hong, Q. Xie, W. Branham, D. Sheehan, and J. Anson, *Development of quantitative structure-activity relationships (QSARs) and their use for priority setting in the testing strategy of endocrine disruptors*, Regulatory Res. Perspect., 1 (2002), pp. 1–16.

[20] Available at www.epa.gov/ncct/dsstox/ (accessed 10/07/07).

[21] H. Fang, W. Tong, L.M. Shi, R. Blair, R. Perkins, W. Branham, B.S. Hass, Q. Xie, S.L. Dial, C.L. Moland, and D.M. Sheehan, *Structure-activity relationships for a large diverse set of natural, synthetic, and environmental estrogens*, Chem. Res. Tox. 14 (2001), pp. 280–294.

[22] R.M. Blair, H. Fang, W.S. Branham, B.S. Hass, S.L. Dial, C.L. Moland, W. Tong, L. Shi, R. Perkins, and D.M. Sheehan, *The estrogen receptor relative binding affinities of 188 natural and xenochemicals: Structural diversity of ligands*, Toxicol. Sci. 54 (2000), pp. 138–153.

[23] W.S. Branham, S.L. Dial, C.L. Moland, B.S. Hass, R.M. Blair, H. Fang, L. Shi, W. Tong, R.G. Perkins, and D.M. Sheehan, *Binding of phytoestrogens and mycoestrogens to the rat uterine estrogen receptor*, J. Nutr. 132 (2002), pp. 658–664.

[24] Available at www.epa.gov/ncct/dsstox/CentralFieldDef.html (accessed 10/07/07).

[25] Available at www.ergweb.com/endocrine/ (accessed 10/07/07).

[26] Available at www.epa.gov/scipoly/oscpendo/ (accessed 10/07/07).

[27] Available at www.vliz.be/projects/Endis/EDNorth.php (accessed 10/07/07).

[28] *Evaluation of possible impacts of endocrine disruptors on the North Sea ecosystem*, Research project MN/DD2/002. Available at www.belspo.be/belspo/fedra/proj. asp?l=en&COD=MN/DD2/002 (accessed 10/07/07).

[29] G.W.A. Milne and J.A. Miller, *The NCI Drug Information System. 1. System overview*, J. Chem. Inf. Comput. Sci. 26 (1986), pp. 154–159.

[30] G.W.A. Milne, M.C. Nicklaus, J.S. Driscoll, S. Wang, and D. Zaharevitz, *National Cancer Institute drug information system 3D database*, J. Chem. Inf. Comput. Sci. 34 (1994), pp. 1219–1224.

[31] Available at http://dtp.nci.nih.gov/webdata.html (accessed 20/07/07).

[32] W.D. Ihlenfeldt, J.H. Voigt, B. Bienfait, F. Oellien, and M.C. Nicklaus, *Enhanced CACTVS browser of the open NCI database*, J. Chem. Inf. Comput. Sci. 42 (2002), pp. 46–57. Available at http://129.43.27.140/ncidb2/ (accessed 20/07/07).

[33] W.D. Ihlenfeldt, Y. Takahashi, S. Abe, and S. Sasaki, *Computation and management of chemical properties in CACTVS: An extensible networked approach toward modularity and flexibility*, J. Chem. Inf. Comput. Sci. 34 (1994), pp. 109–116.

[34] V. Poroikov, D. Akimov, E. Shabelnikova, and D. Filimonov, *Top 200 medicines: Can new actions be discovered through computer-aided prediction?*, SAR QSAR Environ. Res. 12 (2001), pp. 327–344.

[35] J. Devillers, N. Marchand-Geneste, J.C. Doré, J.M. Porcher, and V. Poroikov, *Endocrine disruption profile analysis of 11,416 chemicals from chemometrical tools*, SAR QSAR Environ. Res. 18 (2007), pp. 181–193.

[36] Available at www.u-bordeaux1.fr/e3dba/ (accessed 10/07/07).

[37] *Current status of test methods for detecting endocrine disruptors: in Vitro estrogen receptor binding assays*, NIH report No. 03-4504 (2002).

[38] Available at www.meti.go.jp/english/report/index.html (accessed 27/03/08).

[39] R. Serafimova, M. Todorov, D. Nedelcheva, T. Palvov, Y. Akahori, M. Nakai, and O. Mekenyan, *QSAR and mechanistic interpretation of estrogen receptor binding*, SAR QSAR Environ. Res. 18 (2007), pp. 389–421. Available at http://download.oasis-lmc. org/Publication81-Table1.pdf (accessed 27/03/08).

[40] Available at http://toxnet.nlm.nih.gov/cgi-bin/sis/htmlgen?HSDB (accessed 25/07/07).

[41] D.S. Wishart, C. Knox, A.C. Guo, S. Shrivastava, M. Hassanali, P. Stothard, Z. Chang, and J. Woolsey, *DrugBank: A comprehensive resource for in silico drug discovery and exploration*, Nucleic Acids Res. 34 (2006), pp. D668–D672. Available at http:// redpoll.pharmacy.ualberta.ca/drugbank/ (accessed 26/07/07).

[42] J.W. Zhang, M. Aizawa, S. Amari, Y. Iwasawa, T. Nakano, and K. Nakata, *Development of KiBank, a database supporting structure-based drug design*, Comput. Biol. Chem. 28 (2004), pp. 401–407. Available at http://kibank.iis.u-tokyo.ac.jp/ (accessed 18/07/07).

[43] K. Nakata, S. Amari, and T. Nakano, *Application of KiBank database*, Chem-Bio Informatics J. 6 (2006), pp. 47–54.

[44] R. Wang, X. Fang, Y. Lu, and S. Wang, *The PDBbind Database: Collection of binding affinities for protein-ligand complexes with known three-dimensional structures*, J. Med. Chem. 47 (2004), pp. 2977–2980.

[45] R. Wang, X. Fang, Y. Lu, C.Y. Yang, and S. Wang, *The PDBbind Database: Methodologies and updates*, J. Med. Chem. 48 (2005), pp. 4111–4119. Available at http://sw16.im.med.umich.edu/databases/pdbbind/index.jsp (accessed 18/07/07).

[46] Available at www.pdb.org/ (accessed 20/07/07).

[47] Available at http://pdsp.med.unc.edu/pdsp.php (accessed 20/07/07).

[48] Available at http://pubchem.ncbi.nlm.nih.gov/ (accessed 20/11/07).

[49] R. Lafont, J. Harmatha, F. Marion-Poll, L. Dinan, and I.D. Wilson, *The Ecdysone Handbook, 3rd edition, on-line*. Available at http://ecdybase.org (accessed 23/07/07).

[50] A.L. Ososki and E.J. Kennelly, *Phytoestrogens: A review of the present state of research*, Phytother. Res. 17 (2003), pp. 845–869.

[51] Available at www.ars-grin.gov/duke/ (accessed 24/07/07).

[52] Available at www.leffingwell.com/plants.htm (accessed 24/07/07).

[53] Available at www.pl.barc.usda.gov/usda_chem/achem_home.cfm (accessed 24/07/07).

[54] Available at www.biam2.org (accessed the 25/07/07).

Index

Printed and bound by CPI Group (UK) Ltd, Croydon, CR0 4YY

18/10/2024

01776262-0017